Distributed Data Fusion for Network-Centric Operations

Edited by

David L. Hall
Chee-Yee Chong
James Llinas
Martin Liggins II

CRC Press
Taylor & Francis Group
Boca Raton London New York

CRC Press is an imprint of the
Taylor & Francis Group, an **informa** business

CRC Press
Taylor & Francis Group
6000 Broken Sound Parkway NW, Suite 300
Boca Raton, FL 33487-2742

First issued in paperback 2017

© 2013 by Taylor & Francis Group, LLC
CRC Press is an imprint of Taylor & Francis Group, an Informa business

No claim to original U.S. Government works

Version Date: 2012928

ISBN 13: 978-1-138-07383-8 (pbk)
ISBN 13: 978-1-4398-5830-1 (hbk)

Contents

Foreword ix
Acknowledgment xiii
Editors... xv
Contributors xvii

Chapter 1   Perspectives on Distributed Data Fusion.. 1

David L. Hall

Chapter 2   Distributed Data Fusion: Overarching Design Concerns
and Some New Approaches... 17

*David Nicholson, Steven Reece, Alex Rogers, Stephen Roberts,
and Nick Jennings*

Chapter 3   Network-Centric Concepts: Impacts to Distributed Fusion
System Design 47

James Llinas

Chapter 4   Distributed Detection in Wireless Sensor Networks.......................... 65

*Pramod K. Varshney, Engin Masazade, Priyadip Ray, and
Ruixin Niu*

Chapter 5   Fundamentals of Distributed Estimation 95

Chee-Yee Chong, Kuo-Chu Chang, and Shozo Mori

Chapter 6   Essence of Distributed Target Tracking:
Track Fusion and Track Association 125

Shozo Mori, Kuo-Chu Chang, and Chee-Yee Chong

Chapter 7   Decentralized Data Fusion: Formulation and Algorithms............... 161

Paul Thompson, Eric Nettleton, and Hugh Durrant-Whyte

Chapter 8   Toward a Theoretical Foundation for Distributed Fusion 199

Ronald Mahler

Chapter 9   Object Classification in a Distributed Environment.........................245

          James Llinas and Chee-Yee Chong

Chapter 10   A Framework for Distributed High-Level Fusion 271

          Subrata Das

Chapter 11   Threat Analysis in Distributed Environments 295

          *Hengameh Irandoust, Abder Benaskeur, Jean Roy, and
          Froduald Kabanza*

Chapter 12   Ontological Structures for Higher Levels of Distributed Fusion 327

          Mieczyslaw M. Kokar, Brian E. Ulicny, and Jakub J. Moskal

Chapter 13   Service-Oriented Architecture for Human-Centric Information
          Fusion 347

          Jeff Rimland

Chapter 14   Nonmyopic Sensor Management...365

          Viswanath Avasarala and Tracy Mullen

Chapter 15   Test and Evaluation of Distributed Data and Information
          Fusion Systems and Processes 379

          James Llinas, Christopher Bowman, and Kedar Sambhoos

Chapter 16   Human Engineering Factors in Distributed and Net-Centric
          Fusion Systems...409

          Ann Bisantz, Michael Jenkins, and Jonathan Pfautz

Chapter 17   Distributed Data and Information Fusion in Visual Sensor
          Networks 435

          *Federico Castanedo, Juan Gomez-Romero, Miguel A. Patricio,
          Jesus Garcia, and Jose M. Molina*

Index..467

Foreword

I am very pleased to provide the Foreword for this timely work on distributed fusion. I have been involved in fusion research for the last 15 years, focused on transforming data to support more effective decision making. During that time, I have relied heavily on the advice of the editors of this book and many of the chapter authors to help set the directions for Army-focused basic and applied information fusion initiatives.

I first met the editors about 12 years ago at an Army-sponsored fusion workshop where it was clear that the issues of increased sensors and data sources, along with the introduction of web-based information architectures, had finally overwhelmed the analysis community. Most of the discussions were focused on the problems, but Dave Hall and Jim Llinas began addressing the solutions. They identified relevant terms and definitions, outlined algorithms for specific fusion tasks, addressed many of the evolving architectural issues, pinpointed key technical barriers, and proposed directions for future research. They clearly were long-time experts in the field; but, more importantly, they were visionary in their recognition of rapidly evolving trends in information management and the impact those trends would have on the field of data fusion. It is, therefore, not at all surprising that this, their latest book (along with colleagues), would be focused on distributed fusion.

While there are numerous texts and handbooks on data fusion in general (many written or edited by the editors and authors of this book), there are two major trends that motivate the need for this work. First, the very concept of defense operations has dramatically changed. Modern military missions include, for example, coalition-based counterinsurgency, counternarcotics, counterterrorism, and peacekeeping operations. In a sense, the questions have become more complex. The focus is less on detecting the physical aspects of an oncoming tank battalion, and more on detecting networks of operatives or anomalous events, and integrating them with sociocultural concepts. The impact is that historical fusion algorithms, with their reliance on large-system, sensor data–driven, centralized techniques, must now accommodate human observers, open source information, and distributed decision-making conducted at lower and lower echelons.

A second key trend is that rapid changes in information technology have enabled mobile information architectures, changing our concept of where and how fusion algorithms will be employed. As of February 2012, there are 5.9 billion mobile subscribers worldwide, 1.2 billion mobile web users, and 10.9 billion application downloads in place. While implementation of service-oriented architecture (SOA) and cloud concepts continues to be problematic in the mobile ad hoc network environments of the military, this mobile access trend clearly sets the vision for the future in the Department of Defense.

The impact of these trends on fusion requirements includes the need for the following:

- Near-real-time computational speed: The traditional concept of having an individual warfighter disconnected throughout a mission, occasionally contacting analysts for updates, is no longer acceptable. Untethered soldiers will require ready access to data sources. This means that fusion algorithms must run at near-real-time speeds and be tailored to the user's situation if they are to be effective in supporting tactical operations. In other words, we must begin replacing the notion of large-scale fusion algorithms with small-scale user-adaptive fusion applications.
- Accommodation of more varied data sources: Users will have the ability not only to access data, but to collect and post imagery, voice clips, text messages, etc., as well. Along with this capability comes increased data volume and complexity. Fusion algorithms must handle structured, semistructured, and unstructured data sources alike to support situation awareness. Further, they must rely on data discovery techniques rather than predetermined deductively framed data access; otherwise, they may overlook important new data sources that could prove critical to the decision process.
- Incorporation of trust and confidence concepts: As data sources and users become more widely varied, fusion algorithms must be able to take into account the uncertainties associated with the use of soft data sources, the application of data to problems outside the original scope of the collection effort, and the potential introduction of accidentally or purposefully misleading sources.

So how do we take fusion algorithms, apply them to more complex problems, using more complex data sources, and still meet near-real-time computing constraints? Distributed processing holds out a potential solution, and that solution fits nicely with the concept of cloud computing, where algorithms are automatically distributed across all available resources. However, in reality we know that our current fusion algorithms do not readily lend themselves to parallel techniques. And so it is particularly appropriate that this book begins to tackle the difficult problems of designing and implementing distributed, decentralized information fusion.

Written in a manner that particularly highlights topics of direct relevance to a Department of Defense reader, this text outlines such critical issues as architectural design and the associated impact of network-centric and SOA concepts; fundamentals of estimation, classification, tracking, and threat analysis and their extensions to decentralized implementation; human-centric techniques for visualization and evaluation; and fundamentals of fusion systems engineering.

As is typical for these editors, the chapters provide a well-organized, thorough review of the field from both a theoretical and applied research perspective. The book will most certainly serve as a useful tool for fusion researchers and practitioners alike as we continue to grapple with the critical issue of ensuring our data collection efforts have a clear and positive impact on mission outcome.

Barbara D. Broome, PhD
Chief, Information Sciences Division
U.S. Army Research Laboratory
Adelphi Laboratory Center
Adelphi, Maryland

Acknowledgment

The authors and editors are grateful to Rita Griffith for her efforts to provide excellent copyediting and logistical and administrative support to ensure the successful completion of this book.

Editors

David L. Hall, PhD, is the dean for the Pennsylvania State University College of Information Sciences and Technology (IST). He also serves as a professor of IST and director of the Center for Network Centric Cognition and Information Fusion (NC2IF). Prior to joining IST, he was an associate director of the Penn State Applied Research Laboratory. In this role, he directed an interdisciplinary team of 175 scientists and engineers in conducting research in information science, navigation research, systems automation, and communications science. Dr. Hall has industrial experience, including serving as director of independent research & development (IR&D) and leading a software signal processing group at Raytheon Corporation (HRB Division), manager of the navigation analysis section at the Computer Sciences Corporation, and staff scientist at MIT Lincoln Laboratory. Dr. Hall is the author of over 200 technical papers and several books, including *Mathematical Techniques in Multisensor Data Fusion* (2004) and *Human-Centered Information Fusion* (2010). He is an IEEE fellow and has received the Department of Defense Joe Mignona Award for his contributions to multisensor data fusion. Dr. Hall has lectured internationally on the topics of multisensor data fusion, artificial intelligence, and research management and technology forecasting.

Chee-Yee Chong, PhD, is a chief scientist at BAE Systems Technology Solutions. He received his SB, SM, and PhD in electrical engineering from the Massachusetts Institute of Technology. He taught at the Georgia Institute of Technology in Atlanta, Georgia, before joining Advanced Decision Systems (ADS), a small, advanced research and development company in California. He continued to lead tracking and fusion research at Booz Allen Hamilton after it acquired ADS, and later at ALPHATECH, which was acquired by BAE Systems. Dr. Chong's research interests include centralized and distributed estimation, target tracking, information fusion, optimization and resource management, and application to real-world problems. He has been involved in distributed fusion research for over 25 years, starting with the Distributed Sensor Networks (DSN) program for the U.S. Defense Advanced Research Projects Agency (DARPA) in the 1980s. He is also the cofounder of the International Society of Information Fusion (ISIF), its president since 2004, and general cochair of the 12th International Conference on Information Fusion since 2009. He has served as an associate editor for the *IEEE Transactions on Automatic Control* and the *International Journal of Information Fusion*. Currently he serves as an area editor for the *Journal of Advances in Information Fusion*. Dr. Chong received the Joseph Mignogna Data Fusion Award from the U.S. Department of Defense Joint Directors of Laboratories Data Fusion Group in 2005.

James Llinas, PhD, is a professor emeritus at the State University of New York at Buffalo. He is also the director emeritus for the Center for Multisource Information Fusion (CMIF), a research center that he started some 20 years ago located at the

University at Buffalo. An expert on data fusion, he coauthored the first integrated book on the subject, *Multisensor Data Fusion*, published by Artech House (1990), and has lectured internationally on the subject for over 20 years. For more than a decade, he has been a technical advisor to the Defense Department's Joint Directors of Laboratories Data Fusion Panel. He was the founding president of the International Society of Information Fusion. His expertise in applying data fusion technology to different problem areas ranges from complex defense and intelligence-system applications to nondefense diagnosis. His current projects involve automated reasoning, distributed data fusion, hard and soft data fusion, information fusion architectures, and the scientific foundation of data correlation. He received a doctorate degree in applied statistics and industrial engineering.

Martin E. Liggins II is an engineer with The MITRE Corporation. He has more than 20 years of research and development experience in industry and with the Department of Defense. He has performed fusion research in a number of areas, including sensor and data fusion, multisensor and multitarget tracking, radar, high-performance computing, and program management. He is the author of more than 30 technical and research papers and is coeditor of the *Handbook of Multisensor Data Fusion*, Second Edition. Liggins has served as the chairman of the National Symposium of Sensor and Data Fusion (1995, 2002, and 2003) and has been an active senior committee member since 1990. He has also been active in the SPIE Aerosense Conference on Signal Processing, Sensor Fusion, and Target Recognition since 1992. He was awarded the Veridian Medal Paper Award in fusion research (2002) and the first Rome Air Development Center Major General John J. Toomay Award for advances in multispectral fusion technology (1989).

Contributors

Viswanath Avasarala
GE Research
Niskayuna, New York

Abder Benaskeur
Defence Research and Development
  Canada—Valcartier
Val-Belair, Quebec, Canada

Ann Bisantz
Department of Industrial and Systems
  Engineering
Center for Multisource Information
  Fusion
State University of New York at Buffalo
Buffalo, New York

Christopher Bowman
Data Fusion and Neural Networks
Broomfield, Colorado

Federico Castanedo
Deusto Institute of Technology—
  Deusto Tech
University of Deusto
Bilbao, Spain

Kuo-Chu Chang
Systems Engineering and Operations
  Research Department
George Mason University
Fairfax, Virginia

Chee-Yee Chong
BAE Systems
Los Altos, California

Subrata Das
Machine Analytics, Inc.
Belmont, Massachusetts

Hugh Durrant-Whyte
National ICT Australia
Sydney, New South Wales, Australia

Jesus Garcia
Department of Computer Science
University Carlos III of Madrid
Madrid, Spain

Juan Gomez-Romero
Department of Computer Science
University Carlos III of Madrid
Madrid, Spain

David L. Hall
College of Information Sciences and
  Technology
The Pennsylvania State University
University Park, Pennsylvania

Hengameh Irandoust
Defence Research and Development
  Canada—Valcartier
Val-Belair, Quebec, Canada

Michael Jenkins
Department of Industrial and Systems
  Engineering
Center for Multisource Information
  Fusion
State University of New York at Buffalo
Buffalo, New York

Nick Jennings
School of Electronics and Computer
Science
University of Southampton
Southampton, United Kingdom

Froduald Kabanza
Departement d'informatique
Université de Sherbrooke
Sherbrooke, Quebec, Canada

Mieczyslaw M. Kokar
Department of Electrical and Computer
Engineering
Northeastern University
Boston, Massachusetts

James Llinas
Department of Industrial and Systems
Engineering
Center for Multisource Information
Fusion
State University of New York at Buffalo
Buffalo, New York

Ronald Mahler
Unified Data Fusion Sciences, Inc.
Eagan, Minnesota

Engin Masazade
Department of Electrical Engineering
and Computer Science
Syracuse University
Syracuse, New York

Jose M. Molina
Department of Computer Science
University Carlos III of Madrid
Madrid, Spain

Shozo Mori
BAE Systems
Los Altos, California

Jakub J. Moskal
VIStology, Inc.
Framingham, Massachusetts

Tracy Mullen
Restek
Bellefonte, Pennsylvania

Eric Nettleton
Australian Centre for Field Robotics
The University of Sydney
Sydney, New South Wales, Australia

David Nicholson
BAE Systems
Bristol, United Kingdom

Ruixin Niu
Department of Electrical and Computer
Engineering
Virginia Commonwealth University
Richmond, Virginia

Miguel A. Patricio
Department of Computer Science
University Carlos III of Madrid
Madrid, Spain

Jonathan Pfautz
Charles River Analytics Inc.
Cambridge, Massachusetts

Priyadip Ray
Department of Electrical and Computer
Engineering
Duke University
Durham, North Carolina

Steven Reece
University of Oxford
Oxford, United Kingdom

Jeff Rimland
College of Information Sciences and
    Technology
The Pennsylvania State University
University Park, Pennsylvania

Stephen Roberts
University of Oxford
Oxford, United Kingdom

Alex Rogers
School of Electronics and Computer
    Science
University of Southampton
Southampton, United Kingdom

Jean Roy
Defence Research and Development
    Canada—Valcartier
Val-Belair, Quebec, Canada

Kedar Sambhoos
Department of Industrial and Systems
    Engineering
Center for Multisource Information
    Fusion
State University of New York at Buffalo
Buffalo, New York

Paul Thompson
Australian Centre for Field Robotics
The University of Sydney
Sydney, New South Wales, Australia

Brian E. Ulicny
VIStology, Inc.
Framingham, Massachusetts

Pramod K. Varshney
Department of Electrical Engineering
    and Computer Science
Syracuse University
Syracuse, New York

1 Perspectives on Distributed Data Fusion

David L. Hall

CONTENTS

1.1 Introduction 1
1.2 Brief History of Data Fusion 2
1.3 JDL Data Fusion Process Model 4
1.4 Process Models for Data Fusion 6
1.5 Changing Landscape: Key Trends Affecting Data Fusion 8
1.6 Implications for Distributed Data Fusion 12
References 13

1.1 INTRODUCTION

Multisensor data fusion has an extensive history and has become a relatively mature discipline. Extensive investments in data fusion, primarily for military applications, have resulted in a number of developments: (1) the widely referenced Joint Directors of Laboratories (JDL) data fusion process model (Kessler et al. 1991, Steinberg et al. 1998, Hall and McMullen 2004); (2) numerous mathematical techniques for data fusion ranging from signal and image processing to state estimation, pattern recognition, and automated reasoning (Bar-Shalom 1990, 1992, Hall and McMullen 2004, Mahler 2007, Das 2008, Liggins et al. 2008); (3) systems engineering guidelines (Bowman and Steinberg 2008); (4) methods for performance assessment (Llinas 2008); and (5) numerous applications (see, for example, the *Annual Proceedings of the International Conference on Information Fusion*). Recent developments in communications networks, smart mobile devices (containing multiple sensors and advanced computing capability), and participatory sensing, however, lead to the need to address distributed data fusion. Changes in information technology (IT) introduces an environment in which traditional sensing/computing networks (e.g., for military command and control (C^2) or intelligence, surveillance, and reconnaissance [ISR]) for well-defined situation awareness are augmented (and sometimes surpassed) by uncontrolled, ad hoc information collection. The emerging concept of participatory sensing is a case in point (Burke et al. 2006). For applications ranging from environmental monitoring to crisis management, to political events, information from ad hoc observers provide a huge source of information (albeit uncalibrated). Examples abound: (1) monitoring of the spread of disease by monitoring Google search terms, (2) estimation of earthquake events using Twitter feeds and specialized websites (U.S. Geological Survey (http://earthquake.usgs.gov) n.d.), (3) monitoring political events (http://ushahidi.com n.d.), (4) citizen crime watch (Lexington-Fayette Urban County Division of Police, see http://crimewatch.lfucg.com n.d.), (5) solicitation of citizens to report newsworthy events (Pitner 2012), and (6) use of citizens for collection of scientific data (Hand 2010). While ad hoc observers and open source information provide a huge potential resource of data and information, the use of such data are subject to many challenges such as establishing pedigree of the data, characterization of the observer(s), trustworthiness of the data, rumor effects, and many others (Hall and Jordan 2010).

Traditional information fusion systems involving user-owned and controlled sensor networks, an established system and information architecture for sensor tasking, data collection, fusion, dissemination, and decision making are being enhanced or replaced by dynamic, ad hoc information collection, dissemination, and fusion concepts. These changes provide both opportunities and challenges. Huge new sources of data are now available via global human observers and sensors feeds available via the web. These data can be accessed and distributed globally. Increasingly capable mobile computing and communications devices provide opportunities for advanced processing algorithms to be implemented at the observing source. The rapid creation of new mobile applications (APPs) may provide new algorithms, cognitive aids, and information access methods "for free." Finally, advances in human–computer interaction (HCI) provide opportunities for new engagement of humans in the fusion process, as observers, participants in the cognition process, and collaborating decision makers. However, with such advances come challenges in design, implementation, and evaluation of distributed fusion systems.

This book addresses four key emerging concepts of distributed data fusion. Chapters 1 through 3 introduce concepts in network centric information fusion including the design of distributed processes. Chapters 4 through 8 address how to perform state estimation (viz., estimation of the position, velocity, and attributes of observed entities) in a distributed environment. Chapters 9 through 12 focus on target/entity identification and on higher level inferences related to situation assessment/awareness and threat assessment. Finally, Chapters 13 through 18 discuss the implementation environment for distributed data fusion including emerging concepts of service-oriented architectures, test and evaluation of distributed fusion systems, and aspects of human engineering for human-centered fusion systems. The remainder of this chapter provides a brief history of data fusion, an introduction to the JDL data fusion process model, a review of related fusion models, a discussion of emerging trends that affect distributed data fusion, and finally a discussion of some perspectives on distributed fusion.

1.2   BRIEF HISTORY OF DATA FUSION

The discipline of information fusion has a long history, beginning in the 1700s with the posthumous publication of Bayes' theorem (1763) on probability and Gauss' development of the method of least squares in 1795 to estimate the orbit of the newly discovered asteroid Ceres using redundant observations (redundant in the mathematical sense meaning more observations than was strictly necessary for a minimum data, initial orbit determination). Subsequently, extensive research has been applied to develop methods for processing data from multiple observers or sensors to estimate the state (viz., position, velocity, attributes, and identity) of entities. Mathematical methods in data fusion (summarized in Kessler et al. [1991], Hall and McMullen [2004], and many other books) span the range from signal and image processing methods to estimation methods, pattern recognition techniques, automated reasoning methods, and many others. Such methods have been developed during the entire time period from 1795 to the present.

A brief list of events in the history of information fusion is provided in the following:

- Publication of Bayes' theorem on probability (1763)
- Gauss' original development of mathematics for state estimation using redundant data (1795)
- Development of statistical pattern recognition methods (e.g., cluster analysis, neural networks, etc.) (early 1900s–1940s)
- Development of radar as a major active sensor for target tracking and identification (1940s)
- Development of the Kalman filter (1960) for sequential estimation
- Implementation of U.S. Space Track system (1961)
- Development of military focused all-source analysis and fusion systems (1970s–present)
- First demonstration of the Advanced Research Project Agency computer network (ARPANET)—the precursor to the Internet (1968)
- First cellular telephone network (1978)
- National Science Foundation Computer Science Network (CSNET) (1981)
- Formation of JDL data fusion subpanel (mid-1980s)
- Creation of JDL process model (1990)
- Tri-Service Data Fusion Symposium (1987)
- Formation of the annual National Symposium on Sensor Fusion (NSSDF) (1988)
- Second generation mobile cell phone systems (early 1990s)
- Commercialization of the Internet (1995)
- Creation of the International Society of Information Fusion (ISIF) (1999)
- Annual ISIF Fusion Conferences (since 1998)
- Emergence of nonmilitary applications (1990s to present), including condition monitoring of complex systems, environmental monitoring, crisis management, medical applications, etc.
- Emergence of participatory sensing to augment physical sensors (1990s)

While basic fusion algorithms have been well known for decades, the routine application of data fusion methods for real-time problems awaited the emergence of advanced sensing systems and computing technologies that allowed semi-automated processing. Automated fusion of data fusion requires a combination of processing algorithms, computers capable of executing the fusion algorithms, deployed sensor systems, communication networks to link the sensors and computing capabilities, and systems engineering methods for effective system design, development, deployment, and test and evaluation. Similarly, the emergence of distributed data fusion systems involving hard (physical sensor) data and human (soft) observations requires a combination of new fusion algorithms, computing capabilities, communications systems, global use of smart phones and computing devices, and the emergence of a net-centric generation who routinely makes observations, tweets, reports, and shares such information via the web.

1.3 JDL DATA FUSION PROCESS MODEL*

In the early 1990s, a number of U.S. DoD large-scale funded efforts were underway to implement data fusion systems. An example was the U.S. Army's All Source Analysis System (ASAS) (Federation of American Scientists [www.fas.org/]). The field of data fusion was emerging as a separate discipline, with limited common understanding of terminology, algorithms, architectures or engineering processes. JDL was an administrative group created to assist in coordinating research across the U.S. Department of Defense laboratories. The JDL established a subgroup to focus on issues related to multisensor data fusion. The formal name was the Joint Directors of Laboratories, Technical Panel for Command, Control and Communications (C³) data fusion subpanel. This subgroup created the JDL data fusion process model (see Figure 1.1). The model was originally published in a briefing (Kessler et al. 1991) to the Office of Naval Intelligence and later presented in papers, used as an organizing concept for books (Hall and McMullen 2004, Liggins et al. 2008), national and

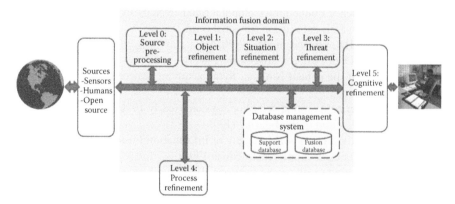

FIGURE 1.1 Top level of JDL data fusion process model. (Adapted from Hall, D.L. and McMullen, S.A.H., *Mathematical Techniques in Multisensor Data Fusion*, Artech House, Norwood, MA, 2004.)

---

* The Joint Directors of Laboratories data fusion process model has been described in multiple references including (1) the original technical report (Kessler et al. 1991) and (2) various textbooks (Waltz and Llinas 1990, Hall and McMullen 2004, Hall and Jordan 2010), review articles (Hall and Llinas 1997), and revisions of the model (Steinberg et al. 1998, Hall et al. 2000, Blasch and Plano 2002). The JDL model has been referenced extensively in books, papers, government solicitations, and tutorials. This section of this chapter is thus not new, but rather a brief summary that paraphrases (and in some cases duplicates) the author's previous writings on this subject.

international conferences, government requests for proposals, and in a few cases government and industrial research organizations. The original briefing (Kessler et al. 1991) presented a hierarchical, three-layer model. The top part of the model is shown in Figure 1.1. For each of the fusion "levels," a second layer identified specific subprocesses and functions, while a third layer identified subfunctions and candidate algorithms to perform those functions. These sublayers are described in Hall and McMullen (2004).

Since its inception, the model has undergone several additions and revisions. The initial model included only the first four levels of fusion processing: object refinement (level 1), situation refinement (level 2), threat refinement (level 3), and process refinement (level 4). Steinberg et al. (1998) extended the model by adding a precursor level of fusion and sought to make the model more broadly applicable beyond military applications. Level 0 fusion involves sensor-based data processing and estimation. Level 0 processing recognized the increasing role of smart sensors and processing at the sensor level. Hall et al. (2000) and, independently, Blasch and Plano (2002) extended the model to include human–computer interaction involving cooperative cognition between a human user and a data fusion system. Other extensions to the data fusion model have been discussed by Llinas, who presents the case for further consideration of current data fusion issues including distributed data fusion systems and ontology-based systems.

The six high-level processes defined in the JDL model are summarized as follows:

1. *Level 0 fusion (data or source preprocessing)* involves processing data from sensors (e.g., signals, images, hyper-spectral images, vector quantities, or scalar data) to prepare the data for subsequent fusion. Examples of data preprocessing include image processing, signal processing, "conditioning" of the data, coordinate transformations (to relate the data from the origin or platform that the sensor is located on to a centralized set of coordinates), filtering, alignment of the data in time or space, and other transformations.
2. *Level 1 fusion (object refinement)* combines data from multiple sensors or sources to obtain the most reliable estimate of the object's location, characteristics, and identity. The term object is usually meant to indicate physical objects such as a vehicle or human. However, we could also fuse data to determine the location and identity of activities, events, or other geographically constrained entities of interest. The issues of object/entity location (estimation) are often discussed separately from the problem of object/entity identification. In real fusion systems, however, these subprocesses are usually integrated.
3. *Level 2 fusion (situation refinement)* uses the results of level 1 processing to develop a contextual interpretation of their meaning. This involves understanding how entities are related to their environment, the relationship among different entities and how they are interrelated. For example, the motion of vehicles in an environment may depend upon factors such as roads, road conditions, terrain, weather, and the presence of other vehicles. The actions of a human in a crowd might be interpreted much differently, than the same human motion and actions in the absence of other

surrounding people. The techniques used for level 2 fusion may involve artificial intelligence, automated reasoning, complex pattern recognition, rule-based reasoning, and many other methods.

4. *Level 3 fusion (threat refinement/impact assessment)* involves projecting the current situation into the future to determine the potential impact or consequences of threats associated with the current situation. Level 3 processing seeks to draw inferences about possible threats, courses of action in response to those perceived threats and how the situation changes based on our changing perceptions. Techniques for level 3 fusion are similar to those used in level 2 processing but also include simulation, prediction, and modeling.

5. *Level 4 fusion (process refinement/resource management)* seeks to improve the fusion process (more accurate, timelier, and more specific). This might be accomplished by redirecting the sensors or information sources, changing the control parameters on the other fusion algorithms or selecting which algorithm or technique is most appropriate to the current situation and available data. The level 4 process involves functions such as sensor modeling, modeling of network communications, computation of measures of performance, and optimization of resource utilization.

6. *Level 5 processing (human–computer interaction/cognitive refinement)* seeks to optimize how the data fusion system interacts with human users. The level 5 process seeks to understand the needs of the human user and respond to those needs by appropriately focusing the fusion system attention on things that are important to the user. Types of functions may include use of advanced displays, search engines, advisory tools, cognitive aids, collaboration tools, and other techniques. This may involve use of traditional HCI functions such as geographical displays, displays of data and overlays, processing input commands, and the use of nonvisual interfaces such as sound or haptic (touch) interfaces.

The originators of the JDL model fully recognized that the JDL levels were an artificial partitioning of the data fusion functions and that the levels overlap. In real systems, fusion is not performed in a sequential (level 0, level 1, …) manner. Instead, the processes are interleaved. For example, in level 1 processing, information about a target's kinematics can provide insight into the target identification and potential threat (level 3). However, this artificial partition of data fusion functions has proven useful for discussion purposes.

1.4 PROCESS MODELS FOR DATA FUSION

There are a number of models that address cognitive and information processes that are related to data fusion. A survey and assessment of these process models was conducted by Hall et al. (2006). A summary of the models (and additional models) is presented in Table 1.1, along with references which describe the models in more detail. Hall et al. (2006) divided the models into two broad categories, data fusion models and decision making models. To a certain extent, this is an arbitrary partitioning but

TABLE 1.1
Summary of Data Fusion Models/Frameworks

Model	Description	References
JDL data fusion process model	A functional model for describing the data fusion process	Kessler et al. (1991) Liggins et al. (2008) Hall and McMullen (2004) Steinberg et al. (1998) Hall et al. (2000) Blasch and Plano (2002)
Functional levels of fusion	An abstraction of input–output functions of the data fusion process—focus on types of data processed and associated techniques appropriate to the data types	Dasarthy (1994)
Transformation of requirements to information processing (TRIP) model	Application of the waterfall development process to data fusion—emphasis on linking inferences to required information and data collection	Kessler and Fabien (2001)
Omnibus model	Adaptation of Boyd's OODA loop for data fusion	Bedworth and O'Brien (2000)
Endsley's model of situational awareness	A cognitive model for situational awareness	Endsley (2003), Endsley et al. (2000)
Three-layer hierarchical model	Three-layer modular approach to data fusion, integrating data at different levels: (1) data level (e.g., signal processing), (2) evidence level (statistical models and decision making), and (3) dynamics level	Thomopoulos (1989)
Behavioral knowledge formalism	Sequence of basic stages of fusion; extraction of a feature vector from data, alignment and association, development of pattern recognition and semantic labels, and linking feature vectors to events	Pau (1988)
Waterfall model	Hierarchical architecture showing flow of data and inferences from data level to decision-making level	Harris et al. (1998)
General data fusion model (DFA) using UML	General data fusion architecture model based on the unified modeling language (UML), using a taxonomy based on definitions of data and variables or tasks	Carvalho et al. (2003)
Unified data fusion (λJDL) model	Model that seeks to unify situation awareness functions, common operating picture, and data fusion	Lambert (1999, 2001)
Recognition primed decision (RPD) making	A naturalistic theory of decision making focused on recognition of perceptual cues and action	Klein (1999), Klein and Zsambok (1997) Kaempf et al. (1996)

(*continued*)

TABLE 1.1 (continued)
Summary of Data Fusion Models/Frameworks

Model	Description	References
Observe, orient, decide, act (OODA) loop	A process model of military decision making based on observing effective commanders; extended by several authors for general situation assessment and decision making	Boyd (1987), Brehmer (2005), Bryant (2006), Rousseau and Breton (2004) Grant (2005)
Salerno's model	A framework that links data sources (categorized by perishability) to perception, comprehension, and projection	Salerno (2002), Salerno et al. (2004)

reflects how these models are referenced in the literature. In addition, models such as the observe–orient–decide–act (OODA) loop have several extensions and variations. Each of these models has advantages and disadvantages related to describing the fusion and decision making process. They are summarized here to indicate the potential variations in how to describe or characterize the process of fusing information to understand an evolving situation and ultimately result in a decision or action. A good discussion of higher level models for data fusion (viz., at the situation awareness and threat assessment levels) is provided by Bosse et al. (2007). It should be noted that the list of models in Table 1.1 is not exhaustive. There are a number of additional models related to specific application domains such as robotics and medicine. It should also be noted that these process models do not explicitly consider the distributed aspect of fusion.

In the domain of military applications and intelligence, the two most utilized models are arguably the JDL data fusion process model summarized in the previous section and Mica Endsley's model of situation awareness (Endsley 2000, Endsley et al. 2003). Because of its extensive use in the situation awareness and cognitive psychology community, it is worth illustrating Endsley's model in Figure 1.2. Endsley's model seeks to line aspects of a cognitive task (illustrated in the top part of the figure) to characteristics on an individual performing the cognition (shown in the bottom part of the figure). Note that the levels in Endsley's model do not correspond to the levels in the JDL model, but rather are meant to model the cognitive processes for situation awareness. Endsley and her colleagues have utilized this model for a variety of DoD applications, performing extensive interviews with operational analysts and knowledge elicitation to identify appropriate techniques for the Endsley levels of fusion. Salerno (2002) and his colleagues (Salerno et al. 2004) have compared the JDL model and Endsley's model and have developed a high-level information functional architecture.

1.5   CHANGING LANDSCAPE: KEY TRENDS AFFECTING DATA FUSION

The context of distributed data fusion involves (1) rapid changes in IT, (2) individual and societal changes impacted and enabled by IT, and (3) the impact of IT as both

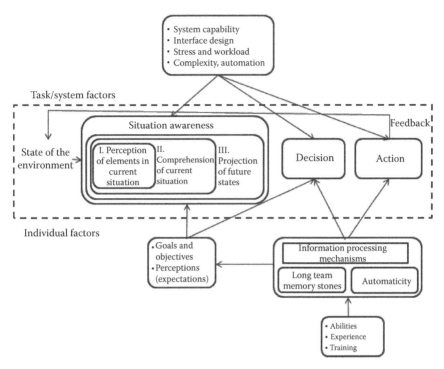

FIGURE 1.2 Endsley's situation awareness model. (Adapted from Endsley, M.R. et al., *Designing for Situation Awareness: An Approach to User-Centered Design*, Taylor & Francis Group, Inc., New York, 2003.)

a cause and solution for global problems. A summary of sample trends is provided in Tables 1.2 through 1.4. The tables list trends related to three main constructs: (1) IT, (2) information, and (3) people. Briefly we see the following trends and associated impacts.

- *Information Technology*—Very rapid changes are occurring in IT, ranging from ubiquitous, persistent surveillance of the entire earth via advanced sensors and human observers, increasingly capable mobile computing devices (via smart phones, embedded "invisible" computers in everyday devices, net-books, notebook computers, etc.), ubiquitous network connectivity with increasing access speeds, and improvements in HCI via multi (human) sensory inputs. This leads to near-universal connectivity among people, a tsunami of data on the web, and access to virtually unlimited computing capability. These have impacts on all aspects of human life and enterprise and certainly affect the concepts and implementation of data fusion systems. A summary of key areas including data collection, mobile computing, and network speed and connectivity is provided in Table 1.2.
- *Information*—The huge increase in available data (including signal, image, video, and text) via sensors and human input leads to major challenges in

TABLE 1.2

Examples of Technology Trends Impacting Data Fusion

Area	Trends and Issues
Data collection	• *Ubiquitous, persistent surveillance*: The capability exists now for worldwide ubiquitous, persistent surveillance. Resources such as national collection systems, long duration unmanned aerial vehicles (UAVs); leave-behind and resident ground-based sensors provide the opportunity for multispectral, multimode surveillance on a 24×7 basis. This allows focused and persistent surveillance about nearly any area of interest. The challenge is how to address the huge avalanche of data to sort through the data to find information of use/interest to generate meaningful knowledge. Such surveillance impacts areas such as environmental monitoring, understanding the distribution and evolution of disease, and crime and terrorism.
	• *New sensors and sensing modalities*: Physical sensors continue to be improved with new modalities of observation, increased sophistication in embedded signal and image processing, increased modes and agility in operation and control, and continuing improvements in sensor-level processing such as semantic meta-data generation, pattern recognition, dynamic sensor performance characterization, target tracking, and adaptive processing.
	• *Open source information*: Websites are available for all sorts of collected information. For example, sites based on reporting and mapping tools (ushahidi.com) provide information on emergency events, political uprisings, etc. Google Street View provides maps and photographs of numerous places around the world with ground level 360° photographs. The photograph sharing site Flickr (flickr.com) contains over 5 billion photographs taken by 10 million active subscribers. Commercial data providers such as DigitalGlobe (digitalglobe.com) provide access to satellite imagery including standard visual images, stereo images, and eight-band spectral images. Data regarding weather information, environmental data, detailed maps, video surveillance cameras, traffic monitoring, and many other types of information are all readily available.
Mobile computing	*Mobile computing* capabilities are rapidly increasing both in functionality, memory, speed, and network interconnectivity. New smart phones have typical specifications that include 4–16 GB memory (expandable to 32 GB), processing speeds in the range from 1 to 1.2 GHz, fourth generation communications speed, and touch screens with 480×800 pixels to 960×640 pixels. Over 1 million open source applications have been developed. The result is incredible hand-carried computing/sensing/communications devices that have proliferated throughout the world.
Network speed and connectivity	*Internet connectivity* is nearing worldwide ubiquity. Original connection via telephone landlines at 60 kilobits per second has changed to connections via television cable coax or fiber optics at typical speeds of 4–6 megabits per second, with additional mobile connection via mobile broadband over terrestrial mobile phone networks, WiFi hotspots in urban areas, and satellite Internet connections. While the United States lags behind other countries, some countries provide connections with speeds of 100 Mbs into homes. Increasingly, mobile devices are sharing and accessing video data via mobile Internet to the extent that video data dominates the data content of the mobile Internet. An excellent site that summarizes the history of the Internet is provided by (zakon.org).

TABLE 1.2 (continued)
Examples of Technology Trends Impacting Data Fusion

Area	Trends and Issues
Cloud computing	*Cloud computing* involves the delivery of computing resources as a product (sharing resources and information over a network), analogous to the concept of the electric grid. The ubiquity of Internet capability allows computing resources (large data storage, sophisticated computer models, large-scale computing capacity, etc.) to be at anyone's fingertips for a fee. Ultimately such concepts could eliminate local IT staff and computers while providing access to unprecedented capability. An example is Wolfram Alpha (wolframalpha.com) which provides free access to large data sets and sophisticated physical and mathematical models.
Human computer interfaces	*Human computing interfaces*: Advances in HCI involve increased fidelity in human access to data as well as multi-(human) sensory methods of interaction. Examples include full-immersion, three-dimensional interfaces and sonification (Ballora 2010) to allow visual and aural pattern recognition, haptic interfaces to provide a sense of touch. The potential exists to create new multisensory, full immersion interfaces that fully engage the sophisticated ability of humans to recognize patterns and detect anomalies.

TABLE 1.3
Examples of Information Trends Impacting Data Fusion

Area	Trends and Issues
Data archiving and distribution	*The exploding digital universe*: According to a 2010 Gartner report, the top three challenges for large enterprises are data growth, followed by system performance and scalability (Harding 2010). In 2007, the digital universe was 2.25×10^{21} bits (281 exobytes); by 2011, it was estimated to grow by a factor of 10. Fast growing data sources include digital TV, surveillance cameras, sensor applications, and social networks. Major issues include how to store, archive, distribute, access, and represent such data (Chute et al. 2008).
Meta-data generation	*Meta-data generation*: Given the enormous amounts of data (signals, images, video) being collected and stored via the *Internet of Things* and human data collection, a challenge involves how to represent the data for subsequent retrieval and use. Significant advances are being made in automated linguistic indexing of pictures (viz., machine-generated semantic labels) with anticipated extensions to signal data and to video data. This would provide the ability to access signal, image, and video data via emerging advanced search engines (e.g., next generation CITESEER type engines [citeseer.ist.psu.edu]).
Hard and soft fusion	*Hard and soft information fusion*: An emerging area in data fusion research is the fusion of hard (traditional physical sensor) data and soft (human observation) data. This topic was first discussed at a Beaver Hollow workshop held in February 2009, hosted by the Center for Multisource Information Fusion (CMIF) (see infofusion.buffalo.edu). The workshop explored issues in the fusion of hard and soft data, characterization of human source data, architecture issues, and even fundamental definitions of the terms hard and soft fusion. Since that workshop, special sessions on hard and soft fusion have been held at the International Society of Information Fusion (ISIF) FUSION 2010 conference and the FUSION 2011 conference.

TABLE 1.4

Examples of People Trends Impacting Data Fusion

Area	Trends and Issues
Digital natives	*Net-generation*: The current "net-generation" of people under the age of 30 have grown up with the Internet, cell phones, social networks, global connectivity, instantly available online resources, and significantly different social outlooks and cognitive approaches than previous generations (see Tapscott 2009). These "digital natives" have different expectations for everything from social interactions to business to problem solving that are having significant impact on all aspects of society. Shirkey (2010) describes some implications of the new era of collaboration which results in projects such as the world's encyclopedia (Wikipedia), shareware software, PatientsLikeMe, Ushahidi, and other dynamic collaborative efforts.
Participatory sensing	*Soft and participatory sensing*: Several developments and trends have provided the opportunity for the creation of a new, worldwide, data collection resource. These include (1) the huge increase in smart phones throughout the world (estimated in 2010 to be greater than 4.6 billion cell phones), (2) the increase in processing capability and sensor "add-ons" to smart phones (including high fidelity cameras, video capability, environmental sensors, etc.), and (3) the emergence of the digital native generation (Palfrey and Gasser 2008) who routinely collect information and share personal information via Twitter, Facebook, and other social sites. This has led to the concept of *participatory sensing*, in which individuals and groups of people actively participate in the collection of information for purposes ranging from crime prevention to scientific studies.

storage, access, archiving, distribution, meta-data generation, and issues such as data pedigree. The ultimate limitation of human attention units (the limited number of people to access data and their limited ability to pay attention to data) will lead to both opportunities and challenges in human–data interaction. Table 1.3 summarizes key areas including data archiving and distribution, meta-data generation, and hard and soft fusion.

- *People*—Finally, changes in IT and availability of information lead to changes in human behavior and expectations. The net-generation (people younger than 30 years) has always had access to the Internet, cell phones, computers, and related technologies. These "digital natives" exhibit different ways of addressing problems, viewpoints on collecting and sharing personal information, ways of establishing distributed social networks, etc. This in turn has implications for education, collaboration, business, and information security. Table 1.4 summarizes the potential impacts of a new generation of digital natives and the emergence of participatory sensing.

1.6   IMPLICATIONS FOR DISTRIBUTED DATA FUSION

As indicated in the previous section, a number of changes in technology, information, and people are impacting and will continue to impact the design and implementation of information fusion systems. Certainly, the proliferation of cell phones (leading to an avalanche of human observations), ubiquitous, high-speed networks, increased mobile computing power, cloud computing, new attitudes of users (based on a digital native outlook), and other factors are impacting data fusion systems. We are seeing the potential for "everyday" fusion systems supporting improved monitoring and operation of automobiles, medical diagnosis, monitoring of the environment, and even smart appliances. It is thus necessary to reconsider traditional data fusion technologies, design, and implementation methods to extend to these new applications and environment. While the changes in technology, information, and people provide increased opportunities, they also enable challenges to traditional thinking about fusion systems. As sensors and sources of information proliferate and new mobile applications become readily available, new challenges will involve (1) calibration and characterization of information sources, (2) establishment of methods to automatically determine the trustworthiness and pedigree of information, (3) the need to automatically generate semantic meta-data to represent signal, image, and video data, (4) how to establish the reliability of open-source software and algorithms, (5) meeting the expectations of increasingly sophisticated users, (6) creation of hierarchies of data and information fusion systems, (7) understanding how to utilize sensor-generated meta-data (e.g., in situ pattern recognition), and (8) robust architectures that span data to knowledge fusion, and many more.

It is hoped that this book will provide some additional insights to begin to address some of these issues.

REFERENCES

Ballora, M. 2010. Beyond visualization: Sonification. In *Human-Centered Information Fusion*, Hall, D. and J. Jordan (eds.), chapter 7. Norwood, MA: Artech House, Inc.

Bar-Shalom, Y. (ed.) 1990. *Multi-Target-Multi-Sensor Tracking Advanced Applications*, vol I. Norwood, MA: Artech House.

Bar-Shalom, Y. (ed.) 1992. *Multi-Target-Multi-Sensor Tracking Advanced Applications*, vol II. Norwood, MA: Artech House.

Bedworth, M. and J. O. O'Brien. 2000. The omnibus model: A new model of data fusion? *IEEE Aerospace and Electronic Systems Magazine,* 15(4), 30–36.

Blasch, E. and S. Plano. 2002. DFIG Level 5 user refinement issues supporting situational assessment reasoning. *Proceedings of SPIE*, vol. 4729, Wyndham, PA, pp. 270–279.

Bosse, E., J. Roy, and S. Wark. 2007. *Concepts, Models and Tools for Information Fusion*. Norwood, MA: Artech House.

Bowman, C. L. and A. N. Steinberg. 2008. Systems engineering approach for implementing data fusion systems. In *Handbook of Multisensor Data Fusion: Theory and Practice*, 2nd edn., M. E. Liggins, D. L. Hall, and J. Llinas (eds.), chapter 22, pp. 561–596. Boca Raton, FL: CRC Press.

Boyd, J. 1987. A discourse on winning and losing. Technical Report, Maxwell AFB, Montgomery, AL.

Brehmer, B. 2005. The dynamic OODA loop: Amalgamating Boyd's OODA loop and the cybernetic approach to command and control. *Proceedings of the 10th International Command and Control Research Technology Symposium*, McLean, VA.

Bryant, D. J. 2006. Rethinking OODA: Toward a modern cognitive framework of command decision making. *Military Psychology*, 18(3), 183.

Burke, J. et al. 2006. Participatory sensing. *Proceedings of WSW'06 at SenSys'06*, October 31, 2006, Boulder, CO.

Carvalho, R., W. Heinzelman, A. Murphy, and C. Coelho. 2003. A general data fusion architecture. *Proceedings of the Sixth International Conference on Information Fusion (Fusion'03)*, July 2003, Cairns, Queensland, Australia, pp. 1465–1472.

Center for MultiSource Information Fusion based at the University of Buffalo. http://www. infofusion.buffalo.edu, owned and maintained by Center for MultiSource Information Fusion, University of Buffalo, June 27, 2012.

Chute, C., A. Manfrediz, S. Minton, D. Reinsal, W. Schlichting, and A. Toncheva. March 2008. The diverse and exploding digital universe: An updated forecast of world wide information growth through 2011. IDC White paper sponsored by EMC.

CiteSeer. http://citeseer.ist.psu.edu/index, CiteSeerX: owned and maintained by The Pennsylvania State University College of Information Sciences and Technology, June 27, 2012.

Das, S. 2008. *High-Level Data Fusion*. Norwood, MA: Artech House.

Dasarthy, B. V. (ed.) 1994. *Decision Fusion*. Washington, DC: IEEE Computer Society.

DigitalGlobe. http://www.digitalglobe.com/, owned and maintained by Digital Globe corporation, June 27, 2012.

Endsley, M. R., B. Bolte, and D. G. Jones. 2003. *Designing for Situation Awareness: An Approach to User-Centered Design*. New York: Taylor & Francis Group, Inc.

Endsley, M. R., L. O. Holder, B. C. Leibrecht, D. C. Garland, R. L. Wampler, and H. D. Matthews. 2000. *Modeling and Measuring Situation Awareness in the Infantry Operational Environment*. Alexandria, VA: U.S. Army Research Institute for the Behavioral and Social Sciences, Infantry Forces Research Unit.

Federation of American Scientists, Intelligence resource Program. http://www.fas.org/irp/ program/process/asas.htm, Federation of American Scientists: Intelligence Resource Program, All Source Analysis System, maintained by Steven Aftergood, updated November 25, 1998.

Flickr. http://www.flickr.com/, owned and maintained by Yahoo, downloaded June 27, 2012.

Grant, T. 2005. Unifying planning and control using an OODA-based architecture. *Proceedings of the 2005 Annual Research Conference of the South African Institute of Computer Scientists and Information Technologists on IT Research in Developing Countries*, Mpumalanga, South Africa, pp. 159–170.

Hall, M. S., S. A. Hall, and T. Tate. 2000. Removing the HCI bottleneck: How the human computer interface (HCI) affects the performance of data fusion systems. *Proceedings of the 2000 MSS National Symposium on Sensor and Data Fusion*, June 2000, San Diego, CA, pp. 89–104.

Hall, D. and J. Jordan. 2010. *Human-Centered Information Fusion*. Norwood, MA: Artech House, Inc.

Hall, D. and J. Llinas. 1997. An introduction to multi-sensor data fusion. *Proceedings of the IEEE*, 85(1), 6–23.

Hall, D. L. and S. A. H. McMullen. 2004. *Mathematical Techniques in Multisensor Data Fusion*. Norwood, MA: Artech House.

Hall, D. et al. 2006. Assessing the JDL model: A survey and analysis of decision and cognitive process models and comparison with the JDL model. *Proceedings of the National Symposium on Sensor Data Fusion*, June 2006, Monterey, CA.

Hand, E. 2010. Citizen science: People power. *Nature* 466, 685–687.

Harding, N. 2010. Gartner: Data storage growth is the top challenge for IT organizations. Posting on IT Knowledge Exchange, November 3, 2010. (see http://itknowledge exchange.techtarget.com/server-farm/gartner-data-storage-growth-is-the-top-challenge-for-it-organizations/)

Harris, C. J., A. Bailey, and T. J. Dodd. 1998. Multi-sensor data fusion in defense and aerospace. *Aeronautical Journal*, 102(1015), 229–244.

Kaempf, G. L., G. Klein, M. L. Thorsden, and S. Wolf. 1996. Decision making in complex naval command-and-control environments. *Human Factors*, 38(2), 220–231.

Kessler, O. et al. November 1991. Functional description of the data fusion process. Report prepared for the Office of Naval Technology Data Fusion Development Strategy, Naval Air Development Center, Warminster, PA.

Kessler, O. and B. Fabien. 2001. Estimation and ISR process integration. Report for the Defense Advanced Projects Research Agency (DARPA), Washington, DC.

Klein, G. A. 1999. *Sources of Power: How People Make Decisions*. Cambridge, MA: MIT Press.

Klein, G. A. and C. E. Zsambok (eds.). 1997. *Naturalistic Decision Making*. Mahwah, NJ: Lawrence Erlbaum Associates, Inc.

Lambert, D. A. 1999. Assessing situations. *Proceedings of the IEEE 1999 Information, Decision and Control*, February 1999, Adelaide, South Australia, Australia, pp. 503–508.

Lambert, D. A. 2001. Situations for situation awareness. *Proceedings of the ISIF Fourth International Conference on Information Fusion, (FUSION 2001)*, August 2001, Montreal, Quebec, Canada, pp. 545–552.

Lexington-Fayette Urban County Division of Police. http://crimewatch.lfucg.com, Lexington-Fayette Urban County of Division Police, Crime Map, owned and maintained by Lexington, Kentucky government, June 29, 2012.

Liggins, M., D. L. Hall, and J. Llinas. 2008. *Handbook of Multisensor Data Fusion*, 2nd edn., Boca Raton, FL: CRC Press.

Llinas, J. 2008. Assessing the performance of multisensor fusion processes. In *Handbook of Multisensor Data Fusion: Theory and Practice*, 2nd edn., M. E. Liggins, D. L. Hall, and J. Llinas (eds.), chapter 25, pp. 655–675. Boca Raton, FL: CRC Press.

Mahler, R. P. S. 2007. *Statistical Multisource-Multi-Target Information Fusion*. Norwood, MA: Artech House.

Palfrey, J. and U. Gasser. 2008. *Born Digital: Understanding the First Generation of Digital Natives*. New York: Basic Books.

Pau, L. F. 1988. Sensor data fusion. *Journal of Intelligent and Robotic Systems*, 1, 103–116.

Pitner, S. 2012. Reporting news with a cell phone, http://handheldjournalism.com/reporting-news-with-a-cell-phone/, March 14, 2010.

Rousseau, R. and R. Breton. 2004. The M-OODA: A model incorporating control functions and teamwork in the OODA loop. *Proceedings of the 2004 Command and Control Research Technology Symposium*, San Diego, CA, pp. 15–17.

Salerno, J. 2002. Information fusion: A high-level architecture overview. *Proceedings of the 5th International Conference on Information Fusion*, Annapolis, MD, pp. 1218–1230.

Salerno, J., M. Hinman, and D. Boulware. 2004. Building a framework for situation awareness. *Proceedings of the 7th International Conference on Information Fusion*, Stockholm, Sweden, pp. 680–686.

Shirkey, C. 2010. *Cognitive Surplus: Creativity and Generosity in a Connected Age*. New York: The Penguin Group.

Steinberg, A. N., C. L. Bowman, and F. E. White. 1998. Revisions to the JDL model. *Joint NATO/IRIS Conference Proceedings*, October 1998, Quebec City, Quebec, Canada.

Tapscott, D. 2009. *Grown Up Digital*. New York: McGraw Hill.

Thomopoulos, S. C. 1989. Sensor integration and data fusion. *Proceedings of SPIE 1189, Sensor Fusion II: Human and Machine Strategies*, November 1989, Philadelphia, PA, pp. 178–191.

U.S. Geological Survey. http://earthquake.usgs.gov, U.S. Geological Survey, earthquake hazards program, U.S. Department of the Interior, June 28, 2012.

Ushahidi. http://ushahidi.com/, Ushahidi, owned and maintained by Ushahidi, a nonprofit technology company, June 29, 2012.

Waltz, E. and J. Llinas. 1990. *Multisensor Data Fusion*. Norwood, MA: Artech House, Inc.

WolframAlpha Computational Knowledge Engine. http://www.wolframalpha.com/, owned and maintained by Wolfram Alpha Corporation, June 29, 2012.

Zakon Group, LLC. http://www.zakon.org/robert/internet/timeline/, Zakon Group Hobbes' Internet Timepline 10.2, by Robert Hobbes Zakon, December 30, 2011.

2 Distributed Data Fusion
Overarching Design Concerns and Some New Approaches

David Nicholson, Steven Reece, Alex Rogers,
Stephen Roberts, and Nick Jennings

CONTENTS

2.1 Introduction 18
    2.1.1 Content 18
2.2 DDF System Concept 18
2.3 DDF Design Concerns 19
2.4 Information Recycling 20
    2.4.1 Bounded Covariance Inflation 20
    2.4.2 Coupling Scalars 22
    2.4.3 Decentralized Tracking Example 23
2.5 Sensor Coordination 24
    2.5.1 Max-Sum Algorithm 25
    2.5.2 Target Tracking Example 27
2.6 Selfish Stakeholders 28
    2.6.1 Problem Description 30
    2.6.2 Valuation Function 31
    2.6.3 Mechanism 33
    2.6.4 Example 34
2.7 Trust and Reputation 34
    2.7.1 Expected Utility of a Contract 35
    2.7.2 Heterogeneous Contracts: Inflated Independent
            Beta Distributions 37
    2.7.3 Heterogeneous Contracts: A Kalman Filter Trust Model 38
    2.7.4 Empirical Evaluation 39
2.8 Future Design Concerns and Opportunities 42
    2.8.1 HAC Design Concerns 42
    2.8.2 HAC Opportunities 43
Acknowledgments 44
References 44

2.1 INTRODUCTION

This chapter exposes some of the design concerns associated with distributed data fusion (DDF) systems and describes how they could be overcome. These concerns arise from the inherent *openness* of DDF systems. Open exchange and fusion of information creates the potential for system degradation as a result of recycling old information, failing to coordinate multiple information sources under the ownership of one or more stakeholders, and failing to recognize untrustworthy information sources. The overarching design concern is how to remove or reduce these problems without compromising the flexibility and scalability benefits of DDF systems.

2.1.1 CONTENT

Section 2.2 introduces the DDF system concept on which this chapter is based. This is a multi-agent system (MAS) in which each agent is a data fusion and decision-making node situated within some larger information fusion network. Section 2.3 introduces four critical design concerns that must be resolved if a multi-agent DDF system is to succeed in practice. Section 2.4 describes the resolution of information recycling concerns with a technique known as bounded covariance inflation (BCI). Section 2.5 reinforces the needs for coordinated actions within a DDF system and describes how this can be achieved with the max-sum algorithm. Section 2.6 describes how the concern of potential selfish actions in a multistakeholder DDF system can be managed by means of computational mechanism design. Finally, Section 2.7 raises the issue of trust and reputation in DDF systems. It describes how a probabilistic model of trust combined with the technique from Section 2.4 can be used to resolve this concern. Each section is highlighted with an example from the familiar domain of target tracking and sensor fusion. Section 2.8 concludes the chapter with some perspectives on the new design challenges that will be raised by future DDF systems that achieve effect by tightly interleaving human and software agent endeavors.

2.2 DDF SYSTEM CONCEPT

The DDF system concept explored in this chapter is illustrated in Figure 2.1. It is composed of autonomous, reactive, and proactive components, referred to as *agents*. These agents filter and fuse data to derive situational information. They interact by exchanging messages over communication links to achieve individual and collective goals. Within this MAS there may be multiple organizational relationships and stakeholders.

Our main focus will be decentralized sensor networks. Each sensor agent is tasked with detecting and tracking multiple targets. Within a region of observation (ROO) an agent is able to estimate the position of targets by making noisy or imprecise measures of their range and bearing. However, in order to better resolve the uncertainty in these position estimates, the agents must acquire target observations from neighboring agents and then fuse these observations with their own.

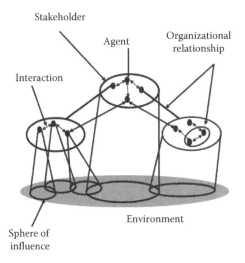

Stakeholder

Agent

Organizational relationship

Interaction

Sphere of influence

Environment

FIGURE 2.1 DDF system concept.

2.3 DDF DESIGN CONCERNS

The DDF design concerns noted in Section 2.1 are explained in more detail in the following:

- *Information recycling.* As DDF networks are dynamic and ad hoc, the information could arrive at any agent from multiple routes. Unless the information is attached with a record of its provenance to eliminate redundancy, there is a risk of recycling common information through the agents' fusion processes. This can give rise to inconsistent situation awareness throughout the system and subsequently to spurious decisions.
- *Sensor coordination.* If each agent in a DDF network determines its next action (e.g., where to look or what to communicate) without considering the actions of the other agents, their collective actions could be highly suboptimal. Unless the network is fully connected with zero propagation delay, the agents will need to explicitly coordinate their actions by communicating with each other until a set of agreed actions is reached.
- *Selfish stakeholders.* In a heterogeneous DDF system the agents may represent distinct stakeholders with different aims and objectives. If they are left to make their own selfish decisions, without any intervention from a system designer, then the overarching DDF system goals are likely to be compromised as the agents will compete for resources.
- *Trust and reputation.* One or more agents in a DDF system may not be trustworthy due to faults, bias, or malice. If these agents are unrecognized, the open nature of DDF systems would permit their false data to propagate to other agents and rapidly pollute the whole system. Thus, agents have to earn their reputations as trustworthy sources as well as estimating the trustworthiness of their information suppliers.

2.4 INFORMATION RECYCLING

Information recycling in a DDF network results in cross-correlation between the estimates of state variables generated by each agent. Ignoring this cross-correlation results in over-confident state estimates, but trying to keep track of cross-correlation requires extra book keeping operations that consume memory and bandwidth. In practice, bounds on the cross-correlation may at least be calculable and conservative estimates may be an acceptable trade-off for preserving the flexibility and scalability benefits of DDF systems. This section introduces the general theory of bounded covariance inflation (BCI) as a viable solution to the information recycling design challenge (Reece and Roberts 2005).

2.4.1 BOUNDED COVARIANCE INFLATION

If $\hat{u}$ is an estimate of the state u then P_{uu}^* is a conservative matrix for the covariance of $\hat{u} - u$ if

$$P_{uu}^* \geq E[\tilde{u}\tilde{u}^T] \quad \text{where } \tilde{u} = \hat{u} - u$$

The symbol $\geq$ denotes positive semi-definite. When u is composed by stacking two vectors, x and y say, with corresponding covariance matrices P_{xx} and P_{yy}, respectively, BCI is the procedure by which P_{uu}^* can be determined from P_{xx} and P_{yy} when the cross-covariance, P_{xy}, between x and y is unknown but bounded

$$[P_{xy} - D_{xy}]^T P_{xx}^{-1}[P_{xy} - D_{xy}] \leq S^2 P_{yy} \tag{2.1}$$

or, equivalently

$$\forall \overline{x}, \overline{y}. \ \left| \overline{x}^T R_{xx}^R (P_{xy} - D_{xy}) R_{yy} \overline{y} \right| \leq S \tag{2.2}$$

for unit vectors $\overline{x}$ and $\overline{y}$, some "centered" matrix D_{xy}, "matrix spread" S and sphering matrices R_{xx} and R_{yy} such that $P_{xx}^{-1} = R_{xx}^T R_{xx}$ and $P_{yy}^{-1} = R_{yy}^T R_{yy}$. When $D_{xy}=0$ then S is the correlation coefficient. In general, we choose D_{xy} so that S is as small as possible (see Figure 2.2).

Given this setup, it is possible to find a proven conservative covariance matrix P_{BCI}^* for all possible joint covariance matrices, P, defined by

$$P = \begin{bmatrix} P_{xx} & P_{xy} \\ P_{xy}^T & P_{yy} \end{bmatrix} \tag{2.3}$$

Define P_{BCI}^*:

$$P_{BCI}^* = \begin{bmatrix} (1 + KS)P_{xx} & D_{xy} \\ D_{xy}^T & \left(1 + \dfrac{S}{K}\right)P_{yy} \end{bmatrix} \tag{2.4}$$

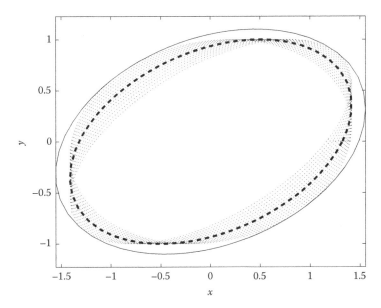

FIGURE 2.2 Family of covariance ellipses (dotted lines) for which $P_{xx}=2$, $P_{yy}=1$, and $0.2<P_{xy}<0.8$. The "centered" ellipse is shown as a thick dashed line. Also shown is a conservative ellipse (solid line) for the family for which $D_{xy}=0.5\times(0.2+0.8)$, $S=0.3$, and $K=1$.

The positive value K is called the inflation factor and is chosen to minimize the overall uncertainty encoded by the covariance matrix. In the remainder of this section we will be concerned with symmetric correlation bounds (i.e., $D_{xy}=0$). BCI with $D_{xy}=0$ effectively replaces two correlated random vectors with two uncorrelated random vectors whose covariance is guaranteed to be conservative with respect to the original vectors.

When x, y, and u are state vectors and u is related to x and y by a linear transform F, thus

$$u = F\begin{pmatrix} x \\ y \end{pmatrix}$$

then a conservative estimate P_{uu}^* of the covariance P_{uu} for $\tilde{u}$ can be obtained from a conservative covariance matrix P^* over $\tilde{x}$ and $\tilde{y}$:

$$\text{If } \hat{u} = F\begin{pmatrix} \hat{x} \\ \hat{y} \end{pmatrix} \text{then } P_{uu}^* = FP^*F^T \tag{2.5}$$

P_{uu}^* is conservative since $P_{uu}^* = FP^*F^T \geq FPF^T = P_{uu}$. Both prediction and estimation fusion operations within the Kalman filter are linear operations for appropriate choices of F. We will now derive the Kalman filter update equation for an inflated covariance matrix.

Assume $\hat{x}$ and $\hat{y}$ are correlated state estimates over the same state space, P_{xx} and P_{yy} are the corresponding covariance matrices and $\hat{u}$ is an estimate obtained by fusing $\hat{x}$ and $\hat{y}$. The inflated covariance for $\tilde{x}$ and $\tilde{y}$ is diagonal and the estimates can be considered to be uncorrelated under the inflated covariance matrix. Therefore a conservative estimate P_{uu} for $\hat{u}$ can be calculated by fusing the random vectors using the Kalman filter and the inflated covariance matrix:

$$P_{uu}^{-1}\hat{x}_u = \frac{P_{xx}^{-1}\hat{x}}{1+KS} + \frac{P_{yy}^{-1}\hat{y}}{1+(S/K)}$$

$$P_{uu}^{-1} = \frac{P_{xx}^{-1}}{1+KS} + \frac{P_{yy}^{-1}}{1+(S/K)}$$

Note that when $S=0$ we recover the information form of the Kalman filter for uncorrelated variables (Durrant-Whyte et al. 2001) and when $S=1$ and $K=\omega/(1-\omega)$ (with $\omega \in [0, 1]$) we recover covariance intersection (Julier and Uhlmann 2001). The next step is to determine upper and lower bounds on cross-correlations.

2.4.2 COUPLING SCALARS

Two correlated estimates $\hat{x}$ and $\hat{y}$ for state vector x and y, respectively, can be decomposed into orthogonal random vectors $\hat{\alpha}$, β_x, and β_y by Gram–Schmidt orthogonalization (Doob 1990)

$$\hat{x} = C_{x\alpha}P_\alpha^{-1}\hat{\alpha}+\beta_x$$
$$\hat{y} = C_{y\alpha}P_\alpha^{-1}\hat{\alpha}+\beta_y \tag{2.6}$$

where $C_{x\alpha}$ and $C_{y\alpha}$ are the cross-covariance between x and α and between y and α, respectively. Since β_x and β_y are orthogonal then

$$P_{xy} = C_{x\alpha}P_\alpha^{-1}C_{\alpha y} \tag{2.7}$$

Thus, the cross-covariance between two random vectors is the information shared between the two vectors projected onto the vector spaces.

To obtain an expression for the minimum cross-correlation bound S in Equation 2.2, first rewrite Equation 2.7

$$P_{xy} = \left[C_{x\alpha}\sqrt{P_\alpha^{-1}}\right]\left[C_{y\alpha}\sqrt{P_\alpha^{-1}}\right]^T$$

and use the Cauchy–Schwarz inequality

$$\left|\bar{x}^T R_{xx}^T P_{xy} R_{yy}\bar{y}\right| \le \sqrt{\text{maxeig}\left[R_{xx}^T C_{x\alpha}P_\alpha^{-1}C_{\alpha x}R_{xx}\right]}$$

$$\times\sqrt{\text{maxeig}\left[R_{yy}^T C_{y\alpha}P_\alpha^{-1}C_{\alpha y}R_{yy}\right]}$$

Comparing the aforementioned inequality with Equation 2.2, we observe that the right-hand side is a known bound for the cross-correlation S. The scalar

$$\Omega = \sqrt{\text{maxeig}\,[R_{xx}^T C_{x\alpha} P_\alpha^{-1} C_{\alpha x} R_{xx}]}$$

is called the coupling scalar for x and is independent of y. The coupling scalars can be calculated locally. Thus, when an agent C receives two messages, one from each of agents A and B, comprising the information vector, matrix, and coupling scalar, a bound on the correlation between the estimates in these messages is

$$S = \Omega_{AC} \times \Omega_{BC}$$

The key point of note is that the cross-correlation between two random vectors can be bounded by the product of just two scalars. This is crucial for limited bandwidth communication applications when bookkeeping messages such as the coupling scalar must be kept to a minimum. The coupling scalar can be interpreted as the fraction of the covariance matrix, which is the correlated part $C_{x\alpha} P_\alpha^{-1} \hat{\alpha}$ of $\hat{x}$. From Equation 2.6 we see that it would be possible to communicate the correlated part and the uncorrelated part β_x of the estimate $\hat{x}$ separately. The receiving agent would then be able to fuse $\hat{x}$ into its own estimate more efficiently than the method described earlier, as only the correlated part of $\hat{x}$ would have to be inflated prior to fusion. However, this alternative approach would involve nearly twice the communication load compared to the coupling scalar approach, which is undesirable in applications where there is limited bandwidth.

2.4.3 Decentralized Tracking Example

In this example three stationary agents track a dynamic process x_t and each agent maintains an estimate of the state of the target using a Kalman filter. All agents have the same behavior model of the target $x_t = x_{t-1} + v_t$ with $v_t \sim N(0, 0.1)$ and they are each able to make a measurement of the target at each time step. The ith agent's observation model is $z_{it} = x_t + \mu_{it}$ with $\mu_{it} \sim N(0, \sigma_i)$ where $\sigma^2 = \{3, 1, 0.1\}$ for the three agents, respectively. Both v and μ are uncorrelated in time and independent of each other and μ_{it} and μ_{jt} are uncorrelated for all $i \neq j$.

The agents communicate intermittently, cycling between agent 1 making contact with agent 3, then agent 3 with agent 2, and then agent 2 with agent 1. A contact takes place each five time intervals. This will correlate the agents' estimates in two ways: through information recycling and because of the fact the agents are modeling the same stochastic process.

Figure 2.3 plots the fused track covariance at each agent for various methods: BCI using both upper and lower cross-correlation bounds, BCI using upper bound only, covariance inflation, and the best and worst possible cases, namely the centralized Kalman filter and local Kalman filters without any communication. BCI is clearly a conservative but consistent performer throughout.

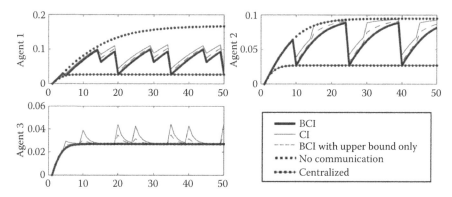

FIGURE 2.3 Agents calculated variances of the track error over time using a variety of methods.

2.5 SENSOR COORDINATION

Sensor coordination presents a fundamental design challenge for DDF systems as often physically distributed devices must act together, under computational and communication constraints, to meet system-wide goals. Consider a wide-area surveillance application in which the sensors are deployed in an ad hoc manner, for example, dropped from a military aircraft or ground vehicle. In this case, the local environment of each sensor, and hence the exact configuration of the network, cannot be determined prior to deployment. The sensors themselves must be equipped with capability to self-organize and coordinate sometime after deployment once the local environment in which they (and their neighbors) find themselves has been determined.

A common feature of these self-organization problems is that the sensors must typically choose between a small number of possible states (e.g., which neighboring sensor to transmit data to, or which sense/sleep schedule to adopt), and the effectiveness of the sensor network as a whole depends not only on the individual choices of state made by each sensor, but on the joint choices of interacting sensors. Thus, to maximize the overall effectiveness of the sensor network, the sensors within the network must typically make coordinated, rather than independent, decisions. Furthermore, this coordinated decision must be performed despite the specific constraints of each individual device (such as limited power, communication, and computational resources), and the fact that each device can typically only communicate with the few other devices in its local neighborhood (due to the use of low-power wireless transceivers, the small form factor of the device and antenna, and the hostile environments in which they are deployed). DDF systems are required to perform coordination without a central coordinator and ensure that the deployed solution scales well as the number of devices within the network increases. The max-sum algorithm is an efficient method by which decentralized sensor coordination can be achieved (Rogers et al. 2011, Waldock and Nicholson 2011).

2.5.1 MAX-SUM ALGORITHM

Consider M sensors and the state of each sensor may be described by a discrete variable x_m. Each sensor interacts locally with a number of other sensors such that the utility of an individual sensor $U_m(\mathbf{x}_m)$ is dependent on its own state and the states of these other sensors (defined by the set $\mathbf{x}_m$). The approach at this stage is generic with no specific assumptions regarding the structure of the individual utility functions.

In this setting, we wish to find the state of each sensor, $\mathbf{x}^*$, such that the sum of the individual sensors' utilities is maximized:

$$\mathbf{x}^* = \arg\max_{\mathbf{x}} \sum_{i=1}^{M} U_i(\mathbf{x}_i) \tag{2.8}$$

Furthermore, in order to enforce a truly decentralized solution, we assume that each sensor only has knowledge of, and can directly communicate with, the few neighboring agents on whose state its own utility depends. In this way, the complexity of the calculation that the sensor performs depends only on the number of neighbors that it has (and not the total size of the network), and thus we can achieve solutions that scale well.

The optimization problem defined by Equation 2.8 is represented as a bipartite factor graph. Specifically, each sensor is decomposed into a variable node that represents its state, and a function node that represents its utility. The function node of each sensor is connected to its own variable node (since its utility depends on its own state) and also to the variable nodes of other sensors whose states impact its utility. For example, we show in Figure 2.4 an example in which three sensors, $\{S_1, S_2, S_3\}$, interact with their immediate neighbors through the overlap of their sensor areas. Figure 2.4c shows the resulting bipartite factor graph in which the sensors are decomposed into function nodes, $\{U_1, U_2, U_3\}$, and variables nodes, $\{x_1, x_2, x_3\}$. The overall function represented by this factor graph is given by

$$U = U_1(x_1, x_2) + U_2(x_1, x_2, x_3) + U_3(x_2, x_3)$$

The max-sum algorithm operates directly on the factor graph representation described earlier. When this graph is cycle-free, the algorithm is guaranteed to converge to the global optimal solution such that it finds the combination of states that

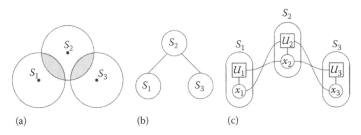

(a)   (b)   (c)

FIGURE 2.4 Sensor network showing (a) the position of three sensors whose fields of view overlap, (b) the sensor interaction graph, and (c) the resulting factor graph with sensors decomposed into function and variable nodes.

maximizes the sum of the sensors' utilities. When applied to cyclic graphs (as is the case here), there is no guarantee of convergence but extensive empirical evidence demonstrates that such family of algorithms generate good approximate solutions. The max-sum algorithm solves this problem in a decentralized manner by specifying messages that should be passed from variable to function nodes and from function nodes to variable nodes. These messages are defined as

- *From variable to function*

$$q_{i \to j}(x_i) = \alpha_{ij} + \sum_{k \in \mathcal{M}_i \setminus j} r_{k \to i}(x_i) \qquad (2.9)$$

where
$\mathcal{M}_i$ is a vector of function indices, indicating which function nodes are connected to variable node i
α_{ij} is a normalizing constant to prevent the messages from increasing endlessly in cyclic graphs

- *From function to variable*

$$r_{j \to i}(x_i) = \max_{\mathbf{x}_j \setminus i} \left[U_j(\mathbf{x}_j) + \sum_{k \in \mathcal{N}_j \setminus i} q_{k \to j}(x_k) \right] \qquad (2.10)$$

where $\mathcal{N}_j$ is a vector of variable indices, indicating which variable nodes are connected to function node j and $\mathbf{x}_j \setminus i \equiv \{x_k : k \infty \mathcal{N}_j \setminus i\}$.

The messages flowing into and out of the variable nodes within the factor graph are functions of a single variable that represent the total utility of the network for each possible value of that variable. At any time during the propagation of these messages, agent i is able to determine which state it should adopt such that the sum over all the agents' utilities is maximized. This is done by locally calculating the function, $z_i(x_i)$, from the messages flowing into agent i's variable node:

$$z_i(x_i) = \sum_{j \in \mathcal{M}_i} r_{j \to i}(x_i) \qquad (2.11)$$

and hence finding $\arg\max_{x_i} z_i(x_i)$.

The messages described earlier may be randomly initialized, and then updated whenever a sensor receives an updated message from a neighboring sensor; there is no need for a strict ordering or synchronization of the messages. In addition, the calculation of the marginal function shown in Equation 2.11 can be performed at any time (using the most recent messages received), and thus sensors have a continuously updated estimate of their optimum state. When the underlying factor graph contains cycles there is no guarantee that the max-sum algorithm will converge; nor that if it does converge it will find the optimal solution. However, extensive empirical evaluation on a number of benchmark coordination problems indicates that it does in fact produce better quality solutions than other state of the art approximate algorithms but at significantly lower computation and communication cost.

Finally, we note that if messages are continuously propagated, and the states of the agents are continuously updated, then the algorithm may be applied to dynamic problems where the interactions between agents, or the utilities resulting from these interactions, may change at any time. For example, within tracking problems where the decentralized coordination algorithm is being used to focus different sensors onto different targets, then the utilities of each sensor are continually changing due to the changing position of targets, and the actions of other sensors. Thus, by continually propagating messages each agent is able to maintain a continuously updated estimate of the state that it should adopt in order to maximize social welfare in this dynamic problem.

2.5.2 Target Tracking Example

In this section the max-sum algorithm is applied to the target tracking example illustrated in Figure 2.5. The system involves three stationary sensors, each of limited observation range shown by the gray areas bounded by the dashed lines. The sensors are initialized with a weak prior over the target positions as well as the targets (factors) they are responsible for maintaining in the factor graph. System performance is measured by the total information in the target tracks.

Three sensor management strategies were implemented: local, centralized, and decentralized. The local strategy selects the sensor (pointing) control parameter that maximizes the total information given by local observations only. The centralized

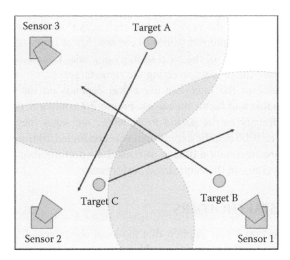

FIGURE 2.5 Example target tracking scenario with three sensors (of limited observation range indicated by the gray shaded areas) and three targets.

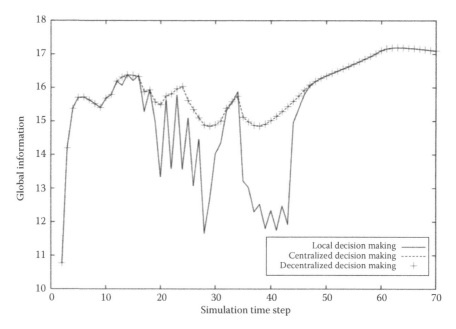

FIGURE 2.6   Performance profile for local, centralized, and decentralized sensor management strategies.

strategy selects control parameters for each sensor based on a brute-force search through all combinations to find the one that yields maximum information. The decentralized strategy solves each of the factors using a brute-force approach and then uses the max-sum algorithm to derive the optimal sensor control parameters that maximize the utility function.

Figure 2.6 displays the performance profile for each strategy. As both targets pass through the center of the environment, each sensor must handover a target to another sensor. The two handover points occur roughly at time steps 18 and 34. At these times the performance of the local strategy degrades since it cannot resolve the conflict that prevents the sensors selecting the same target.

The performance of the max-sum algorithm depends on the time allowed to exchange the variable and factor messages. Figure 2.7 compares performance with the centralized strategy as the period to exchange messages (negotiation time) is adjusted from 50 to 1000 ms (the experiment was conducted 20 times for each negotiation time). As the negotiation time is increased the performance of the decentralized strategy converges on the centralized performance.

2.6   SELFISH STAKEHOLDERS

In the previous section it was implicit that the local objectives of the sensor agents were aligned with the global objective. This situation is best modeled as a cooperative MAS problem in which the agents are designed to work toward the global

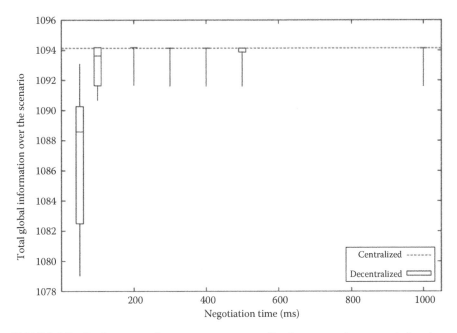

FIGURE 2.7   Performance of max-sum sensor coordination approach as negotiation time is varied.

objective of the system. This can be achieved, as we have seen, through the max-sum algorithm. In this section we consider the situation in which different stakeholders may be responsible for each sensor (or group of sensors). For example, in a disaster relief application, different governmental and nongovernmental organizations must share information gathered by their sensors to help coordinate an effective response. The sensors are now operating in a competitive rather than a cooperative environment. As such, they will attempt to optimize their own gain at a cost to the overall performance of the system. Given this, the challenge is to design a system such that desirable system-wide properties emerge from the interaction between its constituent (selfish) agents (Dash et al. 2005, Rogers et al. 2006).

Computational mechanism design offers a principled framework with which to design systems that exhibit desirable global properties, despite the selfish actions and goals of the constituent parts. It is an extension of the economic field of mechanism design and addresses the additional challenges imposed by a computational setting (i.e., agents that are computationally limited, communication that is not cost or error free, and settings that are open and dynamic). At its core, is the notion that agents hold or require valued items, and are seeking to maximize their own utility through the exchange of these items. In the real world, these items may be goods or services, and thus they will have real monetary value. In the sensor network scenario, information offers a principled currency or valuation metric. It can be applied in any context where sensors make and exchange imprecise observations and thus must deal with uncertainty.

2.6.1 PROBLEM DESCRIPTION

We consider a scenario where a number of sensors are tasked with detecting targets. The sensors each have a partial and inaccurate view of the world and need to communicate with each other in order to increase this accuracy. The "view of the world" in this case is a view of the target passing in the region that the sensors are monitoring. The communication network that the sensors use is constrained by a limited bandwidth. Thus, there is a need to globally decide on how to optimally allocate this bandwidth in order to best satisfy the sensors' overall goal of forming an accurate view of the world.

In more detail, each sensor has two regions that they consider. There is a ROO in which they can observe targets and a region of interest (ROI) they wish to monitor. Figure 2.8 depicts a typical instance of a scenario where the ROI of sensor 1 is shown and there is a target within this ROI. We can observe that agent 1 can already know about this event in its ROI since this overlaps (as it usually does) with its ROO. However, due to noise inherent in the measurement process, agent 1 will have some uncertainty in its observation (e.g., the position, type or speed of the target may be described by a probability distribution rather than an absolute value). Agent 1 can however decrease this uncertainty by gaining data about the target from other agents, namely agents 3 and 5 (which also have the target in their ROO). However, if agent 1 can only receive data from one of these two agents due to bandwidth limitations, it will then have to decide as to which agent to gain the data from. This decision making process is further complicated if the other agents also have to make similar decisions. Thus, different flows of data (i.e., descriptions of which sensors will transmit data and along which path this data will flow) will yield different results in terms of the total reduction of the uncertainty (or the equivalent increase of information). Given this, the high level representation of our problem is then to allocate the flow of data within the bandwidth constraints imposed by the communication network so as to optimize the overall gain in information each sensor has about its ROI.

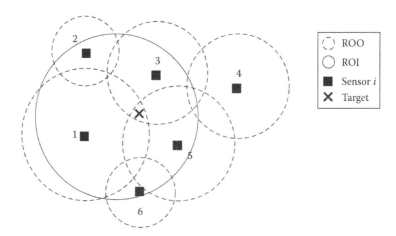

FIGURE 2.8 A multisensor network target tracking scenario.

We tackle this problem by first modeling it as a MAS. Each sensor is then viewed as an agent i, within a set of agents $\mathcal{I}$, which has data and a function x_i that characterizes the accuracy of this data. The data have a size and thus a bandwidth requirement of bw_i. We consider the simplest communication protocol that exhibits a bandwidth constraint, and thus we assume a broadcast protocol whereby any sensor can simultaneously transmit to all other sensors. The total bandwidth available for the transmission of the data is such that only a subset of the sensors can actually transmit their data.

In order to characterize this problem, we first need to make a few assumptions about the scenario:

- The time taken in calculating the allocation of data and in communicating between agents is small compared to the time taken for another target to appear. This allows us time frames where the mechanism can be implemented.
- The agents have perfect and common knowledge about the sensor-network topology and their neighbors. This removes the problem of neighbor discovery in communication systems. These assumptions thus permit us to concentrate solely on the issue of allocating the flow of data under the bandwidth constraints. We now need a way for each agent to value the data received from different agents based around the measure of data accuracy, x_i.

2.6.2 VALUATION FUNCTION

We develop a suitable valuation function based on the information form of the Kalman filter. Now, in the standard Kalman filter, observations are of the form $z(t) = H(t)y(t) + n(t)$, where $y(t)$ is the state of the system at time t, $H(t)$ is the linear observation model and $n(t)$ is a zero mean random variable drawn from a normal distribution with variance R. The covariance update component, $P^{-1}(t|t)$, of the information form of the Kalman filter for N observations is

$$P^{-1}(t \mid t) = P^{-1}(t \mid t-1) + \sum_{j=1}^{N} H^T(j)R^{-1}(j)H(j) \qquad (2.12)$$

The summation in the above expression represents the decrease in covariance and thus the gain in information at time t when all the N observations are fused. In the case of our problem, the value of receiving data from another agent can thus be represented by the gain in information resulting from this observation.

In order to achieve an efficient allocation, this gain in information must be calculated from the measure of the data prior to fusion. Thus, we can represent the measure of accuracy of data x_j, as its covariance, which is calculated from the covariance of its observation $R(j)$:

$$x_j = H^T(j)R^{-1}(j)H(j) \qquad (2.13)$$

Thus, the gain in information of agent i, when all relevant data are transmitted to it and fused, can be expressed as a sum of this measure of accuracy provided by each of the other agents:

$$v_i(\mathbf{x}) = x_i + \sum_{j \in -i} x_j \tag{2.14}$$

where $-i = \mathcal{I} \backslash i$

Equations 2.13 and 2.14 thus cast our valuation function. However, we need to modify this so as to incorporate the characteristics of our scenario. This is because all observations may not fall in an agent's ROI and furthermore an agent may not be able to receive all the data as a result of the bandwidth constraints of the communication network. Defining α_{ij} as the probability that the data observed by agent j is relevant to agent i, and a vector $\mathbf{f}$ as describing the flow of data in the network, then the expected valuation is

$$\bar{v}_i(\mathbf{x}, f) = x_i + \sum_{j \in -i} f_{ij} \alpha_{ij} x_j$$

By slight abuse of notation, we shall hereafter refer to the expected valuation $\bar{v}_i(.)$ as $v(.)$.

From the valuation function, we can observe that the valuation of an agent i depends on x_j, which are signals measured by other agents. There are two conditions that are necessary in order to achieve an efficient allocation when considering selfish agents (Jehiel and Moldovanu 2001). Firstly

$$\frac{\partial v_i(\mathbf{x}, f)}{\partial x_j} = 0 \quad \forall i, j \in \mathcal{I}$$

and secondly

$$\frac{\partial v_i(\mathbf{x}, f)}{\partial x_i} > \frac{\partial v_j(\mathbf{x}, f)}{\partial x_i} \quad \forall i, j \in \mathcal{I}, i \neq j$$

The first condition is automatically satisfied in our case since new data cannot decrease information. In the case of the second condition, it implies that we need to restrict an agent's ROI to its ROO. Otherwise, there may be an event outside its ROO that falls in its ROI such that data from another agent has a greater effect on its utility than its own data. This condition is necessary because otherwise selfish agents may profitably lie about their observed data and derive from it positive utility. Furthermore, the overlap between the agents, ROOs must be such that this condition is satisfied (i.e., $\sum_{j \in -i} \alpha_{ij} < 1$). This means that the ROO of any agent cannot be entirely overlapped by the ROO of other agents (i.e., no agent is redundant).

2.6.3 Mechanism

The aim is to ensure that the global bandwidth resource is used efficiently, that is, ensure that given the limited bandwidth, the information gain of the entire network is maximized. Thus, a mechanism is imposed whereby sensors are called upon to privately reveal the information content of observations to an auctioneer. This auctioneer then allocates the limited bandwidth of the communication network to those sensors whose observations will yield the highest system-wide information gain. However, since each sensor is individually attempting to maximize its information regarding its own ROI, with a simple mechanism there is an opportunity for a sensor to behave strategically (e.g., by understating the information content of its own observations, in an attempt to ensure that bandwidth is allocated to other sensors whose observations it can make use of or by overstating it, in order to deny bandwidth to other sensors).

Such strategic behavior is generally undesirable since it reduces the overall efficiency of the network and is computationally expensive for the individual sensors. Thus, we focus onto a subclass of mechanisms that are said to be strategy-proof or incentive compatible (Dash et al. 2003). That is, within the mechanism, the sensors have a dominant strategy (one which they should adopt regardless of the behavior of other sensors) to truthfully reveal their private information regarding the value of observations to the auctioneer. The mechanism proceeds as follows:

- Each agent i transmits to a central auctioneer its valuation function $v_i(f, \mathbf{x})$ for all the possible allocations of the information flow $f \in \mathcal{F}$, where $\mathcal{F}$ is the set of all feasible flows.
- Each agent i also transmits its observed signal $\hat{x}_i$.
- The center then computes the optimal allocation f_0^* which is calculated as

$$f_0^* = \arg \max_{f \in \mathcal{F}} \left(\sum_{i \in \mathcal{I}} v_i(f, \hat{\mathbf{x}}) \right)$$

- The center also calculates the payment r_i made by each agent i. To do this, the center first finds the m next best allocations as the signal x_i is decreased until the presence of i makes no difference to the allocations. That is, find allocations $f_1^* \ldots f_m^*$ and the signal values z_i^l such that

$$Z_i^l = \inf \left\{ y_i : \sum_{i \in \mathcal{I}} v_i\left(f_l^*, y_i, x_{-i}\right) = \sum_{i \in \mathcal{I}} v_i\left(f_{l+1}^*, y_i, x_{-i}\right) \right\}$$

(where each allocation f_i^* is different) until

$$z_i^m = \inf \left\{ y_i : \sum_{i \in \mathcal{I}} v_i(f_{m-1}^*, y_i, x_{-i}) = \sum_{i \in \mathcal{I}} v_i(f_m^*, y_i, x_{-i}) \right\}$$

where the allocation f_m^* is the optimal allocation when i does not exist, that is

$$f_m^* = \arg\max_{f \in \mathcal{F}} \sum_{j \in \mathcal{I} \backslash i} v_j(f, \mathbf{x})$$

Then the payment to buyer i is

$$r_i = \sum_{l=0}^{m-1} \left[\sum_{j \in \mathcal{I} \backslash i} v_j(f_l^*, z_i^l, x_{-i}) - \sum_{j \in \mathcal{I} \backslash i} v_j(f_{l+1}^*, z_i^l, x_{-i}) \right]$$

The earlier scheme rests upon making an agent derive a utility equal to the marginal contribution that its presence makes to the whole system of agents. Thus the additional part of this mechanism is to take into account the effect that an agent's signal x_i has on the overall utility of the system. This mechanism is general in that it can also be applied to the case of independent valuations. In our scenario, such valuations arise when the ROOs of the sensors do not overlap, and the agents are simply collecting, rather than combining, observations.

2.6.4 EXAMPLE

The mechanism was applied to a simulated sensor fusion problem which allowed the allocation of bandwidth and results of the auction process to be tracked. Figure 2.9 shows the system running.

At the specific instance in time shown in Figure 2.9, the bandwidth is severely limited. Thus, although a target falls into the ROI of both sensors 2 and 3, there is insufficient bandwidth for these sensors to exchange observations (allocated bandwidth is indicated by the thick lines between sensors). The value of information that each sensor receives from other sensors and the payments that they receive in exchange for transmitting their own observations are shown in the bar-graph at the bottom right of the display (note that sensors 1 and 4 both have negative payments since they are currently receiving more information than they are transmitting; indeed, sensor 4 is transmitting no information at all). When sensors truthfully reveal the information content of their observations, they maximize their individual information gain and maintain their budget of currency (shown on the right of the display). However, a sensor that does not adopt this strategy (due to faulty, strategic, or malicious behavior), will not achieve these aims and its budget will gradually be depleted. Such sensors can be recognized and removed from the network, thus incentivizing the truthful reporting that is necessary to ensure that the constrained bandwidth of the sensor network is allocated to achieve the system-wide goal of maximizing the information gain of the entire sensor network.

2.7 TRUST AND REPUTATION

The role of computational models of trust, within MAS in particular and open distributed systems in general, is generating a great deal of research interest. In such

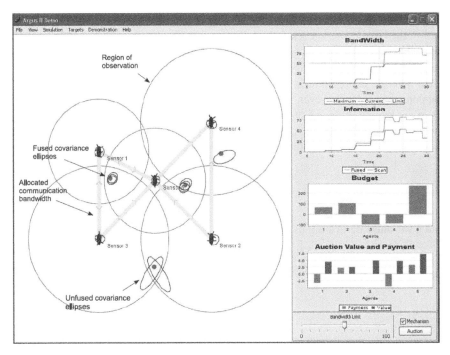

FIGURE 2.9 Example sensor network system showing the auction allocation in process and the resulting communication allocation.

systems, agents must typically choose between interaction partners, and in this context trust can be viewed to provide a means for agents to represent and estimate the reliability with which these interaction partners will fulfill their commitments. Effective trust models should allow agents to (a) estimate the trustworthiness of a supplier as they acquire direct experience, (b) express their uncertainty regarding this estimate, (c) exchange their estimates as reputation reports, and (d) filter and fuse these reputation reports with their own direct experience to yield more accurate estimates (Reece et al. 2007a,b).

This section develops a probabilistic model of computational trust that allows agents to exchange and combine reputation reports over heterogeneous, correlated multidimensional contracts. Specifically, it considers the case of an agent attempting to procure a bundle of services (e.g., audio, video, and data services) that are subject to correlated quality of service failures (e.g., due to use of shared resources or infrastructure), and where the direct experience of other agents within the system consists of contracts over different combinations of these services.

2.7.1 Expected Utility of a Contract

Consider an agent attempting to procure a bundle of services from a single supplier. In order to make a rational decision, or to negotiate a price for this bundle, the agent must estimate the expected utility of a contract with this supplier. Thus, we denote the outcome of a contract as a vector, X, that indicates whether or not each service within the bundle was successfully delivered (e.g., $X = \{o_a = 1, o_b = 0, o_c = 0, \ldots\}$ indicates that service a was successfully delivered, while services b and c were not). If $u(o_a = 1)$ is the marginal utility that the agent derives if service a is successfully delivered, then the expected utility of the agent will depend on the probability that this happens, $p(o_a = 1)$. However, neither the probabilities nor the correlations between them are known to the agent and thus it must use observations of previous contract outcomes to determine a distribution over their possible values. It can then determine an expectation of the expected utility of the contract:

$$E\left[E\left[U\right]\right] = \hat{p}(X)^T U(X) \tag{2.15}$$

and a variance, describing its uncertainty:

$$\mathrm{var}\left(E\left[U\right]\right) = U(X)^T P(X) U(X) \tag{2.16}$$

where

$$U(X) = \begin{pmatrix} u(o_a = 1) \\ u(o_b = 1) \\ u(o_c = 1) \\ \vdots \end{pmatrix}$$

Thus, the agent's estimate of the expected utility is dependent on a trust estimate composed of two expressions: a vector estimate of the probability that each service is successfully delivered

$$\hat{p}(X) = \begin{pmatrix} \hat{p}(o_a = 1) \\ \hat{p}(o_b = 1) \\ \hat{p}(o_c = 1) \\ \vdots \end{pmatrix}$$

and a covariance matrix that describes the uncertainty and correlations in these estimates:

$$P(X) = \begin{pmatrix} V_a & C_{ab} & C_{ac} & \cdots \\ C_{ab} & V_b & C_{bc} & \cdots \\ C_{ac} & C_{bc} & V_c & \cdots \\ \vdots & \vdots & \vdots & \end{pmatrix}$$

where

the diagonal terms, V_a, V_b, and V_c, represent the uncertainties in $p(o_a = 1)$, $p(o_b = 1)$, and $p(o_c = 1)$

the off-diagonal terms C_{ab}, C_{ac}, and C_{bc} represent the correlations between these probabilities

A formalism using the Dirichlet distribution allows an agent to calculate both $\hat{p}(X)$ and $P(X)$ from its direct experience of previous contract outcomes (Reece et al. 2007b). Within this formalism an agent that has observed N contract outcomes in total simply records, for each pair of services (e.g., a and b), the number of times that both were successfully delivered, n_{11}^{ab}, the number of times both were delivered unsuccessfully, n_{00}^{ab}, and both combinations in which one was delivered successfully, and the other unsuccessfully delivered, n_{01}^{ab} and n_{10}^{ab}. These counts over contract outcomes can be communicated as reputation reports, and these reputation reports can be combined by simply aggregating the counts. However, this formalism is limited to the case that contract observations are homogeneous (i.e., all agents observe contracts over the same dimension). Thus, we next consider two formalisms that address the more general case where contract observations are heterogeneous: a simple benchmark formalism using independent beta distributions (with covariance inflation) and a full formalism that uses the Kalman filter.

2.7.2 HETEROGENEOUS CONTRACTS: INFLATED INDEPENDENT BETA DISTRIBUTIONS

We can provide a reasonable benchmark formalism for dealing with heterogeneous contracts through a simple extension of a single dimensional trust model. That is, we do not explicitly represent the correlations between the services within the bundle, but rather, we use independent beta distributions to represent each individual service. Thus, if an agent has direct experience of N previous contract outcomes, in which service a was successfully delivered n_a times, then the trust estimate, $\hat{p}(X)$, can simply be calculated using the standard result from the beta distribution that

$$\hat{p}(o_a = 1) = \frac{n_a + 1}{N + 2} \tag{2.17}$$

Similarly we can calculate the diagonal terms of the covariance matrix, $P(X)$, by again using the standard result from the beta distribution that

$$V_a = \frac{(n_a + 1)(N - n_a + 1)}{(N + 2)^2 (N + 3)} \tag{2.18}$$

Finally, rather than explicitly calculating the off-diagonal elements of the covariance matrix, we can employ the covariance inflation method from Section 2.4 to derive a conservative covariance matrix by simply setting the off-diagonal elements to zero, and multiplying the diagonal variance terms by the number of dimensions in the state vector, X. Thus in the case of two services we have

$$P(X) = \begin{pmatrix} 2V_a & 0 \\ 0 & 2V_b \end{pmatrix}$$

We now develop a more sophisticated approach using the Kalman filter to fuse heterogeneous estimates containing correlation information.

2.7.3 HETEROGENEOUS CONTRACTS: A KALMAN FILTER TRUST MODEL

The Kalman filter trust model operates by fusing an agent's prior trust estimate (calculated from an agent's own direct experience of previous contract outcomes) with reputation reports that are received from other agents in order to give a posterior trust estimate. As described earlier, these trust estimates are represented by a vector, $\hat{p}(X)$, and a covariance matrix, $P(X)$, and the standard form of the Kalman filter provides two equations to update these:

$$\hat{p}_{\text{posterior}} = \hat{p}_{\text{prior}} + K(o - \hat{p}_{\text{prior}})$$

$$P_{\text{posterior}} = (I - K)P_{\text{prior}}$$

where K is the Kalman gain

$$K = P_{\text{prior}}(P_{\text{prior}} + R)^{-1}$$

and o is an observation with covariance R, that together represent the reputation reports received from other agents.

Now, when we have heterogeneous contracts, one or more dimensions of either the prior estimate or the reputation reports may be missing. Within the Kalman filter framework we can simply represent these missing contract observations by setting the corresponding diagonal elements of the covariance matrix to infinity. By doing this we are effectively saying that the estimate for this contract part has no certainty.

In fact, performing these matrix operations involving infinity can be problematic. We can avoid this by using the information form of the Kalman filter whereby an estimate is represented by its precision, Y, which is the inverse of the correlation matrix (i.e., $Y = P(X)^{-1}$), and its information estimate, $\hat{y}$, which is the product of the precision and the state estimate (i.e., $\hat{y} = P(X)^{-1}\hat{p}(X)$).

In this case, the missing information can be represented by inserting zeros into the precision matrix, and as before, the Kalman filter allows us to combine reputation reports with prior beliefs to yield a posterior information estimate and precision matrix:

$$\hat{y}_{\text{posterior}} = \hat{y}_{\text{prior}} + \hat{y}_0$$

$$Y_{\text{posterior}} = Y_{\text{prior}} + Y_0$$

where $Y_0 = R^{-1}$ and $\hat{y}_0 = R^{-1}o$. The information form of the Kalman filter is particularly useful within MAS since reputation reports from multiple agents are simply added (in any order) to an agent's prior estimate. However, the two forms are exactly equivalent, and we can easily switch between the two.

Thus, having presented the Kalman filter in the context of a computational trust model, we describe how an agent's prior estimate is calculated from its own direct experience, and how other agents can communicate reputation reports calculated from their own direct experience.

The prior belief of the agent is represented by a trust estimate, $\hat{p}(X)$, and a covariance matrix, P(X). These can be calculated from an agent's direct experience using the Dirichlet formalism noted earlier. More specifically $\hat{p}(X)$ and the diagonal elements of $P(X)$ are calculated from the counts of contract outcomes (as per Equations 2.17 and 2.18), while the full details of the Dirichlet distribution are required to calculate the off-diagonal terms of $P(X)$ (Reece et al. 2007a). The prior explicitly represents the correlations over the subset of services for which the agent has directly observed previous contract outcomes. When the agent has no direct experience of some services, it may simply insert infinity into the relevant diagonal element of $P(X)$ to reflect this lack of information (or alternatively insert zero into Y if the information form of the Kalman filter is being used).

The Kalman filter fuses a prior estimate with an observation, o, whose covariance is R. In our computational trust model, o and R together represent a reputation report and are calculated from the direct experience of the originating agent. This calculation is different from that which generates $\hat{p}(X)$ and $P(X)$, since the covariance R describes the variability of o about the true probabilities, $p(X)$, while the covariance $P(X)$ describes the variability of $p(X)$ about the estimate $\hat{p}(X)$. This is a subtle but important difference. Calculating o is straightforward since it is a vector estimate of the probability that each service is successfully delivered (i.e., $o = \{o_a, o_b, o_c, \ldots\}$). It is calculated from an agent's previous contract outcomes, and thus if the agent has observed N contracts in total, and service a was successfully delivered in n_a of these, then $o_a = n_a/N$. Note that due to the reasons described earlier, this expression is different from that shown in Equation 2.17.

Calculating R is more complex. Since we are using the Kalman filter with a Dirichlet distribution (rather than the more common Gaussian distribution), the covariance, R, is itself dependent upon the probabilities that each service is successfully delivered, $p(X)$. These probabilities are not known; indeed, these are what we are attempting to estimate. However, the beauty of the Kalman filter lies in its flexibility and we need not worry about finding R exactly. Provided that we can find a conservative matrix, R^*, to use in place of R, we can guarantee that our estimates will remain consistent. We can build such a conservative covariance matrix for R from an agent's direct experience and the method of covariance inflation described in Section 2.4.

2.7.4 EMPIRICAL EVALUATION

To evaluate the effectiveness of the trust formalisms just described, we present simulation results in which ten agents, each with their own direct experience of a supplier that provides two services, participate within a reputation system. We assume that one of these agents is attempting to evaluate the trustworthiness of the supplier in order to calculate the expected utility of interacting with it. As such, the agent must fuse its own direct experience with reputation reports received from the other nine agents.

In each simulation run, contract outcomes are drawn from an arbitrary joint distribution that induces correlations between the services. The contract outcomes are randomly allocated such that some agents observe both services, while others observe just one service. We apply the trust formalisms to calculate posterior trust estimates and then calculate two metrics. The first is a scalar measure of the information content of the trust estimate; a standard way of measuring the uncertainty encoded within the covariance matrix (Bar-Shalom et al. 2001). More specifically, we calculate the determinant of the inverse of the covariance matrix

$$I = \det(P(X)^{-1})$$

and note that the greater the information content, the more precise $\hat{p}(X)$ will be. The second metric measures the normalized error of the estimate:

$$E = \left[\hat{p}(X) - p(X) \right]^T P(X)^{-1} [\hat{p}(X) - p(X)]$$

We perform 1000 repeated simulation runs and calculate the expectation of these two metrics (and the standard error in these expectations). We note that the expectation of the normalized error is commonly termed the normalized standard error and it describes the consistency of the estimate. A consistent estimate has a normalized standard error less than the cardinality of the trust estimate; two in this case. A normalized standard error much less than this value indicates that the covariance matrix is too conservative.

In Figure 2.10 we present these results (with the standard error in the expected values shown as error bars) as the number of contract observations ranges from 10 to 400. We note that the information content of the trust estimates generated by the Kalman filter formalism far exceeds that of those generated using inflated independent beta distributions (typically by a factor of 3). By explicitly representing the correlations between the services, our formalism generates more precise trust estimates. This increased precision is not realized at the cost of producing inconsistent estimates; the normalized standard error of both formalisms is less than two, and thus they both generate consistent estimates. Finally, we note that as the number of contracts increases, the Kalman filter encodes more precise correlation information, and the difference between the formalisms also increases.

Table 2.1 illustrates the effect that the precision of the trust estimate has on an agent's estimate of the expected utility of a contract (calculated using the relationships shown in Equations 2.15 and 2.16 in an example setting where $u(o_a = 1) = 2$ and $u(o_b = 1) = 6$). While both formalisms generate estimates of expected utility close to the true distribution, the more precise covariance matrix of the Kalman filter results in a better estimate of the standard deviation of the expected utility (while that of the inflated independent beta distribution is approximately double the true value).

In summary, we have developed a trust formalism based on the Kalman filter that represents trust as a vector estimate of the probability that each service will be successfully delivered, and a covariance matrix that describes the uncertainty

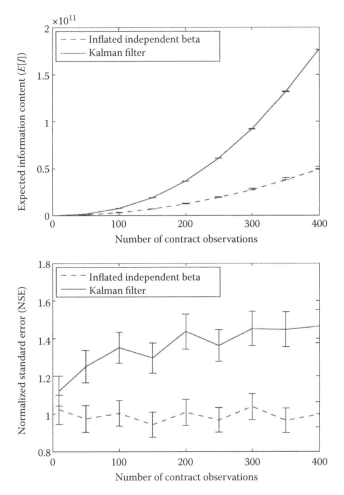

FIGURE 2.10 Comparison of the expected information content, $E[I]$, and normalized standard error (NSE) for trust formalisms using the Kalman filter and independent beta distributions.

TABLE 2.1

Estimated Expected Utility and Its Standard Deviation Calculated from an Agent's Posterior Trust Estimate

Method	$E\big[E[U]\big] \pm \sqrt{\mathrm{Var}\big(E[U]\big)}$
True distribution	5.80 ± 0.27
Inflated independent beta	5.86 ± 0.53
Kalman filter	5.82 ± 0.34

and correlations between these probabilities. We have described how the agents' direct experiences of contract outcomes can be represented and combined within this formalism, and we have empirically demonstrated that the formalism provides significantly better trustworthiness estimates than the alternative of using separate single-dimensional trust models for each separate service (where information regarding the correlation between each estimate is lost).

2.8 FUTURE DESIGN CONCERNS AND OPPORTUNITIES

As cheap sensing and computation increasingly pervades the world around us, it will profoundly change the ways in which we work with computers. Rather than issuing instructions to passive machines, we will increasingly work in partnership with highly interconnected computational components (agents) that are able to act autonomously and intelligently. Humans and software agents will continually and flexibly establish a range of collaborative relationships with one another, forming *human-agent collectives* (*HACs*) to meet their individual and collective goals.

This vision of people and computational agents operating at a global scale raises a very significant design concern that must be faced as we shift to becoming increasingly dependent on systems that interweave human and computational endeavor. As systems based on HACs grow in scale, complexity, and temporal extent, we will increasingly require a principled science that allows us to reason about the computational and human aspects of these systems if we are to avoid developments that are unsafe, unreliable, and lack the appropriate safeguards to ensure societal acceptance.

2.8.1 HAC DESIGN CONCERNS

The global scale and decentralized nature of HACs mean that control and information will be widely dispersed between a large number of potentially self-interested, actors with different aims, objectives, and availabilities. These features of HAC raise the following design challenges:

- Understand how to provide *flexible autonomy* that will allow agents to sometimes take actions in a completely autonomous way without reference to their human owner, while at other times being guided by much closer human involvement in key decisions
- Discover the means by which groups of agents and humans can exhibit *agile teaming* and come together on an ad hoc basis in order to achieve a goal that none of the individuals can achieve in isolation and then disband once the cooperative action has been successful
- Elaborate the principles of *incentive engineering* in which actors' rewards are designed in such a way that the actions that the participants are encouraged to take, when amalgamated, generate socially desirable outcomes
- Design and develop an *accountable information infrastructure* that can provide a step change in situational awareness by blending sensor and crowd generated content in a robust and reliable way, and developing mechanisms that allow its veracity and accuracy to be confirmed and audited

How to solve these challenges and establish the new science needed to understand, build, and apply HACs in the real world is still very much a subject in its infancy. It is sure to draw on the DDF methods described in this book, but they will need to be enriched with insights, understanding, and methodology resulting from a broader multidisciplinary approach involving artificial intelligence, agent-based computing, machine learning, decentralized information systems, participatory systems, and ubiquitous computing.

2.8.2 HAC Opportunities

If these challenges can be solved they will help us meet some of the key societal challenges of sustainability, inclusion, and safety that are crucial to our future. To conclude, let us consider three application domains that are expected to be significant beneficiaries:

Disaster response: Effective disaster response requires rescue services to make critical decisions in the face of an uncertain and rapidly changing situation. We aim to develop systems that allow first responders and software agents to work effectively together in such situations to collect the best possible information from the environment (though diverse sources such as CCTV feeds, UAVs, and crowd generated content), in order to most effectively manage and coordinate the various rescue resources available. Key technologies to achieve these aims include (i) decentralized coordination algorithms that can effectively allocate resources in the absence of centralized control, (ii) methodologies to flexibly handle autonomy so that the decisions that are autonomously made by software agents can be continuously changed as needs arise, and (iii) the ability to track the provenance of information and decisions such that previous decisions can be updated as new information comes to light.

Smart grid: Developing a modern electricity grid where information flows in both directions between consumers and producers is critical to achieving worldwide carbon reduction targets. HACs are an essential part of this vision, for example, the use of agents (or "energy avatars") that are capable of continuously monitoring, predicting, and feeding back information about energy generation and consumption within the grid, in order to satisfy individuals' preferences for cost, carbon, and comfort. Some requirements in support of these aims are (i) coalition formation algorithms that allow multiple self-interested parties such as renewable generators to come together with consumers to create virtual power plants that can more effectively manage the intermittent nature of these energy sources, (ii) algorithms to generate effective short term predictions of demand and supply to allow the optimization of energy use, and (iii) accountable information infrastructure to ensure the information provided to users on their smart meters is easily understandable, credible, and auditable for billing purposes.

Citizen science: Scientific research projects are increasing turning to citizen scientists to help solve problems that defy conventional computational approaches, for example, the Zooniverse projects in astronomy (zooniverse.org). These projects require approaches that allow such problems to be solved at scale, making full use of the skills, preferences, and capabilities of the volunteer participants. To make effective use of volunteer participants within these settings there is a need to develop

(i) algorithms to model and predict the accuracy and trustworthiness of citizen generated content, (ii) methodologies and data models that allow us to track and reason about the provenance of information collected in this way, and (iii) mechanisms that allow us to target which volunteers are asked which questions based on learned models of their capabilities.

ACKNOWLEDGMENTS

The research described in Sections 2.4 and 2.5 was undertaken as part of the ARGUS II DARP project. This was a collaborative project involving BAE Systems, QinetiQ, Rolls-Royce, the University of Oxford, and the University of Southampton. It was funded by the industrial partners together with the EPSRC, MOD, and DTI. The research described in Section 2.6 was funded by the Systems Engineering for Autonomous Systems Defence Technology Centre. The research described in Section 2.7 was undertaken as part of the ALADDIN project and was jointly funded by a BAE Systems and EPSRC strategic partnership (EP/C548051/1). The research described in Section 2.8 is being undertaken as part of the ORCHID project funded by EPSRC.

REFERENCES

Bar-Shalom, Y., X.-R. Li, and T. Kirubarajan. 2001. *Estimation with Applications to Tracking and Navigation.* Hoboken, NJ: Wiley Interscience.

Dash, R.K., D.C. Parkes, and N.R. Jennings. 2003. Computational mechanism design: A call to arms. *IEEE Intelligent Systems*, 18(6): 40–47.

Dash, R., A. Rogers, S. Reece, S.J. Roberts, and N.R. Jennings. 2005. Constrained bandwidth allocation in multi-sensor information fusion: A mechanism design approach. *Proceedings of the Eighth International Conference on Information Fusion*, Philadelphia, PA.

Doob, J.L. 1990. *Stochastic Processes.* New York: Wiley.

Durrant-Whyte, H.F., M. Stevens, and E. Nettleton. 2001. Data fusion in decentralized sensor networks. *Proceedings of the Fourth International Conference on Information Fusion*, Montreal, Quebec, Canada, pp. 302–307.

Jehiel, P. and B. Moldovanu. 2001. Efficient design with interdependent evaluations. *Econometica,* 69(5): 1237–1259.

Julier, S.J. and J.K. Uhlmann. 2001. General decentralized data fusion with covariance intersection (CI). In *Handbook of Multisensor Data Fusion*, D.L. Hall and J. Llinas, eds., Boca Raton, FL: CRC Press, Chapter 12.

Reece, S. and S.J. Roberts. 2005. Robust, low-bandwidth, multi-vehicle mapping. *Proceedings of the Eighth International Conference on Information Fusion*, Philadelphia, PA.

Reece, S., A. Rogers, S.J. Roberts, and N.R. Jennings. 2007a. A multi-dimensional trust model for heterogeneous contract observations. *Proceedings of 22nd AAAI Conference on Artificial Intelligence*, Vancouver, British Columbia, Canada, pp. 128–135.

Reece, S., A. Rogers, S.J. Roberts, and N.R. Jennings. 2007b. Rumours and reputation: Evaluating multi-dimensional trust within a decentralized reputation system. *Proceedings of Sixth International Joint Conference on Autonomous Agents and Multi-Agent Systems (AAMAS-07)*, Honolulu, HI, pp. 1063–1070.

Rogers, A., R. Dash, N.R. Jennings, S. Reece, and S.J. Roberts. 2006. Computational mechanism design for information fusion within sensor networks. *Proceedings of the Ninth International Conference on Information Fusion*, Florence, Italy.

Rogers, A., A. Farinelli, R. Stranders, and N.R. Jennings. 2011. Bounded approximate decentralized coordination via the max-sum algorithm. *Artificial Intelligence*, 175(2): 730–759.

Waldock, A. and D. Nicholson. 2011. A framework for cooperative control applied to a distributed sensor network. *The Computer Journal*, 54(3): 471–481.

Zooniverse. n.d. Real Science Online. www.zooniverse.org (accessed on March 16, 2012.)

3 Network-Centric Concepts

Impacts to Distributed Fusion System Design

James Llinas

CONTENTS

3.1 Introduction 47
3.2 Value Chain Concepts 48
3.3 Value Chain Process 49
3.4 Value of Information in Decision-Making 51
3.5 Role of Fusion 52
3.6 Sense-Making 53
3.7 Nature and Processes of Sense-Making 54
3.8 Role of Fusion 57
3.9 Self-Organization and Self-Synchronization in the Value Chain 60
3.10 Complexity in Sense-Making and Command and Control 61
3.11 Summary 63
References 63

3.1 INTRODUCTION

The history of network-centric concepts in the United States can be said to go back at least to the mid-1980s when the U.S. Defense Department was reorganized under the Goldwater–Nichols Act of 1986 that imputed the notions of "jointness" onto U.S. defense and military operations. Ten years later U.S. Admiral William Owens, in a paper for the Institute of National Strategic Studies at the National Defense University, wrote on the concept of "The Emerging U.S. System of Systems" (Owens 1995) as the foundation of the "Revolution in Military Affairs," involving the extensive use of (and dependency on) information in a layered system framework connecting various military operational functions. A sequence of publications evolved that introduced the notions of net-centricity and eventually the military notion of network-centric warfare (NCW), in which the strong informational dependency persisted. In the networked case, which allows (or should allow) extensive sharing of information, the argument was that NCW enabled the following operational advantages:

- A robustly networked force improves information sharing.
- Information sharing enhances the quality of information and shared situational awareness.
- Shared situational awareness enables collaboration and self-synchronization and enhances sustainability and speed of command.
- These, in turn, dramatically increase mission effectiveness.

In these arguments, combat power is seen to be dependent on information. Related to these ideas, Evans and Wurster (2000) introduced the concepts of information richness and seek to explain how the Internet has changed the economics of *information reach* and the ability of information to create value. In this work, they defined information richness as an aggregate measure of the quality of information and *information reach* as an aggregate measure of the degree that information is shared. Alberts et al. (2001) add the parameter of "quality of interaction" to these factors as influencing the ability to create value, in this case combat value. So it can be argued, following these developments, that combat power and mission effectiveness depend on information quality, information "share-ability," and the nature of interaction among people using information. In a network environment, every node has an opportunity to create information but also to modify it (say, improve its quality), send it forward to other nodes (expedite the sharing of that information), and the people at that node can interact with the information in a way that exploits it for task purposes. Thus, there is the potential for a "chain" of effects that impacts the overall combat value in such a system of systems; i.e., a "value chain" is a latent construct in any information network.

3.2 VALUE CHAIN CONCEPTS

The term "value chain" is cited in the various open works on NCW or network-enabled capability (Alberts et al. 2001), but other sources suggest the term was coined by Michael Porter in 1985 (Porter 1985). The concept is an abstraction related to business processes that operate on a product as part of the product development, and the notion that each process should add value to the product. It seems to be a concept primarily useful for strategic planning that exposes the cost and value drivers at each stage of product development as a basis for analyzing and discerning the best trade-off choices to make toward optimization of value and minimization of cost.

The term has been extended by the business community to apply to broadly based, multi-organizational processes under the phrase "value network," which seems to be particularly applicable to service industries and processes involving nontangible components and products. It is generally presumed in the discussions about value networks that there is a dedicated and altruistic intent among the collaborators to fully cooperate through synchronized interactions toward the single purpose of product value optimization. Clearly, inter-agent communication is crucial to realizing the benefits of a value network (as argued in Alberts et al. [2001], where the "quality of interaction" parameter is introduced, as previously noted), and the overall system can be and usually is complex and exhibiting a variety of inter-agent dependencies, not unlike the complexities in a social network.

The value chain in the NCW case is descriptive of the interdependencies among, and value contributions of, the links from network-centric organizations and improved (value-adding) information processes—and information products— to more effective mission outcomes. As will be discussed later, there are two core assertions that underlie this concept: (1) that the collaborative framework that the net infrastructure provides will improve the quality of *organic (individual-node)* information, and (2) that the same net infrastructure will provide for *improved shareability* of information, in turn leading to more creative, agile, and timely situation assessments and decision making. As noted earlier for the business case, here too there is an assumption of an altruistic imperative and that the network nodes are cooperatively working toward a common goal. This is not unreasonable as an ideal goal but its realization is likely to depend on the specifics of given mission applications and the usual effects of the "fog of war," and mission risks and urgencies in the defense or military context. Even among friendly forces, it is not always the case that the entire force is pulling in the same direction due to localized and random factors.

Also, no small part of the realization of the potential of NCW and the promise of the value chain process will be the willingness of the military to commit to the underlying open, cooperative, and proactive degree of information-sharing that these concepts depend on. As pointed out by Alberts and Hayes (2003), it was not too long ago that the phrase "Knowledge is Power" was employed to convey the notion that possession and control of information (i.e., making it scarce and not sharing it) was a means to achieve power and control. This paradigm thus argues for the control and caching of information, rather than sharing it and generally making it available. In part, these contrasting views relate to the economics of information availability in the general sense as well as the cost of sharing it. With the emergence of the web and the dramatic reductions in the availability-costs of extensive amounts of information and in the marked reductions of all types of networking costs comes the push for a new paradigm that factors sharing into the value-adding processes rather than purposefully resisting it. Of course, this will require a degree of revolution in the way "information-age forces" are structured and in the way they interoperate and in particular how they share information. Military organizations will need to go well beyond the current centralized planning-decentralized execution paradigm to the structures discussed in Alberts and Hayes (2003) to realize much more organizational agility and to empower those at the edge of organizations to decide about information sharing and action-taking.

3.3   VALUE CHAIN PROCESS

Determination of whether the asserted benefits of the tenets of NCW and in particular those of the value chain can be realized begins with understanding the degree to which a force is in fact networked or connected. As is well known, connectivity at the information level is the result of a multilayered process; it begins with the physical connection layer (wires, fiber, transmitters/receivers) but goes well beyond that layer and in the military environment of course involves multilayer security issues and accessibility controls to information. We note the important requirement that to exploit and fuse shared information one must have to have been sent it from somewhere in the network, which in turn depends on what we call "information-sharing strategies (ISS)," those protocols or policies that define who sends what to whom, how often, and in what format. And as has been mentioned earlier, effective and efficient collaboration also presumes the unified focus and altruistic intentions of those nodes in a network that can contribute to the improved problem solution actually doing so, even under combat duress and confusion.

The NCW literature has various diagrammatic representations of the value chain; here we use a simple construct in Figure 3.1 depicting the process and its important components and functions, showing how value is built up in the course of "good" network operations. The figure shows that the first requirement to enable NCW is connectivity via some type of network infrastructure. Shared observational data, data fusion, and information management, done well, lead to significantly improved situational awareness, which when properly shared and integrated into a (possibly-new paradigm of) command and control (C2) and decision-making environment have the potential to yield measurable improvements in mission effectiveness. Closely related to the concept of the value chain is the "conceptual framework" of NCW, depicted here again using the diagram from the Network Centric Operations Conceptual Framework report prepared by Evidence Based Research, Inc. (2003) as Figure 3.2.

Most of Figure 3.2 is, first of all, all about information and its flow in the network but it is (toward the bottom) also about the use of the information in decision-making and action-taking. Important themes in this framework revolve around a few special words and the implied functions: quality—sharing—degree—synchronization. Also a new term appears: "sense-making." Notice also that many of these terms and the associated functions happen to "a degree," and ideally should be measurable through the development of relevant metrics; more is said on this in Chapter 17. Finally, not shown here but important to note in any case is that certain functions are in certain

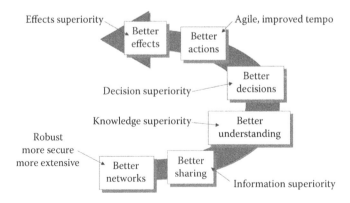

FIGURE 3.1 Network-centric value chain concept diagram.

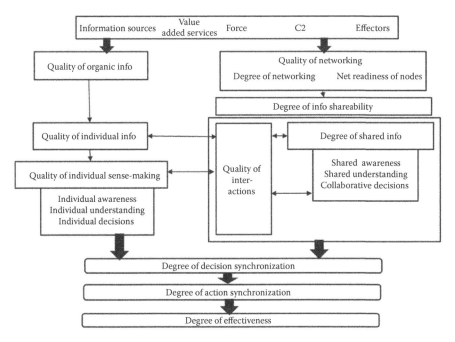

FIGURE 3.2   NCW conceptual framework. (From Evidence Based Research, Inc., Network Centric Operations Conceptual Framework Version 1.0. 2003, Report prepared for Office of Force Transformation, November 2003.)

domains, across the physical, informational, cognitive, and social categories; some involve more than one domain. Everything begins here with improvements in the quality of information at some node; nothing of quality can happen across nodes if the individual nodes have nothing to offer.

3.4   VALUE OF INFORMATION IN DECISION-MAKING

One way to measure the quality of information at a node is by its contribution to both the local and team or network-level decision-making and action-taking that results from employing that information. Usually, the outcomes of actions taken in the context of estimated situational states may be assigned values or *utilities*, which represent the relative desirability of outcomes. This type of approach is typical for cases where rational decision-making and choice-making is appropriate. However there are many modern-day problems, e.g., asymmetric problems, that do not lend themselves to the rational choice, rational decision-making paradigm. If we denote a possible situation state as s, as an instance of S, and utility of action a, given s as U(a,s), we can describe the expected utility as

$$E\{U(s)\} = \sum_{s \subset S} P(s)U(a,s) \qquad (3.1)$$

where P(s) is the probability of any situation s. The situation, s, however is typically unobservable in a direct sense and can be treated here as being estimated by a fusion process on the basis of observable evidence e (of all possible observational data or other evidence, E). That is, the fusion process, assuming that it has a multi-hypothesis capability, produces the distribution of estimated situations P (s|e). If the maximum utility is that associated with taking the optimal action, then we have the maximum of the expectation as follows:

$$\text{Max}\left[E\{U(S\mid e)\}\right] = \max a \subset A \sum_{s \subset S} P(s\mid e)U(a,s) \tag{3.2}$$

If we want to gauge the value of any observable evidence or information e, assuming that what is being shared in the network is observational data or measurements, then we can marginalize over the possible values of e as

$$\text{Max}[E\{U(S\mid E)\}] = \sum_{e \subset E} P(e)\text{Max}[E\{U(S\mid e)\}] \tag{3.3}$$

The value of any observable informational element can then be computed as the difference in maximum utility when the information is included in the above vice excluding it. A similar calculation could be done if what are shared are situational estimates by using slight variations of these equations, using the marginal value of any situational estimate s. The viability and ability to implement calculations of this type will vary from case to case, but some type of quality measures are needed to drive the value-chain process; as footnoted previously, the Network Centric Operations Conceptual Framework report (Evidence Based Research, Inc. 2003) has a rather thorough characterization of a holistic approach to measuring the various "ilities" associated with the value chain process.

3.5   ROLE OF FUSION (1)

It is important here to make a "fusion" remark in light of the implications of Figure 3.2. Any fusion node can only fuse two things: that information which is available to it organically—i.e., information over which it has control, such as locally managed sensor devices—and that information which comes to it *somehow* from the network. Notice the emphasis on "somehow"; it is only through the aforementioned ISS that some type of information flows to a node from the network. Such flows can be the result of a multiplicity of interwoven ISSs such as broadcasts from some nodes, responses to service requests from other nodes, and flows from nodes that the receiving node subscribes to, or yet other flow patterns driven by specified protocols. But it is emphasized that the nature of "non-organic" fusion that can happen at a node is only the result of the synthesis of any such directed or requested (and responsive) information flows, which in turn are the result of defined protocols and policies. A related remark is that fusion can be (should be, if well designed) a contributor to the quality of information and quality of sense-making and understanding, both at the individual or nodal level as well as at the shared level. It could be also argued that the "Level 4: Process Refinement" function of the fusion process could contribute to the nature of the information sharing and other inter-nodal interactions in a positive way, depending on the control authority aspects of how the network is managed.

Further, fusion process design is often spoken of as impacted by "push" requirements—those requirements driven by the input-side, and "pull" requirements, driven by the user-side. The network environment influences both of these requirements-sets in possibly many ways. It can be said that the information flow in the network can be characterized as both delayed and out-of-observation-time-order, and probably Poisson in arrival-rate distribution, all of which could potentially affect fusion algorithm and process operations. New user patterns involving self-organizing and self-synchronizing organizational dynamics will also likely affect how fused information products should optimally be constructed and delivered for use.

3.6   SENSE-MAKING

Following the flow of Figure 3.2, "sense-making" is a process and desired capability at both the individual node level and at the network level. It can be individualized to a person in which case the process would be largely cognitive with some degree of automated support at the individual level. For any netted level of sense-making capability whether within a sub-network at a node or across nodes, the sense-making process relies largely on patterns of collaboration and information exchange. As might be expected, the sense-making term seems to have a number of nominated definitions; a few are offered here to give a sampling:

- Sense-making as making sense of uncertainties in environments through interaction (Weick 1969).
- Sense-making encompasses the range of cognitive activities undertaken by individuals, teams, organizations, and indeed societies to develop awareness and understanding and to relate this understanding to a feasible action space (Alberts 2002).
- Sense-making is defined as the process of creating situation awareness in situations of uncertainty (Leedom 2001).
- Sense-making consists of a set of activities or processes in the cognitive and social domains that begins on the edge of the information domain with the perception of available information and ends prior to taking action(s) that are meant to create effects in any or all of the domains (Alberts and Hayes 2006).

One common theme through the definitions seems to be the notion of dealing with and clarifying an estimated world view while dealing with uncertainty, anomalies, and contradictions. Sieck et al. (2007) depict individualized sense-making as a six-step frame-building process (frames associated to mental representations in this approach), involving sub-processes that seek a frame, and elaborate, question, compare, reframe, and preserve the frame in an iterative process. Each step involves some type of adjudication or reconciliation process to deal with classes of complexity or uncertainty and ambiguity. In this process then, the drive to reduce uncertainty may not be immediately helpful since part of the sense-making process can be to understand the implications of uncertainty and ambiguity. The problem spaces addressed by sense-making processes involve an incomplete understanding of reality and are thus ontologically incomplete; they are also epistemologically incomplete in that available knowledge models are not adequate to describe the observed phenomena. Table 3.1 from Zack (1999) offers a characterization of types of ignorance that sense-making must deal with.

As pointed out in the sense-making literature (McCaskey 1982), the sense-making process is not constrained by the usual models and assumptions of rational decision-making, and a generalized maximization of a type of a utility-type function on the part of the decision-maker. Modern-day adversaries can be expected to act "irrationally" at least by certain standards, and certain arguments suggest that friendly decision-makers need to be equally "irrational" in their decision-making processes. Uncertainty reduction and optimization methods work well in support of the rational choice/rational decision-making model but may warrant reexamination as part of a sense-making process involving a collaborative situation assessment process that is constructing a subjective view of an unknowable, dynamic world and largely dealing with overt deception, equivocal information and the reconciliation of alternative views among the networked decision-making team. The use of bounded rationality models helps in this regard but such models are not the same as the typical descriptions of sense-making. In the sense-making case, it could be said that the networked group is constructing *an interpretation of some complex reality sufficient to achieve a state of commitment to that interpretation and the decisions and actions that may result from it.* This notion interacts with the concept of self-organizing teams in that the sense-making process is a logical precursor to a team setting its own goals and objectives for both action-taking and information-seeking. It could be said that a team can only be labeled as self-organizing if it dynamically sets its own goals and objectives. Commanders then need to limit themselves to presenting the team with an ambiguous challenge rather than defining terms of reference, etc.; whether traditional militaries can adapt to this process is to be seen. Moreover, most fusion processes operate on what could be called explicit information and to varying degrees may not exploit tacit knowledge and contextual information.

McCaskey (1982) offers the list shown in Table 3.2 of types of problems and questions that sense-making type processes are intended to address. It could be said that these are problems involving degrees of bewilderment for analysts or decision-makers. The term "wicked" has also been used to typify such problems involving contradicting information, discrepancies, etc., and the need for problem-solvers to significantly change their mindsets and shed historical preconceptions; see Rittel and Webber (1973).

3.7   NATURE AND PROCESSES OF SENSE-MAKING

Sense-making is sometimes labeled as "constructive reality" and a process that is action-centered and retrospective. This is similar to what some in the fusion community have called "stimulative intelligence," which involves taking actions to stimulate an adversary to an action that is either observable or that aids in clarifying a hypothesis. Such strategies will generally be more successful at the physical level, e.g., when trying to cause actions that manipulate physical objects, but both harder

TABLE 3.1
Forms of Ignorance

Form of Ignorance	Definition	Corrective Response
Uncertainty	Uncertainty is defined as not having sufficient information to describe a current state or to forecast future states, preferred outcomes, or the actions needed to achieve them. Uncertainty can be defined in degrees (e.g., in terms of probability); however, the context of uncertainty is well-defined and meaningful to decision-makers.	Uncertainty can be reduced be acquiring additional information relevant to the problem context. Uncertainty can be tolerated by using assumptions to fill in missing information, or by developing agile responses that can accommodate critical areas of uncertainty.
Complexity	Complexity is defined as being faced with a situation made of an inter-related set of variables, solutions, and stakeholders – each individually understood, but together with exceed the processing capacity of the individual, the team or organization to synthesize. Complexity is defined relative to available experience and expertise: what is complex for one individual might be easily understood by another.	Complexity can be accommodated by breaking problems down into manageable pieces (division of labor). However, this requires the addition of management overhead and the means to bring together the appropriate experts to synthesize the various pieces back into an integrated whole.
Ambiguity	Ambiguity is defined as the inability to make sense out of a situation, regardless of available information. Ambiguity arises when faced with novelty or situations that do not correspond to past experience. Here, what is lacking is not information but the experience and expertise to correctly frame and interpret the information.	Ambiguity can be resolved by acquiring new sources of expertise and/or allowing iterative cycles of collaboration among experts and stakeholders to create new interpretations of the situation. Such collaboration requires well-established social networks for success.
Equivocality	Equivocality is defined as having multiple –equally plausible- interpretations of the same information. Here, interpretations may differ along one or more dimensions; descriptive criteria, problem boundary, relevance of specific underlying factors, multiple stakeholders who each have a vested interest in characterizing the current situation, forecasting its implications, and developing response actions.	As with ambiguity, equivocality can be resolved through iterative cycles of interpretation, discussion, and negotiation among experts and stakeholders. This process can occur either democratically or in authoritative fashion, depending upon the relative influence of each stakeholder and the presence/absence of an overall decision authority.

Source: Zack, M. H., *Knowledge Directions*, 1, 36, 1999.

TABLE 3.2
Sense-Making Problem Characteristics

Category	Characteristics
Nature of the problem	The nature of the problem has shifted from the known (e.g., simple problem) to the unknown (e.g., wicked problem)
	Overall guidance and directions received from functional experts and stakeholders does not set forth a clear and consistent set of goals that address the present operational situation
	Time and other resource constraints necessitate trade-offs among competing goals and operational requirements
Nature of the information	The ability to effectively collect, interpret, and organize information becomes problematic because of the volume of available information or the reliability of this information
	There exist multiple, conflicting interpretations of the available information as different experts or stakeholders each apply their unique perspectives and expertise
	The operational situation appears to present decision-makers with a seemingly inconsistent pattern of features, relationships, or demands
	Functional experts and stakeholders employ symbols and metaphors to articulate their perspective, but these symbols and metaphors are not consistently understood by others
Nature of the decision-makers and stakeholders	Functional experts and stakeholders differ in terms of the underlying values, political goals, or emotional reactions
	Various relevant players lack a clear and consistent assignment of roles and responsibilities
	Decision-makers lack a clear and consistent set of success measures for judging operational progress and adjusting future decisions and actions
	Key decision-makers, functional experts, and stakeholders change as a function of the evolving operational situation

Source: Adapted from McCaskey, M.B., *The Executive Challenge: Managing Change and Ambiguity*, Pitman Publishers, Marshfield, MA, 1982.

to define and execute and likely less successful at the informational and cognitive levels which are both fundamentally more difficult to manipulate and to observe. The sense-making processes are emergent and adaptive but are trying to be kept within a linear inferencing framework. It is also characterized by the problem-solvers' reluctance to simplify interpretations and a reluctance to dispense with information that doesn't fit nominated hypotheses; these teams are also characterized by having a commitment to resilience. With the process involving frequent adaptation, it can also be appreciated that most characterizations of sense-making describe the need for a knowledge management function that keeps track of the dynamics in nominated hypotheses and associated knowledge models to prevent thrashing and a failure to converge. Leedom (2004) shows the diagram of Figure 3.3 to convey the hybrid combination of linear and emergent processes working together in a mission/operational-tempo-based temporal context.

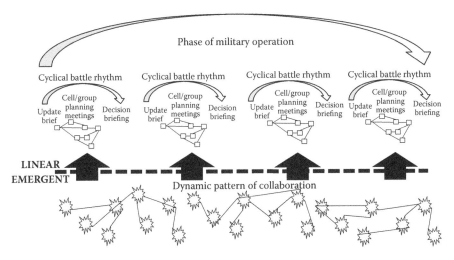

FIGURE 3.3   Sense-making dynamics. (From Leedom, D.K., The analytic representation of sensemaking and knowledge management within a military C2 organization, Air Force Research Laboratory Human Effectiveness Directorate Report AFRL-HE-WP-TR-2004-0083, 2004.)

Weick (1995) depicts the sense-making process as having four functional components as shown in Table 3.3.

Positional arguing involves disparate functional experts coming together in a "community of interest" to develop a shared understanding of the problem space and to nominate actions that will aid in confirming current hypotheses or in aiding the inferencing process. Plausible expectations from the decided actions are formed by the key leaders of the team in the form of projected outcomes or events. Behavioral commitment, as indicated earlier, is action-based and is in a sense a way to help focus the sense-making process on particular components of the problem space for which a leader is committed to a course of action (reflects "commander's intent"). Environmental manipulation is about those actions that are taken to help develop the "constructed reality" that forms the framework of interpretation of the group.

3.8   ROLE OF FUSION (2)

Understanding sense-making and the role for computer-based information fusion processing requires in part an understanding of the various types of information and knowledge involved with sense-making. In Leedom (2004), the knowledge sources described are codified information and knowledge, tacit knowledge, and social knowledge. Clearly, the knowledge coming from the output of an information fusion process falls into the codified knowledge domain. Information sources that are employed by a fusion process will mostly fall into the codified information domain. Among such sources, it can be argued that one particular important information source in this paradigm is that of contextual information. It has been said that "Sense-making is about contextual reality. It is built out of vague questions, muddy answers, and negotiated agreements that attempt to reduce confusion" (van Laere et al. 2007). Context is also a slippery word and has varying interpretations; it can be

TABLE 3.3
Sense-Making Process Characterization

Sense-Making Process	What This Process Entails	Why This Is an Essential Component of Sense-Making
Positional arguing (belief-based)	Various functional experts and/or stakeholders within the team or organization present their perspectives or positions in an attempt to shape the constructed problem framework As part of this collaborative process, each individual attempts to change or expand the knowledge state of others until there exists a commonly shared understanding of how each of the relevant problem elements and potential solution paths fit together in a cohesive whole Sometimes referred to as debative cooperation	Whenever teams or organizations face wicked problems, the major challenge is constructing an appropriate problem framework within which to shape the resulting decisions Wicked problems—including their relevant threats and opportunities—will often be viewed differently by each expert or stakeholders, dependent upon their roles and tacit knowledge
Plausible expectation (belief-based)	Key leaders express their expectation of certain outcomes, events, or future states in order to focus the attention and thinking of their supporting team or organizational members Expectations link belief to action in as much as constructed futures implicitly require certain actions or accomplishments that must be planned and executed by the team or organization Expectations reflect constructed futures that evolve over time to conform with unfolding events and states	The efficiency of sense-making within a team or organization depends upon its leaders focusing the attention and thinking of its members Part of the responsibilities of a leader are to construct a vision for the team or organization out of many possible futures Linking thoughts, teams, and accomplishment is a powerful motivational mechanism for shaping the decision behaviors of others
Behavioral commitment (action-based)	Key leaders demonstrate explicit, public, irrevocable commitment to specific plans and actions in order to further shape and focus the attention and thinking of their supporting team and organizational members Commitment is expressed in the form of approved plans and orders issued to subordinate elements	Individuals, teams, and organizations try hardest to build meaning and understanding around those actions to which they are committed to Prior to leaders expressing commitment, all types of perceptions, experiences, and positions within the team or organization are loosely coupled to an evolving situation

TABLE 3.3 (continued)
Sense-Making Process Characterization

Sense-Making Process	What This Process Entails	Why This Is an Essential Component of Sense-Making
	Commitment serves to provide a team or organization with purpose, order, and value	Commitment transforms unorganized perceptions, experience, and positions into a more orderly and purposeful team
Environmental manipulation (action-based)	Teams and organizations selectively act within their operational environment to conform that environment to their constructed reality	Sense-making is more than merely the passive interpretation of the operational environment as given; it involves the active constitution of a workable reality within which a team or organization operate
	Manipulation reflects the role of the team or organization in actively shaping the future	
	Manipulation can take the form of pre-emptive actions taken to shape the problem space even before that problem space is completely understood	Sense-making links beliefs and action together within an understandable framework; hence, the construction of a reality can involve both hypothesis building and action taking

Source: Weick, K.E., *Sensemaking in Organizations*, Sage Publications, Thousand Oaks, CA, 1995.

difficult to distinguish it from "situation" and tricky to discuss the interplay of situation and context. Contextual information, necessary to the determination of a context, can be seen to have two roles: (1) an "a priori" role where it is proactively designed into some fusion-based estimation algorithm—in this case the algorithm designed is able to prespecify what contextual information is relevant to the estimation process, and integrate it into the algorithm design (using terrain information in ground target tracking is one example), and (2) an "a posteriori" role, where contextual information is drawn upon to clarify or constrain an estimate that has been separately developed, i.e., contextual information is used after the fact of an externally asserted inference for the purpose of improved interpretation. In the latter case a type of "relevance filter" has to be designed to select, retrieve, and employ the pertinent contextual information for clarification purposes. The employment of contextual information, which can be relatively static but also dynamic (weather, e.g.), in the sense-making process adds a layer of complexity and also opens the process to various biasing effects.

What seems to be needed to support the sense-making process is a type of non-monotonic logic; one appealing model is the abductive process, which pursues plausibility rather than accuracy (Lundberg 2000). Another way to view this is that we apply abduction when there is a lack of dependable causal models as typically driven by the traditionally deductive data fusion frameworks, i.e., when only "symptoms" are available and plausible causes have to be developed. However, there is likely no single inferencing process that can be argued as the foundation of sense-making; an inferencing toolkit is probably a better model.

There are various important messages for the information fusion community in reviewing the characterizations of sense-making:

- One is that sense-making and rational decision-making will in many cases need to coexist—they are each appropriate to different problem classes, and will very likely require different data fusion processes to support them.
- Another is that the fusion community needs to make a determination of whether it is possible for fusion processing as it is known today to fit into or be extended in some way to support the sense-making process.
- But the fusion community also needs to reflect on and develop a new model for fusion as supportive of sense-making per se, and what the new functional model of that process should be and what the technological challenges are toward implementing that model.

3.9 SELF-ORGANIZATION AND SELF-SYNCHRONIZATION IN THE VALUE CHAIN

The problem framework that gives rise to the need for a sense-making process can be said to form one of the drivers that fosters the need for self-organization of an operating unit: a sense of tension or difference, misunderstanding, or under-determination where meaning is in dispute. This tension necessarily or at least naturally leads to a need for communication and the new type of social dynamic that sense-making is. Hammond and Sanders (2002) argue that the dialogic creation of meaning (one could say sense-making) is a self-organizing process. They suggest that it is the tension between disorder created by randomness and order imposed by shared meaning that drives the need to communicate. However, while communicative activity aids in creating meaning and order in the face of equivocal information, the communication processes create disorder at the same time. What happens is that as the group begins converging on a problem solution, new directions begin to emerge in a kind of convergent-emergent tension. This engenders a bit of a twist on the sense-making process characterized as only convergent to a consensus; it is likely that in the confusing, equivocal environments that sense-making is designed for that divergent factors will enter into the process. Wheatley (1992) describes this as a productive localized "chaos" that enables the opportunity for participants to let go of previous assumptions and seek "out of the box" solutions.

As regards self-synchronization, the mostly widely quoted definition of self-synchronization related to NCW comes from Cebrowski (Cebrowski and Garstka 1998): "Self-synchronization is the ability of a well-informed force to organize and synchronize warfare activities from the bottom-up. The organizing principles are unity of effort, clearly articulated commander's intent, and carefully crafted rules of engagement. Self-synchronization is enabled by a high level of [knowledge of] one's own forces, enemy forces, and all appropriate elements of the operating environment. It overcomes the loss of combat power inherent in top-down command directed synchronization characteristics of more conventional doctrine and converts combat from a step function to a high-speed continuum." A simpler definition of self-synchronization (Costanza 2003) is "the ability of a well-informed force to organize and coordinate complex warfare from the bottom up."

It is usually considered that the "self" in "self-synchronization" implies the ability of an agent to arrange the timing aspects of its own activities without the influence or control of other agents, implying a sense of independence. In terms of analysis and decision-making style, to be independent an agent needs to be proactive in his actions otherwise he may be captive to the reactions driven by the adversary. Other factors necessary for enabling self-synchronization include maintaining an awareness of commander's intent at all times, i.e., operating within that mind-set, and being able to dynamically prioritize activities. It is of course not usual that an agent acts strictly alone, so the notion of "self" in realistic cases relates to a kind of collective self-synchronization, and each agent in such collectives must be thinking synergistically, having a willingness to share resources and power. It also implies that such agents are synergistic communicators—empathetic listeners that understand the basic needs of a collaborator that enable achieving actions which are truly helpful to both agents, rather than compromises coming from negotiation-type communications. In the end, the self-synchronizing collective molds itself to the tasks and operations at hand; the molding forces are a kind of shaping context of people, problems, and resources. These factors are not unlike the "seven habits of highly successful people" that Covey (1990) sets as imperatives, e.g., being proactive, operating with an end in mind (e.g., commander's intent), having priorities, thinking synergistically, and seeking first to understand.

3.10   COMPLEXITY IN SENSE-MAKING AND COMMAND AND CONTROL

Self-organization and self-synchronization are easy to talk about but very difficult to execute in the best way. Part of the rationale regarding the need for such agile behavior comes from the "Law of Requisite Variety" of Moffat (2003), where it follows from cybernetic arguments that to properly control a complex system (the dynamic asymmetric battlefield), the variety of the controller function (the number of accessible states which it can occupy) must match the variety of the combat system itself. In other words, the control system itself, here the C2 (human-based) organization, has to be complex. This Law of Requisite Variety implies that the control system must exhibit great agility in dealing with the dynamics and complexity of combat involving hybrid teams. But that agility must be controlled to some degree else it can result in chaotic behavior. According to Moffat (2003), "the representation of the C2 process must reflect two different mechanisms. The first is the lower level interaction of simple rules or algorithms, which generate the required system variety. The second is the need to damp these by a top-down C2 process focused on campaign objectives." In a broad sense, the relationships between complex concepts and the behavior of an "information age force" are characterized as shown in Table 3.4.

Thus, it is not surprising to see considerable literature discussing the NCW sense-making and C2 processes as modeled by a complex adaptive system (CAS). If a CAS model is appropriate, then there is a need to understand CASs well enough to predict their macro-level behavior, a result of nonlinear micro-level behaviors. A related goal is to design and construct a CAS-based C2 process having a desired,

TABLE 3.4

Relations between Complexity Factors and Force Factors

Complexity Concept	Information Age Force
Nonlinear interaction	Combat forces composed of a large number of nonlinearly interacting parts
Decentralized control	There is no master "oracle" dictating the actions of each and every combatant
Self-organization	Local action, which often appears chaotic, induces long-range order
Nonequilibrium order	Military conflicts, by their nature, proceed far from equilibrium. Correlation of local effects is key
Adaptation	Combat forces must continually adapt and coevolve in a changing environment
Collectivist dynamics	There is continual feedback between the behavior of combatants and the command structure

Source: Moffat, J., *Complexity Theory and Network Centric Warfare*, CCRP Press, Washington, DC, 2003.

or perhaps bounded, emergent behavior with a theoretical understanding that the emergent behavior will be most fit for a particular C2 or mission objective. The CAS/C2 literature speaks of the C2 process as ideally operating "on the edge of chaos"; i.e., within the favorable, predictable macro-behavioral bounds of the inherent CAS C2 process, but not tipping into chaotic behavior.

Since information fusion processes are information-providing processes into such decision-making and C2 operations, it is then important for fusion process designers to understand that they are supplying information into this nonlinear decision-support environment. One way to study such interdependencies is via the multi-agent systems construct, and probably the most research in CAS for C2 has been along these lines. Some of the notable examples of using intelligent agents to study emergent behavior in warfare are the Irreducible Semi-Autonomous Adaptive Combat (ISAAC) works, and the Enhanced ISAAC Neural Simulation Toolkit (EINSTein), from the U.S. Marine Corps Combat Development Command (MCCDC) as part of their Project Albert research (Ilachinski, 1999). There are yet other efforts that have employed the agent paradigm for such research (Hummel et al., 2005, Yang et al., 2005, Lauren 2000). These test beds have been used for a wide variety of research studies that have aided in developing insights into the behaviors and performance of CASs. Other methods have been applied to explore the CAS-data fusion interdependency, but overall, the research and thus design knowledge is limited; this is considered a robust area for needed research.

Regarding other methods, Urken (2011) has studied "error-resilient data fusion" (ERDF) processes, in which the contributors to the formation of a composite situational estimate employ voting procedures. In the ERDF approach, the properties of the systems used to represent and aggregate votes produce a high probability of producing what Urken calls "error resilient collective outcomes" (ERCOs). When such a voting process produces a reliable ERCO, neither outstanding votes or data, nor unelapsed time, will change the collective inference, yielding a robust result or situational interpretation. So ERCO results provide a basis for ignoring uncollected critical data and enabling agents to take immediate action to adapt to changes in their environment. Alternate approaches to dealing with CAS aspects for both fusion and network design have been put forward in a limited body of work, such as the biologically inspired strategies described in Urken (2011) and Ferro and Pioggia (2009). However, by and large, the information fusion community has not developed an organized research strategy to explore the nature of fusion functions and processes in the context of CAS.

3.11 SUMMARY

It is anticipated that not only the military but extensive business and civil systems will be operating in a network-centric context from the point of view of the under-lying informational infrastructure. There are clearly advantages to employing net-worked systems but there is little doubt that there are also system design trade-off issues regarding the formation of the physical network and perhaps the even more important issue of how the network is used. In the value chain characterization, one can to some degree build in ways to improve information quality and sharing through mandated processes and protocols, but the intermodal interactions and human inputs and controls also play into the overall effectiveness equation. If the sense-making and CAS paradigms indeed apply toward modeling such interactions, the information fusion community will need to better study and understand how to design fusion processes to operate in these highly adaptive and nonlinear user environments. The implications of these new models of "sense-making," consensus-formation, convergent–emergent interpretation dynamics, productive local chaos, etc., on the requirements for data fusion process design and development are likely to be rather revolutionary.

REFERENCES

Alberts, D. S. 2002. Information age transformation. *CCRP Program Monograph* online, http://www.dodccrp.org/html4/books_downloads.html

Alberts, D. S., J. J. Garstka, R. E. Hayes, and D. T. Signori. 2001. *Understanding Information Age Warfare*. Washington, DC: CCRP Publications.

Alberts, D. S. and R. E. Hayes. 2003. Power to the edge. *CCRP Program Monograph* online, http://www.dodccrp.org/html4/books_downloads.html

Alberts, D. S. and R. E. Hayes. 2006. Understanding command and control. *CCRP Program Monograph* online, http://www.dodccrp.org/html4/books_downloads.html

Cebrowski, A. K. and J. J. Garstka. January 1998. Network centric earfare: Its origins and future. *U.S. Naval Institute Proceedings*, 124(1): 35.

Costanza, C. D. May, 2003. *Self-Synchronization, the Future Joint Force and the United States Army's Objective Force*. Monograph, Fort Leavenworth, KS: School of Advanced Military Studies United States Army Command and General Staff College.

Covey, S. 1990. *The Seven Habits of Highly Effective People: Powerful Lessons in Personal Change*. New York: Fireside Publishers.

Evans, P. and T. S. Wurster. 2000. *Blown to Bits: How the New Economics of Information Transforms Strategy*. Boston, MA: Harvard Business Press.

Evidence Based Research, Inc. November 2003. Network centric operations conceptual frame-work version 1.0. 2003. Report prepared for Office of Force Transformation, http://www.oft.osd.mil/library/library_files/document_353_NCO%20CF%20Version%20 1.0%20(FINAL).doc

Ferro M. and G. Pioggia. 2009. A biologically based framework for distributed sensory fusion and data processing. In: *Sensor and Data Fusion*, N. Milisavljevic (Ed.). Vienna, Austria: InTech.

Hammond, S. C. and M. L. Sanders. 2002. Dialogue as social self-organization: An introduction. *Emergence: Complexity and Organization* 4(4): 7–24.

Hummel, J. R., J. H. Christiansen, C. M. Macal, and M. J. North. White paper, 2005. The development of complex adaptive systems based decision support systems. Decision and Information Sciences Division, Argonne National Laboratory, Argonne, IL.

Ilachinski, A. 1997. Irreducible semi-autonomous adaptive combat (ISAAC): An artificial-life approach to land combat. *Military Operations Research* 5(3): 29.

Ilachinski, A. February 1999. Towards a science of experimental complexity: An artificial-life approach to modeling warfare. Special issue of *Kybernetes Journal*.

van Laere, J., M. Nilsson, and T. Ziemke. 2007. Implications of a Weickian perspective on decision-making for information fusion research and practice. *10th International Conference on Information Fusion*, Quebec City, Quebec, Canada.

Lauren, M. K. 2000. Modeling combat using fractals and the statistics of scaling systems. *Military Operations Research* 5(3): 47–58.

Leedom, D. K. 2001. *Final Report: Sensemaking Symposium.* (Technical Report prepared under contract for Office of Assistant Secretary of Defense for Command, Control, Communications & Intelligence.) Vienna, VA: Evidence Based Research. Inc., http://www.dodccrp.org/files/sensemaking_final_report.pdf

Leedom, D. K. 2004. The analytic representation of sensemaking and knowledge management within a military C2 organization. Air Force Research Laboratory Human Effectiveness Directorate Report AFRL-HE-WP-TR-2004-0083.

Lundberg, C. G. 2000. Made sense and remembered sense: Sensemaking through abduction. *Journal of Economic Psychology* 21(6): 691–709.

McCaskey, M. B. 1982. *The Executive Challenge: Managing Change and Ambiguity.* Marshfield, MA: Pitman Publishers.

Moffat, J. 2003. *Complexity Theory and Network Centric Warfare.* Washington, DC: CCRP Press.

Owens, W. A. May 1995. The emerging system of systems. *U.S. Naval Institute Proceedings* (121): 36–39.

Porter, M. E. 1985. *Competitive Advantage: Creating and Sustaining Superior Performance.* New York: The Free Press.

Rittel, H. and M. Webber. 1973. Dilemmas in a general theory of planning. In: *Policy Sciences*, Vol. 4, pp. 155–169. Amsterdam, the Netherlands: Elsevier Scientific Publishing Company, Inc. [Reprinted in N. Cross (Ed.). 1984. *Developments in Design Methodology.* Chichester, U.K.: John Wiley & Sons.]

Sieck, W. R. et al. May 2007. FOCUS: A model of sensemaking, Technical Report 1200, http://www.au.af.mil/au/awc/awcgate/army/tr1200.pdf

Urken, A. B. 2011. Voting theory, data fusion, and explanations of social behavior. Paper from the *AAAI 2011 Spring Symposium*, pp. 29–34. Stanford, CA.

Weick, K. E. 1969. *The Social Psychology of Organizing.* Reading, MA: Addison-Wesley.

Weick, K. E. 1995. *Sensemaking in Organizations.* Thousand Oaks, CA: Sage Publications.

Wheatley, M. 1992. *Leadership and the New Science: Learning about Organization from an Orderly Universe.* San Francisco, CA: Berrett-Koehler.

Yang, A., H. A. Abbass, R. Sarker, and M. Barlow. 2005. *Network Centric Multi-Agent Systems: A Novel Architecture.* TR-ALAR-200504004. The Artificial Life and Adaptive Robotics Laboratory, School of Information Technology and Electrical Engineering, University of New South Wales, Kensington, New South Wales, Australia.

Zack, M. H. 1999. Managing organizational ignorance. *Knowledge Directions* 1: 36–49.

4 Distributed Detection in Wireless Sensor Networks

Pramod K. Varshney, Engin Masazade,
Priyadip Ray, and Ruixin Niu

CONTENTS

4.1 Introduction 65
4.2 Distributed Detection over Ideal Communication Channels 67
    4.2.1 Bayesian Formulation 69
    4.2.2 Neyman–Pearson Formulation 70
    4.2.3 Design of Fusion Rules 71
    4.2.4 Asymptotic Regime 73
    4.2.5 Counting Rule 74
    4.2.6 False Discovery Rate–Based Sensor Decision Rules 76
        4.2.6.1 Review of Multiple Comparison Problems in Statistics 77
        4.2.6.2 Algorithm to Control FDR 78
        4.2.6.3 Design Guidelines for Distributed Detection Systems 79
    4.2.7 Correlated Decisions 81
4.3 Distributed Detection over Nonideal Communication Channels 84
    4.3.1 Distributed Detection with Partial Channel State Information 87
    4.3.2 Distributed Detection with No Channel State Information 88
4.4 Conclusions 89
References 90

4.1 INTRODUCTION

There are many practical situations in which one is faced with a decision-making problem. Based on observations regarding a certain phenomenon, a particular course of action needs to be employed from a set of possible options. Decision-making structures are found in many real-world situations that include financial institutions, air-traffic control, oil exploration, medical diagnosis, military command and control, electric power networks, weather prediction, and industrial organizations. In conventional decision-making scenarios, a sensor transmits its raw observation to a processor where optimal detection is carried out based on conventional statistical techniques. The branch of statistics dealing with these types of problems is known as statistical decision theory or hypothesis testing. In the context of radar and communication theory, it is known as detection theory [1–4]. More recently, the trend is to employ multiple sensors to observe a phenomenon. For decision making, raw observations from all the sensors can be transmitted to a central processor where an optimum decision rule can be designed based on conventional detection theory. However, centralized processing based on raw observations from multiple sensors is neither efficient nor necessary. It may consume excessive energy and bandwidth in communications and may impose a heavy computation burden at the central processor.

In distributed detection [1,5,6], multiple detectors (sensors) work collaboratively to distinguish between two or more hypotheses. In a binary distributed detection problem, the objective might be the determination of the absence or presence of a signal of interest, or in a multiple hypothesis testing problem, the objective might be the classification of multiple signals or targets. Local sensors can carry out preliminary processing of data and only communicate with each other and/or the central processing unit called the fusion center with the most informative information relevant to the global objective. As we describe later in the chapter, the global objective might be the minimization of detection error probability or maximization of probability of detection given a fixed false alarm rate constraint. Deployment of multiple sensors for signal detection improves system survivability, results in improved detection performance or in a shorter decision time to attain a prespecified performance level. From the signal processing perspective, two inherently different problems need to be considered for the distributed detection system: the design of the decision rule at the fusion center (often referred to as the fusion rule), which strives for an optimal system performance using compressed input from distributed sensors, and the design of local sensor signal processing algorithms. These two problems are intertwined with each other and they need to be jointly solved to optimize a prescribed performance criterion.

Recently, wireless sensor networks (WSNs) have gained much attention and interest and have become a very active research area. Due to their flexibility, enhanced surveillance coverage, robustness, mobility, and cost effectiveness, WSNs have found wide applications in areas such as military surveillance, and environmental monitoring. Usually, a WSN consists of a large number of low-cost and low-power sensors, which are deployed in the environment to collect observations from an event of interest. Each sensor preprocesses and extracts information from the raw observations and has the ability to communicate with other sensor nodes or the fusion center via wireless channels. The fusion center processes all the sensor data and arrives at a global inference. The detection ability of a WSN is crucial for various applications. As an example, in a surveillance scenario, the presence or absence of a target is usually determined before its attributes, such as its position or velocity, are estimated. For WSNs, the classical distributed detection framework needs to be reconsidered by taking into account the important features and limitations of sensors and the wireless channels between the sensors and the fusion center. Since a WSN has stringent resource availability in terms of power and/or bandwidth, the design of appropriate distributed detection algorithm should satisfy the resource constraints of the WSN. Furthermore, error-free transmission of sensor measurements to the fusion center over wireless channels may require high transmission power and/or powerful error correction codes which might be prohibitive for sensors with limited power and processing capabilities. Therefore, channel impairments should be taken into account in the design of distributed detection systems. A recent survey [7] summarizes the results on distributed detection, estimation, and tracking in WSNs with a special emphasis on solutions that take into account the communication network connecting the sensors and the resource constraints at the sensors.

The remainder of the chapter is organized as follows. In Section 4.2, under the conditional independence assumption, we first introduce the conventional design of decision rules at the local sensors and at the fusion center to optimize detection performance, under the Bayesian and Neyman–Pearson (NP) criteria. In many practical scenarios, it may be difficult to obtain the optimal decision rules which require information about the performance of individual sensors. Hence, decision rules that do not require this information are desirable. Later in this section, we discuss false discovery rate (FDR)-based decision fusion which does not require the knowledge of the local sensor parameters while employing nonidentical decision thresholds at each sensor. In Section 4.3, we investigate the decision fusion problem, where the channels between the sensors and the fusion center are subject to fading and noise. We review channel aware decision fusion algorithms with different degrees of channel state information. Finally, in Section 4.4, a summary of the chapter is presented and some open challenging issues for distributed detection are addressed.

4.2 DISTRIBUTED DETECTION OVER IDEAL COMMUNICATION CHANNELS

When there are two possible sets of action, the problem is a binary hypothesis testing problem. We label the two possible choices as H_0 and H_1. Hypothesis H_0 usually represents the absence of an object or event and Hypothesis H_1 corresponds to its presence. If there are M hypotheses with $M > 2$, it is a multiple hypothesis testing problem or M-ary detection problem. In this chapter, we focus on the binary hypothesis testing problem. More detailed treatment for the multiple hypothesis testing problem can be found in the literature [8–13].

In the hypothesis testing problem, the source or event of interest is not directly observable. Corresponding to each hypothesis, an observation (a set of observations), which is a random variable (vector) in the observation space is generated according to some probabilistic law. Let us assume that there are K sensors in the WSN and the observation at each of the K sensors, z_k, corresponds to either of the two hypotheses

$$H_0 \sim p_0(\theta)$$
$$H_1 \sim p_1(\theta)$$

(4.1)

where $p_0(\theta)$ and $p_1(\theta)$ are the pdfs under H_0 and H_1, respectively. More specifically, if the problem is to detect the absence or presence of the signal of interest, the received observation at each sensor has the form

$$z_k = \begin{cases} n_k & \text{Under } H_0 \\ \theta + n_k & \text{Under } H_1 \end{cases} \tag{4.2}$$

where

θ represents the parameter vector that characterizes the hypothesis H_1
n_k represents the noise

By examining the observation, we try to infer which hypothesis is the correct one based on a certain decision rule. Usually, a decision rule partitions the observation space into decision regions corresponding to the different hypotheses. The hypothesis corresponding to the decision region where the observation falls is declared true. Whenever a decision does not match the true hypothesis, an error occurs. To obtain the fewest errors (or least cost), the decision rule plays an important role and should be designed according to the optimization criterion in use.

Parallel configuration, as shown in Figure 4.1, is the most common topological structure that has been studied quite extensively in the literature. In parallel topology, the sensors do not communicate with each other and there is no feedback from the fusion center to any sensor. Sensors either transmit their measurements z_k's directly to the fusion center or send a quantized version of their local measurements defined by the mapping rule $u_k = \gamma_k(z_k) k \in \{1, 2, \ldots, K\}$. Based on the received information $\mathbf{u} = [u_1, \ldots, u_K]$, the fusion center arrives at the global decision $u_0 = \gamma_0(\mathbf{u})$ that favors either H_1 (decides $u_0 = 1$) or H_0 (decides $u_0 = 0$). The goal is to obtain the optimal set of decision rules $\Gamma = (\gamma_0, \gamma_1, \ldots, \gamma_K)$ according to the objective function under consideration which can be formulated according to Bayesian formulation or NP formulation. For general network structures, the optimal solution to the distributed detection problem, i.e., the optimal decision rules ($\gamma_1, \ldots, \gamma_K$), is NP-complete [14–16]. Nonetheless, under the conditional independence assumption the optimum solution becomes tractable.

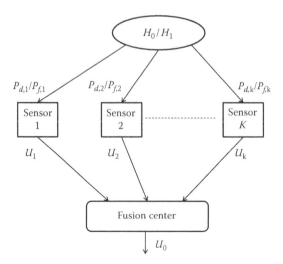

FIGURE 4.1   Parallel configuration.

The conditional independence assumption implies that the joint density of the observations obeys

$$p(z_1,\ldots,z_K \mid H_j) = \prod_{k=1}^{K} p(z_k \mid H_j), \quad \text{for } j = 0,1 \qquad (4.3)$$

Consider a scenario in which the observations at the sensors are conditionally independent as well as identically distributed. The symmetry in the problem suggests that the decision rules at the sensors should be identical. But counterexamples have been found in which nonidentical decision rules are optimal [16–19]. In the following sections, the decision rules at local sensors and the fusion center are designed according to Bayesian and NP formulations for the parallel configuration.

4.2.1 BAYESIAN FORMULATION

Let the vector of sensor decisions be denoted as $\mathbf{u} = [u_1, \ldots, u_K]$ so that the conditional densities under the two hypotheses are $p(\mathbf{u}|H_0)$ and $p(\mathbf{u}|H_1)$ respectively. The observations are generated from these conditional densities which are assumed known. The a priori probabilities of the two hypotheses denoted by $P(H_0)$ and $P(H_1)$ are assumed to be known. In the binary hypothesis testing problem, four possible actions can occur. Let $C_{i,j}$, $i \in \{0, 1\}, j \in \{0, 1\}$ represent the cost of declaring H_i true when H_j is present. The Bayes risk function is given by

$$\mathcal{R} = \sum_{i=0}^{1}\sum_{j=0}^{1} C_{i,j} P(H_j) P(\text{Decide } H_i \mid H_j \text{ is present})$$

$$= \sum_{i=0}^{1}\sum_{j=0}^{1} C_{i,j} P(H_j) \int_{\mathcal{U}_i} p(\mathbf{u} \mid H_j)d\mathbf{u} \qquad (4.4)$$

where $\mathcal{U}_i$ is the decision region corresponding to hypothesis H_i which is declared true for any observation falling in the region $\mathcal{U}_i$. Let $\mathcal{U}$ be the entire observation space so that $\mathcal{U} = \mathcal{U}_0 \cup \mathcal{U}_1$ and $\mathcal{U}_0 \cap \mathcal{U}_1 = \varnothing$.

If $C_{0,0} = C_{1,1} = 0$ and $C_{0,1} = C_{1,0} = 1$, we have the minimum probability of error criterion, i.e., $R = P_e = P(u_0 = 1|H_0)P_0 + P(u_0 = 0|H_1)P_1$. The probability of error is given by

$$P_e = P(H_0)P_F + P(H_1)(1 - P_D) \qquad (4.5)$$

where
$P_F = P(u_0 = 1|H_0)$ denotes the probability of false alarm
$P_D = P(u_0 = 1|H_1)$ denotes the probability of detection

Given the vector of local sensor decisions, $\mathbf{u}$, the probability of error is expressed as

$$P_e = P(H_0)P(u_0 = 1 \mid H_0) + P(H_1)(1 - P(u_0 = 1 \mid H_1)) \qquad (4.6)$$

which can be written as

$$P_e = P(H_1) + P(u_0 = 1 | \mathbf{u})[P(H_0)P(\mathbf{u} | H_0) - P(H_1)P(\mathbf{u} | H_1)]$$

P_e is minimized if

$$P(u_0 = 1 | \mathbf{u}) = 0 \quad \text{when } [P(H_0)P(\mathbf{u} | H_0) - P(H_1)P(\mathbf{u} | H_1)] > 0$$
$$P(u_0 = 1 | \mathbf{u}) = 1 \quad \text{when } [P(H_0)P(\mathbf{u} | H_0) - P(H_1)P(\mathbf{u} | H_1)] < 0$$

(4.7)

The earlier property leads to the following likelihood ratio test (LRT) at the fusion center [1]:

$$\frac{P(\mathbf{u} | H_1)}{P(\mathbf{u} | H_0)} = \prod_{k=1}^{K} \frac{p(u_k | H_1)}{p(u_k | H_0)} \underset{u_0=0}{\overset{u_0=1}{\gtrless}} \frac{P(H_0)}{P(H_1)}$$

(4.8)

The quantity on the left-hand side is known as the likelihood ratio and the quantity on the right-hand side is the threshold. Let

$$\mathbf{u}^i = [u_1, \ldots, u_{i-1}, u_{i+1}, \ldots, u_K],$$

$$A(\mathbf{u}^i) = P(u_0 = 1 | \mathbf{u}^{i1}) - P(u_0 = 1 | \mathbf{u}^{i0})$$

$$\mathbf{u}^{ij} = [u_1, \ldots, u_{i-1}, u_i = j, u_{i+1}, \ldots, u_K], \quad j = 0,1$$

and $C_F = P_0(C_{10} - C_{00})$, $C_D = (1 - P_0)(C_{01} - C_{11})$. Then, the LRT at each sensor has the form

$$\frac{p(z_i | H_1)}{p(z_i | H_0)} \underset{u_i=0}{\overset{u_i=1}{\gtrless}} \frac{\sum_{\mathbf{u}^i} C_F A(\mathbf{u}^i) \prod_{k=1,k\neq i}^{K} P(u_k | H_0)}{\sum_{\mathbf{u}^i} C_D A(\mathbf{u}^i) \prod_{k=1,k\neq i}^{K} P(u_k | H_1)} \quad \text{for } i = 1, \ldots, K$$

(4.9)

Conditional independence assumption and establishing the optimality of LRT at local sensors does not completely solve the problem. Note that the LRT thresholds at the sensors are coupled with each other which affect the system performance in an interdependent manner. Almost invariably used for finding the local sensor thresholds is the so called person-by-person optimization (PBPO) approach, where each sensor's threshold is optimized assuming fixed decision rules at all other sensors and the fusion center [20]. Unfortunately, the PBPO algorithm does not necessarily lead to a global optimal solution and may only lead to a local minimum of the solution space. Multiple initializations may be needed to obtain global optimum.

4.2.2 NEYMAN–PEARSON FORMULATION

The NP formulation of the distributed detection problem can be stated as follows: Let α be a prescribed bound on the global probability of false alarm such that $P_F = P(u_0 = 1|H_0) \leq \alpha$. Then the problem is to find (optimum) local and global decision rules that maximize the probability of detection $P_D = P(u_0 = 1|H_1)$ given $P_F = P(u_0 = 1|H_0) \leq \alpha$.

Under the conditional independence assumption, the mapping rules at the sensors as well as the decision rule at the fusion center are threshold rules based on the appropriate likelihood ratios [21,22]:

$$\frac{p(z_k \mid H_1)}{p(z_k \mid H_0)} \begin{cases} > t_k, & \text{then } u_k = 1 \\ = t_k, & \text{then } u_k = 1 \quad \text{with probability } \epsilon_k \\ < t_k, & \text{then } u_k = 0 \end{cases} \qquad (4.10)$$

for $k = 1, \ldots, K$, and

$$\prod_{k=1}^{K} \frac{P(u_k \mid H_1)}{P(u_k \mid H_0)} \begin{cases} > \lambda_0, & \text{decide } H_1 \text{ or set } u_0 = 1 \\ = \lambda_0, & \text{randomly decide } H_1 \text{ with probability } \epsilon \\ < \lambda_0, & \text{decide } H_0 \text{ or set } u_0 = 0 \end{cases} \qquad (4.11)$$

If the likelihood ratio in (4.10) is a continuous random variable with no point mass, then the randomization is unnecessary and ϵ_k can be assumed to be zero without losing optimality. The threshold λ_0 in (4.11) as well as the local thresholds t_k in (4.10) need to be determined so as to maximize P_D for a given $P_F = \alpha$. This can still be quite difficult even though the local decision rules and the global fusion rule are LRTs [1]. Since (4.11) is known to be a monotone fusion rule, one can solve for the set of optimal local thresholds $\{t_k, i = 1, \ldots, K\}$ for a given monotone fusion rule and compute the corresponding P_D. One can then successively consider other possible monotone fusion rules and obtain the corresponding detection probabilities. The final optimal solution is the one monotone fusion rule and the corresponding local decision rules that provide the largest P_D. An iterative gradient method was proposed in [23] to find the thresholds satisfying the preassigned false alarm probability. Finding the optimal solution in this fashion is possible only for very small values of N. The complexity increases with N, because (1) the number of monotone rules grows exponentially with N, and (2) finding the optimal $\{t_k, i = 1, \ldots, K\}$ for a given fusion rule is an optimization problem involving an $N - 1$ dimensional search (it is one dimension less than N because of the constraint $P_F = \alpha$).

4.2.3 DESIGN OF FUSION RULES

Given the local detectors, the problem is to determine the fusion rule to combine local decisions optimally. Let us first consider the case where local detectors make only hard decisions, i.e., u_k can take only two values 0 or 1 corresponding to the two hypotheses H_0 and H_1. Then, the fusion rule is essentially a logical function with K binary inputs and one binary output. There are 2^{2^K} possible fusion rules in general and an exhaustive search strategy is not feasible for large K.

Let $P_{f,k}$ and $P_{d,k}$ denote the probabilities of false alarm and detection of sensor k, respectively, i.e., $P_{f,k} = P(u_k = 1|H_0)$ and $P_{d,k} = P(u_k = 1|H_1)$. According to (4.8) and (4.11), the optimum fusion rule is given by the LRT:

$$\prod_{k=1}^{K} \frac{P(u_k \mid H_1)}{P(u_k \mid H_0)} \underset{u_0=0}{\overset{u_0=1}{\gtrless}} \lambda \tag{4.12}$$

Here, λ is determined by the optimization criterion in use. The left-hand side of (4.12) can be written as

$$\prod_{k=1}^{K} \frac{P(u_k \mid H_1)}{P(u_k \mid H_0)} = \prod_{k=1}^{K} \left(\frac{P(u_k = 1 \mid H_1)}{P(u_k = 1 \mid H_0)} \right)^{u_k} \left(\frac{P(u_k = 0 \mid H_1)}{P(u_k = 0 \mid H_0)} \right)^{1-u_k}$$

$$= \prod_{k=1}^{K} \left(\frac{P_{d,k}}{P_{f,k}} \right)^{u_k} \left(\frac{1 - P_{d,k}}{1 - P_{f,k}} \right)^{1-u_k} \tag{4.13}$$

Taking the logarithm of both sides of (4.12), we have the Chair–Varshney fusion rule [24]

$$\sum_{k=1}^{K} \left[u_k \log \frac{P_{d,k}}{P_{f,k}} + (1 - u_k) \log \frac{1 - P_{d,k}}{1 - P_{f,k}} \right] \underset{u_0=0}{\overset{u_0=1}{\gtrless}} \log \lambda \tag{4.14}$$

This rule can also be expressed as

$$\sum_{k=1}^{K} \left[\log \frac{P_{d,k}(1 - P_{f,k})}{P_{f,k}(1 - P_{d,k})} \right] u_k \underset{u_0=0}{\overset{u_0=1}{\gtrless}} \log \lambda + \sum_{k=1}^{K} \log \frac{1 - P_{f,k}}{1 - P_{d,k}} \tag{4.15}$$

Thus, the optimum fusion rule can be implemented by forming a weighted sum of the incoming local decisions and comparing it with a threshold. The weights and the threshold are determined by the local probabilities of detection and false alarm. If the local decisions have the same statistics, i.e., $P_{f,k} = P_{f,l}$ and $P_{d,k} = P_{d,l}$ for $k \neq l$, the Chair–Varshney fusion rule reduces to a T-out-of-K form or a counting rule, i.e., the global decision $u_0 = 1$ if T or more sensor decisions are one. This structure of the fusion rule reduces the computational complexity considerably.

So far, we have assumed that the parameters characterizing a hypothesis, θ, are fixed and known leading to the conditional independence assumption. In many situations, these parameters can take unknown values or a range of values. Such hypotheses are called composite hypotheses and the corresponding detection problem is known as composite hypothesis testing. If θ is characterized as a random vector with known probability densities under the two hypotheses, the LRT can be extended to composite hypothesis testing in a straightforward manner:

$$\Lambda(\mathbf{u}) = \frac{\displaystyle\int_{\Theta_1} p(\mathbf{u} \mid \boldsymbol{\theta}, H_1) p(\boldsymbol{\theta} \mid H_1) d\boldsymbol{\theta}}{\displaystyle\int_{\Theta_0} p(\mathbf{u} \mid \boldsymbol{\theta}, H_0) p(\boldsymbol{\theta} \mid H_0) d\boldsymbol{\theta}} \overset{u_0=1}{\underset{u_0=0}{\gtrless}} \eta \qquad (4.16)$$

If $\boldsymbol{\theta}$ is nonrandom, i.e., fixed but unknown constant, one would like to be able to obtain uniformly most powerful (UMP) results for an optimum scheme based on an NP test. If a UMP test does not exist, we can use the maximum likelihood estimates of its value under the two hypotheses as the true values in an LRT, resulting in the so-called generalized likelihood ratio test (GLRT):

$$\Lambda(\mathbf{u}) = \frac{\max_{\boldsymbol{\theta} \in \Theta_1} p(\mathbf{u} \mid \boldsymbol{\theta}, H_1)}{\max_{\boldsymbol{\theta} \in \Theta_0} p(\mathbf{u} \mid \boldsymbol{\theta}, H_0)} \overset{u_0=1}{\underset{u_0=0}{\gtrless}} \eta \qquad (4.17)$$

Note that the optimum NP or Bayesian detectors involve an LRT as in (4.12). Although the NP and Bayesian detectors are optimum in the sense of maximizing P_D for a fixed P_F, and minimizing the Bayes risk, the associated LRTs require the complete knowledge of the pdfs $p(\mathbf{u}|H_1)$ and $p(\mathbf{u}|H_0)$ which may not always be available in a practical application. Also, there are many detection problems where the exact form of the LRT is too complicated to implement. Therefore, simpler and more robust suboptimal detectors are used in numerous applications [25]. For some suboptimal detectors, the detection performance can be improved by adding an independent noise to the observations under certain conditions which is known as stochastic resonance (SR) noise [26]. The work in [27] first discusses the improvability of the detection performance by adding SR noise given a suboptimal fixed detector. If the performance can be improved, then the best noise type is determined in order to maximize P_D without increasing P_F. The work in [28] discusses variable detectors.

In this chapter, we have focused on fixed-sample-size detection problems for the parallel architecture. Solutions for arbitrary topologies such as serial [1,29–31] and tree have been derived and are discussed in [32–34]. In fixed-sample-size detection, the fusion center arrives at a decision after receiving the entire set of sensor observations or decisions. Sequential detectors may choose to stop at any time and make a final decision or continue to take additional observations [35–39]. Moreover, in consensus-based detection [40–42], which requires no fusion center, sensors first collect sufficient observations over a period of time. Then, subsequently they run the consensus algorithm to fuse their local log likelihood ratios.

4.2.4 ASYMPTOTIC REGIME

In this section, we describe some results when the number of sensors becomes very large, i.e., we discuss some asymptotic results. It has been shown that identical decision rules are optimal in the asymptotic regime where the number of sensors increases to infinity [16,43]. In other words, the identical decision rule assumption often results in little or no loss of optimality. Therefore, identical local decision rules are frequently assumed in many situations, which reduces the computational complexity considerably.

For any reasonable collection of decision rules Γ, the probability of error at the fusion center goes to zero exponentially as the number of sensors K grows unbounded. It is then adequate to compare decision rules based on their exponential rate of convergence to zero:

$$\lim_{K \to \infty} \frac{\log P_e(\Gamma)}{K} \qquad (4.18)$$

It was shown that for the binary hypothesis testing problem, use of identical local decision rules for all the sensor nodes is asymptotically optimal in terms of the error exponent [43]. In [44], the exact asymptotics of the minimum error probabilities achieved by the optimal parallel fusion network and the system obtained by imposing the identical decision rule constraint was investigated. It was shown analytically that the restriction of identical decision rules leads to little or no loss of performance. Asymptotic regimes applied to distributed detection are convenient because they capture the dominating behaviors of large systems. This leads to valuable insights into the problem structure and its solution.

In the asymptotic regime, it has been shown in [45] that if there exists a binary quantization function γ_b whose Chernoff information exceeds half of the information contained in an unquantized observation, then transmitting binary decisions from sensors to the fusion center becomes optimal. The requirement is fulfilled by many practical applications [46] such as the problem of detecting deterministic signals in Gaussian noise and the problem of detecting fluctuating signals in Gaussian noise using a square-law detector. In these scenarios, the gain offered by having more sensor nodes outperforms the benefits of getting detailed information from each sensor.

4.2.5 COUNTING RULE

Most of the results discussed so far on distributed detection are based on the assumption that the local sensors' detection performances, namely, either the local sensors' signal to noise ratio (SNR) or their probability of detection and false alarm rate, are known to the fusion center. For a WSN consisting of passive sensors, it might be very difficult to estimate local sensors' performances via experiments because sensors' distances from the signal of interest might be unknown to the fusion center and to the local sensors. Even if the local sensors can somehow estimate their detection performances in real time, it can be still very expensive to transmit them to the fusion center, especially for a WSN with very limited system resources. Hence, the knowledge of the local sensors' performances cannot be taken for granted and a fusion rule that does not require local sensors' performances is highly preferable. Without the knowledge of local sensors' detection performances and their positions, an approach at the fusion center is to treat every sensor equally. An intuitive solution is to use the total number of "1"s as a statistic since the information about which sensor reports a "1" is of little use to the fusion center. In [47–49], a counting-based fusion rule is proposed, which uses the total number of detections ("1"s) transmitted from local sensors as the statistic,

$$\Lambda(\mathbf{u}) = \sum_{k=1}^{K} u_k \underset{u_0=0}{\overset{u_0=1}{\gtrless}} T \tag{4.19}$$

where T is the threshold at the fusion center, which can be decided by a prespecified probability of false alarm P_F. This fusion rule is called the counting rule. It is an attractive solution, since it is quite simple to implement, and achieves very good detection performance in a WSN with randomly and densely deployed low-cost sensor nodes.

The performance of a distributed detection system that is the probability of false alarm and the probability of detection at the fusion center needs to be calculated from

$$P_F = P(u_0 = 1 \mid H_0) = P(\Lambda(\mathbf{u}) > \eta \mid H_0)$$
$$P_D = P(u_0 = 1 \mid H_1) = P(\Lambda(\mathbf{u}) > \eta \mid H_1) \tag{4.20}$$

which requires the probability density function of the test statistic $\Lambda(\mathbf{u})$. For the counting rule as in (4.19), under hypothesis H_0, the total number of detections $\Lambda = \sum_{k=1}^{K} u_k$ follows a binomial distribution. For a given threshold T, the false alarm rate can be calculated as follows:

$$P_F = \sum_{k=T}^{K} \binom{K}{k} P_f^k (1 - P_f)^{N-k} \tag{4.21}$$

where $P_{f,1} = \cdots = P_{f,K} = P_f$. For the sensing model in (4.2) where θ is fixed and known, the detection probability can be obtained from

$$P_D = \sum_{k=T}^{K} \binom{K}{k} P_d^k (1 - P_d)^{N-k} \tag{4.22}$$

where all the sensors use identical decision thresholds. In many practical scenarios, while computing P_D, decisions are not independent of each other under hypothesis H_1, since the decisions are all dependent on the target and sensors coordinates which can also be random variables. For such cases, several approximations for computing the distribution of $\Lambda(\mathbf{u})$ under H_1 can be found in [47–49].

The calculation of P_D and P_F may become difficult since it requires the probability density function of the decision rule $\Lambda(\mathbf{u})$. Deflection coefficient is a useful performance measure when the statistical properties of the received measurements are limited to moments up to a given order as

$$D(\Lambda) = \frac{(E[\Lambda \mid H_1] - E[\Lambda \mid H_0])^2}{\mathrm{Var}(\Lambda \mid H_0)} \tag{4.23}$$

which requires the first two moments of the decision test statistic Λ.

Our previous survey [50] also summarizes the decision fusion results based on identical decision rules at each sensor. Next, we summarize FDR-based decision fusion which uses nonidentical decision thresholds at each sensor.

4.2.6 FALSE DISCOVERY RATE–BASED SENSOR DECISION RULES

Let us consider a detection scenario where the sensors which are located within the target's finite radius of influence receive identical target signal and the rest of the sensors do not receive any target signal. This "disk" target signal model may be applied to scenarios such as oil or chemical leaks [51] or to approximate more general electromagnetic or acoustic target models. Though this is a very simple model, it clearly captures the scenario where the sensors in the network receive nonidentical target signals (all sensors receive identical target signal has been the primary assumption in the distributed detection literature). As mentioned earlier, design of the optimum local and global decision rules for such problems is very difficult. Earlier related work [47,49] assumes that all the sensors use an identical local threshold for an LRT to obtain a local decision. Since the probability of detection of each sensor is unknown due to unknown target and sensor location, the optimal Chair–Varshney fusion rule cannot be used for this problem. An intuitive choice is to constrain the fusion center decision statistic to be linear in the total number of local detections, i.e., employ the "count" as the statistic, and perform a threshold test to obtain the global decision. This approach may also be viewed as performing multiple hypotheses tests (each sensor performing a binary hypothesis test locally)* and the fusion center using the results of these tests (i.e., the outcome of the local hypotheses tests) to come up with a global decision. Therefore, the detection problem essentially reduces to obtaining the optimal set of the two design parameters, the local and global decision thresholds. Hence, from here on we will use the terms "decision rules" and "decision thresholds" interchangeably in this article. Note that optimization of distributed detection systems where the local sensor SNRs may be unknown has been investigated in [52–54]. However, the optimization techniques in [52–54] require the knowledge or an estimate of the local sensor SNRs. Note that, the estimation of the local sensor SNRs is very difficult as it is a function of the sensor and target location which is generally unknown. In [55], the authors propose a detection scheme based on the control of FDR, which employs nonidentical local sensor decision rules without increasing the total number of design parameters. Also, the FDR-based detection strategy proposed in [55] does not require an estimate of the local sensor SNRs. The FDR-based scheme is discussed in some detail in this section.

---

* Note that in this section, multiple hypotheses tests indicate multiple binary hypothesis tests and a formal definition is provided in the next section. In the previous sections, we use multiple hypotheses testing to indicate M-ary tests.

Since FDR was first proposed in the context of multiple hypotheses problems (also known as multiple comparison problems [MCPs]) in statistics, we next provide a brief review of MCPs.

4.2.6.1 Review of Multiple Comparison Problems in Statistics

Multiple comparisons refer to multiple simultaneous hypothesis tests. When a family of tests is conducted, it is often meaningful to define an error measure for the family instead for the individual tests. One of the most common measures is the family-wise error rate (FWER) [56], defined as the probability of committing any type I error or false alarm. If the error rate for each test is α then the FWER α_F for k tests is given by

$$\alpha_F = P(V \geq 1) = 1 - (1 - \alpha)^k \tag{4.24}$$

where V is defined in Table 4.1. As can be seen from Equation 4.24 for a single comparison, $\alpha_F = \alpha$. When the number of comparisons increases, α remains constant but α_F increases. This is a fundamental problem of MCPs and classical multiple comparison procedures aim to control this error measure. A method to control FWER, known as the Bonferroni procedure, controls the FWER in the *strong* sense, i.e., under all conditions. The method is based on the Bonferroni inequality, which says that the probability of the union of a number of events is less than or equal to the sum of their individual probabilities:

$$P(A_1 \cup A_2 \cup \cdots \cup A_k) \leq \sum_{i=1}^{k} P(A_i) \tag{4.25}$$

Hence, if each individual test is performed at the probability of false alarm $\alpha^* = \alpha_F/k$, the FWER for the family of tests is maintained at α_F. But this procedure is very conservative and results in significantly reduced probability of detection (reduced power). A radically different and more liberal approach proposed by Benjamini and Hochberg [57] controls FDR, defined as the fraction of false rejections among those hypotheses rejected. Table 4.1 defines some terms leading to the definition of FWER and FDR for a binary hypothesis testing problem involving two hypotheses H_0 and H_1.

FDR is defined as the expected ratio of the number of false alarms (declared H_1 when H_0 is true) to the total number of detections (consisting of both true and false

TABLE 4.1
Notations to Define FDR

	Declared H_0	Declared H_1	Total
H_0 True	U	V	K_0
H_1 True	T	S	$K - K_0$
Total	$K - R$	R	N

detections). The fraction of false alarms to the total number of detections can be viewed through the random variable defined as

$$Q = \begin{cases} \dfrac{V}{V+S}, & \text{if } V+S \neq 0 \\ 0, & \text{if } V+S = 0 \end{cases} \tag{4.26}$$

FDR (Q_e) is defined to be the expectation of Q,

$$Q_e = E(Q) \tag{4.27}$$

Along with this metric, Benjamini and Hochberg [57] also proposed the following algorithm to control FDR for multiple comparisons.

4.2.6.2 Algorithm to Control FDR

Suppose $p_1, p_2, \ldots, p_K$ are the p-values for K tests and $p_{(1)}, p_{(2)}, \ldots, p_{(k)}$ denote the ordered p-values. The p-value for an observation s_k is defined as

$$p_k = \int_{s_k}^{\infty} f_0(s) ds \tag{4.28}$$

where $f_0(s)$ is the probability density function of the observation under H_0.

The algorithm by Benjamini and Hochberg [57] which keeps the FDR below a value γ, is provided as follows:

1. Calculate the p-values of all the observations and arrange them in ascending order.
2. Let d be the largest k for which $p_{(k)} \leq k\gamma/K$.
3. Declare all observations corresponding to $p_{(k)}$, $k=1, \ldots, d$, as H_1.

Under the assumption of independence of test statistics corresponding to the true null hypotheses (H_0), this procedure controls the FDR at γ. It has also been proved later in [58], that this same procedure also controls the FDR when the test statistics have positive regression dependency on each of the test statistics corresponding to the true null hypothesis. Note that the FDR-based decision-making system looks for the largest index $k=d$ such that $p_{(d)} \leq d\gamma/K$. There may be other indices $k=l$, where $l < d$ for which the condition $p_{(l)} \leq l\gamma/K$ may be true, but the FDR-based decision system looks for the largest value of k for which this is true. The reason behind this, as discussed in [57], is to achieve the largest probability of detection while constraining the FDR to less than or equal to γ. A detailed proof for the control of FDR by this algorithm is provided in [57]. It should also be noted that the assumption of independence of test statistics corresponding to the false null hypotheses (H_1) is not needed for the proof of the theorem.

As the ordering of p-values is required for the FDR control procedure described in [57], the procedure conventionally needs centralized processing. For the distributed detection problem considered earlier, the sensors can only send one bit to the fusion center and hence a distributed ordering scheme is necessary. A decentralized FDR procedure has been proposed in [55] which requires only one-bit communication capability for each sensor and achieves the same performance as the centralized Benjamini–Hochberg procedure. The maximum communication cost for the entire network is less than or equal to K bits per detection round, where K is the total number of sensors in the network.

An important property of FDR is now presented in the following proposition [57].

Proposition 1

If all MCP hypotheses are true H_0s, i.e., $K_0=K$, control of FDR is equivalent to the control of FWER. However, if some of the MCP hypotheses are true H_1s, i.e., $K_0 \leq K$, the FDR is smaller than or equal to FWER.

As seen from Proposition 1, FDR is the expectation of a ratio and hence the control of FDR is more liberal compared to the control of FWER in general, and as the number of true H_1s increases, the local detection probability increases. Also, as seen from the algorithm provided earlier, the control of FDR results in a data dependent rejection region (decision region) unlike conventional statistical tests where the rejection region is fixed a priori. This characteristic of FDR, as illustrated next, is the primary motivation behind the control of FDR for distributed detection to design local decision thresholds.

4.2.6.3 Design Guidelines for Distributed Detection Systems

Based on the earlier discussion on MCPs, if K sensors employ an identical decision threshold equal to τ (or p-value threshold of $Q(\tau)$*), the FWER is controlled at a value of $NQ(\tau)$ under all conditions. However, an FDR-based threshold selection scheme, with FDR parameter γ, will result in control of the FWER to γ when there is no target in the ROI, i.e., all MCP hypotheses are true H_0s. In the presence of a target, i.e., when some MCP hypotheses are true H_1s, as seen from Proposition 1, the FWER is greater than the FDR. Thus, when there is no target, an FDR-based scheme may be designed to control the FWER at any arbitrary level. But the same scheme, in the presence of a target, is more liberal (in the sense of permitting more local detections) at the cost of higher FWER. Hence, the total number of detections (irrespective of whether they are true or false local detections) over the sensor field, increases significantly in the presence of a target compared to an identical threshold scheme. Thus, the control of FDR provides better separation of the probability mass functions (pmfs) of the "count" under the global hypothesis G_0 (target absent in ROI) and the global hypothesis G_1 (target present in ROI) compared to a scheme that controls the FWER. Here by "better separation" it is implied that for the FDR-based

---

* The Q function is the complementary distribution function of the standard Gaussian, which is defined as $Q(y) = 1/\sqrt{2\pi} \int_y^\infty \exp(-z^2/2)dz$.

detection scheme, it is likely that the distance (quantifiable in terms of metrics such as the deflection coefficient) between the pmfs of the "count" under hypotheses G_0 and G_1 will be more compared to an identical threshold approach.

As discussed earlier, the two* design parameters for the distributed detection system are the local sensor decision threshold parameter (γ for FDR-based strategy) and the global decision threshold parameter, denoted by T. For any observed count $\Delta \in Z$ (Z denotes the set of integers $[0, \ldots, K]$), the binary hypotheses testing problem at the fusion center is given by

$$G_0 : P(\Delta = i; G_0) = p_0(\Delta): \quad \text{Target absent}$$

$$G_1 : P(\Delta = i; G_1) = p_1(\Delta): \quad \text{Target present}$$

(4.29)

If $T(\Delta)$ is the decision statistic, the optimal test under the NP criterion is given by a randomized decision rule which chooses the hypothesis G_1 with probability $\delta_T(\Delta)$, where

$$\delta_T(\Delta) = \begin{cases} 1, & \text{if } T(\Delta) > T \\ \kappa, & \text{if } T(\Delta) = T \\ 0, & \text{if } T(\Delta) < T \end{cases}$$

(4.30)

where

    T is the global threshold
    κ is the randomization parameter
    $T(\Delta)$ is the likelihood ratio

However, for the problem considered here, the optimal NP detector is very complex. Hence, a simplified detector is adopted in which the test statistics is linear in "count," i.e., $T(\Delta) = \Delta$. The threshold T and the randomization constant κ are chosen such that the system-wide probability of false alarm is controlled. The system-wide probability of false alarm P_{FA} for this simplified detector is given by

$$P_{FA} = P(\Delta > T; G_0) + \kappa P(\Delta = T; G_0)$$

(4.31)

The system-wide probability of detection P_D for this simplified detector is given by

$$P_D = P(\Delta > T; G_1) + \kappa P(\Delta = T; G_1)$$

(4.32)

For the FDR-based detector, for any arbitrary FDR parameter γ, the parameters T and κ are selected such that the system-level probability of false alarm is constrained. The system-level probability of false alarm for a threshold T and randomization constant κ is given by [55]

$$P_{FA} = \sum_{k=T+1}^{K} \binom{K}{k} (1-\gamma) \left(\frac{k\gamma}{K} \right)^k \left(1 - \frac{k\gamma}{K} \right)^{K-k-1} + \kappa \binom{K}{T} (1-\gamma) \left(\frac{k\gamma}{K} \right)^T \left(1 - \frac{k\gamma}{K} \right)^{K-T-1}$$

(4.33)

---

* Note that due to discrete global test statistics, a third design parameter is the randomization constant.

Also, for any arbitrary FDR parameter γ, T and κ, the system-wide probability of detection is given by [55]

$$P_D = \sum_{T+1}^{K} P(\Delta = k; G_1) + \kappa P(\Delta = T; G_1) \tag{4.34}$$

where $P(\Delta = k; G_1)$ is the probability of observing "count" k for a target present in the ROI [55]. For large K [55], the system-wide probability of detection may be approximated by

$$P_D \approx Q\left(\frac{T - K\overline{p_d}}{\sqrt{K\overline{p_d}(1 - \overline{p_d})}}\right) \tag{4.35}$$

where $\overline{p_d}$ is the average probability of detection for a sensor.

The choice of the optimum FDR parameter γ, where optimality is with respect to system-level detection performance, is a difficult problem. Receiver operating characteristic (ROC)-based optimization procedures to obtain the best γ or τ is computationally prohibitive. A computationally less intensive approach is to obtain γ or τ via optimization of the deflection coefficient. Under Gaussian assumptions, it is known that maximizing the deflection coefficient maximizes the detection performance [59] in terms of the ROC. Though, under non-Gaussian conditions, there is no general result showing that larger deflection coefficient achieves better performance in terms of ROC curves. It is, however, intuitive that increased deflection coefficient generally implies greater separation between $P(\Delta; G_0)$ and $P(\Delta; G_1)$ and hence is likely to lead to better detector design. Hence, the FDR parameter γ is set at a value such that the deflection coefficient is maximized. A comparative detection performance for an FDR-based scheme and an identical threshold scheme is shown in Figure 4.2. It is observed that the FDR-based detection approach shows significant improvement in performance over the classically used identical decision threshold approach.

4.2.7 CORRELATED DECISIONS

An important result in distributed detection is that for the classical framework, LRTs at the local sensors are optimal if observations are conditionally independent given each hypothesis [16]. This property drastically reduces the search space for an optimal set of local decision rules. Although the resulting problem is not necessarily easy, it is amenable to analysis in many contexts. In general, it is reasonable to assume conditional independence across sensor nodes if the uncertainty comes mainly from device and ambient noise. However, it does not necessarily hold for arbitrary sensor systems. For instance, when sensors lie in close proximity of one another, we expect their observations to be strongly correlated. If the observed signal is random in nature or the sensors are subject to common external noise, conditional independence assumption may also fail. Without the conditional independence assumption, the joint density of the observations, given the hypothesis, cannot be

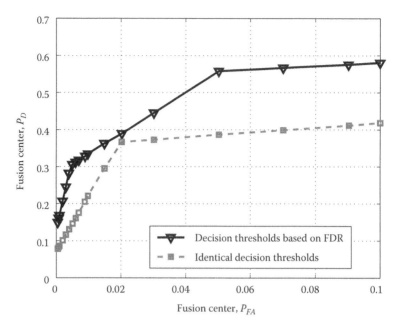

FIGURE 4.2 Detection performance comparison of FDR-based scheme and identical threshold scheme.

written as the product of the marginal densities, as in (4.3). The optimal tests at the sensors are no longer of the threshold type based solely on the likelihood ratio of the observations at the individual sensors. In general, finding the optimal solution to the distributed detection problem becomes intractable [14]. Distributed detection with conditionally dependent observations is known to be a challenging problem in decentralized inference.

One may restrict attention to the set of likelihood ratio–based tests and employ algorithms to determine the best solution from this restricted set. The resulting system may yield acceptable performance. This approach has been adopted in [60] where detection of known and unknown signals in correlated noise was considered. For the case of two sensors observing a shift-in-mean of Gaussian data, Chen and Papamarcou [61] develop sufficient conditions for the optimality of each sensor implementing a local LRT. Aalo and Viswanathan [62] assume local LRTs at multiple sensors and study the effect of correlated noise on the performance of a distributed detection system. The detection of a known signal in additive Gaussian and Laplacian noise is considered. System performance deteriorates when the correlation increases. In [63], two correlation models are considered. In one, the correlation coefficient between any two sensors decreases geometrically as the sensor separation increases. In the other model, the correlation coefficient between any two sensors is a constant. Asymptotic performance with Gaussian noise when the number of sensors goes to infinity is examined. In [64], Blum et al. study distributed detection of known signals in correlated non-Gaussian noise, where the noise is restricted to be circularly symmetric. Lin and Blum examine two-sensor distributed detection of known signals in correlated t-distributed noise in [65]. Simulation results show that in some specific cases the optimum local decision rules are better than LRTs. A distributed M-ary hypothesis testing problem when observations are correlated is examined from a numerical perspective in [66]. Willett et al. study the two detector case with dependent Gaussian observations, the simplest meaningful problem one can consider, in [67]. They discover that the nature of the local decision rules can be quite complicated. The recent work presented in [68] proposes a new framework for distributed detection under conditionally dependent observations which builds a hierarchical conditional independence model. Through the introduction of a hidden variable that induces conditional independence among the sensor observations, the proposed model unifies distributed detection with dependent or independent observations.

Constraining the local sensor decision rules to be suboptimal binary quantizers for the dependent observations problem, improvement in the global detection performance can still be attained by taking into account the correlation of local decisions while designing the fusion rule. Towards this end, design of fusion rules using correlated decisions has been proposed in [69,70]. In [69], Drakopoulos and Lee have developed an optimum fusion rule based on the NP criterion for correlated decisions assuming that the correlation coefficients between the sensor decisions are known and local sensor thresholds generating the correlated decisions are given. Using a special correlation structure, they studied the performance of the detection system versus the degree of correlation and showed how the performance advantage obtained by using a large number of sensors degrades as the degree of correlation between local decisions increases. In [70], the authors employed the Bahadur–Lazarsfeld series expansion of probability density functions to derive the optimum fusion rule for correlated local decisions. By using the Bahadur–Lazarsfeld expansion of probability density functions, the pdf of local correlated binary decisions can be represented by the pdf of independent random variables multiplied by a correlation factor. In many practical situations, conditional correlation coefficients beyond a certain order can be assumed to be zero. Thus, computation of the optimal fusion rule becomes less burdensome. When all the conditional correlation coefficients are zero, the optimal fusion rule reduces to the Chair–Varshney rule. Here, the implementation of the fusion rule was carried out assuming that the joint density of sensor observations is multivariate Gaussian, which takes into consideration the linear dependence of sensor observations by using the Pearson-correlation coefficient in the covariance matrix. An implicit assumption is that individual sensor observations are also Gaussian distributed.

In many applications, the dependence can get manifested in many different nonlinear ways. As a result, more general descriptors of correlation than the Pearson correlation coefficient, which only characterizes linear dependence, may be required [71]. Moreover, the marginal distributions of sensor observations characterizing their univariate statistics may also not be identical. Here, emphasis should be laid on the fact that multivariate density (or mass) functions do not necessarily exist for arbitrary marginal density (or mass) functions. In other words, given arbitrary marginal distributions, their joint distribution function cannot be written in a straightforward manner.

An interesting approach for the fusion of correlated decisions, that does not necessarily require prior information about the joint statistics of the sensor observations or decisions, is described next. Its novelty lies in the usage of *copula theory* [72]. The application of copula theory is widespread in the fields of econometrics and finance. However, its use for signal processing applications has been quite limited. The authors in [73,74] employ copula theory for signal detection problems involving correlated observations as well as for heterogeneous sensors observing a common scene. For the fusion of correlated decisions, copula theory does not require prior information about the joint statistics of the sensor observations or decisions and constructs the joint statistics based on a copula selection procedure. Note that the copula function–based fusion will fail to perform better than the Chair–Varshney rule if the constructed joint distribution using a particular parametric copula function does not adequately model the underlying joint distribution of the sensor observations. Therefore, training is necessary in order to select the best copula function. The topic of copula function selection for the distributed detection problem is considered in [75].

4.3 DISTRIBUTED DETECTION OVER NONIDEAL COMMUNICATION CHANNELS

For systems employing high SNR and/or effective channel error correction coding, communication may have extremely low error rates and can be assumed lossless, meaning that the local decisions can be transmitted to the fusion center without errors. On the other hand, the lossless communication assumption should be subject to careful scrutiny in WSNs. Increasing power and/or employing powerful error correction codes may not always be possible because of the stringent resources of WSNs. Furthermore, in a hostile environment, the power of transmitted signal should be kept to a minimum to attain a low probability of intercept/detection (LPI/LPD). Therefore, it may be necessary in many situations to tolerate the loss during data transmission to some extent. To overcome this loss, it is highly desirable to integrate the communication and decision fusion functions intelligently to achieve an acceptable system performance without spending extra system resources. This motivates the study of fusion of local decisions corrupted during the transmission process due to channel fading/noise impairment.

The model for a distributed detection system in the presence of fading channels is illustrated in Figure 4.3. Decisions at local sensors, denoted by u_k for $k = 1, \ldots, K$, are transmitted over parallel channels that are assumed to undergo independent fading. In this section, we consider a discrete-time Rayleigh flat fading channel with a stationary and ergodic complex gain of $h_k e^{j\phi_k}$ between the kth sensor and the fusion center. Note that h_k and ϕ_k denote the fading envelope and the phase of the channel, respectively. It is assumed that the channel gain remains constant during the transmission of a decision and channels are independent of each other. We further simplify the analysis by assuming binary signaling and replace $u_k \in \{0, 1\}$ by $s_k \in \{-1, 1\}$, so that the effect of the fading channel reduces to a real scalar multiplication for phase coherent reception. The phase coherent reception can be either accomplished through limited training for stationary channels, or, at a small cost of SNR degradation, by employing

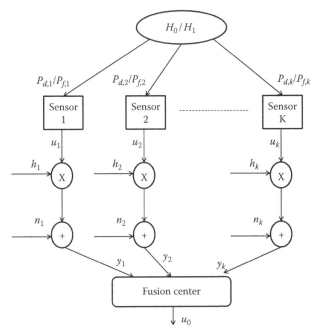

FIGURE 4.3 Parallel fusion model in the presence of fading and noisy channels between local sensors and the fusion center. u_k is the binary decision made by the kth sensor, h_k is the fading channel gain, n_k is a zero-mean Gaussian random variable with variance σ^2, and y_k is the observation received by the fusion center from the kth sensor, where $k \in \{1, \ldots, K\}$.

differential encoding for fast fading channels which results in the same signal model. The received signal model for sensor k is illustrated as

$$\tilde{y}_k = h_k e^{j\phi_k} s_k + v_k \qquad (4.36)$$

where v_k is a zero-mean complex Gaussian noise with independent real and imaginary parts having identical variance σ_n^2, i.e., $\mathbf{CN}(0, 2\sigma_n^2)$. Note that the notation $\mathbf{CN}$ represents complex Gaussian distribution. Without loss of generality, we make the assumption of Rayleigh fading channels with unit power, i.e., $h_k e^{j\phi_k} \sim \mathbf{CN}(0,1)$, therefore $E[h_k^2] = 1$. Using the knowledge of the channel phase at the receiver, the observation model at the fusion center for the kth sensor can be obtained as

$$y_k = h_k s_k + n_k \qquad (4.37)$$

Since v_k follows a circularly symmetric complex Gaussian distribution, the noise term $n_k \triangleq Re\{v_k e^{-j\phi_k}\}$ is real WGN with variance σ_n^2, i.e., $n_k \sim \mathcal{N}(0, \sigma_n^2)$.

Optimal Likelihood Ratio–Based Fusion Rule: By assuming instantaneous channel state knowledge regarding the fading channel and the local sensor performance indices, i.e., the $P_{d,k}$ and $P_{f,k}$ values, the optimal likelihood ratio (LR)-based fusion rule has been derived in [76], with the fusion statistic (LR) given by

$$\Lambda(\mathbf{y}) = \log \left[\frac{p(\mathbf{y} \mid H_1)}{p(\mathbf{y} \mid H_0)} \right]$$

$$= \sum_{k=1}^{K} \log \left[\frac{P_{d,k} \exp\left(-(y_k - h_k)^2 / 2\sigma_n^2\right) + (1 - P_{d,k}) \exp\left(-(y_k + h_k)^2 / 2\sigma_n^2\right)}{P_{f,k} \exp\left(-(y_k - h_k)^2 / 2\sigma_n^2\right) + (1 - P_{f,k}) \exp\left(-(y_k + h_k)^2 / 2\sigma_n^2\right)} \right] \quad (4.38)$$

where $\mathbf{y} = [y_1, \ldots, y_K]^T$ is a vector containing data received from all the K sensors. Note that, this fusion rule requires both local sensor performance indices and instantaneous CSI. Given exact channel state information and under conditional independence assumption under both hypotheses, the distribution of the optimal LR-based fusion statistic is given in [77]. Several suboptimum fusion rules that relax the requirements on a priori knowledge have also been proposed in [76].

Chair–Varshney Fusion Rule: In [76], the Chair–Varshney fusion statistic [24] has been shown to be a high-SNR approximation to (4.38)

$$\Lambda_1 = \sum_{sign(y_k)=1} \log \frac{P_{d,k}}{P_{f,k}} + \sum_{sign(y_k)=-1} \log \frac{1 - P_{d,k}}{1 - P_{f,k}} \quad (4.39)$$

where Λ_1 does not require any knowledge regarding the channel gain but does require $P_{d,k}$ and $P_{f,k}$ for all k. The probability distribution of the Chair–Varshney statistic, which is very helpful for performance analysis, has also been shown in [78]. This approach may suffer significant performance loss at low to moderate channel SNR.

Maximum Ratio Combining (MRC) Fusion Rule: It has been shown in [76] that for small values of channel SNR, Λ in (4.38) reduces to

$$\hat{\Lambda}_2 = \sum_{k=1}^{K} (P_{d,k} - P_{f,k}) h_k y_k \quad (4.40)$$

Further, if the local sensors are identical, i.e., $P_{d,k} = P_D$ and $P_{f,k} = P_F$ for all ks, then Λ further reduces to a form analogous to an MRC statistic:

$$\Lambda_2 = \frac{1}{K} \sum_{k=1}^{K} h_k y_k \quad (4.41)$$

Λ_2 in (4.41) does not require the knowledge of P_d and P_f provided $P_d - P_f > 0$. Knowledge of the channel gain is, however, required.

Equal Gain Combining (EGC) Fusion Rule: Motivated by the fact that Λ_2 resembles an MRC statistic for diversity combining, a third alternative in the form of an EGC has been proposed, which requires minimum amount of information:

$$\Lambda_3 = \frac{1}{K} \sum_{k=1}^{K} y_k \quad (4.42)$$

Interestingly enough, Λ_3 outperforms both Λ_1 and Λ_2 for a wide range of SNR in terms of its detection performance [76].

4.3.1 DISTRIBUTED DETECTION WITH PARTIAL CHANNEL STATE INFORMATION

The optimal LR-based fusion rule presented in Equation 4.38 requires instantaneous CSI, i.e., h_k and ϕ_k, for all the sensors in the WSN. However, for a WSN with very limited resources (energy and bandwidth), it is prohibitive to spend resources on estimating the channel gain every time a local sensor sends its decision to the fusion center. Thus, it is imperative to avoid channel estimation and conserve resources at the possible cost of relatively small performance degradation. This is the reasoning behind the exploration of new fusion rules that do not require instantaneous channel gains, h_k. In many WSN scenarios, the statistics of the fading (random) channel and the additive Gaussian noise can be estimated in advance, and used as prior information. It is the goal to develop a new LR-based fusion rule with only the prior information regarding the channel statistics instead of the instantaneous CSI.

Under hypothesis H_j, we have

$$p(y_k \mid H_j) = \sum_{u_k} [p(u_k \mid H_j)p(y_k \mid s_k)]$$

$$= P(u_k = 1 \mid H_j)p(y_k \mid s_k = 1) + P(u_k = 0 \mid H_j)p(y_k \mid s_k = -1)$$

and

$$p(y_k \mid s_k) = \int_0^\infty p(y_k \mid h_k, s_k)f(h_k)\,dh_k \tag{4.43}$$

By assuming a Rayleigh fading channel with unit power (i.e., $E[h_k^2] = 1$), the pdf of h_k is

$$p(h_k) = 2h_k e^{-h_k^2}, \quad h_k \ge 0 \tag{4.44}$$

and

$$p(y_k \mid h_k, s_k) = \frac{1}{\sqrt{2\pi}\sigma_n}\exp\left(-\frac{(y_k - h_k s_k)^2}{2\sigma_n^2}\right) \tag{4.45}$$

Then, the log LR based on the knowledge of channel statistics and local detection performance indices is expressed as [78]

$$\Lambda_4 = \log\left[\frac{f(\mathbf{y} \mid H_1)}{f(\mathbf{y} \mid H_0)}\right]$$

$$= \sum_{k=1}^{K} \log\left\{\frac{1 + [P_{d,k} - Q(ay_k)]\sqrt{2\pi}ay_k e^{(ay_k)^2/2}}{1 + [P_{f,k} - Q(ay_k)]\sqrt{2\pi}ay_k e^{(ay_k)^2/2}}\right\} \tag{4.46}$$

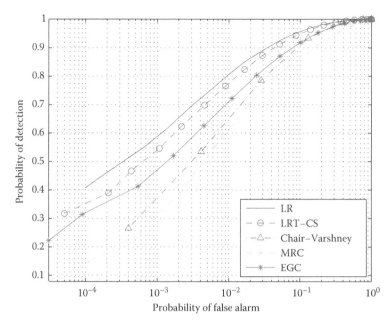

FIGURE 4.4   ROC curves for various fusion statistics for the Rayleigh fading channel with average channel SNR=4 dB. There are $k=8$ sensors with $P_{d,k}=0.6$ and $P_{f,k}=0.05$.

where $a = 1/\left(\sigma_n\sqrt{1+2\sigma_n^2}\right)$. As shown in Figure 4.4, the optimal LR-based fusion rule provides the best detection performance, however it requires instantaneous gain of the channel. On the other hand, its performance can be approached closely by the LRT fusion rule with partial channel knowledge (LRT-CS). The performance of the LRT-CS fusion rule is slightly worse than the optimal LR-based fusion rule with instantaneous channel gains and is better than the three suboptimal schemes.

4.3.2 Distributed Detection with No Channel State Information

Acquiring phase information of transmission channels can be costly as it typically requires training overhead. This overhead may be substantial for time-selective fading channels when mobile sensors are involved or the fusion center is constantly moving. Thus, incoherent-detection-based decision fusion rule has been introduced in Ref. [79]. In the incoherent case, the fusion statistics are based on the received envelope, or equivalently, the received power from each sensor. Denoting $r_k = |y_k|^2$, given the channel state information h_k, the signal power for the kth channel output is given by

$$p(r_k \mid h_k, u_k = 0) = \frac{1}{2\sigma_n^2}\exp\left(-\frac{r_k}{2\sigma_n^2}\right)$$

$$p(r_k \mid h_k, u_k = 1) = \frac{1}{2\sigma_n^2}I_0\left(\frac{h_k}{\sigma_n^2}\sqrt{r_k}\right)\exp\left(-\frac{h_k^2 + r_k}{2\sigma_n^2}\right)$$

(4.47)

where $I_0(.)$ is the zeroth-order modified Bessel function of the first kind. Given $p(h_k)$ as in Equation 4.44,

$$p(r_k \mid u_k = 0) = \frac{1}{2\sigma_n^2} \exp\left(-\frac{r_k}{2\sigma_n^2}\right)$$

$$p(r_k \mid u_k = 1) = \frac{1}{1+2\sigma_n^2} \exp\left(-\frac{r_k}{1+2\sigma_n^2}\right)$$

(4.48)

Then the LLR (log-likelihood ratio) can be given as

$$\Lambda(\mathbf{r}) = \log\left[\frac{p(\mathbf{r}\mid H_1)}{p(\mathbf{r}\mid H_0)}\right]$$

$$= \sum_{k=1}^{K} \log\left[\frac{P_{d,k}(1/(1+2\sigma_n^2))\exp\left(-r_k/(1+2\sigma_n^2)\right)+(1-P_{d,k})(1/2\sigma_n^2)\exp\left(-r_k/2\sigma_n^2\right)}{P_{f,k}(1/(1+2\sigma_n^2))\exp\left(-r_k/(1+2\sigma_n^2)\right)+(1-P_{f,k})(1/2\sigma_n^2)\exp\left(-r_k/2\sigma_n^2\right)}\right]$$

(4.49)

For the case of known fading statistics, Ricean and Nakagami fading channels have also been considered in [79]. In this section, we have investigated channel aware decision fusion algorithms with different degrees of channel state information for single-hop networks [76–79]. Extensions to multi-hop WSNs can be found in [80,81], while channel-optimized local quantizer design methods are provided in [82–84]. To counter sensor or channel failures, robust binary quantizer design has been proposed in [85]. Channel aware distributed detection has also been studied in the context of cooperative relay networks [86,87].

4.4 CONCLUSIONS

In this section, we summarize and further discuss distributed detection and decision fusion for a multi-sensor system. In a conventional distributed detection framework, it is assumed that local sensors' performance indices are known and communication channels between the sensors and fusion center are perfect. Under these assumptions, the design for optimal decision fusion rule at the fusion center and the optimal local decision rules at sensors was discussed under Bayesian and NP criteria. For a WSN consisting of passive sensors, it might be very difficult to estimate local sensors' performance indices and it can be very expensive to transmit them to the fusion center. Counting rule is an intuitive solution which uses the total number of "1"s as a decision statistic since the information about which sensor reports a "1" is of little use to the fusion center. Recent research shows that FDR-based decision fusion with nonidentical thresholds can substantially improve the detection performance as compared to counting rule with identical thresholds.

In a WSN setting with severe constraints on energy, bandwidth, and delay, transmitting sensor decisions to the fusion center over error free channels may become unrealistic since error free transmission may require high transmission power and/ or powerful error correction codes. Therefore, channel impairments should be taken into account in the design of distributed detection systems. Channel aware decision fusion algorithms where each has different degrees of channel state information have been reviewed.

For distributed detection in WSNs, in [55], it has been assumed that the communication channels between the sensors and the fusion center are perfect. It will be interesting to study the effect of imperfect communication channels on the detection performance of the proposed FDR-based framework. Also, the FDR framework has been proposed for the detection of a single target in the ROI. Extension of the FDR framework to detection of multiple targets in the ROI is an interesting and challenging research problem. It is also assumed that every sensor has identical noise power. Extension of the proposed framework to include the scenario of nonidentical noise power at each sensor is an interesting research problem.

Dense deployment of sensors in the WSN introduces redundancy in coverage, so selecting a subset of sensors may still provide information with the desired quality. Adaptive sensor management policies can be applied in distributed detection which select a subset of active sensors or distribute the available resources among the informative sensors while meeting the application requirements in terms of quality of service [36].

In this chapter, we have focused on parallel decision fusion architecture where sensors transmit their observations directly to the fusion center. For serial decision fusion, the information processing dealing with distributed data in the context of accurate signal detection and energy-efficient routing is currently emerging as a fruitful research area [88,89].

REFERENCES

1. P.K. Varshney, *Distributed Detection and Data Fusion*, Springer, New York, 1997.
2. H.L. Van Trees, *Detection, Estimation and Modulation Theory*, Vol. 1, Wiley, New York, 1968.
3. H.V. Poor, *An Introduction to Signal Detection and Estimation*, Springer-Verlag, New York, 1988.
4. C.W. Helstrom, *Elements of Signal Detection and Estimation*, Prentice-Hall, Englewood Cliffs, NJ, 1995.
5. R. Viswanathan and P.K. Varshney, Distributed detection with multiple sensors: Part I—Fundamentals, *Proceedings of the IEEE*, 85(1), 54–63, January 1997.
6. R.S. Blum, S.A. Kassam, and H.V. Poor, Distributed detection with multiple sensors: Part II—Advanced topics, *Proceedings of the IEEE*, 85(1), 64–79, January 1997.
7. V. Veeravalli and P.K. Varshney, Distributed inference in wireless sensor networks, *Philosophical Transactions of the Royal Society*, 370(1958), 100–117, January 2012.
8. J.P. Shaffer, Multiple hypothesis testing, *Annual Review of Psychology*, 46(1), 561–584, 1995.
9. M. Schwartz, W.R. Bennett, and S. Stein, *Communication Systems and Techniques*, Wiley, New York, 1995.

10. B. Eisenberg, Multihypothesis problems, in *Handbook of Sequential Analysis*, B.K Ghosh and P.K. Sen, Eds. New York, Marcel Dekker, Vol. 118, pp. 229–244, 1991.

11. X. Zhu, Y. Yuan, C. Rorres, and M. Kam, Distributed M-ary hypothesis testing with binary local decisions, *Information Fusion*, 5(3), 157–167, 2004.

12. Q. Zhang and P.K. Varshney, Decentralized M-ary detection via hierarchical binary decision fusion, *Information Fusion*, 2(1), 3–16, 2001.

13. C.W. Baum and V.V. Veeravalli, A sequential procedure for multihypothesis testing, *IEEE Transactions on Information Theory*, 40(6), 1994–2007, 1994.

14. J. Tsitsiklis and M. Athans, On the complexity of decentralized decision making and detection problems, *IEEE Transactions on Automatic Control*, 30, 440–446, May 1985.

15. N.S.V. Rao, Computational complexity issues in synthesis of simple distributed detection networks, *IEEE Transactions on Systems, Man, Cybernetics*, 21, 1071–1081, September/October 1991.

16. J.N. Tsitsiklis, Decentralized detection, in *Advances in Statistical Signal Processing*, H.V. Poor and J.B. Thomas, Eds. JAI Press, Greenwich, CT, 1993.

17. J.N. Tsitsiklis, On threshold rules in decentralized detection, in *Proceedings of the 25th IEEE Conference on Decision and Control*, Athens, Greece, 1986, pp. 232–236.

18. P. Willet and D. Warren, Decentralized detection: When are identical sensors identical, in *Proceedings Conference on Information Science and Systems*, Princeton, NJ, 1991, pp. 287–292.

19. M. Cherikh and P.B. Kantor, Counterexamples in distributed detection, *IEEE Transactions on Information Theory*, 38, 162–165, January 1992.

20. Z.B. Tang, K.R. Pattipati, and D. Kleinman, An algorithm for determining the detection thresholds in a distributed detection problem, *IEEE Transactions on Systems, Man, and Cybernetics*, 21, 231–237, January/February 1991.

21. A.R. Reibman, *Performance and Fault-Tolerance of Distributed Detection Networks*, PhD thesis, Duke University, Durham, NC, 1987.

22. S.C.A. Thomopoulos, R. Viswanathan, and D.K. Bougoulias, Optimal distributed decision fusion, *IEEE Transactions on Aerospace and Electronic Systems*, 25, 761–765, September 1989.

23. C.W. Helstrom, Gradient algorithms for quantization levels in distributed detection systems, *IEEE Transactions on Aerospace and Electronic Systems*, 31, 390–398, January 1995.

24. Z. Chair and P.K. Varshney, Optimal data fusion in multiple sensor detection systems, *IEEE Transactions on Aerospace and Electronic Systems*, 22, 98–101, January 1986.

25. J.B. Thomas, Nonparametric detection, *Proceedings of the IEEE*, 58(5), 623–631, 1970.

26. S. Kay, Can detectability be improved by adding noise? *IEEE Signal Processing Letters*, 7(1), 8–10, January 2000.

27. H. Chen, P.K. Varshney, S.M. Kay, and J.H. Michels, Theory of the stochastic resonance effect in signal detection: Part I; fixed detectors, *IEEE Transactions on Signal Processing*, 55(7), 3172–3184, July 2007.

28. H. Chen and P.K. Varshney, Theory of the stochastic resonance effect in signal detection: Part II; variable detectors, *IEEE Transactions on Signal Processing*, 56(10), 5031–5041, October 2008.

29. P.F. Swaszek, On the performance of serial networks in distributed detection, *IEEE Transactions on Aerospace and Electronic Systems*, 29(1), 254–260, January 1993.

30. Z.B. Tang, K.R. Pattipati, and D.L. Kleinman, Optimization of detection networks. I. Tandem structures, *IEEE Transactions on Systems, Man and Cybernetics*, 21(5), 1044–1059, September/October 1991.

31. W.P. Tay, J.N. Tsitsiklis, and M.Z. Win, On the subexponential decay of detection error probabilities in long tandems, *IEEE Transactions on Information Theory*, 54(10), 4767–4771, October 2008.

32. Z.-B. Tang, K.R. Pattipati, and D.L. Kleinman, Optimization of detection networks. II. Tree structures, *IEEE Transactions on Systems, Man and Cybernetics*, 23(1), 211–221, January/February 1993.

33. W.P. Tay, J.N. Tsitsiklis, and M.Z. Win, On the impact of node failures and unreliable communications in dense sensor networks, *IEEE Transactions on Signal Processing*, 56(6), 2535–2546, June 2008.

34. W.P. Tay, J.N. Tsitsiklis, and M.Z. Win, Bayesian detection in bounded height tree networks, *IEEE Transactions on Signal Processing*, 57(10), 4042–4051, October 2009.

35. Q. Zou, S. Zheng, and A.H. Sayed, Cooperative sensing via sequential detection, *IEEE Transactions on Signal Processing*, 58(12), 6266–6283, December 2010.

36. Q. Cheng, P.K. Varshney, K.G. Mehrotra, and C.K. Mohan, Bandwidth management in distributed sequential detection, *IEEE Transactions on Information Theory*, 51(8), 2954–2961, August 2005.

37. H. Chen, P.K. Varshney, and J.H. Michels, Improving sequential detection performance via stochastic resonance, *IEEE Signal Processing Letters*, 15, 685–688, 2008.

38. V.V. Veeravalli, Decentralized quickest change detection, *IEEE Transactions on Information Theory*, 47(4), 1657–1665, May 2001.

39. R. Niu and P.K. Varshney, Sampling schemes for sequential detection with dependent data, *IEEE Transactions on Signal Processing*, 58(3), 1469–1481, March 2010.

40. D. Bajovic, D. Jakovetic, J. Xavier, B. Sinopoli, and J.M.F. Moura, Distributed detection via Gaussian running consensus: Large deviations asymptotic analysis, *IEEE Transactions on Signal Processing*, 59(9), 4381–4396, September 2011.

41. Z. Li, F.R. Yu, and M. Huang, A distributed consensus-based cooperative spectrum-sensing scheme in cognitive radios, *IEEE Transactions on Vehicular Technology*, 59(1), 383–393, January 2010.

42. S. Stankovic, N. Ilic, M.S. Stankovic, and K.H. Johansson, Distributed change detection based on a consensus algorithm, *IEEE Transactions on Signal Processing*, 59(12), 5686–5697, December 2011.

43. J.N. Tsitsiklis, Decentralized detection with a large number of sensors, *Mathematics of Control, Signals, and Systems*, 1, 167–182, 1988.

44. P. Chen and A. Papamarcou, New asymptotic results in parallel distributed detection, *IEEE Transactions on Information Theory*, 39(6), 1847–1863, November 1993.

45. J. Chamberland and V.V. Veeravalli, Decentralized detection in sensor networks, *IEEE Transactions on Signal Processing*, 51, 407–416, February 2003.

46. J.F. Chamberland and V.V. Veeravalli, Asymptotic results for decentralized detection in power constrained wireless sensor networks, *IEEE Journal on Selected Areas in Communications*, 22(6), 1007–1015, August 2004.

47. R. Niu, P.K. Varshney, and Q. Cheng, Distributed detection in a large wireless sensor network, *International Journal on Information Fusion*, 7(4), 380–394, December 2006.

48. R. Niu and P.K. Varshney, Distributed detection and fusion in a large wireless sensor network of random size, *EURASIP Journal on Wireless Communications and Networking*, 5(4), 462–472, September 2005.

49. R. Niu and P.K. Varshney, Performance analysis of distributed detection in a random sensor field, *IEEE Transactions on Signal Processing*, 56(1), 339–349, January 2008.

50. Q. Cheng, R. Niu, A. Sundaresan, and P.K. Varshney, Distributed detection and decision fusion with applications to wireless sensor networks, *Integrated Tracking, Classification, and Sensor Management: Theory and Applications*, Wiley/IEEE, June 2012.

51. B. Krishnamachari and S. Iyengar, Distributed Bayesian algorithms for fault-tolerant event region detection in wireless sensor networks, *IEEE Transactions on Computers*, 53(3), 241–250, March 2004.

52. F. Gini, F. Lombardini, and L. Verrazzani, Decentralized CFAR detection with binary integration in weibull clutter, *IEEE Transactions on Aerospace and Electronic Systems*, 33(2), 396–407, April 1997.

53. F. Gini, F. Lombardini, and L. Verrazzani, Decentralised detection strategies under communication constraints, *IEE Proceedings—Radar, Sonar and Navigation*, 145(4), 199–208, August 1998.

54. F. Gini, F. Lombardini, and P.K. Varshney, On distributed signal detection with multiple local free parameters, *IEEE Transactions on Aerospace and Electronic Systems*, 35(4), 1457–1466, October 1999.

55. P. Ray and P.K. Varshney, False discovery rate based sensor decision rules for the network-wide distributed detection problem, *IEEE Transactions on Aerospace and Electronic Systems*, 47(3), 1785–1799, July 2011.

56. E.L. Lehman and J.P. Romano, *Testing Statistical Hypotheses*, Springer, New York, 3rd edn., 2008.

57. Y. Benjamini and Y. Hochberg, Controlling the false discovery rate: A practical and powerful approach to multiple testing, *Journal of the Royal Statistical Society, Series B*, 57(1), 289–300, 1995.

58. Y. Benjamini and D. Yekutieli, The control of the false discovery rate in multiple testing under dependency, *Annals of Statistics*, 29, 1165–1188, 2001.

59. B. Picinbono, On deflection as a performance criterion in detection, *IEEE Transactions on Aerospace and Electronic Systems*, 31(3), 1072–1081, July 1995.

60. G.S. Lauer and N.R. Sandell Jr., Distributed detection with waveform observations: Correlated observation processes, in *Proceedings of the 1982 American Controls Conference*, Arlington, VA, 1982, Vol. 2, pp. 812–819.

61. P. Chen and A. Papamarcou, Likelihood ratio partitions for distributed signal detection in correlated Gaussian noise, in *Proceedings of IEEE International Symposium on Information Theory*, Whistler, Canada, Septemper 1995, p. 118.

62. V. Aalo and R. Viswanathan, On distributed detection with correlated sensors: Two examples, *IEEE Transactions on Aerospace and Electronic Systems*, 25, 414–421, May 1989.

63. V. Aalo and R. Viswanathan, Asymptotic performance of a distributed detection system in correlated Gaussian noise, *IEEE Transactions on Signal Processing*, 40, 211–213, January 1992.

64. R. Blum, P. Willett, and P. Swaszek, Distributed detection of known signals in nonGaussian noise which is dependent from sensor to sensor, in *Proceedings of Conference of the Information Sciences and Systems*, Baltimore, MD, March 1997, pp. 825–830.

65. X. Lin and R. Blum, Numerical solutions for optimal distributed detection of known signals in dependent t-distributed noise: The two-sensor problem, in *Proceedings of the Asilomar Conference on Signals, Systems, and Computers*, Pacific Grove, CA, November 1998, pp. 613–617.

66. Z. Tang, K. Pattipati, and D. Kleinman, A distributed M-ary hypothesis testing problem with correlated observations, *IEEE Transactions on Automatic Control*, 37, 1042–1046, July 1992.

67. P.K. Willett, P.F. Swaszek, and R.S. Blum, The good, bad, and ugly: Distributed detection of a known signal in dependent Gaussian noise, *IEEE Transactions on Signal Processing*, 48, 3266–3279, December 2000.

68. H. Chen, P.K. Varshney, and B. Chen, A novel framework for distributed detection with dependent observations, *IEEE Transactions on Signal Processing*, 60(3), 1409–1419, March 2012.

69. E. Drakopoulos and C.-C. Lee, Optimum multisensor fusion of correlated local decisions, *IEEE Transactions on Aerospace and Electronic Systems*, 27(4), 593–606, July 1991.

70. M. Kam, Q. Zhu, and W.S. Gray, Optimal data fusion of correlated local decisions in multiple sensor detection systems, *IEEE Transactions on Aerospace and Electronic Systems*, 28(3), 916–920, July 1992.

71. D.D. Mari and S. Kotz, *Correlation and Dependence*, Imperial College Press, London, U.K., 2001.

72. R.B. Nelsen, *An Introduction to Copulas*, Springer-Verlag, New York, 1999.

73. A. Sundaresan, P.K. Varshney, and N.S.V. Rao, Copula-based fusion of correlated decisions, *IEEE Transactions on Aerospace and Electronic Systems*, 47(1), 454–471, 2011.

74. S.G. Iyengar, P.K. Varshney, and T. Damarla, A parametric copula-based framework for hypothesis testing using heterogeneous data, *IEEE Transactions on Signal Processing*, 59(5), 2308–2319, May 2011.

75. A. Sundaresan, Detection and source location estimation of random signal sources using sensor networks, PhD thesis, Syracuse University, Syracuse, New York, 2010.

76. B. Chen, R. Jiang, T. Kasetkasem, and P.K. Varshney, Channel aware decision fusion for wireless sensor networks, *IEEE Transactions on Signal Processing*, 52, 3454–3458, December 2004.

77. I. Bahceci, G. Al-Regib, and Y. Altunbasak, Parallel distributed detection for wireless sensor networks: Performance analysis and design, in *IEEE Global Telecommunications Conference, GLOBECOM*, St. Louis, MO, 2005. IEEE, Piscataway, NJ, Vol. 4, p. 5.

78. R. Niu, B. Chen, and P. K. Varshney, Fusion of decisions transmitted over Rayleigh fading channels in wireless sensor networks, *IEEE Transactions on Signal Processing*, 54(3), 1018–1027, March 2006.

79. R. Jiang and B. Chen, Fusion of censored decisions in wireless sensor networks, *IEEE Transactions on Wireless Communications*, 4(6), 2668–2673, November 2005.

80. Y. Lin, B. Chen, and P.K. Varshney, Decision fusion rules in multi-hop wireless sensor networks, *IEEE Transactions on AES*, 51, 475–488, April 2005.

81. I. Bahceci, G. Al-Regib, and Y. Altunbasak, Serial distributed detection for wireless sensor networks, in *International Symposium on Information Theory, ISIT*, Adelaide, Australia, 2005, pp. 830–834.

82. B. Chen and P.K. Willett, On the optimality of likelihood ratio test for local sensor decision rules in the presence of non-ideal channels, *IEEE Transactions on Information Theory*, 51(2), 693–699, 2005.

83. B. Liu and B. Chen, Channel optimized quantizers for decentralized detection in wireless sensor networks, *IEEE Transactions on Information Theory*, 52, 3349–3358, July 2006.

84. B. Liu and B. Chen, Decentralized detection in wireless sensor networks with channel fading statistics, *EURASIP Journal on Wireless Communications and Networking*, 2007, 11, January 2007.

85. Y. Lin, B. Chen, and B. Suter, Robust binary quantizers for distributed detection, *IEEE Transactions on Wireless Communications*, 6(6), 2172–2181, June 2007.

86. B. Liu, B. Chen, and R.S. Blum, Minimum error probability cooperative relay design, *IEEE Transactions on Signal Processing*, 55(2), 656–664, February 2007.

87. H. Chen, P.K. Varshney, and B. Chen, Cooperative relay for decentralized detection, in *Proceedings of the 2008 IEEE International Conference on Acoustics, Speech and Signal Processing*, Las Vegas, NV, March 2008, pp. 2293–2296.

88. Y. Yang, R.S. Blum, and B.M. Sadler, Energy-efficient routing for signal detection in wireless sensor networks, *IEEE Transactions on Signal Processing*, 57(6), 2050–2063, 2009.

89. Y. Sung, S. Misra, L. Tong, and A. Ephremides, Cooperative routing for distributed detection in large sensor networks, *IEEE Journal on Selected Areas in Communications*, 25(2), 471–483, 2007.

5 Fundamentals of Distributed Estimation

Chee-Yee Chong, Kuo-Chu Chang , and Shozo Mori

CONTENTS

5.1 Introduction 96
5.2 Distributed Estimation Architectures..97
    5.2.1 Fusion Architecture Graph 98
        5.2.1.1 Singly Connected Fusion Architectures 98
        5.2.1.2 Multiply Connected Fusion Architectures.........................99
    5.2.2 Information Graph 100
        5.2.2.1 Singly Connected Information Graphs for Singly
                  Connected Fusion Architectures.......................................101
        5.2.2.2 Multiply Connection Information Graphs for
                  Hierarchical Fusion..101
        5.2.2.3 Information Graph for Distributed Architectures.............103
    5.2.3 Information Communicated and Common Prior Knowledge 104
    5.2.4 Selecting Appropriate Architectures 104
5.3 Bayesian Distributed Fusion Algorithm 105
    5.3.1 Bayesian Distributed Estimation Problem and Solution..................105
    5.3.2 Bayesian Distributed Fusion for Gaussian Random Vectors 107
5.4 Optimal Bayesian Distributed Fusion for Different Architectures 109
    5.4.1 Hierarchical Architecture 109
        5.4.1.1 Hierarchical Fusion without Feedback 109
        5.4.1.2 Hierarchical Fusion with Feedback 110
    5.4.2 Arbitrary Distributed Fusion Architecture 110
5.5 Suboptimal Bayesian Distributed Fusion Algorithms 111
    5.5.1 Naïve Fusion 112
    5.5.2 Channel Filter Fusion 112
    5.5.3 Chernoff Fusion 113
    5.5.4 Bhattacharyya Fusion 114
5.6 Distributed Estimation for Gaussian Distributions or Estimates
with Error Covariances..115
    5.6.1 Maximum A Posteriori Fusion or Best Least Unbiased Estimate....115
    5.6.2 Cross-Covariance Fusion..116
5.7 Distributed Estimation for Object Tracking 117
    5.7.1 Deterministic Dynamics..117
    5.7.2 Nondeterministic Dynamics..118

        5.7.2.1   Augmented State Vector and Approximation 119
        5.7.2.2   Using Cross-Covariance at a Single Time 119
5.8   Distributed Estimation for Object Classification 119
        5.8.1   Distributed Object Classification Architectures 119
        5.8.2   Distributed Classification Algorithms 120
5.9   Summary 121
5.10 Bibliographic Notes 121
References 122

5.1   INTRODUCTION

Many applications such as target tracking, robotics, and manufacturing have increasingly used multiple sensor or data sources to provide information. Multiple sensors provide better coverage than a single sensor, either over a larger geographical area or broader spectrum. By generating more measurements, they can improve detection and false alarm performance. Improved accuracy (location, classification) can also result from viewing or phenomenological diversity provided by multiple sensors. For example, similar sensors that are not co-located can provide more accurate measurements on target location by exploiting different viewing angles, while dissimilar sensors such as radar and optical can observe different features for better object recognition.

   The measurements from multiple sensors can be processed or fused at a central site or multiple sites. The centralized fusion architecture requires communicating all the measurements to a single site and is theoretically optimal because the information in the measurements is not degraded by any intermediate processing. When the sensors are geographically distributed, it may make sense to also distribute the processing, with each processing site responsible for the measurements from one or more sensors. These sites can communicate their results to other fusion sites for further processing. The distributed fusion architecture has many advantages such as lower bandwidth by communicating processing results rather than measurements, availability of processing results for local functions such as sensor management, distribution of the processing load to multiple sites, and less vulnerability because there is no single point of failure. Furthermore, a properly designed distributed fusion system can provide modularity and scalability for rapid incorporation of more sensors.

   Because of these advantages, there are many examples of distributed fusion systems including net-centric military systems, robotics teams, and wireless sensor networks, where centralized processing is not practical. However, many technical issues need to be addressed for distributed fusion systems to achieve high performance. The first issue is selecting the appropriate fusion architecture that connects sensors with the processors or agents at the fusion sites and how the data are shared with other sites in the network. The fusion architecture also specifies the information flow between the agents. The second issue is how the data should be processed by each agent to provide the best performance. For example, a fusion agent has to recognize when common information occurs in any received data to avoid double counting when fusing the data.

This chapter presents the fundamental concepts for distributed data fusion. In particular, we focus on the estimation problem where the goal of fusion is to compute an estimate of the state from measurements collected by multiple sensors. The state may be continuous and time varying such as the position and velocity of a vehicle in object tracking. It may also be discrete and static such as the class of an object in object classification. We focus on estimation to exclude discussions on data association issues that are important in object tracking. These issues will be discussed in Chapter 6.

The rest of this chapter is structured as follows. Section 5.2 discusses distributed fusion architectures, their advantages and disadvantages, the use of information graph to represent information flow, and selection of an appropriate architecture. Section 5.3 presents the Bayesian fusion equation for combining two probability functions, or their means and covariances. Section 5.4 shows how the information graph can be used to keep track of information flow in a distributed estimation system and how it can be used to derive fusion equations for various fusion architectures. Section 5.5 discusses some suboptimal but practical approaches that are based on approximations of the optimal approach. Section 5.6 presents algorithms for fusing estimates characterized by means and covariances. Section 5.7 discusses distributed fusion for object tracking when the state is continuous and time-varying. Section 5.8 discusses distributed fusion for object classification when the state is a discrete and static random variable. Section 5.9 provides a summary, and Section 5.10 contains some bibliographic notes.

Much has been published on distributed estimation over the last three decades with summaries provided in Chong et al. (1990) and Liggins and Chang (2009). Our discussion focuses on algorithms that are non-iterative, i.e., we will not address the consensus problem (Teneketzis and Varaiya 1988, Olfati-Saber 2005). We also view decentralized estimation (Durrant-Whyte et al. 1990) as a special case of distributed estimation. Furthermore, we sometimes use fusion and estimation to mean the same thing, and consider conditional probability (density) as a form of estimate.

5.2 DISTRIBUTED ESTIMATION ARCHITECTURES

The basic components of a distributed estimation system are sensors, processors (estimation or fusion agents), and users. Sensors generate measurements or data on the objects of interest. The measurements contain information on the object state such as position, velocity, or class. Estimation or fusion agents process sensor data or results received from other fusion agents to generate better estimates. Users are the consumers of the fusion results. A user can be the controller in a robotic system or a commander in a surveillance system. In a distributed fusion or estimation system, there are multiple sensors, processors, and users. These components are usually distributed geographically and connected together by a communication network.

The fusion architecture (Chong 1998) consists of three components. At the system level, the communication graph represents network connectivity between the components. When sensors collect measurements and processors fuse estimates at multiple times, the information graph represents the detailed information flow from the sensors to the processors. Finally, the information content communicated also has to be specified.

5.2.1 FUSION ARCHITECTURE GRAPH

The fusion architecture graph represents the connectivity of the fusion system as determined by the communication network. The nodes of the graph represent the sensors and processors, and the directed edges are the communication paths between the components. There are two main types of system architectures based on the number of communication paths from a particular sensor to the processor.

5.2.1.1 Singly Connected Fusion Architectures

In a singly connected fusion architecture, there is a single path between any sensor–processor pair. Figure 5.1 shows four examples of singly connected fusion architectures—centralized, decoupled, replicated centralized, and hierarchical without feedback.

In the centralized architecture, measurements from all sensors are sent to a single fusion site or agent to be processed. Theoretically this architecture produces the best performance since there is no information loss. However, centralization implies high communication load over the network, high processing load at the fusion site, and low survivability due to a single point of failure. The decoupled architecture partitions the sensors into multiple sets with a fusion site responsible for each set. This architecture is appropriate when there is a natural partitioning of the sensors so that the sensors in the same set can help each other but those outside the set provide little additional information. This architecture has the lowest computation and communication requirements. However, the performance can be poor if the sensors cannot be partitioned easily. In the replicated centralized architecture, multiple fusion sites process data from overlapping sets of sensors. There is no communication among the fusion sites. This architecture has high performance and reliability due to the

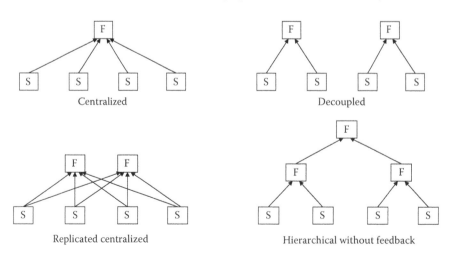

FIGURE 5.1 Singly connected fusion architectures.

multiple sites processing the same data. However, it also has high communication and processing costs.

These three architectures do not allow communication among the fusion sites. Thus there is a single information path from a sensor to a fusion site. This allows the use of simple fusion algorithms since the double counting or rumor propagation problem does not exist. These architectures are useful since they serve as benchmarks for comparing the performance of other distributed fusion architectures.

In the hierarchical (without feedback) architecture, the fusion sites are arranged in a hierarchy with the low-level fusion sites processing sensor data to form local estimates. These estimates are sent to a high-level fusion site to be combined. In order to realize the benefit of reduced communication, the communication rate from the low-level site to the high level should be lower than the sensor observation rate. As compared to the centralized architecture, the hierarchical architecture has the advantage of lower communication, lower processing cost when the low-level site processes data from a smaller set of sensors, and increased reliability. However, multiple information paths can occur if the sensors and fusion sites collect measurements and process at multiple times.

5.2.1.2 Multiply Connected Fusion Architectures

In a multiply connected fusion architecture, there are multiple communication paths between a pair of sensor and processor. Figure 5.2 shows four examples of multiply connected fusion architectures—hierarchical with sensor sharing, hierarchical with feedback, peer-to-peer, and broadcast.

In the hierarchical with sensor sharing architecture, the measurements from one sensor are processed by multiple fusion sites. This makes sense when that sensor is particularly powerful. However, high-level fusion is difficult because the common information from that sensor cannot be removed easily. In the hierarchical with feedback

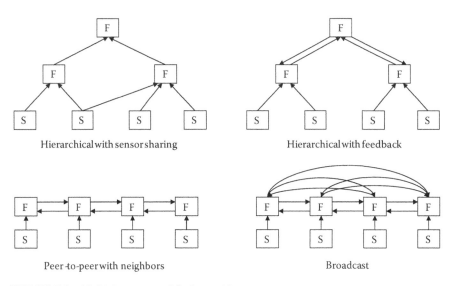

FIGURE 5.2   Multiply connected fusion architectures.

architecture, the accuracy of the local estimates can be enhanced by feeding back high-level estimates (which include information from more sensors) to the low level where the data are to be combined. In this feedback architecture, information flows in both directions, from low level to high level and also from high level to low level.

In the peer-to-peer architecture, a fusion agent has two-way communication with another fusion agent (with only neighbors in a decentralized architecture). In the broadcast architecture, a fusion agent broadcasts its results to multiple fusion agents, who can also broadcast their own results. These two are examples of fully distributed architectures where the communication is dynamic and not specified a priori. For example, a fusion site may send its results to another fusion site depending on the results or in response to a request for information from another site. Such architectures can adapt dynamically to the current situation. In general, multiply connected fusion architectures are more robust against failures, but algorithms are more difficult to develop because of the multiple information paths.

5.2.2 INFORMATION GRAPH

The fusion architecture graph characterizes information paths at a high level. It does not describe how each measurement or fusion result flows through the system, and particularly it does not portray the effects of time between updates or communications due to repeated sensor observations and fusion processing. In particular, the architectures in Figures 5.1 and 5.2 do not represent the relationship between the estimates and the sensor data at different times, which is needed in order to identify the common information to avoid double counting or data incest. The information graph (Chong et al. 1982, 1983, 1985, 1986, 1987, Chong and Mori 2004) represents the detailed information flow and transactions in a fusion architecture specified by communication paths. It also supports the development of optimal and suboptimal fusion algorithms. A similar graph model can be found in McLaughlin et al. (2004, 2005).

The nodes in the information graph represent information events. The observation node represents the observation event of a sensor at a specific time; the fusion node represents a fusion event at a fusion site at a specific time. There are two main types of fusion events: fusion of sensor observation with the local fusion result, and fusion of the processing results from other sites with the local results.

The directed edges or links represent the communication between information nodes. Note that the observation node is a leaf node with no predecessors and its successor nodes are always fusion nodes. The predecessor node of a fusion node may be an observation node or another fusion node. A fusion node may have other fusion nodes as successors or no successor nodes.

The edges in the graph can be used to trace the information available to a node. A directed path from Node A to Node B means that Node B has access to the information at Node A, and in general each node has access to the information of its predecessor nodes. The specific information available depends on what is communicated. Sensor data are transmitted from an observation node to a fusion node but usually estimates are communicated between fusion nodes. If the estimate is the sufficient statistics, then the maximum information at a node consists of the sensor data based on all its ancestor observation nodes.

A main problem in distributed fusion is identifying the common information shared by two estimates that need to be fused. The information graph provides a useful tool to discover the source of this common information. If two fusion nodes have common ancestors, then the estimates at these nodes contain the information of the common ancestors. If two fusion nodes have no common predecessor, there is no sharing of information except for the prior. The following are some examples of information graphs.

5.2.2.1 Singly Connected Information Graphs for Singly Connected Fusion Architectures

Figure 5.3 shows the information graphs for the centralized, replicated centralized, and decoupled fusion architectures of Figure 5.1. Note that time is now represented explicitly. These information paths are singly connected because there is only one information path from each observation node to a fusion node.

5.2.2.2 Multiply Connection Information Graphs for Hierarchical Fusion

Figure 5.4 shows the information graph for hierarchical fusion without feedback architecture. Even though the fusion architecture graph is singly connected when there is no feedback from the high-level site, the information graph (on the left) is multiply connected due to repeated communication and fusion. For example, both fusion nodes H and L have the predecessor node $\bar{L}$, that is, the information at $\bar{L}$ is included in the information at H and L. Thus fusion of H and L have to make sure that the common information of $\bar{L}$ is not double counted. This multiply connected information graph can be transformed into a singly connected graph by modifying the processing and communication strategies. One approach is to have the local fusion site send only new information since the last time it communicated with the high-level fusion site. This is equivalent to deleting the edge at the local site

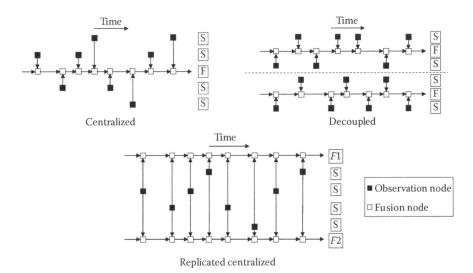

FIGURE 5.3 Singly connected information graph for singly connected fusion architectures.

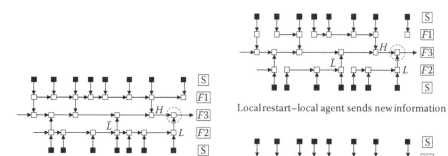

FIGURE 5.4   Information graph for hierarchical fusion without feedback.

after each communication (local restart and sending new information in Figure 5.4). Effectively, the local fusion site has a separate estimator whose output is communicated. The other approach of getting a singly connected graph is not to allow memory at the high-level fusion site. Then the fusion nodes will only have observation nodes from each sensor (global restart and no memory in Figure 5.4).

Figure 5.5 shows the information graph for hierarchical fusion with feedback. As in hierarchical fusion without feedback, the multiply-connected information graph for high level fusion can be converted to a singly connected network if the local

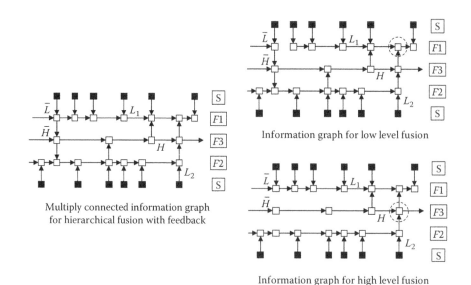

FIGURE 5.5   Information graph for hierarchical fusion with feedback.

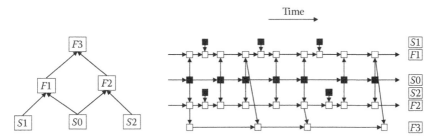

FIGURE 5.6   Information graph for hierarchical fusion with common sensor.

fusion site sends fusion results that do not rely on feedback or the sensor observations before the last communication. Effectively, the local site keeps two sets of books—an optimal estimate for local use based on all sensor observations and an estimate for communication based only on new local observations received since the last communication. Similarly, the low-level fusion site can obtain a singly connected fusion graph by deleting the appropriate edges.

Figure 5.6 shows the information graph for hierarchical fusion with a common sensor. In this case, the information graph is inherently multiply connected and it is difficult to convert it into a singly connected information graph.

5.2.2.3   Information Graph for Distributed Architectures

The information graphs for general distributed architectures are usually multiply connected because of the possible communication paths. However, it is sometimes possible to convert them to singly connected information graphs by designing the appropriate information exchange. Figure 5.7 shows how the information graphs for peer-to-peer and broadcast fusion architectures can be made singly connected if

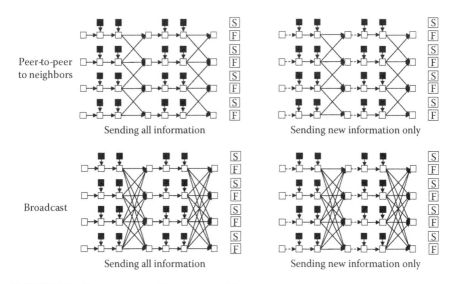

FIGURE 5.7   Peer-to-peer and broadcast architectures.

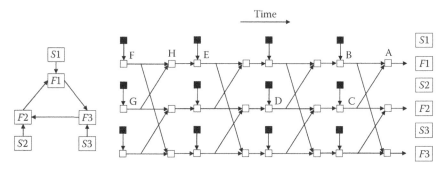

FIGURE 5.8   Information graph for cyclic architecture.

each local fusion site only communicates the new information received since the last communication. The dotted lines are information paths that are only maintained for generating the local estimates but not for communication.

The information graph can become complicated if the fusion sites only communicate their fusion results. Figure 5.8 shows a cyclic fusion architecture where site 1 sends its local fusion result to site 3, site 3 to site 2, and site 2 to site 1. It is difficult to identity new information because of the loopy communication in this architecture. From the information graph, the most recent common predecessors of B and C are D and E. The common predecessors of D and E are the nodes F and G, which have the same information as H.

5.2.3   INFORMATION COMMUNICATED AND COMMON PRIOR KNOWLEDGE

Defining the distributed fusion architecture also requires specifying the type of data communicated. The data can be the sensor measurements collected by the site or processing results, which can be estimates or probabilities of the state. Choosing what to communicate is a tradeoff between bandwidth and amount of information. Communicating measurements require the most bandwidth but provide the most information. Processing is easy for the fusion site receiving the measurements because the measurement errors are generally independent. Effectively, each fusion site performs centralized fusion of measurements and the information graph is singly connected.

When processing results such as estimates or probabilities are communicated, additional information is frequently needed for optimal fusion by the receiver. For example, network topology or information pedigree is needed to construct the information graph to identify the common information. Optimal fusion may also require knowing other estimates. When such information or sufficient statistics is not available, fusion can only be suboptimal. For example, optimal fusion for tracking objects with nonzero process noise requires knowledge of state estimates at multiple previous times. The fusion will be suboptimal when the state is only known at the current time.

5.2.4   SELECTING APPROPRIATE ARCHITECTURES

The fusion architecture has a significant impact on the development and performance of the distributed fusion system. A fusion architecture can be evaluated by the amount of information generated, communication bandwidth, algorithm complexity, and robustness. The following are some general guidelines for selecting fusion architectures:

- Use all sensor data for optimal performance. A fusion site should have access to as much sensor data as possible and a fusion node in the information graph should include information on all observation nodes (ancestors) that can be communicated to the fusion node.
- Compress sensor data for efficient communication. Less communication bandwidth is needed if the information in multiple observation nodes can be captured by a single intermediate fusion node. However, compression may result in information loss and introduce multiply connected information paths.
- Find architectures with singly connected information paths. Then the information to be fused will not contain common information and the fusion algorithm will be relatively simple.
- Use redundant paths for robustness/survivability. Each observation node should have multiple paths to reach a fusion node. However, redundancy may result in more processing/communication cost and/or more complicated fusion algorithms.

5.3 BAYESIAN DISTRIBUTED FUSION ALGORITHM

The goal of distributed estimation is to generate an "optimal" estimate for each fusion site given the information available to the fusion site. We assume that local estimates (or probabilities) and not measurements are communicated to the fusion site. The advantage is local use of estimates and lower bandwidth due to data compression.

When measurements are communicated, as in centralized fusion, the fusion algorithm can exploit the independent measurement errors or the conditional independence of the measurements given the state or variable to be estimated. When only local estimates are shared across the network, this conditional independence may be lost due to common information resulting from prior communication. In some cases, the "state" may not be large enough due to internal variables not included in the estimates. These are issues that have to be addressed in developing distributed fusion algorithms.

The following sections will develop the optimal Bayesian distributed fusion algorithm for a general object state. For object tracking, this state is a temporal sequence of states (e.g., position, velocity) at each time, and the observation is a temporal sequences) of measurements, e.g., range, angle for a radar. For object classification, the state is the object class and attributes such as object size and the observations are observed features such as measured length.

5.3.1 BAYESIAN DISTRIBUTED ESTIMATION PROBLEM AND SOLUTION

Let x be the state to be estimated. The state may be a continuous random variable such as the position and velocity of an object or a discrete random variable such as the class of an object. Let $p(x)$ be the prior probability density function for a continuous variable or the probability distribution for a discrete variable.

Suppose the measurement sets at two fusion nodes (as in the information graph), node 1 and node 2, are

$$Z_1 = \{z_{11}, z_{12}, z_{13}, \ldots\}$$
$$Z_2 = \{z_{21}, z_{22}, z_{23}, \ldots\}$$

(5.1)

These measurements may come from multiple sensors at different times or the same time. Assume the measurements are conditionally independent given x, i.e.,

$$p(z_{ij}, \ldots, z_{mn}, \ldots \mid x) = p(z_{ij} \mid x) \cdots p(z_{mn} \mid x) \cdots$$

(5.2)

This assumption is valid if the measurement errors are independent across sensors and over time.

The fusion nodes compute the local posterior conditional probabilities $p(x|Z_1)$ and $p(x|Z_2)$. The goal of distributed estimation is to compute the posterior conditional probability $p(x|Z_1 \cup Z_2)$ given all the measurements $Z_1 \cup Z_2$.

The fused information set $Z_1 \cup Z_2$ is the union of each node's private information and the common information (Figure 5.9), i.e.,

$$Z_1 \cup Z_2 = (Z_1 \setminus Z_2) \cup (Z_2 \setminus Z_1) \cup (Z_1 \cap Z_2)$$

(5.3)

where \ denotes set difference. Then, the assumption (5.2) of conditional independence of the measurements given the state implies that

$$p(Z_1 \cup Z_2 \mid x) = p(Z_1 \setminus Z_2 \mid x) p(Z_2 \setminus Z_1 \mid x) p(Z_1 \cap Z_2 \mid x)$$
$$= \frac{p(Z_1 \mid x) p(Z_2 \mid x)}{p(Z_1 \cap Z_2 \mid x)}$$

(5.4)

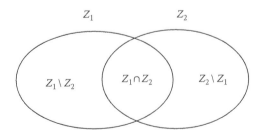

FIGURE 5.9   Decomposition into private and common information.

Bayes rule leads to the Bayesian distributed fusion equation

$$p(x \mid Z_1 \cup Z_2) = \frac{p(Z_1 \mid x)p(Z_2 \mid x)p(x)}{p(Z_1 \cap Z_2 \mid x)p(Z_1 \cup Z_2)}$$

$$= C^{-1} \frac{p(x \mid Z_1)p(x \mid Z_2)}{p(x \mid Z_1 \cap Z_2)} \tag{5.5}$$

where the normalizing constant is given by

$$C = \frac{p(Z_1 \cap Z_2)p(Z_1 \cup Z_2)}{p(Z_1)p(Z_2)} \tag{5.6}$$

This Bayesian distributed fusion equation states that the fused posterior probability $p(x \mid Z_1 \cup Z_2)$ is the product of the local probabilities $p(x \mid Z_1)$ and $p(x \mid Z_2)$, divided by the common probability $p(x \mid Z_1 \cap Z_2)$, which is included in each of the local probabilities.

The Bayesian fusion equation can be used to derive optimal fusion equations for the state of interest as long as the measurements are conditionally independent given the state. The key is identifying the common information that has to be removed to avoid double counting. This common information is usually a prior probability or estimate or the information shared during the last communication. Thus, fusion requires knowing the common probability $p(x \mid Z_1 \cap Z_2)$ in addition to $p(x \mid Z_1)$ and $p(x \mid Z_2)$.

5.3.2 BAYESIAN DISTRIBUTED FUSION FOR GAUSSIAN RANDOM VECTORS

Suppose the state x is a Gaussian random vector with known mean and covariance, and the measurements are also Gaussian with zero mean errors and known covariance. Then the local estimates are Gaussian random vectors with means $\hat{x}_i$ and covariances P_i. The fused estimate is also Gaussian with mean $\hat{x}_{1\cup 2}$ and covariance $P_{1\cup 2}$. Then the fusion equation (5.5) becomes

$$P_{1\cup 2}^{-1}\hat{x}_{1\cup 2} = P_1^{-1}\hat{x}_1 + P_2^{-1}\hat{x}_2 - P_{1\cap 2}^{-1}\hat{x}_{1\cap 2} \tag{5.7}$$

$$P_{1\cup 2}^{-1} = P_1^{-1} + P_2^{-1} - P_{1\cap 2}^{-1} \tag{5.8}$$

where $\hat{x}_{1\cap 2}$ and $P_{1\cap 2}$ are the mean and covariance of the state estimate given the common information.

Equations 5.7 and 5.8 are the information matrix form of the fusion equations because the inverse of the covariance matrix is the information matrix. Equation 5.8 states that the information matrix of the fused estimate is the sum of the information matrices of the local estimates minus the information matrix of the common estimate. Equation 5.7 states that the information of the fused estimate is the sum of the local information minus the common information. As in the general case, optimal fusion requires knowing the $\hat{x}_{1\cap 2}$ and $P_{1\cap 2}$ in addition to the estimates and covariances to be fused.

The information matrix fusion equations can be derived directly from the information filter equations (Chong 1979). Suppose each fusion node $i = 1, 2$, has the observation equation

$$Z_i = H_i x + v_i \tag{5.9}$$

where
H_i is the observation matrix
v_i is a zero mean independent observation noise with error covariance R_i

Then the information filter form of the estimate $\hat{x}_i$ is given by

$$P_i^{-1}\hat{x}_i = \bar{P}^{-1}\bar{x} + H_i'R_i^{-1}Z_i \tag{5.10}$$

with error covariance given by

$$P_i^{-1} = \bar{P}^{-1} + H_i'R_i^{-1}H_i \tag{5.11}$$

where $\bar{x}$ and $\bar{P}$ are the mean and covariance of x. Given the measurements Z_1 and Z_2, the optimal estimate $\hat{x}$ and its error covariance P are given by the information filter equations

$$P^{-1}\hat{x} = \bar{P}^{-1}\bar{x} + H'R^{-1}Z \tag{5.12}$$

$$P^{-1} = \bar{P}^{-1} + H'R^{-1}H \tag{5.13}$$

where the measurement vector Z, observation matrix H, and noise covariance matrix R are

$$Z = \begin{bmatrix} Z_1 \\ Z_2 \end{bmatrix} \quad H = \begin{bmatrix} H_1 \\ H_2 \end{bmatrix} \quad R = \begin{bmatrix} R_1 & 0 \\ 0 & R_2 \end{bmatrix} \tag{5.14}$$

Since

$$H'R^{-1}Z = H_1'R_1^{-1}Z_1 + H_2'R_2^{-1}Z_2 \tag{5.15}$$

$$H'R^{-1}H = H_1'R_1^{-1}H_1 + H_2'R_2^{-1}H_2 \tag{5.16}$$

Equations 5.12 and 5.13 become

$$P^{-1}\hat{x} = \bar{P}^{-1}\bar{x} + H'_1 R_1^{-1} Z_1 + H'_2 R_2^{-1} Z_2 \tag{5.17}$$

$$P^{-1} = \bar{P}^{-1} + H'_1 R_1^{-1} H_1 + H'_2 R_2^{-1} H_2 \tag{5.18}$$

These are the information fusion equations used in Durrant-Whyte et al. (1990) to fuse measurements communicated by the fusion agents. Substituting into (5.10) and (5.11) produce the information matrix fusion equations similar to Equations 5.7 and 5.8

$$P^{-1}\hat{x} = P_1^{-1}\hat{x}_1 + P_2^{-1}\hat{x}_2 - \bar{P}^{-1}\bar{x} \tag{5.19}$$

$$P^{-1} = P_1^{-1} + P_2^{-1} - \bar{P}^{-1} \tag{5.20}$$

5.4   OPTIMAL BAYESIAN DISTRIBUTED FUSION FOR DIFFERENT ARCHITECTURES

The Bayesian distributed fusion equation assumes a hierarchical architecture with no feedback. Furthermore, the local estimates are only fused once by the fusion agent. However, with the help of the information graph, this equation (in either general or linear form) can be used to derive optimal fusion equations for complex architectures and identify the information that needs to be communicated in addition to the estimates to be fused. It is also trivial to extend to fusing multiple local estimates. The following sections contain some examples.

5.4.1   HIERARCHICAL ARCHITECTURE

5.4.1.1   Hierarchical Fusion without Feedback

Consider the example of Figure 5.4 with $F3$ as the fusion site. When there is no feedback from the high level, the common information in the received estimate $p(x|Z_L)$ and the current estimate $p(x|Z_H)$ is the estimate $p(x|Z_{\bar{L}})$ last communicated from the low-level fusion site F2. From (5.5), the fused estimate or probability function is given by

$$p(x \mid Z_H \cup Z_L) = C^{-1} \frac{p(x \mid Z_H) p(x \mid Z_L)}{p(x \mid Z_{\bar{L}})} \tag{5.21}$$

When the probability distribution is Gaussian, applying Equations 5.7 and 5.8 yields

$$P_{H \cup L}^{-1} = P_H^{-1} + P_L^{-1} - P_{\bar{L}}^{-1} \tag{5.22}$$

$$P_{H \cup L}^{-1} \hat{x}_{H \cup L} = P_H^{-1} \hat{x}_H + P_L^{-1} \hat{x}_L - P_{\bar{L}}^{-1} \hat{x}_{\bar{L}} \tag{5.23}$$

where the subscripts represent the information nodes.

5.4.1.2   Hierarchical Fusion with Feedback

For hierarchical architecture with feedback in Figure 5.5, fusion takes place at both levels. For fusion at the low level node $F1$, the common predecessor of L_1 and H is $\bar{L}$, the fusion node of the last communication from low level to high level. For fusion of H at the high-level node $F3$ with the estimate at L_2 from $F2$, the common predecessor is $\bar{H}$, the fusion node of the last communication from high level to low level. For low-level fusion, the common information shared (from the information graph) is the last estimate sent to the high level. Thus the fusion equation for low level is

$$p(x \mid Z_H \cup Z_{L_1}) = C^{-1} \frac{p(x \mid Z_H) p(x \mid Z_{L_1})}{p(x \mid Z_{\bar{L}})} \tag{5.24}$$

Similarly, the high-level fusion equation is

$$p(x \mid Z_H \cup Z_{L_2}) = C^{-1} \frac{p(x \mid Z_H) p(x \mid Z_{L_2})}{p(x \mid Z_{\bar{H}})} \tag{5.25}$$

When the variables are Gaussian, the low-level fusion equations are

$$P_{H \cup L_1}^{-1} = P_H^{-1} + P_{L_1}^{-1} - P_{\bar{L}}^{-1} \tag{5.26}$$

$$P_{H \cup L_1}^{-1} \hat{x}_{H \cup L_1} = P_H^{-1} \hat{x}_H + P_{L_1}^{-1} \hat{x}_{L_1} - P_{\bar{L}}^{-1} \hat{x}_{\bar{L}} \tag{5.27}$$

and the high-level fusion equations are

$$P_{H \cup L_2}^{-1} = P_H^{-1} + P_{L_2}^{-1} - P_{\bar{H}}^{-1} \tag{5.28}$$

$$P_{H \cup L_2}^{-1} \hat{x}_{H \cup L_2} = P_H^{-1} \hat{x}_H + P_{L_2}^{-1} \hat{x}_{L_2} - P_{\bar{H}}^{-1} \hat{x}_{\bar{H}} \tag{5.29}$$

5.4.2   ARBITRARY DISTRIBUTED FUSION ARCHITECTURE

The optimal fusion algorithm for arbitrary distributed fusion architectures is found by repeated application of the Bayesian fusion equation (5.5). The algorithm starts by identifying the common predecessor nodes of the information nodes whose estimates are to be fused. If there is only one common predecessor node, then the information at that node becomes the $p(Z_1 \cap Z_2)$ in the denominator of (5.5). If there are multiple common predecessor nodes, then (5.5) is used again to compute $p(Z_1 \cap Z_2)$ terms of the information (probability) at these nodes and the information at their common predecessor nodes. The process is repeated until each conditional probability involves only one information node. Thus the fusion equation for the general fusion architecture consists of a product of probabilities representing information to be fused and divisions representing redundant information to be removed. The general fusion equation has the form (Chong et al. 1987, 1990)

$$p\left(x \mid \bigcup_{i=1}^{N} Z_i\right) = C^{-1} \prod_{j \in J} p(x \mid Z_j)^{\alpha(j)} \tag{5.30}$$

where
   J is a set of predecessor nodes of the fusion node
   C is a normalizing constant
   $\alpha(\cdot)$ is either $+1$ or -1 depending on whether information is to be added or deleted
For Gaussian case, the fusion equations are

$$P^{-1}\hat{x} = \sum_{j \in J} \alpha(j) P_j^{-1} \hat{x}_j \tag{5.31}$$

$$P^{-1} = \sum_{j \in J} \alpha(j) P_j^{-1} \tag{5.32}$$

The hierarchical fusion equations discussed earlier are special cases of these equations. For the cyclic architecture of Figure 5.8, repeated application of the fusion equation results in the following equation:

$$p(x \mid Z_A) = C^{-1} \frac{p(x \mid Z_B) p(x \mid Z_C)}{p(x \mid Z_{D \cup E})}$$

$$= C^{-1} \frac{p(x \mid Z_B) p(x \mid Z_C) p(x \mid Z_{F \cup G})}{p(x \mid Z_D) p(x \mid Z_E)}$$

$$= C^{-1} \frac{p(x \mid Z_B) p(x \mid Z_C) p(x \mid Z_H)}{p(x \mid Z_D) p(x \mid Z_E)} \tag{5.33}$$

The equations for the Gaussian case are

$$P_A^{-1}\hat{x}_A = P_B^{-1}\hat{x}_B + P_C^{-1}\hat{x}_C - P_D^{-1}\hat{x}_D - P_E^{-1}\hat{x}_E + P_H^{-1}\hat{x}_H \tag{5.34}$$

$$P_A^{-1} = P_B^{-1} + P_C^{-1} - P_D^{-1} - P_E^{-1} + P_H^{-1} \tag{5.35}$$

5.5   SUBOPTIMAL BAYESIAN DISTRIBUTED FUSION ALGORITHMS

The optimal distributed fusion algorithm described in the previous section is based upon identifying and removing redundant information using the information graph. When the bandwidth does not support communication of information pedigree, such as in ad hoc wireless sensor networks, the relevant part of the information graph cannot be constructed by the fusion node. Even if the information pedigree can be communicated, in a dynamic network with possible failures and adaptive communication strategies, the optimal distributed fusion algorithm may not be practical due to the long pedigree information needed for de-correlation. This section presents several practical and scalable algorithms (Chang et al. 2010) based on approximations of the optimal algorithms (5.30) through (5.32). To simplify the notation, we again focus on the fusion of two information nodes with either probability functions given by $p_1(x)$ and $p_2(x)$, or estimates $\hat{x}_1$ and $\hat{x}_2$ with error covariances P_1 and P_2. The fusion result is represented by probability function $p(x)$ or an estimate $\hat{x}$ with error covariance P.

5.5.1 NAÏVE FUSION

Naïve fusion ignores the dependence in the information to be fused or the denominator in the optimal Bayesian fusion equation (5.5). Thus the naïve fusion algorithm is

$$p(x) = C^{-1} p_1(x) p_2(x) \qquad (5.36)$$

where C is the normalizing constant. For Gaussian case, the common information is similarly ignored in Equations 5.7 and 5.8, resulting in the following equations for the fused state estimate and error covariance

$$P^{-1} = P_1^{-1} + P_2^{-1}$$
$$P^{-1}\hat{x} = P_1^{-1}\hat{x}_1 + P_2^{-1}\hat{x}_2 \qquad (5.37)$$

By not subtracting the prior information matrix (inverse of the prior covariance matrix), the computed fused error covariance is smaller than the true error covariance, resulting in an estimate of naïve fusion that may be overconfident.

The naïve fusion equation for the Gaussian case is sometimes called the convex combination equation because it can be shown that the fused estimate is given by

$$\hat{x} = P_2(P_1 + P_2)^{-1}\hat{x}_1 + P_1(P_1 + P_2)^{-1}\hat{x}_2 \qquad (5.38)$$

For the cyclic architecture of Figure 5.8, naïve fusion only retains $p(x|Z_B)$ and $p(x|Z_C)$.

5.5.2 CHANNEL FILTER FUSION

The channel filter (Grime and Durrant-Whyte 1994, Nicholson et al. 2001, Bourgault and Durrant-Whyte 2004) can be viewed as a first-order approximation of the optimal fusion algorithm. The distributed estimation system consists of a number of channels with each defined by a pair of transmitting and receiving nodes. In the channel filter, the fusion node keeps track of the communication history for all the information nodes that it receives data from. When it receives a new estimate to be fused from a node, it retrieves the more recent estimate from that node and considers it as the only common information to be removed, ignoring earlier information nodes that may have contributed to the common information. In that sense, the channel filter can be considered as a first-order approximation to the optimal information graph approach.

Specifically, the channel filter fusion equation is given as

$$p(x) = C^{-1} \frac{p_1(x)p_2(x)}{\bar{p}(x)} \qquad (5.39)$$

where

    C is a normalizing constant

    $\bar{p}(x)$ is the probability function received from the same channel at the previous communication time and is the common "prior information" to be removed in the fusion formula, with mean $\bar{x}$ and covariance $\bar{p}$ when Gaussian.

When both $p_1(x)$ and $p_2(x)$ are Gaussian with means and covariances $\hat{x}_1$, P_1 and $\hat{x}_2$, P_2 respectively, the fused state estimate and corresponding error covariance are given by

$$P^{-1} = P_1^{-1} + P_2^{-1} - \bar{P}^{-1} \qquad (5.40)$$

$$P^{-1}\hat{x} = P_1^{-1}\hat{x}_1 + P_2^{-1}\hat{x}_2 - \bar{P}^{-1}\bar{x} \qquad (5.41)$$

The first-order approximation of channel filter fusion is suboptimal because it does not account for all common information shared by the estimates to be fused. However, it may only be slightly suboptimal if the time between when that redundancy occurred and the current processing time is relatively long. For the cyclic architecture of Figure 5.8, channel filter approximates (5.33) by the following

$$p(x \mid Z_A) = C^{-1} \frac{p(x \mid Z_B)p(x \mid Z_C)}{p(x \mid Z_D)} \qquad (5.42)$$

and ignores the other terms in the optimal fusion equation. Similarly, the fusion equations for the Gaussian case become

$$P_A^{-1}\hat{x}_A = P_B^{-1}\hat{x}_B + P_C^{-1}\hat{x}_C - P_D^{-1}\hat{x}_D \qquad (5.43)$$

$$P_A^{-1} = P_B^{-1} + P_C^{-1} - P_D^{-1} \qquad (5.44)$$

5.5.3 CHERNOFF FUSION

Chernoff information fusion also ignores completely the dependence in the information to be fused. However, instead of assigning equal weights as in naïve fusion, the fusion formula allows different weights for the probabilities to be fused, resulting in

$$p(x) = C^{-1}p_1^w(x)p_2^{1-w}(x) \tag{5.45}$$

where $w \in [0\ 1]$ is an appropriate parameter which minimizes a chosen criterion. The fusion algorithm is called Chernoff fusion when the criterion to be minimized is the Chernoff information (Cover and Thomas 1991) defined by the normalizing constant C. It can be shown that the resulting fused probability function that minimizes the Chernoff information is the one "halfway" between the two original densities in terms of the Kullback Leibler distance (Cover and Thomas 1991). In the case when both $p_1(x)$ and $p_2(x)$ are Gaussian, the resulting fused density is also Gaussian with mean and covariance given by

$$P^{-1} = wP_1^{-1} + (1-w)P_2^{-1} \tag{5.46}$$

$$P^{-1}\hat{x} = wP_1^{-1}\hat{x}_1 + (1-w)P_2^{-1}\hat{x}_2 \tag{5.47}$$

This formula is identical to the covariance intersection (CI) fusion technique (Chong and Mori 2001, Nicholson et al. 2001, 2002, Hurley 2002, Julier 2006, Julier et al. 2006). Therefore, the CI technique can be considered as a special case of (5.45). In theory, Chernoff fusion can be used to combine any two arbitrary probabilities in a log-linear fashion. However, the resulting fused probability may not preserve the same form as the original ones. Also in general, obtaining the proper weighting parameter to satisfy a certain criterion may involve extensive search or computation.

5.5.4 BHATTACHARYYA FUSION

Bhattacharyya fusion is a special case of Chernoff fusion (5.45), when the parameter w is set to be 0.5. Then the normalizing constant of (5.45) becomes $B = \int \sqrt{p_1(x)p_2(x)}dx$, which is the Bhattacharyya bound. The fusion algorithm is

$$p(x) = B^{-1}\sqrt{p_1(x)p_2(x)} \tag{5.48}$$

When both $p_1(x)$ and $p_2(x)$ are Gaussian, the fusion equation can be written as

$$P^{-1} = \frac{1}{2}(P_1^{-1} + P_2^{-1}) \tag{5.49}$$

$$P^{-1}\hat{x} = \frac{1}{2}(P_1^{-1}\hat{x}_1 + P_2^{-1}\hat{x}_2)$$

or

$$\hat{x} = (P_1^{-1} + P_2^{-1})^{-1}(P_1^{-1}\hat{x}_1 + P_2^{-1}\hat{x}_2) \tag{5.50}$$

Therefore, Bhattacharyya fusion is similar to naïve fusion for the Gaussian case. However, the resulting fused covariance is twice as big as that of naïve fusion. Note that the fusion equation can be rewritten as

$$P^{-1} = \frac{1}{2}(P_1^{-1} + P_2^{-1}) = (P_1^{-1} + P_2^{-1}) - \frac{1}{2}(P_1^{-1} + P_2^{-1}) \tag{5.51}$$

$$P^{-1}\hat{x} = \frac{1}{2}(P_1^{-1}\hat{x}_1 + P_2^{-1}\hat{x}_2)$$

$$= (P_1^{-1}\hat{x}_1 + P_2^{-1}\hat{x}_2) - \frac{1}{2}(P_1^{-1}\hat{x}_1 + P_2^{-1}\hat{x}_2) \tag{5.52}$$

This formula replaces the common prior information of (5.40) and (5.41) for channel filter by the average of the two sets of information to be fused, namely, $\bar{P}^{-1} \Leftarrow \frac{1}{2}(P_1^{-1} + P_2^{-1})$ and $\bar{P}^{-1}\bar{x} \Leftarrow \frac{1}{2}(P_1^{-1}\hat{x}_1 + P_2^{-1}\hat{x}_2)$. In other words, instead of removing the common prior information from the previous communication as in the channel filter case, the common information of Bhattacharyya fusion is approximated by the "average" of the two locally available information sets.

5.6 DISTRIBUTED ESTIMATION FOR GAUSSIAN DISTRIBUTIONS OR ESTIMATES WITH ERROR COVARIANCES

In Section 5.5, we presented several suboptimal algorithms that avoid the exact identification and removal of redundant information using the information graph. These algorithms can be viewed as approximations of the optimal fusion algorithm for general probability functions. This section presents fusion algorithms that are optimal according to some criteria when the information to be fused is either Gaussian or can be represented by estimates with error covariances.

In the following, we assume that the state to be estimated has mean $\bar{x}$ and covariance $\bar{P}$, the estimates to be fused are $\hat{x}_1$ and $\hat{x}_2$ with error covariances P_1 and P_2, and cross-covariance $P_{12} = P_{21}'$. Note that in addition to the common prior $\bar{x}$ and $\bar{P}$, there is additional dependence between $\hat{x}_1$ and $\hat{x}_2$ represented by the cross-covariance $P_{12} = P_{21}'$. Thus removing the common prior alone is not sufficient for generating the best fused estimate.

5.6.1 MAXIMUM A POSTERIORI FUSION OR BEST LEAST UNBIASED ESTIMATE

Let $z = [\hat{x}_1' \ \hat{x}_2']'$ be the augmented vector of the estimates to be fused. Assume z and x are jointly Gaussian with mean $\bar{z}$ and $\bar{x}$, with covariances

$$P_{xz} = P_{zx}' \triangleq E\{(x - \bar{x})(z - \bar{z})'\} \tag{5.53}$$

$$P_{zz} \triangleq E\{(z - \bar{z})(z - \bar{z})'\} \tag{5.54}$$

Then given z, $p(x|z)$ is also Gaussian with mean and covariance given by (Anderson and Moore 1979)

$$\hat{x} = \bar{x} + P_{xz}P_{zz}^{-1}(z - \bar{z}) \tag{5.55}$$

$$P = \bar{P} - P_{xz}P_{zz}^{-1}P_{zx} \tag{5.56}$$

Note that (5.55) is also the maximum a posteriori (MAP) estimate (Mori et al. 2002, Chang et al. 2004) and can be expressed as

$$\hat{x} = \bar{x} + W_1(\hat{x}_1 - \bar{x}) + W_2(\hat{x}_2 - \bar{x}) = W_0\bar{x} + W_1\hat{x}_1 + W_2\hat{x}_2 \tag{5.57}$$

with $W_0 = I - W_1 - W_2$ and $P_{0i} = E((x - \bar{x})(\hat{x}_i - \bar{x})')$ for $i = 1, 2$, where

$$[W_1 \ W_2] = P_{xz}P_{zz}^{-1} = [P_{01} \ P_{02}]\begin{bmatrix} P_1 & P_{12} \\ P_{21} & P_2 \end{bmatrix}^{-1} \tag{5.58}$$

If $\hat{x}_1$ and $\hat{x}_2$ are not jointly Gaussian but the moments are known, (5.57) is the best linear unbiased estimate (BLUE) (Zhu and Li 1999, Li et al. 2003). Note that the MAP estimate or BLUE requires more information for its calculation. In addition to the common prior $\bar{x}$ and $\bar{P}$, and the estimates $\hat{x}_1$ and $\hat{x}_2$ with error covariances P_1 and P_2, it also requires the cross-covariances $P_{12} = P'_{21}$ between the estimates, and the cross-covariances P_{01} and P_{02} between the estimates and the state. If the estimates $\hat{x}_1$ and $\hat{x}_2$ are generated from measurements with independent errors, (5.57) and (5.58) reduce to the standard fusion equations of (5.19) and (5.20).

5.6.2 CROSS-COVARIANCE FUSION

The cross-covariance fusion rule (Bar-Shalom and Campo 1986) considers explicitly the cross-covariance of the local estimates to be fused. The fusion rule is given by

$$\hat{x} = W_1\hat{x}_1 + W_2\hat{x}_2 \tag{5.59}$$

where

$$W_i = (P_j - P_{ji})(P_1 + P_2 - P_{12} - P_{21})^{-1} \tag{5.60}$$

for $i = 1, 2$ with $j = 3 - i$. Since $W_1 + W_2 = I$, the fused estimate is unbiased if the local estimates are also unbiased. It can be shown that Equation 5.59 maximizes the classical likelihood function $p(\hat{x}_1, \hat{x}_2|x)$ with x viewed as a parameter. Thus, the cross-covariance fusion rule is also the maximum likelihood fusion rule. As shown in (Chang et al. 1997), Equation 5.59 is the unique solution of the BLUE without a priori *information*, i.e., the linear solution obtained without using a priori information (initial condition). This follows from the fact the MAP estimates becomes the maximum likelihood estimate when the prior covariance becomes very large.

If we ignore the cross covariance P_{ij}, (5.60) becomes, for $i=1, 2$ with $j=3-i$,

$$W_i = P_j(P_1 + P_2)^{-1} = (P_1^{-1} + P_2^{-1})^{-1} P_i^{-1} \qquad (5.61)$$

which is the fusion rule obtained by treating the two estimates $\hat{x}_1$ and $\hat{x}_2$ as if they were two conditionally independent observations of x. This is again the *convex combination rule*.

Since $\det\left(\begin{bmatrix} P_1 & P_{12} \\ P_{21} & P_2 \end{bmatrix}\right) = \det(P_1 - P_{12}P_2^{-1}P_{21})\det(P_2)$, ignoring the cross covariance as $P_{12}=0$ increases the size of the ellipsoid defined by the joint covariance matrix $\begin{bmatrix} P_1 & P_{12} \\ P_{21} & P_2 \end{bmatrix}$. Thus, the simplified fusion rule (5.61) is obtained by inflating the joint covariance matrix.

5.7 DISTRIBUTED ESTIMATION FOR OBJECT TRACKING

In this section, we discuss how the general approach for distributed estimation can support object tracking (Liggins et al. 1997, Chong et al. 2000). Multi-object tracking involves two steps: associating measurements to form object tracks, and estimating the states of the objects given the tracks. Our discussion will focus on single object state estimation or filtering. The association problem in object tracking will be addressed in Chapter 6.

For object state estimation, the state of the object is a random process that evolves according to a dynamic model given by the transition probability $p(x_{k+1}|x_k)$, where x_k is the state of the object at time t_k. Measurements are generated from the state according to a measurement model $p(z_k|x_k)$. The objective of object state estimation is to generate the estimate of the state, $p(x_k|Z_k)$, given the cumulative measurements $Z_k=(z_0, z_1, ..., z_k)$. Recursive state estimation or filtering consists of two steps: predicting $p(x_k|Z_k)$ to the time of the next measurement to obtain $p(x_{k+1}|Z_k)$ and updating with the current measurement to generate $p(x_{k+1}|Z_{k+1})$. Since the prediction step uses only the object dynamic model and does not depend on measurements, distributed estimation focuses on the update step.

We assume a hierarchical fusion architecture to discuss the approach. Each low-level fusion agent i generates an updated estimate of the object state given its local measurements $p(x_k|Z_{ik})$ where $Z_{ik}=(z_{i0}, z_{i1}, ..., z_{ik})$. The high-level fusion site or agent combines the low-level (updated) estimates to form the fused estimate $p(x_k|Z_k)$ where $Z_k=Z_{1k} \cap Z_{2k}$.

5.7.1 DETERMINISTIC DYNAMICS

An object is said to have deterministic dynamics if its future state is determined completely by the current state, i.e., the state transition probability is a delta function. Ballistic missiles and space objects are examples of objects with deterministic dynamics. It can be easily shown that conditional independence of the measurements z_{ik} given x_k for all i and k implies conditional independence of the cumulative measurements Z_{ik} given x_k for all i and k, i.e.,

$$p(Z_{1k}, Z_{2k} \mid x_k) = p(Z_{1k} \mid x_k) p(Z_{2k} \mid x_k) \tag{5.62}$$

Thus the Bayesian distributed fusion equation can be used and

$$p(x_k \mid Z_{1k}, Z_{2k}) = C^{-1} \frac{p(x_k \mid Z_{1k}) p(x_k \mid Z_{2k})}{p(x_k \mid Z_{k-1})} \tag{5.63}$$

where

    C is a normalizing constant
    $p(x_k \mid Z_{k-1})$ is the common prior that can be extrapolated from $p(x_{k-1} \mid Z_{k-1})$ provided
        by the fusion site

When the random variables are Gaussian, the fusion equations are

$$P_{k|k}^{-1} \hat{x}_{k|k} = P_{1,k|k}^{-1} \hat{x}_{1,k|k} + P_{2,k|k}^{-1} \hat{x}_{2,k|k} - P_{k|k-1}^{-1} \hat{x}_{k|k-1} \tag{5.64}$$

$$P_{k|k}^{-1} = P_{1,k|k}^{-1} + P_{2,k|k}^{-1} - P_{k|k-1}^{-1} \tag{5.65}$$

where $\hat{x}_{i,k|l}$ and $P_{i,k|l}$ are the estimate and error covariance of x_k given Z_{il} and $\hat{x}_{k|l}$ and $P_{k|l}$ are the fused estimate and error covariance given x_k and Z_l.

    If there is no feedback from the fusion site, the fusion equation is

$$p(x_k \mid Z_{1k}, Z_{2k}) = C^{-1} \frac{p(x_k \mid Z_{1k}) p(x_k \mid Z_{2k})}{p(x_k \mid Z_{1,k-1}) p(x_k \mid Z_{2,k-1})} p(x_k \mid Z_{k-1}) \tag{5.66}$$

When the random variables are Gaussian, the fusion equations become

$$P_{k|k}^{-1} \hat{x}_{k|k} = P_{1,k|k}^{-1} \hat{x}_{1,k|k} - P_{1,k|k-1}^{-1} \hat{x}_{1,k|k-1} + P_{2,k|k}^{-1} \hat{x}_{2,k|k} - P_{2,k|k-1}^{-1} \hat{x}_{2,k|k-1} + P_{k|k-1}^{-1} \hat{x}_{k|k-1} \tag{5.67}$$

$$P_{k|k}^{-1} = P_{1,k|k}^{-1} - P_{1,k|k-1}^{-1} + P_{2,k|k}^{-1} - P_{2,k|k-1}^{-1} + P_{k|k-1}^{-1} \tag{5.68}$$

For deterministic object dynamics, the fusion equations reconstruct the optimal centralized estimate independent of number of sensor revisits between fusion times. This is not the case for nondeterministic object dynamics.

5.7.2 Nondeterministic Dynamics

When the object has nondeterministic dynamics, the cumulative measurements Z_{ik} are no longer conditionally independent given x_k. Effectively, the common process noise or nondeterministic dynamics destroys the conditional independence. Then the fusion equations (5.63) through (5.68) are no longer optimal or exact unless the low-level fusion agents communicate with the high-level agent after each observation time. For hierarchical fusion with feedback, the high-level fusion agent also has to send the fused estimate back to the local agents after each fusion.

5.7.2.1   Augmented State Vector and Approximation

Let $X_k = [x_0', x_1', \ldots, x_k']$ be the augmented state vector consisting of the states at multiple observation times. Then the cumulative measurements Z_{ik} are conditionally independent given X_k, and the optimal fusion equations are (5.63) through (5.68) with x_k replaced by X_k. However, this approach may not be practical because the probability density functions or covariance matrices involve high dimensions.

5.7.2.2   Using Cross-Covariance at a Single Time

For problems that can be represented by Gaussian distributions or means and covariances, the approach of Section 5.6 can be used to handle the conditional dependence due to nondeterministic dynamics. Specifically, let $\hat{x}_{1,k|k}$ and $\hat{x}_{2,k|k}$ be the estimates to be fused with error covariances $P_{1,k|k}$ and $P_{2,k|k}$, cross-covariance $P_{12,k|k} = P_{21,k|k}'$, and common prior $\hat{x}_{k|k-1}$ with covariance $P_{k|k-1}$. Then the MAP, BLUE, or cross-covariance fusion rules can be used by replacing $\hat{x}_i$, P_i, P_{12}, $\bar{x}$, and $\bar{P}$ with $\hat{x}_{i,k|k}$, $P_{i,k|k}$, $P_{12,k|k}$, $\hat{x}_{k|k-1}$ and $P_{k|k-1}$ respectively in the fusion equations. Chapter 6 has a comparison of the different fusion rules for nondeterministic dynamics.

5.8   DISTRIBUTED ESTIMATION FOR OBJECT CLASSIFICATION

The general fusion approach in Section 5.3 can be used for distributed object classification (Chong and Mori 2005) where the state of interest is a discrete and constant random variable representing the object class. When the conditional independence assumption is satisfied, optimal distributed object classification can be performed using (5.5) of Section 5.3. In general, selecting object class as the state will not satisfy the conditional independence assumption because measurements containing class information may also depend on other variables such as viewing angles. In the following, we will consider hierarchical fusion at a single time to focus on the information that should be used in fusion. More complicated communication patterns will require checking for common information and removing it, using approximate algorithms if necessary. Chapter 9 contains a more detailed discussion on distributed object classification.

5.8.1   DISTRIBUTED OBJECT CLASSIFICATION ARCHITECTURES

The common fusion architectures for object classification are centralized measurement fusion, decision fusion, and probability fusion. In centralized measurement fusion, measurements containing object class information are fused at a central site. This architecture is theoretically optimal because the central site has access to all the measurements but requires the most communication. In decision fusion, each local site performs classification using the local measurements and sends the decision to the fusion site. Decisions require very little bandwidth to communicate but may not contain enough information for generating a good decision after fusion. Thus we will

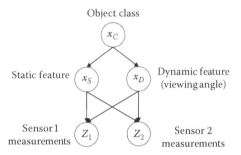

FIGURE 5.10 Bayesian network for object classification.

focus on probability fusion, in which each local site generates a conditional probability of the object class from the local measurements, and the fusion site combines the conditional probabilities to form the centralized conditional probability.

The key to high-performance probability fusion is determining the state used for generating the probability. In general, the object class is not sufficient as a state for optimal fusion because the measurements may depend on other object attributes in addition to the class (Chong and Mori 2004). Consider the example in Figure 5.10 where a Bayesian network is used to show that the measurements z_1 and z_2 depend on the object class x_C through the static object attribute x_S such as size and the dynamic attribute x_D such as viewing angle. As shown in Figure 5.10, the measurements are conditionally independent given x_S and x_D, i.e.,

$$p(z_1, z_2 \mid x_S, x_D) = p(z_1 \mid x_S, x_D) p(z_2 \mid x_S, x_D) \tag{5.69}$$

but not conditionally independent given only x_C, i.e.,

$$p(z_1, z_2 \mid x_C) = \int p(z_1, z_2, x_S, x_D \mid x_C) \, dx_S dx_D$$

$$= \int p(z_1, z_2 \mid x_S, x_D) p(x_S, x_D \mid x_C) \, dx_S dx_D \neq p(z_1 \mid x_C) p(z_2 \mid x_C) \tag{5.70}$$

Thus, for optimal distributed fusion, the state to be communicated should be x_S and x_D.

5.8.2 DISTRIBUTED CLASSIFICATION ALGORITHMS

For optimal distributed classification, the object state in the probabilities should make the measurements conditionally independent. For the example in Figure 5.10, the state consists of the static attribute x_S and the dynamic attribute x_D. Then the optimal fusion equation is

$$p(x_S, x_D \mid z_1, z_2) = C^{-1} \frac{p(x_S, x_D \mid z_1) p(x_S, x_D \mid z_2)}{p(x_S, x_D)} \tag{5.71}$$

From this, the object class probability can be computed as

$$p(x_C \mid z_1, z_2) = \int p(x_C \mid x_S, x_D) p(x_S, x_D \mid z_1, z_2) dx_S dx_D \qquad (5.72)$$

When only the probabilities of the object class are communicated, naïve fusion can be used to obtain an approximate solution

$$p(x_C \mid z_1, z_2) = C^{-1} \frac{p(x_C \mid z_1) p(x_C \mid z_2)}{p(x_C)} \qquad (5.73)$$

Note that in the case the information ignored is not the common prior due to communication but the states that lead to conditional independence.

5.9 SUMMARY

This chapter presented the fundamental concepts for distributed estimation, which are crucial for developing distributed fusion algorithms. We discussed various distributed fusion architectures, their advantages and disadvantages, the use of information graph to represent information flow, and selection of an appropriate architecture. We presented the Bayesian fusion equation for combining two probability functions, and the equation for estimates given by means and covariances. The Bayesian fusion equation, when used with the information graph, can be used to derive fusion equations for various fusion architectures. Since the fusion equation can be complicated, requiring pedigree or network information for complicated architectures, it is necessary to approximate the optimal algorithm with suboptimal algorithms for implementation in real systems. When the estimates to be fused are Gaussian or can be characterized by means and covariances, there are several linear combination rules such as MAP, BLUE, and cross-covariance fusion. We also showed that the distributed estimation approach can be used for object tracking and object classification.

5.10 BIBLIOGRAPHIC NOTES

Research in distributed estimation started around 1980 and addresses the problem of reconstructing the optimal estimate from the local estimates (Chong 1979, Speyer 1979, Willsky et al. 1982, Castanon and Teneketzis 1985). A general distributed estimation approach (Chong et al. 1982, 1983, 1985, 1987) for arbitrary architectures was investigated under the Distributed Sensor Networks (DSN) program sponsored by the Defense Advanced Research Projects Agency (DARPA). By using the information graph to track information flow in the system, the optimal fusion algorithm avoids double counting of information or data incest. The DSN program also developed general distributed tracking algorithms (Chong et al. 1986, 1990). Around 1990, researchers in the United Kingdom and Australia developed similar decentralized fusion algorithms (Durrant-Whyte et al. 1990, Grime and Durrant-Whyte 1994) that avoid data incest, and CI algorithms (Nicholson et al. 2001, 2002) to address unknown correlation between local estimates to be fused. Bar-Shalom and Campo (1986) developed the first fusion algorithm that uses the cross-covariance between the local estimates. This paper was followed by the BLUE fusion algorithm (Zhu and Li 1999, Li et al. 2003) and the MAP fusion rule (Mori et al. 2002, Chang et al. 2004). The last two papers also contain performance evaluation of fusion algorithms, along with Chong and Mori (2001) and Chang et al. (2010).

REFERENCES

Anderson, B. D. O. and J. B. Moore. 1979. *Optimal Filtering*. Englewood Cliffs, NJ: Prentice-Hall.

Bar-Shalom, Y. and L. Campo. 1986. The effects of the common process noise on the two-sensor fused-track covariance. *IEEE Transactions on Aerospace and Electronic Systems*, 22: 803–805.

Bourgault, F. and H. F. Durrant-Whyte. 2004. Communication in general decentralized filter and the coordinated search strategy. In *Proceedings of the 7th International Conference on Information Fusion*, Stockholm, Sweden.

Castanon, D. and D. Teneketzis. 1985. Distributed estimation algorithms for nonlinear systems. *IEEE Transactions on Automatic Control*, 30: 418–425.

Chang, K. C., C. Y. Chong, and S. Mori. 2010. Analytical and computational evaluation of scalable fusion algorithms. *IEEE Transactions on Aerospace and Electronic Systems*, 46: 2022–2034.

Chang, K. C., R. K. Saha, and Y. Bar-Shalom. 1997. On optimal track-to-track fusion. *IEEE Transactions on Aerospace and Electronic Systems*, 33: 1271–1276.

Chang, K. C., T. Zhi, S. Mori, and C. Y. Chong. 2004. Performance evaluation for MAP state estimate fusion. *IEEE Transactions on Aerospace and Electronic Systems*, 40: 706–714.

Chong, C. Y. 1979. Hierarchical estimation. In *Proceedings of MIT/ONR Workshop on C3*, Monterey, CA.

Chong, C. Y. 1998. Distributed architectures for data fusion. In *Proceedings of the 1st International Conference on Multisource Multisensor Information Fusion*, Las Vegas, NV.

Chong, C. Y., K. C. Chang, and S. Mori. 1986. Distributed tracking in distributed sensor networks. In *Proceedings of 1986 American Control Conference*, Seattle, WA.

Chong, C. Y. and S. Mori. 2001. Convex combination and covariance intersection algorithms in distributed fusion. In *Proceedings of the 4th International Conference on Information Fusion*, Montréal, Québec, Canada.

Chong, C. Y. and S. Mori. 2004. Graphical models for nonlinear distributed estimation. In *Proceedings of the 7th International Conference on Information Fusion*, Stockholm, Sweden.

Chong, C. Y. and S. Mori. 2005. Distributed fusion and communication management for target identification. In *Proceedings of the 8th International Conference on Information Fusion*, Philadelphia, PA.

Chong, C. Y., S. Mori, and K. C. Chang. 1985. Information fusion in distributed sensor networks. In *Proceedings of 1985 American Control Conference*, Boston, MA.

Chong, C. Y., S. Mori, and K. C. Chang. 1987. Adaptive distributed estimation. In *Proceedings of 26th IEEE Conference on Decision and Control*, Los Angeles, CA.

Chong, C. Y., S. Mori, and K. C. Chang. 1990. Distributed multitarget multisensor tracking. In *Multitarget Multi-Sensor Tracking: Advanced Applications*, ed. Y. Bar-Shalom, pp. 247–295. Norwood, MA: Artech House.

Chong, C. Y., S. Mori, K. C. Chang, and W. H. Barker. 2000. Architectures and algorithms for track association and fusion. *IEEE Aerospace and Electronic Systems Magazine*, 15: 5–13.

Chong, C. Y., S. Mori, E. Tse, and R. P. Wishner. 1982. Distributed estimation in distributed sensor networks. In *Proceedings of 1982 American Control Conference*, Arlington, VA.

Chong, C. Y., E. Tse, and S. Mori. 1983. Distributed estimation in networks. In *Proceedings of 1983 American Control Conference*, San Francisco, CA.

Cover, T. M. and J. A. Thomas. 1991. *Elements of Information Theory*. New York: Wiley.

Durrant-Whyte, H. F., B. S. Y. Rao, and H. Hu. 1990. Toward a fully decentralized architecture for multi-sensor data fusion. In *Proceedings of 1990 IEEE International Conference on Robotics and Automation*, Cincinnati, OH.

Grime, S. and H. Durrant-Whyte (eds.). 1994. Communication in decentralized systems. In *IFAC Control Engineering Practice*. Oxford, U.K.: Pergamon Press.

Hurley, M. 2002. An information-theoretic justification for covariance intersection and its generalization. In *Proceedings of the 5th International Conference on Information Fusion*, Annapolis, MD.

Julier, S. J. 2006. An empirical study into the use of Chernoff information for robust, distributed fusion of Gaussian mixture models. In *Proceedings of the 9th International Conference on Information Fusion*, Florence, Italy.

Julier, S. J., J. K. Uhlmann, J. Walters, R. Mittu, and K. Palaniappan. 2006. The challenge of scalable and distributed fusion of disparate sources of information. In *Proceedings of SPIE Conference on Multisensor, Multisource Information Fusion: Architectures, Algorithms, and Applications*, Vol. 6242, Kissimmee, FL.

Li, X., R. Y. Zhu, J. Wang, and C. Han. 2003. Optimal linear estimation fusion—Part I: Unified fusion rules. *IEEE Transactions on Information Theory*, 49: 2192–2208.

Liggins, M. E. and K. C. Chang. 2009. Distributed fusion architectures, algorithms, and performance within a network-centric architecture. In *Handbook of Multisensor Data Fusion: Theory and Practice*, eds. M. E. Liggins, D. H. Hall, and J. Llinas. Boca Raton, FL: CRC Press.

Liggins, II, M. E., C. Y. Chong, I. Kadar, M. G. Alford, V. Vannicola, and S. Thomopoulos. 1997. Distributed fusion architectures and algorithms for target tracking. *Proceedings of IEEE*, 85: 95–107.

McLaughlin, S. P., R. J. Evans, and V. Krishnamurthy. 2005. A graph theoretic approach to data incest management in network centric warfare. In *Proceedings of the 8th International Conference on Information Fusion*, Philadelphia, PA.

McLaughlin, S. P., V. Krishnamurthy, and R. J. Evans. 2004. Bayesian network model for data incest in a distributed sensor network. In *Proceedings of the 7th International Conference on Information Fusion*, Stockholm, Sweden.

Mori, S., W. H. Barker, C. Y. Chong, and K. C. Chang. 2002. Track association and track fusion with non-deterministic target dynamics. *IEEE Transactions on Aerospace and Electronic Systems*, 38: 659–668.

Nicholson, D., S. J. Julier, and J. K. Uhlmann. 2001. DDF: An evaluation of covariance intersection. In *Proceedings of the 4th International Conference on Information Fusion*, Montréal, Québec, Canada.

Nicholson, D., C. M. Lloyd, S. J. Julier, and J. K. Uhlmann. 2002. Scalable distributed data fusion. In *Proceedings of the 5th International Conference on Information Fusion*, Annapolis, MD.

Olfati-Saber, R. 2005. Distributed Kalman filter with embedded consensus filters. In *Proceedings of the 44th IEEE Conference on Decision and Control*, Seville, Spain.

Speyer, J. L. 1979. Computation and transmission requirements for a decentralized linear-quadratic-Gaussian control problem. *IEEE Transactions on Automatic Control*, 24: 266–269.

Teneketzis, D. and P. Varaiya. 1988. Consensus in distributed estimation. In *Advances in Statistical Signal Processing*, ed. H. V. Poor, pp. 361–386. Greenwich, CT: JAI Press.

Willsky, A., M. Bello, D. Castanon, B. Levy, and G. Verghese. 1982. Combining and updating of local estimates along sets of one-dimensional tracks. *IEEE Transactions on Automatic Control*, 27: 799–813.

Zhu, Y. and X. R. Li. 1999. Best linear unbiased estimation fusion. In *Proceedings of the 2nd International Conference on Information Fusion*, Sunnyvale, CA.

6 Essence of Distributed Target Tracking
Track Fusion and Track Association

Shozo Mori, Kuo-Chu Chang, and Chee-Yee Chong

CONTENTS

6.1 Introduction 125
6.2 Track Fusion 127
    6.2.1 One-Time Track Fusion 127
        6.2.1.1 One-Time Track Fusion Rules 129
        6.2.1.2 Calculation of Cross-Covariance Matrix 133
        6.2.1.3 Covariance Intersection Methods 134
        6.2.1.4 Optimality of Track Fusion 135
        6.2.1.5 Performance Comparison of One-Time Track
                  Fusion Rules 137
    6.2.2 Repeated Track Fusion 143
        6.2.2.1 Repeated Track Fusion without Feedback 144
        6.2.2.2 Repeated Track Fusion with Feedback 148
6.3 Track Association 150
    6.3.1 Track Association Problem Definition 151
    6.3.2 Track Association Metrics 152
    6.3.3 Comparison of Track Association Metrics 155
6.4 Conclusions 156
References 157

6.1 INTRODUCTION

This chapter describes an important, practical, widely studied application of the distributed estimation theories described in Chapter 5, i.e., distributed target tracking. Multiple-target tracking problems can be viewed as an extension of classical dynamical state estimation problems, or *filtering* problems (Wiener 1949, Kalman 1960, Kalman and Bucy 1960, Anderson and Moore 1979), to estimate the states of generally moving physical entities. An essence of the extension is from single-target problems to multiple-target problems with an unknown number of targets, without a priori target identification, where any observation originates from any one of the modeled targets, or an object of no interest (i.e., clutter, false alarms, etc.) (Blackman 1986, Bar-Shalom and Fortmann 1988, Bar-Shalom and Li 1993, Blackman and Popoli 1999, Bar-Shalom et al. 2001, 2011). In short, multiple-target tracking problems are dynamical state estimation problems with data association problems.

As any other information processing system, given a set of sources of information, i.e., sensors, optimal or near-optimal state estimates are obtained by *central processing*, i.e., by centrally processing all the relevant information provided by all the sensors. However, in many large-scale system designs, an alternative processing architecture, i.e., *distributed processing*, is preferred because of the lack of single-point-of-failure, generally reduced communication requirements, and possible minimization of processing bottlenecks, as discussed in the previous chapter. This preference is particularly prevalent for multiple-target tracking problems, mainly because of often severely heavy information-processing requirements for solving data association problems. In distributed tracking systems, the data association requirements are typically divided into (i) *local data association* where sensor measurements are *correlated* together into *local* (or *sensor*) *tracks*, and (ii) *global processing* where local tracks are *associated* and *fused* together into a set of *global* (or *system*) *tracks*. In this way, the processing and the communication loads may be system-wide balanced.

For this reason, the studies of distributed tracking started almost at the same time when the multiple-target tracking itself began to be studied. We can cite a pioneering work (Singer and Kanyuck 1971) and two seminal papers (Bar-Shalom 1981) and (Bar-Shalom and Campo 1986), which cover two essences of distributed tracking, i.e., *track association* and *track fusion*. As mentioned in the previous chapter, the studies of the track association and fusion problems were formulated and solved in the framework of distributed estimation problems (Chong et al. 1985, 1987, Liggins II et al. 1997), with general sensor and information networks. Since then, the amount of the literature on track fusion has exploded (Hashemipour et al. 1988, Durrant-Whyte et al. 1990, Belkin et al. 1993, Lobbia and Kent 1994, Drummond 1997a, Miller et al. 1998, Zhu and Li 1999, Li et al. 2003), and many others.

As described in Drummond (1997a), Liggins II et al. (1997), Chong et al. (2000), Moore and Blaire (2000), Dunham et al. (2004), and Liggins II and Chang (2009), many distributed tracking systems, both military and civilian, have been developed and operated, system engineering studies, mainly of so-called *fusion architecture* studies, have been conducted, and performance of various functions and algorithms has been examined. Recently, the topics of the distributed target tracking have been migrated into the area of the robotics (Durrant-Whyte et al. 1990) and the distributed large-scale sensor networks (Iyengar and Brook 2005).

Instead of covering the entire areas concerning the distributed tracking, this chapter revisits its two essences, i.e., track association and track fusion, in terms of track fusion *rules* and track association *metrics*. We will describe as many rules and metrics that have been proposed and examine them, as quantitatively as possible.

To do this, we need to limit our scope using simple abstract mathematical models. However, we will try to cover as many practical factors as possible: consequences of one-time versus repeated information exchanges, fusion of information from similar versus dissimilar sensors, target maneuverability, a priori position and velocity uncertainty, and target density. We also choose a minimum complexity of system architecture, i.e., a two-sensor, or two-station system, by which we can isolate the two essential problems, i.e., the track association and fusion, to enable clear comparison of many algorithms. In this way, we can discuss key design factors for the distributed tracking, i.e., fusion with or without feedback, and the effect of the depth of memory of the past informational transactions, etc.

The rest of this chapter is divided into two major sections: Section 6.2 describes representative track fusion rules, and numerically compares the performance, under a set of prescribed variations of track fusion environments and designs. Section 6.3 examines a simple one-time track-to-track association and compares the performance using various track-to-track association metrics.

6.2   TRACK FUSION

Although track association is prerequisite to track fusion, we will discuss track fusion first in this section before discussing track association in the next section. Despite a large volume of works on track fusion, the studies on the track association are still rather sparse comparing with the studies on the track fusion.

We will first consider a simple, *one-time* track-to-track fusion problem in Section 6.2.1 and more complicated cases where track fusion is done *repeatedly* in Section 6.2.2.

6.2.1   ONE-TIME TRACK FUSION

Suppose that two sensors, $i = 1, 2$, have been observing the same target as

$$y_{ik} = H_{ik}x(t_{ik}) + \eta_{ik} \tag{6.1}$$

at time t_{ik}, for $k = 1, \ldots, N_i$, such that $t_{i1} < \cdots < t_{iN_i}$, where each measurement error η_{ik} is an independent zero-mean Gaussian random vector with covariance matrix* $R_{ik} = \mathbb{E}\left(\eta_{ik}\eta_{ik}^T\right)$ and H_{ik} is an observation matrix with appropriate dimensions. $x(\cdot)$ in (6.1) is the target state process defined by

$$\frac{d}{dt}x(t) = A_t x(t) + B_t \dot{w}(t) \tag{6.2}^\dagger$$

---

* By X^T we mean the transpose of a vector or matrix X. $\mathbb{E}$ is the conditional or unconditional mathematical expectation operator.
† More precisely, (6.2) is meant to be a stochastic differential equation, $dx(t) = Ax(t)dt + Bdw(t)$, with unit-intensity Wiener process $(w(t))_{t \in [t_0, \infty)}$. We assume $x(t_0)$, $(w(t))_{t \in [t_0, \infty)}$, and $((\eta_{ik})_{k=1}^{N_i})_{i=1}^2$ are all independent from each other.

on $[t_0, \infty)$ with $t_0 \leq \min\{t_{11}, t_{21}\}$ on a Euclidean target state space E, with a unit-intensity vector white noise process $(\dot{w}_t)_{t\in[t_0,\infty)}$, and by the initial state $x(t_0)$, a Gaussian vector with mean $\bar{x}_0$ and covariance matrix $\bar{V}_0$, i.e.,* $P(x(t_0)) = g(x(t_0) - \bar{x}_0; \bar{V}_0)$.

We assume the local data processor, for each sensor, $i = 1, 2$, produces the local estimate $\hat{x}_t = \mathbb{E}\left(x(t_F) \middle| (y_{ik})_{k=1}^{N_i}\right)$, which is the conditional expectation[†] of the target state $x(t_F)$ at a common fusion time $t_F \geq \max\{t_{1N_1}, t_{2N_2}\}$, conditioned by the local data $(y_{ik})_{k=1}^{N_i}$, together with estimation error covariance matrix $V_i = \mathbb{E}((\hat{x}_i - x(t_F))(\hat{x}_i - x(t_F))^T)$ that we assume is strictly positive definite, i.e., $P(x(t_F) | (y_{ik})_{k=1}^{N_i}) = g(x(t_F) - x_i; V_i)$. Our track fusion problem is then defined as the problem of generating a "good" estimate $\hat{x}_F$ of the target state $x(t_F)$ as a function of the local estimates $\hat{x}_1$ and $\hat{x}_2$.

The joint probability density function of the two local estimation errors can then be written as

$$P(\hat{x}_1 - x(t_F),\ \hat{x}_2 - x(t_F)) = g\left(\begin{bmatrix} \hat{x}_1 - x(t_F) \\ \hat{x}_2 - x(t_F) \end{bmatrix}; \begin{bmatrix} V_1 & V_{12} \\ V_{21} & V_2 \end{bmatrix}\right) \tag{6.3}$$

We need to consider the cross-covariance matrix, $V_{12} = \mathbb{E}((\hat{x}_1 - x(t_F))(\hat{x}_2 - x(t_F))^T)$ and $V_{21} = V_{12}^T$, in (6.3), because the initial condition $x(t_0) = x_0$ and the process noise $(w(t))_{t\in[t_0,\infty)}$ in (6.2) both commonly affect the two estimates, $\hat{x}_1$ and $\hat{x}_2$.

Let the local estimation errors be denoted by $\tilde{x}_i \overset{\text{def}}{=} \hat{x}_i - x(t_F)$, $i = 1, 2$. Then, we should immediately recognize the following three facts:

1. For each i, the estimation error $\tilde{x}_i$ is independent (orthogonal) to the state estimate $\hat{x}_i$.
2. The two estimation error vectors, $\tilde{x}_1$ and $\tilde{x}_2$, are correlated.
3. Each estimation error $\tilde{x}_i$ is *not* necessarily independent of the target state $x(t_F)$.

Although (1) is the basic fact of the linear Gaussian estimation (cf., e.g., Anderson and Moore 1979), (2) and (3) are the distinct characteristics of the track fusion problems, which prevent us from treating the two local estimates as if they were two independent sensor measurements of the target state $x(t_F)$. As mentioned earlier, (2) originated from the common use of the initial state condition and the process noise while (3) is simply due to the fact that $\hat{x}_i$ is the processed result, correlated to the initial condition $x(t_0) = x_0$, and hence $\tilde{x}_i$ is correlated to $x(t_F)$.

Some of the track fusion rules described subsequently can be used for track fusion problems with nonlinear target dynamics and nonlinear observation models. In such a case, (6.3) may be considered as a Gaussian approximation of a non-Gaussian joint estimation error probability distribution.

---

* For this chapter, we use P and p as the generic symbols for conditional or unconditional probability density or mass function, and g as the generic zero-mean Gaussian density function, i.e., $g(\xi;V)^{\text{def}} = \det(2\pi V)^{-1/2} \exp(-(1/2)\xi^T V^{-1}\xi)$.

† In this chapter, we use any conditioning in the strict Bayesian sense, e.g., $P(x|y) = P(x, y)/P(y)$. $(y_{ik})_{k=1}^{N_i}$ is shorthand for a finite sequence $(y_{i1}, y_{i2}, \ldots, y_{iN_i})$.

6.2.1.1 One-Time Track Fusion Rules

In the following, for the sake of simplicity, we will drop the time index and replace $x(t_F)$ by x, $E(x(t_F))$ by $\bar{x}$, and write $\bar{V} = \mathbb{E}((x - \bar{x})(x - \bar{x})^T)$. All the rules described in this section are in the form of the linear combination

$$\hat{x}_F = W_0\,\bar{x} + W_1 x_1 + W_2 x_2 \tag{6.4}$$

with $W_0 + W_1 + W_2 = I$ (unbiasedness), where all the weight matrices are constant and independent of sensor data, $(y_{ik})_{k=1}^{N_i}$, $i = 1, 2$, either as a conscious choice, or as a consequence of the linear-Gaussian assumptions. The estimation error covariance matrix V_F can therefore be evaluated by

$$V_F = \begin{bmatrix} W_0 & W_1 & W_2 \end{bmatrix} \begin{bmatrix} \bar{V} & V_{01} & V_{02} \\ V_{01}^T & V_1 & V_{12} \\ V_{02}^T & V_{12}^T & V_2 \end{bmatrix} \begin{bmatrix} W_0^T \\ W_1^T \\ W_2^T \end{bmatrix} \tag{6.5}$$

The covariance matrix V_i is provided with each local state estimator, $i = 1, 2$, and the a priori state variance $\bar{V}$ at the fusion time t_F is given by

$$\bar{V} = \Phi(t_F, t_0)\bar{V}_0 \Phi(t_F, t_0)^T + Q(t_F, t_0) \tag{6.6}$$

where $\Phi(t, \tau)$ is the fundamental solution matrix of $(A_t)_{t \in [t_0, \infty)}$, defined by a matrix differential equation $(\partial/\partial t)\Phi(t, \tau) = A_t \Phi(t, \tau)$ with $\Phi(\tau, \tau) = I$, and $\dot{Q}(\cdot, \cdot)$ is defined by

$$Q(t_2, t_1) = \int_{t_1}^{t_2} \Phi(t_2, \tau) B_\tau B_\tau^T \Phi(t_2, \tau)^T \, dt \tag{6.7}$$

for any $t_0 \leq t_1 \leq t_2$. Later, Section 6.2.1.2 shows how to calculate the cross-covariance V_{12} between the two local state estimation errors, $\tilde{x}_1$ and $\tilde{x}_2$, as well as the cross-covariance V_{0i} between the state a priori expectation error, $\bar{x} - x$, and the local state estimation error $\tilde{x}_i$, $i = 1, 2$.

Some of the fusion rules described later in this section *declare* the estimation error covariance matrix V_F by itself, assuming implicitly or explicitly that some of the statistics, e.g., V_{12} or V_{0i}, are not available when fusing the two local estimates, $\hat{x}_1$ and $\hat{x}_2$. In that case, the declared covariance matrix V_F may not be the *true* one defined by (6.5). We will call the declared V_F *honest* (*consistent*) if it coincides with the one calculated by (6.5), *pessimistic* if it is generally larger, and *optimistic* if smaller.

6.2.1.1.1 Bar-Shalom–Campo and Speyer Fusion Rules

The Bar-Shalom–Campo fusion rule, described in a seminal paper (Bar-Shalom and Campo 1986), is defined by the weights, $W_0 = 0$, and

$$W_i = (V_j - V_{ji})(V_1 + V_2 - V_{12} - V_{21})^{-1} \tag{6.8}$$

for $i=1, 2$ with $j=3-i$. Since $W_1+W_2=I$, the unbiasedness of the local estimates implies the unbiasedness of the fused estimate $\hat{x}_F$. As shown in Li et al. (2003), this fusion rule is obtained as a unique solution $x=\hat{x}_F$ that maximizes the *likelihood function* $L(x|\hat{x}_1, \hat{x}_2)$ defined as $L(x|\hat{x}_1, \hat{x}_2)=p(\hat{x}_1 - x, \hat{x}_2 - x)$, where $p(\tilde{x}_1, \tilde{x}_2)$ is the joint probability density function of the two local estimation errors, $\tilde{x}_1$ and $\tilde{x}_2$.

We should note that this likelihood function $L(x|\hat{x}_1,\hat{x}_2)$ is not the likelihood function $P(\hat{x}_1, \hat{x}_2|x)$ in the strict Bayesian sense, i.e., the conditional joint probability density function of the *data*, $\hat{x}_1$ and $\hat{x}_2$ given the *true state* x, because the estimation errors, $(\tilde{x}_1, \tilde{x}_2)$, are not independent of the target state $x=x(t_F)$. Nonetheless, $L(x|\hat{x}_1, \hat{x}_2)=p(\hat{x}_1-x, \hat{x}_2-x)$ is certainly qualified as a likelihood function of x in the classical statistics sense, i.e., a joint probability density function of $\hat{x}_1$ and $\hat{x}_2$, when we consider x as a *constant parameter to be determined*. Two other fusion rules based on the conditional expectation and the likelihood function, both in the strict Bayesian sense, will be described later in this section. Both of those rules, as well as the Bar-Shalom–Campo rule, use the cross-covariance matrix V_{12}, generated by the common factors, i.e., the common initial condition and the common process noise.

If we ignore the cross-covariance V_{ij}, (6.8) becomes, for $i=1, 2$ with $j=3-i$,

$$W_i = V_j(V_1 + V_2)^{-1} = (V_1^{-1} + V_2^{-1})^{-1}V_i^{-1} \tag{6.9}$$

which is the fusion rule obtained by treating two estimates $\hat{x}_1$ and $\hat{x}_2$ as if they were two conditionally independent observations of x. Since the gain matrices (6.9) are obtained by normalizing two positive definition matrices V_i or V_i^{-1} to have $W_1 + W_2 = I$, we may call (6.9) the *simple convex combination rule*, with some caution for not confusing this with the *covariance intersection fusion rules* described later. It is also called the *naïve fusion* rule in Chang et al. (2008). We call this simplified rule the *Speyer fusion rule*, because this fusion rule seems to have appeared for the first time as Equation (22) of Speyer (1979).

Since $\det\left(\begin{bmatrix} V_1 & V_{12} \\ V_{12}^T & V_2 \end{bmatrix}\right) = \det(V_1 - V_{12}V_2^{-1}V_{12}^T)\det(V_2)$, ignoring the cross-variance

as $V_{12}=0$ means an increase in the ellipsoidal area defined by the joint covariance matrix $\begin{bmatrix} V_1 & V_{12} \\ V_{12}^T & V_2 \end{bmatrix}$. In that sense, we may say the simplified fusion rule (6.9) is obtained by inflating the joint covariance matrix. Using either fusion rule, the fused estimate is unbiased in the sense $E(\hat{x}_F-x)=0$. For the Bar-Shalom–Campo rule, the declared fused estimation error covariance matrix, $V_F = V_1 - (V_1 - V_{12})(V_1 + V_2 - V_{12} - V_{12}^T)^{-1}(V_1 - V_{12}^T)$, is *honest* (or *consistent*), while, ignoring the cross-covariance, $V_F = (V_1^{-1} + V_2^{-1})^{-1} = V_1 - V_1(V_1 + V_2)^{-1}V_1$ for the Speyer rule is not honest and generally optimistic.

6.2.1.1.2 Tracklet Fusion Rule

For each sensor $i=1, 2$, let $\hat{p}_i(x) = P\left(x\left|(y_{ij})_{j=1}^{N_i}\right.\right)$. Then, as shown in Chong (1979), the *tracklet fusion rule* to obtain the fused probability density function $\hat{p}_F$ by fusing $\hat{p}_1$ and $\hat{p}_2$ can be written as $\hat{p}_F(x)=C^{-1}\hat{p}_1(x)\hat{p}_2(x)/\bar{p}(x)$, with the a priori probability density $\bar{p}(x)$ and the normalizing constant C. This fusion rule can be applied to any probability (generally non-Gaussian) distributions on any appropriate target space E as long as the densities and the integral are all well defined. In our linear-Gaussian case, as shown in Chong et al. (1983,1986,1990), etc., the tracklet fusion rule is defined by

$$W_i = V_F V_i^{-1}, i = 1, 2; \quad W_0 = -V_F \bar{V}^{-1} = I - W_1 - W_2 \tag{6.10}$$

with the declared fused estimate error covariance matrix $V_F = (V_1^{-1} + V_2^{-1} - \bar{V})^{-1}$, where $\bar{V} = E((\bar{x} - x)(\bar{x} - x)^T)$ is the a priori covariance matrix.

Unfortunately, the tracklet fusion rule may not be *exact* in the sense, $\hat{p}_F(x) = P(x | (y_{1j})_{j=1}^{N_1}, (y_{2j})_{j=1}^{N_2})$, unless the target dynamics are deterministic, i.e., $B_t \equiv 0$ in (6.2). Nonetheless, the extrapolation of the a priori covariance matrix by (6.6) takes the effects of the process noise into account. However, the declared fused estimate error covariance matrix $V_F = (V_1^{-1} + V_2^{-1} - \bar{V})^{-1}$ is often not honest and generally optimistic.

The fusion rule (6.10) can be rewritten as

$$\begin{cases} V_F^{-1} x_F = \bar{V}^{-1} \bar{x} + \tilde{V}_1^{-1} z_1 + \tilde{V}_2^{-1} z_2 \\ V_F^{-1} = \bar{V}^{-1} + \tilde{V}_1^{-1} + \tilde{V}_2^{-1} \end{cases} \tag{6.11}$$

with, for $i = 1, 2$,

$$\begin{cases} \tilde{V}_i^{-1} z_i = V_i^{-1} \hat{x}_i - \bar{V}^{-1} \bar{x} \\ \tilde{V}_i^{-1} = V_i^{-1} - \bar{V}^{-1} \end{cases} \tag{6.12}$$

Equation 6.11 appears as a Kalman filter update equation that updates the state estimate by the two conditionally independent measurements, z_1 and z_2. Equation 6.12 can be interpreted as the *decorrelation* of the two local estimates, $\hat{x}_1$ and $\hat{x}_2$, by removing the prior information represented by the pair $(\bar{x}, \bar{V})$ from $\hat{x}_1$ and $\hat{x}_2$. The decorrelated estimates, z_1 and z_2, defined by Equation 6.12, are called the *equivalent measurements*, or the *pseudo-measurements*, or the state estimates of *tracklets* (or a track segment, a portion of a track, small enough represented by a single Gaussian distribution but large enough to have such a full-state representation, defined by $(z_i, \tilde{V}_i)$) (Belkin et al. 1993, Lobbia and Kent 1994, Drummond 1997a, 1997b), etc. This is the reason why we call this rule the tracklet fusion rule. Because Equation 6.11 is in the information matrix form of Kalman filter, a distributed track fusion algorithm using Equation 6.11 with Equation 6.12 is sometimes called *information filter* or *information matrix filter* (Chang et al. 2002).

The decorrelation formula (6.12) also gives us a convenient way of representing a tracklet or a track segment by a pair $(z_i, \tilde{V}_i)$ of the equivalent measurement and its measurement error covariance matrix, or equivalently $\left(\tilde{V}_i^{-1} z_i, \tilde{V}_i^{-1} \right)$. From this point of view, the distributed track fusion algorithm that use this pair, $(z_i, \tilde{V}_i)$ or $\left(\tilde{V}_i^{-1} z_i, \tilde{V}_i^{-1} \right)$

to represent approximately the conditionally independent unit of information, is called the *Channel filter* in Durrant-Whyte et al. (1990) and Rao et al. (1993).

6.2.1.1.3 Minimum-Variance (MV) Fusion Rule

Let $\hat{x}_F = \mathbb{E}(x|\hat{x}_1, \hat{x}_2)$, i.e., the conditional expectation of the target state $x = x(t_F)$ at the fusion time t_F, given the two local state estimates, $\hat{x}_1$ and $\hat{x}_2$. It is well known (cf., e.g., Rhodes 1971) that the estimate $\hat{x}_F$ minimizes the expected estimation error* $\mathbb{E}(\|\hat{x} - x\|^2|\hat{x}_1, \hat{x}_2)$ among all the estimates $\hat{x}$, as defined as any measurable functions of $\hat{x}_1$ and $\hat{x}_2$. Because of the Gaussianness, the fused estimate, $\hat{x}_F = \mathbb{E}(x|\hat{x}_1, \hat{x}_2)$, is also the maximum a posteriori (MAP) estimate of x conditioned by $\hat{x}_1$ and $\hat{x}_2$, and is given by (6.4) using $\begin{bmatrix} W_1 & W_2 \end{bmatrix} = V_{xz}V_{zz}^{-1}$, and $W_0 = I - W_1 - W_2$, with

$$\begin{cases} V_{zz} = \begin{bmatrix} V_1 + \overline{V} - V_{01}^T - V_{01} & V_{12} + \overline{V} - V_{01}^T - V_{02} \\ V_{12}^T + \overline{V} - V_{02}^T - V_{01} & V_{22} + \overline{V} - V_{02}^T - V_{02} \end{bmatrix} \\ V_{xz} = \begin{bmatrix} \overline{V} - V_{01} & \overline{V} - V_{02} \end{bmatrix} \end{cases} \tag{6.13}$$

where

V_{zz} is the self-covariance matrix of $z = \begin{bmatrix} \hat{x}_1^T & \hat{x}_2^T \end{bmatrix}^T$

V_{xz} is the cross-covariance between x and z

We should note that, since $\hat{x}_i - \overline{x} = \hat{x}_i - x + x - \overline{x} = \tilde{x}_i - (\overline{x} - x)$, $i = 1, 2$, we have $\mathbb{E}\left(\left(\hat{x}_i - \overline{x} \right)\left(\hat{x}_j - \overline{x} \right)^T \right) = V_{ij} + \overline{V} - V_{0i}^T - V_{0j}^T$, for $i, j = 1, 2$, with $V_i = V_{ii}$, and $E((x - \overline{x}) (\hat{x}_i - \overline{x})^T) = \overline{V} - V_{0i}$, for $i = 1, 2$. Therefore, while the Bar-Shalom–Compo rule considers only the correlation caused by the common process noise, and what the tracklet rule considers explicitly only the correlation caused by the use of the common a priori information, the MV rule considers both and provides the optimal estimate as the conditional expectation given $z = \begin{bmatrix} \hat{x}_1^T & \hat{x}_2^T \end{bmatrix}^T$. The declared fused estimate error covariance matrix, $V_F = \overline{V} - V_{xz}V_{zz}^{-1}V_{xz}^T$, is honest.

As shown in Zhu and Li (1999) and Li et al. (2003), this MV fusion rule is also the *best linear unbiased estimate (BLUE)* by choosing the best weights (W_0, W_1, W_2) to minimize the estimation error variance, under constraint $W_0 + W_1 + W_1 = I$, and hence we may call it the *BLUE fusion rule*. The Bar-Shalom–Campo rule is obtained as the BLUE rule with more restriction, i.e., by the minimization with respect to (W_1, W_2), with the constraints $W_0 = 0$ and $W_1 + W_2 = I$.

6.2.1.1.4 Bayesian Maximum-Likelihood Fusion (BML) Rule

Define $z = \begin{bmatrix} \hat{x}_1^T & \hat{x}_2^T \end{bmatrix}^T$ as before. Consider $P(z|x)$, which is the conditional probability density of $z = \begin{bmatrix} \hat{x}_1^T & \hat{x}_2^T \end{bmatrix}^T$ (data) given x (state to be estimated), i.e., the likelihood function in the strict Bayesian sense. Reversing the roles of x and z, we have $P(z|x) = g(z - \hat{z}; \hat{V}_{zz})$ with

---

* By $\|\cdot\|$, we mean the standard Euclidean norm, i.e., $\|x\| = \sqrt{x^T x}$ for any vector x.

$$\begin{cases} \hat{z} = \bar{z} + V_{zx}\bar{V}^{-1}(x - \bar{x}) \\ \hat{V} = V_{zz} - V_{zx}\bar{V}^{-1}V_{xz} \end{cases} \tag{6.14}$$

Hence, the likelihood function $P(z|x)$ (as a function of x) is maximized at

$$\hat{x}_F = \bar{x} + \bar{V}\left(V_{xz}\hat{V}_{zz}^{-1}V_{zx}\right)^{-1}V_{xz}\hat{V}_{zz}^{-1}(z - \bar{z}) \tag{6.15}$$

Thus the maximum likelihood estimate of the target state x given the local estimates $\hat{x}_1$ and $\hat{x}_2$ can be expressed by (6.4) with the weight matrices calculated by $[W_1 \ W_2] = MV_{xz}\hat{V}_{zz}^{-1}$ with $M = \bar{V}(V_{xz}\hat{V}_{zz}^{-1}V_{xz}^T)^{-1}$ with $W_0 = I - W_1 - W_2$, instead of the MV fusion weights $[W_1 \ W_2] = V_{xz}V_{zz}^{-1}$.

We may say the likelihood function $P(z|x)$ is the likelihood function in the strict Bayesian sense, and hence we call the fusion rule defined by (6.15) the *Bayesian Maximum Likelihood Rule*, or BML Rule.

6.2.1.2 Calculation of Cross-Covariance Matrix

Besides the Speyer (simple convex combination) fusion rule and the tracklet fusion rule, it is necessary to calculate the cross-covariance matrix between the estimation errors, $\tilde{x}_1$ and $\tilde{x}_2$, of local estimates, $\hat{x}_1$ and $\hat{x}_2$. The calculation was described in Bar-Shalom (1981), for the synchronous sensor case, which can be easily extended to nonsynchronous cases as shown in Mori et al. (2002). To do this, let $T = \bigcup_{i=1}^{2} \{t_{ik}\}_{k=1}^{N_i}$ be the union of the observation times of the two sensors, $(T_k)_{k=1}^{N}$ be the unique enumeration of T such that $T_1 < T_2 < \ldots < T_N$, and $I_k = \{i \in \{1,2\} | T_k = t_{ik'} \text{ for some } k'\}$ for every k.

Let $\bar{V}_{12k}$ and $\hat{V}_{12k}$ be the cross-covariance matrices between estimation errors of the state estimates of $x(T_k)$ based on $\{y_{1k}|t_{1k} \le T_{k-1}\}$ and $\{y_{2k}|t_{2k} \le T_{k-1}\}$, and between those by $\{y_{1k}|t_{1k} \le T_k\}$ and $\{y_{2k}|t_{2k} \le T_k\}$, respectively. Then we have

$$\bar{V}_{12k} = \Phi(T_k, T_{k-1})\hat{V}_{12(k-1)}\Phi(T_k, T_{k-1})^T + Q(T_k, T_{k-1}) \tag{6.16}$$

and

$$\hat{V}_{12k} = \begin{cases} (I - K_{1k'}K_{1k'})\bar{V}_{12k} & \text{if } I_k = \{1\} \text{ and } t_{1k'} = T_k \\ \bar{V}_{12k}(I - K_{2k'}K_{2k'})^T & \text{if } I_k = \{2\} \text{ and } t_{2k'} = T_k \\ (I - K_{1k'}K_{1k'})\bar{V}_{12k}(I - K_{2k''}K_{2k''})^T & \text{if } I_k = \{1,2\} \text{ and } t_{1k'} = t_{2k''} = T_k \end{cases} \tag{6.17}$$

together with an appropriate initial condition, where $K_{1k'}$ and $K_{2k'}$ are the Kalman filter gain matrices used by sensors 1 and 2 to process $y_{1k'}$ and $y_{2k''}$, respectively. The cross-covariance matrix V_{12} between $\tilde{x}_1$ and $\tilde{x}_2$ can be obtained at the end of this recursion, with an extra extrapolation (6.16) (at the end) if necessary.

The cross-covariance matrices $\hat{V}_{0ik}$, between the estimation error of the state estimate of $x(T_k)$ conditioned on $\{y_{ik}|t_{ik} \le T_k\}$ and the a priori extrapolation error $\mathbb{E}(x(T_k) - x(T_k))$, can be calculated similarly by the extrapolation equation (6.16), and the second update equation of (6.17), to obtain V_{01} and V_{02} (which are necessary for calculating the MV and BML fusion rules). The local estimation error covariance matrices, V_1 and V_2, are of course provided by the local Kalman filters, while the a priori extrapolation error covariance matrix $\bar{V}$ is calculated by (6.6).

6.2.1.3   Covariance Intersection Methods

The covariance intersection (CI) method was introduced as a method of fusing two estimate-covariance pairs, $(\hat{x}_1, V_1)$ and $(\hat{x}_2, V_2)$ when the cross-covariance V_{12} of the estimation errors is not known or available. The CI approach is a heuristic approach to adjust the commonly used the simple weighting, i.e., the Speyer fusion rule (6.9), as

$$
\begin{cases}
V_F^{-1}\hat{x}_F = \alpha V_1^{-1}\ \hat{x}_1 + \left(1-\alpha\right)V_2^{-1}\hat{x}_2 \\
V_F^{-1} = \alpha V_1^{-1} + \left(1-\alpha\right)V_2^{-1}
\end{cases}
\tag{6.18}
$$

with a fixed scalar parameter $\alpha \in [0, 1]$, i.e., (6.4) with $W_0 = 0$, $W_1 = \alpha V_F V_1^{-1}$, and $W_2 = \left(1-\alpha\right)V_F V_2^{-1}$. The term "covariance intersection" originates from the fact that the ellipsoid* $\left\{ x \in E \middle| \|x\|^2_{V_F^{-1}} \le \chi^2 \right\}$ is included in the intersection of two ellipsoids, $\left\{ x \in E \middle| \|x\|^2_{V_i^{-1}} \le \chi^2 \right\}$, $i = 1, 2$, for any given $\chi^2 > 0$ (Nicholson et al. 2001, 2002, Julier et al. 2006). But the terminology may appear rather confusing because the ellipsoid $\left\{ x \in E \middle| \| x - \hat{x}_F\|^2_{V_F^{-1}} \le \chi^2 \right\}$ is not necessarily contained in the intersection $\bigcap_{i=1}^{2} \left\{ x \in E \middle| \| x - \hat{x}_i\|^2_{V_i^{-1}} \le \chi^2 \right\}$.

The CI rule (6.18) can be viewed as a Gaussian case of the fusion rule,

$$
\hat{p}_F(x) = \frac{p_1(x)^\alpha\, p_2(x)^{1-\alpha}}{\displaystyle\int_E p_1(x')^\alpha\, p_2(x')^{1-\alpha}dx'}
\tag{6.19}
$$

which is called the *Chernoff fusion rule* in Hurley (2002) and Julier (2006), because the denominator of the right-hand side of (6.19) is known as *Chernoff information* (Cover and Thomas 2006). There are several proposals about how to choose the parameter $\alpha \in [0, 1]$. A couple of choices for the scalar weight α are shown below.

6.2.1.3.1   Shannon Fusion Rule

Consider the continuous-random-variable version of the entropy, known as the differential entropy or the continuous entropy of the fused probability distribution,

---

* By $\|\cdot\|_A$, we mean the norm on any Euclidean space defined by a positive definite symmetric matrix A as $\|x\|_A \overset{\text{def}}{=} \sqrt{x^T A x}$ for each vector x.

$$H(\hat{p}_F) = -\int_E \ln\left(\hat{p}_F(x)\right)\hat{p}_F(x)\,dx \tag{6.20}$$

The fusion rule that minimizes (6.20) can be called the *minimum entropy fusion* or *Shannon rule*. In the case where $\hat{p}_F$ is Gaussian with the CI covariance matrix V_F, we have $H(\hat{p}_F) = (1/2)(\ln(\det(2\pi V_F)) + \dim(E))$, the minimization of which becomes the minimization of the determinant. The resultant fusion rule (6.18) is called the *Shannon rule* in Chang et al. (2008).

6.2.1.3.2 Chen–Arambel–Mehra Fusion Rule

This fusion rule is defined as the one that minimizes the estimation error mean square, i.e., the trace of the fused covariance matrix V_F defined in (6.18), $\mathrm{tr}\left(V_F\right) = \mathrm{tr}\left(\left(\alpha V_1^{-1} + (1-\alpha)V_2^{-1}\right)^{-1}\right)$, as a function of $\alpha \in [0, 1]$. Let $\hat{\alpha} \in [0, 1]$ such that the trace $\mathrm{tr}(V_F)$ is minimized at $\alpha = \hat{\alpha}$. Chen et al. (2002) show a very interesting interpretation of this optimal $\hat{\alpha}$, i.e., the corresponding CI fusion gain matrix pair, $W_1 = \hat{\alpha} V_F V_1^{-1}$ and $W_2 = (1-\hat{\alpha})V_F V_2^{-1}$, is an optimal solution that minimizes

$$h(W_1 W_2) = \sqrt{\mathrm{tr}W_1 V_1 W_1^T} + \sqrt{\mathrm{tr}W_2 V_2 W_2^T} \tag{6.21}$$

subject to the unbiasedness condition $W_1 + W_2 = I$. We call the convex intersection with this $\hat{\alpha}$ the *Chen–Arambel–Mehra fusion rule*.

6.2.1.4 Optimality of Track Fusion

The track fusion rules described so far obtain an target state estimate that is the "best" in some sense, e.g., maximum likelihood, maximum a posteriori (MAP), minimum variance, etc., given the two local target state estimates, either explicitly or implicitly, under the assumption that the local estimates are optimal in the usual sense, i.e., the outputs of the local Kalman filters. However, because the conditioning uses only the local estimates that are not sufficient statistics when the target dynamics are nondeterministic, e.g., with process noise in the target model to account for target maneuvers, the performance of the fused estimate is generally inferior* to that of the central processing using all the raw sensor data, $\left(\left(y_{ik}\right)_{k=1}^{N_i}\right)_{i=1}^{2}$. For this reason, the performance of a track fusion rules should be compared with that of central processing, rather than the MAP (or the MV) fusion rule, whose optimality is also limited to the conditioning by the local sate estimates, $\hat{x}_1$ and $\hat{x}_2$.

Furthermore, there may be one more compelling reason why the performance of every track fusion rule must be compared with the centralized tracking performance. That is because, for the first time in the long history of track fusion studies, it was recently shown that the reconstruction of the globally optimal state estimate only by fusing or combining the local estimates is possible. Although such reconstruction can be obviously done in cases with deterministic dynamics or the full-rate communication, it is remarkable to see that the fusion

---

* Except for some extreme cases such as the cases where the local agents send out the local estimates after every synchronized observation (i.e., the full-communication-rate cases).

rule in Koch (2008, 2009), Govaers and Koch (2010, 2011), which we may call the *Koch–Govaers fusion rule*, can achieve the global optimality with any asynchronous communication and arbitrary communication rate to achieve the global optimality after each communication.

Using the notation in Section 6.2.1.2, the Koch–Govaers fusion rule requires local estimates, $\left((\bar{x}_{ik})_{k=1}^{N} \right)_{i=1}^{2}$ and $\left((\hat{x}_{ik})_{k=1}^{N} \right)_{i=1}^{2}$, each $\bar{x}_{ik}$ paired with the covariance matrix $\bar{V}_{k}$, and each $\hat{x}_{ik}$ paired with the covariance matrix $\hat{V}_{ik}$, to satisfy

$$P\left(x(T_k) \middle| \left\{ y_{ik'} \middle| t_{ik'} \le T_{k-1}, i = 1,2 \right\} \right) = C^{-1} \prod_{i=1}^{2} g\left(x(T_k) - \bar{x}_{ik}; \bar{V}_{k} \right) \qquad (6.22)$$

and

$$P\left(x(T_k) \middle| \left\{ y_{ik'} \middle| t_{ik'} \le T_{k}, i = 1,2 \right\} \right) = C'^{-1} \prod_{i=1}^{2} g\left(x(T_k) - \hat{x}_{ik}; \hat{V}_{ik} \right) \qquad (6.23)$$

The recent series of papers (Koch 2009, Govaers and Koch 2010, 2011) show that those two requirements can be satisfied by the extrapolation step

$$\begin{cases} \bar{x}_{ik} = 2\Phi(T_k, T_{k-1}) \left(\sum_{i'=1}^{2} \hat{V}_{i'(k-1)}^{-1} \right) \hat{V}_{i(k-1)} \hat{x}_{i(k-1)} \\ \bar{V}_{k} = 2\left(\Phi(T_k, T_{k-1}) \left(\sum_{i'=1}^{2} \hat{V}_{i'(k-1)}^{-1} \right)^{-1} \Phi(T_k, T_{k-1})^{T} + Q\Phi(T_k, T_{k-1}) \right) \end{cases} \qquad (6.24)$$

and the updating step

$$\begin{cases} \hat{x}_{ik} = \bar{x}_{ik} + K'_{ik} y_{ik'} - H_{ik'} \bar{x}_{ik} \\ \hat{V}_{ik} = \left(I - K'_{ik} H_{ik'} \right) \bar{V}_{k} \end{cases} \text{if } T_k = t_{ik'} \quad \begin{cases} \hat{x}_{ik} = \bar{x}_{ik} \\ \hat{V}_{ik} = \bar{V}_{k} \end{cases} \qquad (6.25)$$

with $K'_{ik} = \bar{V}_{k-1} H_{ik'}^{T} \left(H_{ik'} \bar{V}_{k-1} H_{ik'}^{T} + R_{ik'} \right)^{-1}$ if $T_k = t_{ik'}$, each for each sensor $i = 1, 2$. The initialization can be done in any way to satisfy the condition, either (6.22) or (6.23), for any appropriate k.

We should note two crucial facts: (i) neither $\left((\bar{x}_{ik})_{k=1}^{N} \right)_{i=1}^{2}$ nor $\left((\hat{x}_{ik})_{k=1}^{N} \right)_{i=1}^{2}$ are necessarily locally optimal estimates in any sense, and (ii) both extrapolation (6.24) and update (6.25) require knowledge of the local variance matrices $(\hat{V}_{ik})_{i=1}^{2}$, not only its own but also those of the other sensor. In other words, global optimality is obtained by sacrificing local optimality, and we need extensive knowledge in terms of covariance matrices of the other local processor. As far as (i) is concerned, however, local optimality, if necessary, can be maintained by locally running the Kalman filter for each sensor in parallel to the estimates $\left((\bar{x}_{ik})_{k=1}^{N} \right)_{i=1}^{2}$ and $\left((\hat{x}_{ik})_{k=1}^{N} \right)_{i=1}^{2}$ defined by (6.24) and (6.25). The latter requirement (ii), however, looks too excessive at the first glance.

We should note, however, that for our linear-Gaussian systems, all the estimation error (self and cross) covariance matrices as well as all the crucial parameters, such as the Kalman filter gain matrices and the innovations variance matrices, are all constant (i.e., not random). Hence, in the probability theory, they are all *known* or a part of the problem definition or the problem statement. In practice, however, the exchange or transfer of such knowledge may require significant communication bandwidth that may or may not be available. Therefore, realistically, we need to assume those parameters may have to be considered as a part of system design parameters, i.e., *off-line information* communicated "beforehand." Otherwise, e.g., communicating each measurement error covariance matrix for every local observation to a fusion center, or to each other local agent, may be equivalent to or exceeds the full measurement communication. Furthermore, whenever the linearity or the Gaussianness is questionable and extended Kalman filters are needed, the covariance matrices may become data-dependent, and hence, at least, some adjustment may become necessary. Thus the feasibility of this "optimal" track fusion algorithm remains to be demonstrated in a practical situation.

6.2.1.5   Performance Comparison of One-Time Track Fusion Rules

In order to characterize various track fusion rules and to compare with each other, we would like to use simple yet realistic examples. For this purpose, we chose a four-dimensional (two-dimensional position, two-dimensional velocity) state space, with the Ornstein–Uhlenbeck model, i.e., $A_t \equiv \begin{bmatrix} 0 & I \\ 0 & -\beta I \end{bmatrix}$, $B_t \equiv \begin{bmatrix} 0 \\ \sqrt{q}I \end{bmatrix}$, and

$\bar{V}_0 = \begin{bmatrix} \sigma_p^2 I & 0 \\ 0 & \sigma_v^2 I \end{bmatrix} (\beta > 0 \text{ and } q = 2\beta\sigma_v^2 > 0)$, and the two-dimensional position-only

observation,* $H_{ik} \equiv [I\ 0]$. The Ornstein–Uhlenbeck model can approximate a realistic target maneuver behavior known as a *random-tour* behavior with β^{-1} as the mean time between two maneuvers or of the length of each constant-velocity leg (Washburn 1969, Vebber 1991). For the sake of simplicity, we use synchronous, uniform sampling (measurements), i.e., $\Delta t \equiv t_{i(k+1)} - t_{ik}$ for $k = 1, \ldots, N$, $t_0 = t_{i1}$, and $t_F = t_{iN}$, for $i = 1, 2$.

It is customary to use the so-called almost constant-velocity model or the small-white-noise model to model target maneuvers, i.e., $\beta = 0$. Since we have chosen the Ornstein–Uhlenbeck model instead, we would provide some explanation. The Ornstein–Uhlenbeck dynamics are usually determined by two parameters, the inverse β of the time constant (which can be considered the mean time between two maneuvers) and the white noise intensity q that drives the variations in the velocity, from the initial condition. However, if the target velocity is, a priori, a stationary process defined by the stochastic differential equation $dv(t) = -\beta v(t)dt + \sqrt{q}dw(t)$ with the stationary covariance matrix $\sigma_v^2 I$, the two parameters are constrained as $q = 2\beta\sigma_v^2$. Using the stationary velocity process with standard deviation σ_v reflects the physical reality of real moving objects, in particular, on ground or on surface

---

* Where I and 0 are the 2×2 identity and zero matrices, respectively.

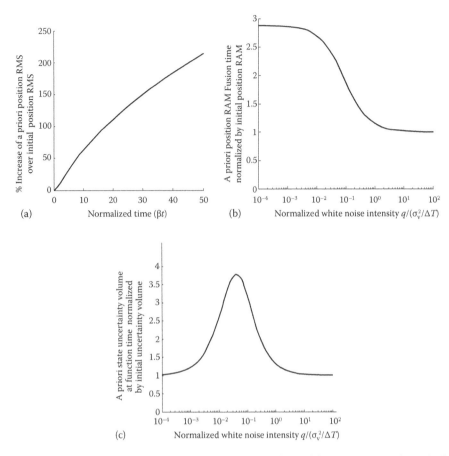

FIGURE 6.1 Characterization of Ornstein–Uhlenbeck model: (a) Increase of a priori position uncertainty in time, (b) a priori position uncertainty as function of normalized white noise intensity, (c) size of state uncertainty ellipsoid as function of normalized white noise intensity.

water or under water. Avoidance of ever increasing a priori velocity uncertainty (contradicting with reality) is the major motivation for using the Ornstein–Uhlenbeck model.

Figure 6.1 presents the key features of the Ornstein–Uhlenbeck model. Figure 6.1a shows the time increase of the a priori root mean square (RMS) position by the Ornstein–Uhlenbeck model, which increases as $\beta^{-1}(1 - e^{-\beta t})\sigma_v$ (which can be approximated as $\sigma_v t$ when β is small, and approaches 0 when β is large) by the velocity uncertainty, and by the white noise intensity, as $\sigma_v\sqrt{2\beta^{-1}t}$ for large β and $\sqrt{(2/3)\beta\sigma_v^2 t^3}$ for small β. The increase as the function of time is generally much slower than the small white noise model. Figure 6.1b shows the a priori positional RMS at a fixed time ΔT as a function of the normalized white noise intensity $q/\left(\sigma_v^2/\Delta T\right)$. We should note that, as $q \downarrow 0$, since $q = 2\beta\sigma_v^2$, we have $\beta \downarrow 0$, i.e., the model approaches a deterministic system, and that, as the white noise intensity increases $q \uparrow \infty$, β increases as $\beta \uparrow \infty$, and a priori position uncertainty approaches to be stationary one.

As $\beta \uparrow \infty$ (hence $q \uparrow \infty$), the average time between maneuvers approaches zero, and it will eventually reach the point where there are so many maneuvers to every direction, the effects cancel each other, resulting in the almost stationary positional RMS. Figure 6.1c shows the volume of the state uncertainty hyper volume at fixed time ΔT (normalized by the initial state hyper volume) as the function of normalized white noise intensity $q/\left(\sigma_v^2/\Delta T\right)$. It is interesting to see that, for both small and large q's (consequently small and large β's), the position-velocity joint uncertainty volume approaches the same volume at the initial time, and the maximum of the volume is attained in the middle. For a small β, the position-velocity cross-covariance makes the state uncertainty volume time-invariant, while for a large β, the position-velocity cross-covariance disappears and both position and velocity covariance matrices become stationary.

6.2.1.5.1 Supplementary Sensor Case

Let us consider two sensors that have almost the same performance characteristics, so that the addition of the second sensor to the first sensor is *supplementary*. As an extreme case in such situations, we assume, with $N = 10$ (track fusion after each sensor accumulates 10 measurements), $R_{ik} \equiv \begin{bmatrix} \sigma_m^2 & 0 \\ 0 & \sigma_m^2 \end{bmatrix}$, $k = 1,\ldots,N, i = 1, 2$, which implies $V_1 = V_2$.

In this extreme case, the Bar-Shalom–Campo rule is reduced to $W_1 = W_2 = (1/2)I$, which is the same as the Speyer rule. In other words, no matter how big the cross-covariance between the two local track estimation errors is, the inter-sensor cross-covariance is *irrelevant* to the fusion rule.

Moreover, because $V_1 = V_2$, any value α in the unit interval [0, 1] provides the same $\hat{V}_F$ for any rule of the CI method. Although the actual fused estimation error covariance may change with the weight α, considering the symmetry, it is reasonable to choose $\alpha = 1/2$, which makes both the Shannon and the Chen–Arambel–Mehra rules[*] become the same as the Bar-Shalom–Campo and the Speyer rules. As is well known, however, the fused estimation error covariance by any CI rule is the same as each local estimation error covariance and is extremely overestimated (overly pessimistic).

Figure 6.2 compares the performance by the four fusion rules: the Bar-Shalom–Campo rule (also Speyer and CI rules), the MV rule, the tracklet rule, and the BML rule, with the centralized tracking performance, when the normalized white noise intensity, $q/\left(\sigma_v^2/\Delta t\right)$, is varied in a wide range. Other key parameters are set as $\sigma_p = 10\sigma_m$ (the initial position standard deviation) and $\sigma_v = 3(\sigma_m/\Delta t)$ (the stationary velocity standard deviation).

First of all, we should note that the deterioration of the estimation performance from centralized tracking is very small, i.e., less than 4% for the Bar-Shalom–Campo,

---

[*] With $\alpha = 1/2$, the denominator of the right hand side of (6.19) becomes $\int_E \sqrt{p_1(x)p_2(x)}dx$, the expression known as the *Bhattacharyya bound*, and hence, we may call covariance intersection fusion rule with $\alpha = 1/2$, *Bhattacharyya fusion rule* (Chang et al. 2008).

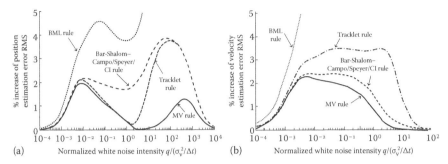

FIGURE 6.2 Percent increase of RMS estimation errors over centralized tracking performance as function of normalized process noise intensity: supplementary sensors: (a) RMS position estimation error and (b) RMS velocity estimation error.

minimum variance (MV), and the tracklet rules, over a wide range of the process noise intensity, which is consistent with the observations reported in Bar-Shalom and Campo (1986) and Mori et al. (2002). The apparent poor performance of the fusion rule using the likelihood function in the strict Bayesian sense, labeled as BML rule, in the figure, is, however, rather surprising. The deterioration of the performance of the BML rule from centralized tracking is within 10%–30% for both the position and the velocity error RMS, for small process noise intensity, when $q < (\sigma_v^2/\Delta t)$. However, when $q > (\sigma_v^2/\Delta t)$, the estimation errors, in particular for the velocity estimates, deteriorate and seem to increase rapidly.

As mentioned in Section 6.2.1.1, the BML rule is defined using the likelihood function in the strict Bayesian sense, i.e., the conditional probability density of the *data* $P(\hat{x}_1, \hat{x}_2|x)$ given the target state x to be estimated. In order for the BML rule to be close to the optimal in the sense of the minimum variance, we must have $\bar{V} \approx V_{xz}(V_{zz} - V_{zx}\bar{V}^{-1}V_{xz})^{-1}V_{zx}$, which is apparently violated for large process noise intensities. As mentioned in Section 6.2.1.1, the Bar-Shalom–Campo rule uses a likelihood function in the classical statistics sense, and apparently, its performance is much better than the ML estimate using the likelihood function in the strict Bayesian sense. In Figure 6.2, the full extent of the BML rule performance is not shown, since its bad performance will otherwise obscure the comparison of the performance of the other three fusion rules.

The MV rule provides optimal performance in terms of the estimation error variance as shown in Figure 6.2. It is however interesting to see that the tracklet rule, which does not use the inter-sensor cross-covariance matrix, shows better performance in terms of positional estimation than the Bar-Shalom–Campo rule, which uses the cross-covariance. But the order of the performance is reversed for the velocity estimation. This trend holds generally true for complementary sensor and repeated fusion cases, as shown later. All three fusion rules, Bar-Shalom–Campo, tracklet, and MV, converge to the performance of centralized tracking performance both when $q \downarrow 0$ and $q \uparrow \infty$, although the Bar-Shalom–Campo rule that does not use the a priori target state information exhibits a small bias as $q \downarrow 0$.

Figure 6.3 shows a similar comparison when we vary the initial state position standard deviation, σ_p, which represents the a priori information, in a wide range.

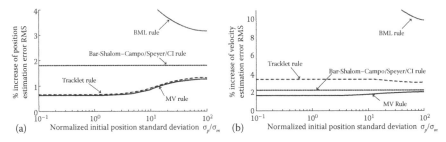

FIGURE 6.3 Percent increase of RMS estimation errors over centralized tracking performance as function of normalized initial position standard deviation—supplementary sensors: (a) RMS position estimation error and (b) RMS velocity estimation error.

For this figure, the process noise intensity and the stationary velocity covariance are kept constant at $q = 0.1\left(\sigma_v^2/\Delta t\right)$ and $\sigma_v = 3(\sigma_m/\Delta t)$, respectively. The Bar-Shalom–Campo rule does not use the initial state (a priori) information, and its performance is invariant with respect to σ_p. Like Figure 6.2 obtained by varying the process noise intensity, the performance of the BML rule using the likelihood function in the strict Bayesian sense (that we may call the Bayesian likelihood function) is noticeably worse than the Bar-Shalom–Campo rule that is a maximum likelihood estimate using a likelihood function in the classical statistics sense. In particular, the estimation error of the BML rule exhibits more than 20% increase in the velocity estimation error RMS over centralized tracking for small initial position uncertainty (small σ_p), although Figure 6.3b does not show that part. Again the position estimation performance by the tracklet rule is better than the Bar-Shalom–Campo rule consistently, and the order of the performance is reversed for the velocity estimation.

We performed similar studies by changing the stationary velocity standard deviation σ_v, and did not observe any significant effects on the performance of any of the fusion rules.

6.2.1.5.2 Complementary Sensor Case

Let us consider cases where two sensors *compensate* with each other, by letting $R_{1k} = \begin{bmatrix} \sigma_m^2 & 0 \\ 0 & 4\sigma_m^2 \end{bmatrix}$ and $R_{2k} = \begin{bmatrix} 4\sigma_m^2 & 0 \\ 0 & \sigma_m^2 \end{bmatrix}$, using the same parameters otherwise, including the Ornstein–Uhlenbeck model. Figure 6.4 shows the changes in estimation performance by several fusion rules due to the process noise intensity.

In this case, the 90° difference in the orientations of the local measurement error covariance matrices, R_{1k} and R_{2k}, is propagated into the local state estimation error covariance matrices, V_1 and V_2, and the state fusion weight matrices, W_1 and W_2, of various fusion rules. In particular, the difference in the behaviors of the fusion rules that use the inter-sensor, cross-covariance matrix V_{12} (Bar-Shalom–Campo and MV rules) and those that do not use it (tracklet, Speyer, and CI rules) becomes visible in Figure 6.4. Nonetheless, like the supplementary sensor case of Figure 6.3, the estimation performance deterioration of the four fusion rules from the centralized tracking performance remains within a very small range, i.e., 4%–5%. In Figure 6.4, we exclude

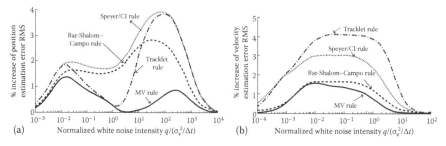

FIGURE 6.4 Percent increase of RMS estimation errors over centralized tracking performance as function of normalized process noise intensity—complementary sensors: (a) RMS position estimation error and (b) RMS velocity estimation error.

the performance of the BML rule using the likelihood in the strict Bayesian sense to prevent its bad performance from obscuring the comparison of other fusion rules.

There are apparently two peaks in the departure of distributed tracking fusion performance from the centralized tracking performance, i.e., the low q peak and the high q peak. For the position estimation performance, the tracklet rule that does not use the V_{12} exhibits better performance over the Bar-Shalom–Campo rule for very small q's (close to deterministic cases). For large q's, on the other hand, the Bar-Shalom–Campo rule using the V_{12} exhibits clear advantages over the other rules that do not use V_{12}. For the velocity estimation performance, the advantage of the Bar-Shalom–Campo rules over others (except for the MV rule) is uniform with respect to the process noise intensity q.

Unlike the supplementary sensor case (where $V_1 = V_2$), the scalar weight α in (6.18) does change the fused estimation error covariance V_F since $V_1 \neq V_2$. However, in our examples, for both supplementary and the complementary cases, since the measurement error covariance matrices are diagonal, all the state estimation (self and cross) covariance matrices are also diagonal. The minimization of the determinant of the CI fused state estimation error covariance matrix $\left(\alpha V_1^{-1} + (1-\alpha)V_2^{-1}\right)^{-1}$ is therefore the same as the minimization of its trace, and both are reduced to the maximization of $\alpha(1-\alpha)$, achieved uniquely at $\alpha = 1/2$. This makes all the CI rules identical to the Speyer rule, i.e., $W_i = \left(V_1^{-1} + V_2^{-1}\right)V_i^{-1}$, $i = 1, 2$. In other words, both Shannon and Chen–Arambel–Mehra rules become the same as the Speyer rule.

Figure 6.5 shows the sensitivity of the four algorithms to the initial state estimation accuracy, i.e., the dependence on the a priori information. Both the Bar-Shalom–Campo and the Speyer rules do not use the a priori information, and hence, only very small secondary effects are visible. Because of the sensors' difference in observability, the effects of including the cross-correlation or not are apparent. Like the case chosen for Figure 6.3, the process noise intensity and the stationary velocity covariance are kept constant at $q = 0.1\left(\sigma_v^2/\Delta t\right)$ and $\sigma_v = 3(\sigma_m/\Delta t)$. With this parameter, as shown in Figure 6.4, the tracklet rule performs better than the Bar-Shalom–Campo rule, for the position estimation, while the opposite is true for the velocity estimation.

Other tendencies are almost identical with those shown in the supplementary sensor case (Figure 6.3). Both Figures 6.4 and 6.5 exhibit the robustness of various track

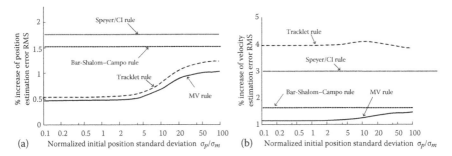

FIGURE 6.5 Percent increase of RMS estimation errors over centralized tracking performance as function of normalized initial position standard deviation—complementary sensors: (a) RMS position estimation error and (b) RMS velocity estimation error.

fusion rules, due to the changes in the key tracking parameters, i.e., the process noise intensity level and the initial state accuracy (except for the BML fusion rule), as well as Figures 6.2 and 6.3.

6.2.2 REPEATED TRACK FUSION

In the previous section, we considered a simple case where track fusion takes place only once to fuse two local state estimates. In this section, we will explore cases where communication between two sensors or to a fusion center is repeated.

Figure 6.6 shows three architectures of distributed tracking systems using two sensors that have their own independent local data processing capabilities. The two sensor systems may act as two completely autonomous systems that exchange data between them, or alternatively, report their processed data to a high level system, which we may call a fusion center. The fusion center may feed fused state estimates back to the two local sensor systems, to improve the performance of the local systems. In this section, we first consider the cases where there is no feedback, and then later, the cases with feedback.

We assume the same linear dynamics of a target to be tracked using two sensors with linear observations, as described in Section 6.2.1. For the sake of simplicity, let us consider only cases where the informational exchange happens synchronously at

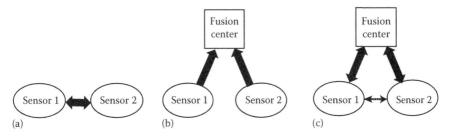

FIGURE 6.6 Three possible architectures using two sensor systems with local data processing capabilities: (a) two-autonomous-sensor distributed system, (b) two-level hierarchical distributed system, and (c) hierarchical system with feedback.

the same time, but repeatedly at t_{F1}, t_{F2},...$(t_0 < t_{F1} < t_{F2} < ...)$. The local estimates of the two sensors and their estimation error covariance matrices will be denoted as $((\hat{x}_{11}, V_{11}), (\hat{x}_{12}, V_{12}))$ at t_{F1}, $((\hat{x}_{21}, V_{21}), (\hat{x}_{22}, V_{22}))$ at t_{F2}, $((\hat{x}_{31}, V_{31}), (\hat{x}_{32}, V_{32}))$ at t_{F3}, and so forth, while the fused state estimate at each t_{Fk}, $k = 1, 2, ...$ will be denoted as $\hat{x}_{Fk}$.

For repeated-fusion cases, with or without feedback, the fusion rules $\hat{x}_{Fk} = \phi_k(\hat{x}_{11}, \hat{x}_{12}; \hat{x}_{21}, \hat{x}_{22}; ...; \hat{x}_{k1}, \hat{x}_{k2})$ (where ϕ_k is a linear or affine function because we are using a linear-Gaussian model) can be categorized as follows:

1. *Memoryless*: $\hat{x}_{Fk} = \phi_k(\hat{x}_{k1}, \hat{x}_{k2})$ uses only the most recent local estimates $(\hat{x}_{k1}, \hat{x}_{k2})$.
2. *Limited memory*: $\hat{x}_{Fk} = \phi_k(\hat{x}_{(k-\ell+1)1}, \hat{x}_{(k-\ell+1)2}; ...; \hat{x}_{k1}, \hat{x}_{k2})$ uses only the most recent ℓ pairs of local estimates.
3. *Full memory*: The full history $(\hat{x}_{11}, \hat{x}_{12}; \hat{x}_{21}, \hat{x}_{22}; ...; \hat{x}_{k1}, \hat{x}_{k2})$ of the past local estimates is used.

We may categorize the Bar-Shalom–Campo, the Speyer, and the CI rules into the memoryless fusion rules, while the MV, and the BML rules can be made to be either limited or full memory rules, and the tracklet fusion rule may become a memoryless or one-step limited memory rule.

6.2.2.1   Repeated Track Fusion without Feedback

Let us first consider the cases where each sensor subsystem maintains the local data processing only with the local data, and does not mix with data from other sensors, while fused state estimates are calculated by fusing the unmixed local estimates. The rationale for not letting the local sensor system use the fused information is that, depending on what fusion rule is used and how fused results are fed back to the local data processing system, the performance of local systems, and eventually of the overall system, may deteriorate, rather than improve, by *contamination* of the otherwise *pure* local data. This data processing can be achieved either by a hierarchical or two-autonomous-system design, as shown in Figure 6.7.

Figure 6.7 shows two information graphs (described in Chapter 5) to illustrate the information flow in track fusion without feedback. The two information graphs are equivalent to each other and describe informational transactions in a

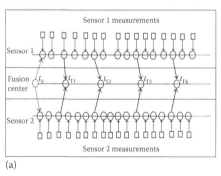

(a)

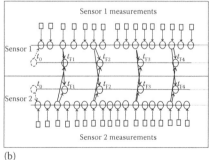

(b)

FIGURE 6.7 Information graphs for processing architectures of two-sensor track fusion without feedback: (a) two sensor and fusion center and (b) two autonomous sensors.

two-sensor-one-fusion-center system and a two-autonomous-sensor system. Dotted lines and circles with dotted lines represent a priori information, and squares represent raw sensor data that are fed into and accumulated in the local sensor data information graph nodes. The data accumulation, represented by horizontal informational flows at the same horizontal position, is represented without arrows in the graph. As shown in the graph, the same data processing can be implemented in either (a) hierarchical architecture or (b) autonomous architecture (or replicated hierarchical architecture). In the latter case, each local system maintains two state estimation filters, one local and one global. The global filter maintained by each local sensor system is sometimes called a *shadow tracker* (Drummond 1997b).

The various fusion rules introduced in Section 6.2.1 can be adapted as follows:

- *Bar-Shalom–Campo, Speyer, and CI fusion rules*: Those rules do not use the a priori information. For repeated track fusion, the a priori information at one fusion time t_{Fk} can be viewed as the information accumulated up to the previous fusion time $t_{F(k-1)}$. Thus these fusion rules ignore this a priori information, and simply combine the latest available local state estimates (i.e., memoryless fusion rules).

- *Minimum Variance (MV) Fusion Rule*: As indicated in Figure 6.7, either fusion center or the fused state filter in a local system accumulates the local estimates as $((\hat{x}_{11}, V_{11}), (\hat{x}_{12}, V_{12}), \ldots, ((\hat{x}_{(k-1)}1, V_{(k-1)}1), (\hat{x}_{(k-1)}2, V_{(k-1)}2))$, at fusion time t_F, in addition to the current pair $((\hat{x}_{k1}, V_{k1}), (\hat{x}_{k2}, V_{k2}))$. Therefore, a general linear estimate $\hat{x}_{Fk}$ of the target state $x(t_F)$ is a linear function of all the available estimates $\left((\hat{x}_{\kappa i}, V_{\kappa i})_{i=1}^2 \right)_{\kappa=1}^k$ plus the a priori information $P(x(t_0))$ (or equivalently $P(x(t_F))$) with mean $\bar{x}_k$ and covariance matrix $\bar{V}_k$. By letting $z = \left((\hat{x}_{\kappa i})_{i=1}^2 \right)_{\kappa=1}^k$ and $x = x(t_{Fk})$ in (6.13), and by calculating the covariance matrices V_{xz} and V_{zz} with augmented dimensions, the MV fusion rule can be expressed as

$$x_{Fk} = W_{k0}\bar{x}_k + \sum_{\kappa=1}^k \sum_{i=1}^2 W_{\kappa i}\hat{x}_{\kappa i} \qquad (6.26)$$

which is a full-memory fusion rule. Note that the calculation of the matrices, V_{xz} and V_{zz}, are not trivial involving many random vectors $\left((\hat{x}_{\kappa i})_{i=2}^2 \right)_{\kappa=1}^k$. Nonetheless, it can be done through a simple extension of the method described in Section 6.2.1.2. We should note that the MV fusion rule is called the *quasi-tracklet fusion method* in Gao and Li (2010).

Replacing the summation $\sum_{\kappa=1}^k$ in (6.26) by $\sum_{\kappa=k-\ell+1}^k$, we have a memoryless ($\ell = 1$) or a limited memory ($\ell \geq 1$) fusion rule. In such a case, the MV rule obtained in that way is the BLUE with respect to the weights $(W_{k0}, W_{k1}, W_{k2}, \ldots, W_{(k-\ell+1)}1, W_{(k-\ell+1)2})$ with the constraint $W_{k0} + \sum_{\kappa=k-\ell+1}^k \sum_{i=1}^2 W_{\kappa i} = I$, while the Bar-Shalom–Campo rule is the BLUE with respect only to (W_{k1}, W_{k2}) with $W_{k1} + W_{k2} = I$.

The BML fusion rule using the likelihood function in the strict Bayesian sense, defined by (6.15) in Section 6.2.1.1 for one-time track fusion, can be extended to the repeated fusion rules without feedback in exactly the same way as the MV rule. However, because of the poor performance that we found for the single-time track fusion, we will exclude the BML fusion rules from our consideration in the rest of this chapter.

6.2.2.1.1 Tracklet Fusion Rule and Decorrelation Method

The tracklet fusion rule defined by (6.10) in Section 6.2.1.1 can be directly translated into the repeated fusion as $W_{k0} = -V_{Fk}\bar{V}_k^{-1}$, $W_{k1} = V_{Fk}V_{k1}^{-1}$, and $W_{k2} = V_{Fk}V_{k2}^{-1}$. In other words, we can apply the one-time tracklet fusion rule (6.10) used to decorrelate the past fused estimate from the most recent pair of the local estimates. Without feedback to the local processing, it can be shown (cf., e.g., Chong et al.1990) that this rule can achieve the performance of the centralized tracker for deterministic target dynamics without process noise, and for non-deterministic target dynamics when the fusion rate is the same as the sensor revisit rate.

Another approach is to *decorrelate* the local estimates between the current estimates $\hat{x}_{ki}$ at the current fusion time t_{Fk}, and the previous fusion time $t_{F(k-1)}$, by rewriting Equation 6.12 as

$$\begin{cases} \tilde{V}_{ki}^{-1}z_{ki} = V_{ki}^{-1}\hat{x}_{ki} - \bar{V}_{ki}^{-1}\bar{x}_{ki} \\ \tilde{V}_{ki}^{-1} = V_{ki}^{-1} - \bar{V}_{ki}^{-1} \end{cases} \tag{6.27}$$

to obtain the decorrelated pair $(z_{ki}, \tilde{V}_{ki})$, for each sensor, $i = 1, 2$. This rule is similar to Equation 6.11 except that the local past estimates in used in decorrelation (Chong 1979). As mentioned in Section 6.2.1.1, the vector, z_{ki}, $i = 1, 2$, obtained this way, is called the *pseudo-measurement* or the *equivalent measurement*, and the measurements between the two consecutive fusion times $t_{F(k-1)}$ and t_F are often called a *tracklet*. The decorrelated pair $(z_{ki}, \tilde{V}_{ki})$, $i = 1, 2$, is then used to obtain the updated fused estimate $\hat{x}_{Fk}$, using the Kalman filter update equations

$$\begin{cases} V_{Fk}^{-1}\hat{x}_{Fk} = \bar{V}_{Fk}^{-1}\bar{x}_{Fk} + \tilde{V}_{k1}^{-1}z_{k1} + \tilde{V}_{k2}^{-1}z_{k2} \\ V_{Fk}^{-1} = \bar{V}_{Fk}^{-1} + \tilde{V}_{k1}^{-1} + \tilde{V}_{k2}^{-1} \end{cases} \tag{6.28}$$

The local prediction $(\bar{x}_{ki}, \bar{V}_{ki})$ and the global prediction $(\bar{x}_{Fk}, \bar{V}_{Fk})$ can be obtained by the extrapolation described in Section 6.2.1.1.

The tracklet fusion rule may viewed as a memoryless rule, since it uses only the most recent pair of local estimates, $(\hat{x}_{k1}, \hat{x}_{k2})$, although it uses the extrapolated a priori state mean. On the other hand, decorrelation of the local estimates uses the extrapolated pair of $(\bar{x}_{k1}, \bar{x}_{k2})$ of the last local estimates $(\hat{x}_{(k-1)1}, \hat{x}_{(k-1)2})$, as well as the extrapolation $\bar{x}_{Fk}$ of the last fused state estimate $\hat{x}_{F(k-1)}$, i.e., a fusion rule with limited (one-step) memory. However, it can be readily shown that the two methods are equivalent to each other only if the target dynamics are deterministic, i.e., $B_t \equiv 0$ in (6.2). However, when the target dynamics are not deterministic, their performance will be different. As mentioned in Section 6.2.1.1, this local estimate decorrelation fusion rule is called the *Channel filter* in Bourgault and Durrant-Whyte (2004).

6.2.2.1.2 Numerical Example of Repeated Track Fusion without Feedback

Figure 6.8 compares the performance of various track fusion rules applied to repeated-fusion-without-feedback case. We used the same simplified model defined in Section 6.2.1.5, i.e., the Ornstein–Uhlenbeck model with the time constant β and $q = 2\beta\sigma_v^2$, with a priori position standard deviation, σ_p and the velocity standard deviation σ_v. Only the complementary case with $R_{1k} = \begin{bmatrix} \sigma_m^2 & 0 \\ 0 & 4\sigma_m^2 \end{bmatrix}$ and $R_{2k} = \begin{bmatrix} 4\sigma_m^2 & 0 \\ 0 & \sigma_m^2 \end{bmatrix}$ is used.

As shown in Section 6.2.1.5, 10 local synchronous measurements are taken between two consecutive fusion times, which are repeated 5 times, at the end of which we evaluate the performance of each repeated track fusion rule by the methods described in Section 6.2.1.2. The performance is shown only for the variation of the normalized process noise intensity. When the initial positional covariance matrix (determined by σ_p) or the stationary velocity covariance matrices (determined by σ_v) is varied, virtually no sensitivity was found due to the relatively long simulation period.

Comparing Figure 6.8 with Figure 6.4 in Section 6.2.1.5, the relative trends of the various fusion rules remain the same for the positional estimation performance, while the deterioration of the velocity estimation performance when the process noises is noticeably smaller in the repeated fusion than the one-time fusion. Since this is a complementary-sensor case, the local estimation error covariance matrices are different, and hence the fusion weights of the Bar-Shalom–Campo and the Speyer rules are different, resulting in some differences in Figure 6.8. However, because of the use of the completely complementary sensors defined by constant measurement error covariance matrices, R_{k1} and R_{k2}, all the CI fusion rules become the same with $\alpha = 1/2$, as in the one-time fusion case (for Figure 6.4) and is identical to the Speyer rule.

As described in (6.26), the full-memory MV fusion rule uses an increasing number of past fused state estimates in the fusion rule as track fusion is repeated, requiring correlation among a larger number of past local estimates. To obtain Figure 6.8, we considered two cases for the number of local estimates used by the MV estimate.

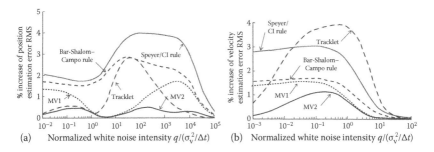

FIGURE 6.8 Percent increase of RMS estimation errors over centralized tracking performance as function of normalized process noise intensity: repeated fusion without feedback—complementary sensors: (a) RMS position estimation error and (b) RMS velocity estimation error.

Case 1 (MV1) uses only the most recent local estimate for each sensor and the fused estimates is $\hat{x}_{Fk} = W_{k0}\bar{x}_k + \sum\limits_{i=2}^{2} W_{ki}\hat{x}_{ki}$ for each sensor and the fused estimate is

$\hat{x}_{Fk} = W_{k0}\bar{x}_k + \sum\limits_{i=2}^{2} \left(W_{ki}\hat{x}_{ki} + W_{(k-1)i}\hat{x}_{(k-1)i} \right)$. As mentioned earlier, the MV fusion rules outperform any other fusion rules that were compared. We should also note that, except for the MV2 fusion rule, all the fusion rules do not exhibit the convergence to the performance of the centralized tracking when the system approaches the deterministic dynamics, i.e., $\beta \downarrow 0$, which may be a general indication of a potential instability associated with repeated fusion without feedback. Nonetheless, as seen in Figure 6.8, the performance deterioration of distributed tracking from centralized tracking by various fusion rules remains relatively very small, i.e., 1%–5%, over a very wide range of the process noise intensity level q. In particular, the performance by the relatively simple fusion rules, i.e., Bar-Shalom–Campo, Speyer, and CI, is found to be very robust.

The tracklet rule shown in Figure 6.8 is in its decorrelation form, which is widely used to decorrelate a sequence of up-stream trackers' outputs that are input into a fusion engine that fuses tracking information from multiple sources, given in terms of state estimates rather than raw sensor measurements. The decorrelation form of the tracklet rule is defined by (6.27) and (6.28). This tracklet fusion rule is practical because it does require inter-sensor local target state estimation error covariance, and the result in Figure 6.8 justifies its use in the cases where fused state estimates are not fed back the local tracks.

6.2.2.2 Repeated Track Fusion with Feedback

It is rather intuitive to expect better state estimation performance, both local and global, by feeding back the fused state estimates to the local tracking agents. However, even using linear models, i.e., a rather idealized version of generally nonlinear real-world systems, such expectation may not be realized, depending on what fusion rule is used. This is the case because, although some fusion rules may perform reasonably well for state estimates at fusion times as shown in Chang et al. (2002), they may declare wrong, generally unreasonably optimistic estimation error covariance matrices, thereby contaminating the performance of the local trackers. This may cause secondary effects such as contamination of the fused state estimates generated later in by local tracking agents and subsequent deterioration of the overall performance. For this reason, repeated fusion without feedback may be preferred in many practical cases.

Repeated track fusion with feedback can be illustrated by the information graph shown in Figure 6.9.

In this figure, feedback is represented by those from the fusion center to the local processors in (a) two-local-sensor-one-fusion-center architecture and by arrows that connect local processing information graph nodes directly in (b) two-autonomous-sensor architecture. In (b), two-autonomous-sensor distributed architecture, each local processing node sends its current state estimate at an agreed upon fusion time to the other node, and upon the receipt of the state estimate from the other node,

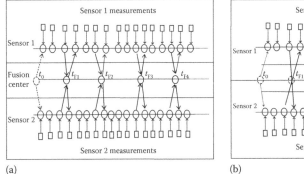
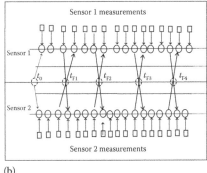

(a)   (b)

FIGURE 6.9 Information graphs for processing architectures of two-sensor track fusion with feedback: (a) two sensor and fusion center and (b) two autonomous sensors.

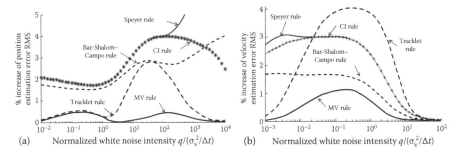

(a)   (b)

FIGURE 6.10 Percent increase of RMS estimation errors over centralized tracking performance as function of normalized process noise intensity: repeated fusion with feedback—complementary sensors: (a) RMS position estimation error and (b) RMS velocity estimation error.

fuses the local and remote state estimates into the global estimate at that moment. Then, until the next fusion time, each local sensor processes only its local data.

Figure 6.10 shows the performance of the various fusion rules adapted to repeated track fusion with feedback, using the same complementary-sensor model used to compare the performance of Figure 6.8. The adaptations are shown below.

6.2.2.2.1 Bar-Shalom–Campo, Speyer, and CI Rules

The same fusion rules are used but the local state estimates and estimation error covariance matrices modified by feedback are used. All the covariance matrices are diagonal due to the use of the same diagonal measurement error covariance matrices R_{ki}'s. Hence, all the CI rules become the same with $\alpha = 1/2$ as shown earlier. However, although the Bar-Shalom–Campo fusion rule provides the honest fused state estimation error covariance, neither the Speyer nor the CI fusion rule does. The Speyer rule ignores the cross-covariance and results in generally optimistic estimation error covariance matrices, whereas the CI rules generally produce grossly pessimistic estimation error covariance matrices, with a typical determinant about four times bigger than that of the actual estimation error, both contaminating the local trackers' performance.

6.2.2.2.2   Tracklet Rule

In this fusion rule, $\hat{x}_{Fk} = \hat{V}_{Fk}(V_{k1}^{-1}\hat{x}_{k1} + V_{k2}^{-1}\hat{x}_{k2} - \bar{V}_{Fk}^{-1}\bar{x}_{Fk})^{-1}$, the a priori global state estimation pair, $(\bar{x}_{Fk}, \bar{V}_{Fk})$, obtained by extrapolating the last fusion result $(\bar{x}_{F(k-1)}, \bar{V}_{F(k-1)})$, is used to eliminate the redundant information contained by the two local state estimates through the feedback and remove double counting. Nonetheless, the declared fused estimation error covariance matrix $\hat{V}_{Fk} = (V_{k1}^{-1} + V_{k2}^{-1} - \bar{V}_{Fk}^{-1})^{-1}$ is not honest and generally optimistic, thus contaminating the local sensor data processing.

6.2.2.2.3   MV Fusion Rule

The MV1 rule as defined in Section 6.2.1.1, which only uses the most recent local estimates as $\hat{x}_{Fk} = W_{k0}\bar{x}_k + \sum_{i=1}^{2} W_{ki}\hat{x}_{ki}$, is used because the most recent estimates contain all the significant updates by the local agents due to feedback of the fused estimation results to the local agents. Any version of the MV fusion rules based on the BLUE principle generates honest estimation error covariance matrices, and hence there will be no contamination propagated through fusion and its feedback to the local processors.

6.2.2.2.4   Numerical Example of Repeated Track Fusion with Feedback

The same simplified linear models with synchronized complementary sensors as those to produce Figure 6.8 are used for Figure 6.10.

   The behavior of the tracklet fusion rule is much more stable than that in fusion without feedback and behaves as a good approximation of the MV rule, which we may consider an *almost optimal distributed fusion algorithm,* as far as positional estimation is concerned. As observed in Figures 6.2 through 6.5, 6.8 and 6.10, however, the simpler rules, i.e., Bar-Shalom–Campo, Speyer, and CI, may provide better velocity estimation performance for a range of the process noise intensity levels q. The MV rule may be improved more by considering linear optimal estimate using longer length of memory. On the other hand, the Bar-Shalom–Campo, the Speyer, and the CI fusion rules do not use the a priori information. In the fusion with feedback case, we can see its consequences in Figure 6.10, although all the variations are within a relatively small margin, i.e., 5%. Therefore, we see again the robustness of the simple fusion rules, despite concerns about the use of information that may be much less than information available at each fusion time, and about the contamination of the local tracker by not honest (either pessimistic or optimistic) state fusion estimation error covariance matrices.

6.3   TRACK ASSOCIATION

Track association is a prerequisite for track fusion in a distributed tracking system. However, in many cases, association is rather obvious, and therefore target state estimation from multiple sensors or track fusion becomes the major problem. On the other hand, when the target density is high, track association becomes a much more important problem than track fusion. As the track density becomes even higher, track association and track fusion can no longer be treated as separate problems. In that case, many local track association hypotheses are possible and equally likely, so that the best local association hypothesis may not provide high-quality tracks for association at the fusion site. Then some form of distributed multiple hypothesis tracking is needed (Chong et al. [1990], Dunham et al. [2004]).

In this section, we treat the situation where the local tracks are good enough so that distributed tracking can be viewed as a two-stage problem, i.e., track association followed by track fusion. The concept of distributed tracking in terms of track association was developed shortly after target tracking started to be investigated with modern estimation theory or filtering theory. Early work includes Singer and Kanyuck (1971) and Yaakov Bar-Shalom (1981).

6.3.1 TRACK ASSOCIATION PROBLEM DEFINITION

We use the same linear-Gaussian model described in Section 6.2.1. We assume, however, a fixed number n of "true" targets represented by the system $\left((x_i(t))_{t \in [t_0, \infty)}\right)_{i=1}^n$ of n replicated stochastic processes on the time interval $[t_0, \infty)$ with the joint initial condition $\Pr ob\left\{x_1(t_0) \in dx_1(t_0),\ldots,x_n(t_0) \in dx_n(t_0)\right\} = \prod_{i=1}^n g\left(x_i(t_0) - \overline{x}_0; V_0\right)dx_i(t_0)$.

Each individual stochastic process $\left(x_i(t)\right)_{t \in [t_0, \infty)}$, $i = 1,\ldots,n$, is defined as in Section 6.2.1, with a system $\left((\dot{w}_i(t))_{t \in [t_0, \infty)}\right)_{i=1}^n$ of white noises, or equivalently Wiener processes $\left((w_i(t))_{t \in [t_0, \infty)}\right)_{i=1}^n$. The target density can be measured by $\gamma_0(x) = ng(x - \overline{x}_0; V_0)$ so that for any measurable subset B in the target state space E the integral $\int_B \gamma_0(x)dx$ is the expected number of targets whose initial condition $x_i(t_0)$ is in the set B.

Instead of assuming the number n of targets to be a known constant, we may assume that n is a random variable. When n is a Poisson random variable, the system $\left(x_i(t_0)\right)_{i=1}^n$ of random vectors in E is a *Poisson point process*. We maintain the constant n assumption for this chapter for the sake of simplicity, because the main purpose of this chapter is to compare various track association metrics.

We assume the following scenario: All the targets are visible by each of the two sensors, i.e., we assume the detection probability (for the local track level) by each sensor is unity. We also assume that there are *no false tracks*. The last assumption is supported by the fact that any track made up solely of false alarms would have been weeded out by the local sensor's tracking. Thus we have n targets that are observed by two sensors, which produces n local tracks through N_i measurements, $i = 1, 2$, prior to a fusion time t_F. Then our goal is to associate two sets of local tracks represented by the n-tuple of state estimates, $\left(\hat{x}_{1i}(t_F)\right)_{i=1}^n$, and $\left(\hat{x}_{2i}(t_F)\right)_{i=1}^n$, at the fusion time t_F, where each estimate $\hat{x}_{ij}$ is associated with the estimation error covariance matrix V_{ij}.

Uncertainty of the association between the true targets and the set of tracks from each sensor, $i = 1, 2$, can be modeled by two independent assignment functions a_i that is a permutation on the set $\{1,\ldots, n\}$. The association hypothesis between the two sets of local tracks is then expressed as $a(i) = \left(a_2 \cdot a_1^{-1}\right)(i)$, i.e., the i-th track from sensor 1 and the $a(i)$-th track from sensor 2 share the same origin. The problem is then the determination of the most likely or most probable association according to an evaluation function that has the general form

$$P(a) = C^{-1} \prod_{i=1}^{n} \ell(i, a(i)) \qquad (6.29)$$

where
  C is the normalizing constant
  $\ell(i, j)$ is the *likelihood* of track i from sensor 1 that shares the same origin (target) as track j from sensor 2

Under an appropriate set of assumptions mentioned earlier, (6.29) becomes the a posteriori probability of the association a conditioned by the set of state estimates of all the tracks from both sensors, with the normalizing constant C. However, in this chapter, we consider (6.29) as the expression that relates the association hypothesis evaluation function to the track association metrics represented by the likelihood function $\ell(i, j)$ or its half negative logarithm, $L(i, j) = -(1/2)\ln(\ell(i, j))$.

6.3.2 TRACK ASSOCIATION METRICS

Using the negative half logarithm $L(i, j) = -(1/2)\ln(\ell(i, j))$, the optimal track association $\hat{a}$ is obtained by minimizing the association cost $f(a) = \sum_{i=1}^{n} L(i, a(i))$. By the track association metrics, we mean the metrics that represent the cost $L(i, j)$ for associating the ith track from sensor 1 and the jth track from sensor 2. Some of the metrics in the following list were originally developed as the metric to be used in the classical chi-square test, but can be considered as an association metric because of its structure.

Singer–Kanyuck metric: In a pioneering paper (Singer and Kanyuck 1971), the usual chi-square metric

$$L(i, j) = \left\| \hat{x}_{1i} - \hat{x}_{2j} \right\|^2_{(V_{1i} + V_{2j})^{-1}} = (\hat{x}_{1i} - \hat{x}_{2j})^T (V_{1i} + V_{2j})^{-1} (\hat{x}_{1i} - \hat{x}_{2j}) \qquad (6.30)$$

is proposed. This metric can be interpreted as the negative half logarithm of

$$\ell(i, j) = \int_E \hat{p}_{1i}(x) \, \hat{p}_{2j}(x) \, dx = \int_E g(x - \hat{x}_{1i}; V_{1i}) g(x - \hat{x}_{2j}; V_{2j}) \, dx = g(x_{1i} - \hat{x}_{2j}; V_{1i} + V_{2j})$$

$$(6.31)$$

when we eliminate the factor $\det(2\pi(V_{1i} + V_{2j}))^{-1/2}$, or its negative half logarithm $\ln(\det(2\pi(V_{1i} + V_{2j})))$, as a constant that appears in the metrics for the other pairs.

Bar-Shalom metric: Yaakov Bar-Shalom proposed the metric

$$L(i,j) = \left\| \hat{x}_{1i} - \hat{x}_{2j} \right\|^2_{\left(V_{1i}+V_{2j}-V_{12ij}-V_{12ij}^T\right)^{-1}} \tag{6.32}$$

in Bar-Shalom (1981) to be used also in a chi-square test for the track association, where V_{12ij} is the cross-covariance between two tracks, track i from sensor 1 and track j from sensor 2, obtained assuming that they originate from the same target. This metric can be interpreted as the negative half logarithm of

$$\ell(i,j) = \int_E g\left(\begin{bmatrix} x - \hat{x}_{1i} \\ x - \hat{x}_{2j} \end{bmatrix}; \begin{bmatrix} V_{1i} & V_{12ij} \\ V_{12ij}^T & V_{2j} \end{bmatrix}\right) dx = g\left(\hat{x}_{1i} - \hat{x}_{2j}; V_{1i} + V_{2j} - V_{12ij} - V_{12ij}^T\right) \tag{6.33}$$

when we ignore the factor $\det\left(2\pi\left(V_{1i} + V_{2j} - V_{12ij} - V_{12ij}^T\right)\right)^{-1/2}$, or its negative half logarithm $\ln\left(\det\left(2\pi\left(V_{1i} + V_{2j} - V_{12ij} - V_{12ij}^T\right)\right)\right)$, as a constant that is to be canceled out.

CI metric: To the best of our knowledge, there is no track association metric based on the CI principle. However, based on the observation on the two metrics described earlier, and on the definition of the CI fusion (6.19), an appropriate track association metric may be defined as

$$L(i,j) = \left\| \hat{x}_{1i} - \hat{x}_{2j} \right\|^2_{\left(\hat{\alpha}_{ij}^{-1}V_{1i}+\left(1-\hat{\alpha}_{ij}\right)^{-1}V_{2j}\right)^{-1}} \tag{6.34}$$

This metric can be interpreted as the negative half logarithm of

$$\ell(i,j) = \int_E \hat{p}_{1i}(x)^{\hat{\alpha}_{ij}} \hat{p}_{2j}(x)^{(1-\hat{\alpha}_{ij})} dx = \int_E g(x - \hat{x}_{1i}; V_{1i})^{\hat{\alpha}_{ij}} g(x - \hat{x}_{2j}; V_{2j})^{(1-\hat{\alpha}_{ij})} dx \tag{6.35}$$

when we ignore the factor $\left(\dfrac{\det(V_{1i})^{(1-\hat{\alpha}_{ij})} \det(V_{2j})^{\hat{\alpha}_{ij}}}{\det\left(\left(1-\hat{\alpha}_{ij}\right)V_{1i} + \hat{\alpha}_{ij}V_{2j}\right)}\right)^{1/2}$, or its negative half logarithm, as a constant that is to be canceled out. The "optimal" weight $\hat{\alpha}_{ij} \in [0,1]$ may be chosen as the one that either maximizes the determinant $\det\left(\alpha V_{1i}^{-1} + (1-\alpha)V_{2j}^{-1}\right)$ (corresponding to the Shannon fusion rule), or minimizes trace $\left(\left(\alpha V_{1i}^{-1} + (1-\alpha)V_{2j}^{-1}\right)^{-1}\right)$ (corresponding to the Chen–Arambel–Mehra fusion rule).

Chong–Mori–Chang metric: Under the assumption that there are no false tracks and missed tracks for the two-sensor track-to-track association, we can show that the Bayesian track association hypothesis evaluation formula is expressed by (6.29) using the track association likelihood given by

$$\ell(i,j) = \int_E \frac{\hat{p}_{1i}(x)\,\hat{p}_{2j}(x)}{\bar{p}(x)}\,dx = \int_E \frac{g\left(x - \hat{x}_{1i}; V_{1i}\right)g\left(x - \hat{x}_{2j}; V_{2j}\right)}{g\left(x - \bar{x}; \bar{V}\right)}\,dx$$

$$= \left(\frac{\det\left(\bar{V}\right)\det\left(\hat{V}_{Fij}\right)}{\det\left(V_{1i}\right)\det\left(V_{2j}\right)}\right)^{1/2}\exp\left(-\frac{1}{2}\left(\left\|\hat{x}_{Fij} - \hat{x}_{1i}\right\|_{V_{1i}^{-1}}^2 + \left\|\hat{x}_F - \hat{x}_{2j}\right\|_{V_{2j}^{-1}}^2 - \left\|\hat{x}_F - \bar{x}\right\|_{\bar{V}^{-1}}^2\right)\right)$$

$$(6.36)$$

where

$\bar{p}(x) = g(x - \bar{x};\,\bar{V})$ is the a priori probability density of the target state $x = x(t_F)$ at the fusion time t_F

$(\hat{x}_{Fij},\,\hat{V}_{Fij})$ is the pair of the fused state estimate and the estimation error covariance matrix, obtained by the track fusion rule, defined by (6.15) in Section 6.2.1.1, i.e.,

$$\begin{cases} \hat{V}_{Fij}^{-1}\hat{x}_{Fij} = V_{1i}^{-1}\hat{x}_{1i} + V_{2j}^{-1}\hat{x}_{2j} - \bar{V}^{-1}\bar{x} \\ \hat{V}_{Fij}^{-1} = V_{1i}^{-1} + V_{2j}^{-1} - \bar{V}^{-1} \end{cases} \tag{6.37}$$

all under the hypothesis that the i-th track from sensor 1 and the j-th track from sensor 2 originate from the same target. Unfortunately, like the tracklet fusion rule (6.15), the last statement is true only when the target dynamics are deterministic, i.e., there is no process noise ($B_t = 0$). Nonetheless, like the tracklet fusion rule, combining with the nondeterministic extrapolation formula, the track association metric of (6.36) can be adapted to the nondeterministic cases by combining with the nondeterministic extrapolation formula. By eliminating the four determinant factors from (6.36) as the factors that can be canceled out, the negative half logarithm of the track likelihood becomes

$$L(i,j) = \left\|\hat{x}_F - \hat{x}_{1i}\right\|_{V_{1i}^{-1}}^2 + \left\|\hat{x}_F - \hat{x}_{2j}\right\|_{V_{2j}^{-1}}^2 - \left\|\hat{x}_F - \bar{x}\right\|_{\bar{V}^{-1}}^2 \tag{6.38}$$

Expanded State metric: This metric is obtained by expanding the target state from the state $x(t_F)$ at the fusion time t_F to the states at multiple times, $(t_1, t_2, \ldots, t_n)$, within the time interval $[t_0, t_F]$. If this set $(t_1, t_2, \ldots, t_n)$ covers all the measurement times by both sensors, we can reformulate the nondeterministic problem defined in Section 6.2.1, as a *static state problem* in which the "static" states are $(x(t_1), \ldots, x(t_n))$ instead of $x(t_F)$. In this way, all the uncertainty generated by the process noise is translated into the cross-covariance among the target states at different times. Then the track association hypothesis evaluation formula (6.29) using the Chong–Mori–Chang metric (6.36) becomes truly the conditional probability of each association hypothesis in the Bayesian sense, from which we can obtain the MAP probability track association hypothesis by solving the classical bipartite assignment problem.

Remarks: Strictly speaking, the use of (6.29) is justified only when the number of targets is known, i.e., when there is no missed target and there are no false tracks.

When missed targets are possible, we may have unpaired local tracks. In such a case, as shown in Mori and Chong (2003) and Ferry (2010), when there may be unpaired local tracks, each track-to-track association must be adjusted according to the estimate of the target density, and when the a priori number of targets is not Poisson, the constant C in (6.19) may depend on the number of paired and unpaired local tracks. The cases where there may be false tracks are theoretically more complicated. We can find a proposal of track-to-track association metric used in such a case in Blackman and Popoli (1999), and a recent theoretical treatment can be found in Mori et al. (2009).

The sensor biases and the track association are closely related, and may not be separable in some cases. In such a case, the track association metric in (6.29) can be modified by the sensor bias probability distribution, as shown in Levedahl (2002), Mori and Chong (2007), and Ferry (2010).

6.3.3 COMPARISON OF TRACK ASSOCIATION METRICS

In order to compare the various track association metrics described in Section 6.3.2, we will examine the track association performance using the evaluation function (6.29) with different track association metrics. A simple linear model, using the Ornstein–Uhlenbeck target dynamics and two complementary sensors described in Section 6.2.1.5, is used for this purpose. The complementary sensor case was chosen to mimic a situation where each local sensor is able to separate the targets relatively well into a set of high-quality local tracks, but there is still significant association uncertainty between the local tracks from both sensors, as illustrated in Figure 6.11.

Figure 6.12 shows the result of this comparison. Unlike the track fusion performance analysis of Section 6.2, there is no obvious analytical method of predicting the track association performance by any of the association metrics described earlier. Therefore, Monte Carlo analysis was conducted. In each run, a random set of 100 targets was generated according to the model described in Section 6.3.1, assuming synchronous observation with the same number of 10 local measurements for each track. The initial position uncertainty standard deviation is 10 times as big as the measurement error, i.e., $\sigma_P = 10\sigma_m$. The figure shows a comparison

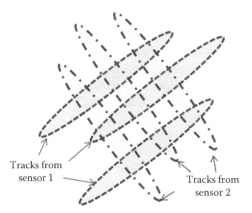

FIGURE 6.11 Local tracks from two complementary sensors.

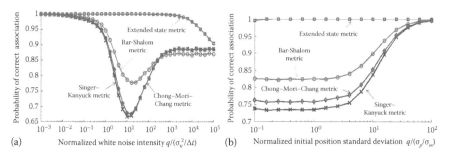

FIGURE 6.12 Track association performance comparison: (a) track association performance as function of normalized process noise intensity and (b) track association performance as function of normalized initial position standard deviation.

of association performance by (1) the Bar-Shalom metric, (2) the Singer–Kanyuck metric, (3) Chong–Mori–Chang metric, and (4) the extended state metric, varying (a) the normalized process noise intensity q and (b) the initial position uncertainty standard deviation σ_p. The complementary sensor case with 90° different sensor measurement error covariance matrices is used for this comparison, resulting in the equal weights for the CI fusion rule, i.e., $\alpha = 1/2$, in (6.18). The corresponding CI track association metric is defined by (6.34) with equal weight $\alpha_{ij} \equiv 1/2$. This weight makes the CI association metric the same as the Singer–Kanyuck metric.

For each run, we examined each target to see whether the tracks originating from that target are correctly associated or not. Then the probability of correct association, as defined as the probability of each track from sensor 1 being assigned to the "correct" track from sensor 2 ("correct" as indicated by the ground truth), was calculated as the number of correctly associated targets over the total number of targets. Each point in the figure was obtained by averaging 1000 samples.

In Figure 6.12a and b, the advantage of using the inter-sensor cross-covariance in the association metric is clearly shown by the better performance of the Bar-Shalom metric over the Singer–Kanyuck or the Chong–Mori–Chang metric that does not use the cross-covariance matrix. The deterioration of the association performance for the middle range of the process noise intensity can be explained by its effect on the joint target state density, shown in Figure 6.1c in Section 6.2.1.5. The use of the a priori state mean by the Chong–Mori–Chang metric results in better association performance by that using the Singer–Kanyuck metric, but the difference is rather small because the 10 local measurements may lessen the effect of the initial condition. The association performance using either of metrics is worse than that of the Bar-Shalom metric that uses the cross-covariance. In almost all situations, the extended state metric exhibits much better association performance than the other association metrics because it considers the state estimates at multiple times, and not just at the fusion time.

6.4 CONCLUSIONS

This chapter has addressed the track fusion and association problems in distributed multiple-target tracking. We have reviewed several track fusion algorithms developed over the last three decades and compared their performance. The use of linear-Gaussian models allows closed form analytical performance evaluation. Simple but realistic target dynamics with the Ornstein–Uhlenbeck model were used to compare the various track fusion rules for one-time fusion, and repeated fusion cases, with and without feedback of the global fused target state estimates to the local tracking agents. Our analysis indicates that even though some fusion rules perform slightly better than others depending on the situation, the performance of the more common fusion rules such as speyer, minimum variance (MV) or BLUE, Bar-Shalom Campo, decorrelation, is only slightly worse (<5%) than that of centralized tracking. The choice of the appropriate fusion rule should depend on factors such as communication requirements, implementation difficulty, and robustness.

Various track association metrics were compared with respect to track association performance for a simple one-time track fusion. For the complementary sensor case, we confirmed clearly better track association performance of the Bar-Shalom metric that considers the cross-covariance between two local tracks hypothesized to originate from the same target, over the Singer–Kanyuck metric or the Chong–Mori–Chang metric that does not use such cross-covariance information. At the same time, the extended state vector for track association, which requires more data and computation, exhibits much better track association performance than any other association metrics. This is not surprising because association tracks with only the state estimates are difficult when the target is maneuvering. An approximate extended state track association metric may be desirable in the case of highly nondeterministic target dynamics.

REFERENCES

Anderson, B. D. O. and J. B. Moore. 1979. *Optimal Filtering*. Englewood Cliffs, NJ: Prentice-Hall.

Bar-Shalom, Y. 1981. On the track-to-track correlation problems. *IEEE Transactions on Automatic Control* AC 26(2): 571–572.

Bar-Shalom, Y. and L. Campo. 1986. The effect of the common process noise on the two-sensor fused-track covariance. *IEEE Transactions on Aerospace and Electronic Systems* AES 22(6): 803–805.

Bar-Shalom, Y. and T. E. Fortmann. 1988. *Tracking and Data Association*. San Diego, CA: Academic Press.

Bar-Shalom, Y. and X. R. Li. 1993. *Estimation and Tracking: Principles, Techniques and Software*. Dedham, MA: Artech House.

Bar-Shalom, Y., X. R. Li, and T. Kirubarajan. 2001. *Estimation with Applications to Tracking and Navigation: Theory, Algorithms, and Software*. New York: John Wiley & Sons.

Bar-Shalom, Y., P. K. Willet, and X. Tian. 2011. *Tracking and Data Fusion: A Handbook of Algorithms*. Storrs, CT: YBS Publishing.

Belkin, B., S. L. Anderson, and K. M. Sommar. 1993. The pseudomeasurement approach to track-to-track data fusion. *Proceedings of the 1993 Joint Service Data Fusion Symposium*, Laurel, MD, pp. 519–538.

Blackman, S. S. 1986. *Multiple-Target Tracking with Radar Application*. Norwood, MA: Artech House.

Blackman, S. S. and R. Popoli. 1999. *Design and Analysis of Modern Tracking Systems*. Norwood, MA: Artech House.

Bourgault, F. and H. F. Durrant-Whyte. 2004. Communication in general decentralized filters and the coordinated search strategy. *Proceedings of the 7th International Conference on Information Fusion*, Stockholm, Sweden, pp. 723–730.

Chang, K. C., C.-Y. Chong, and S. Mori. 2008. On scalable distributed sensor fusion. *Proceedings of the 11th International Conference on Information Fusion*, Cologne, Germany.

Chang, K. C., R. K. Saha, and Y. Bar-Shalom. 1997. On optimal track-to-track fusion. *IEEE Transactions on Aerospace and Electronic Systems* 33(4): 1271–1276.

Chang, K. C., Z. Tian, and R. K. Saha. 2002. Performance evaluation of track fusion with information matrix filter. *IEEE Transactions on Aerospace and Electronic Systems* 38(2): 455–466.

Chen, L., P. O. Arambel, and R. K. Mehra. 2002. Estimation under unknown correlation: Covariance intersection revised. *IEEE Transactions on Automatic Control* 47(11): 1879–1882.

Chong, C. Y. 1979. Hierarchical estimation. *Proceedings of the MIT/ONR Workshop on C3*, Monterey, CA.

Chong, C. Y., E. Tse, and S. Mori. 1983. Distributed estimation in network. *Proceedings of the 83 American Control Conference*, San Francisco, CA.

Chong, C. Y. and S. Mori. 2001. Convex combination and covariance intersection algorithms in distributed fusion. *Proceedings of the 4th International Conference in Information Fusion*, Montréal, Québec, Canada.

Chong, C.-Y., S. Mori, and K. C. Chang. 1985. Information fusion in distributed sensor networks. *Proceedings of the 1985 American Control Conference*, Boston, MA, pp. 830–835.

Chong, C. Y., K. C. Chang, and S. Mori. 1986. Distributed tracking in distributed sensor networks. *Proceedings of the 1986 American Control Conference*, Seattle, WA.

Chong, C. Y., S. Mori, and K. C. Chang. 1987. Adaptive distributed estimation. *Proceedings of the 26th IEEE Conference on Decision and Control*, Los Angeles, CA, pp. 2233–2238.

Chong, C. Y., S. Mori, and K. C. Chang. 1990. Distributed multitarget multisensor tracking (Chapter 8). In *Multitarget-Multisensor Tracking: Advanced Applications*. Y. Bar-Shalom (Ed.), pp. 247–295, Norwood, MA: Artech House.

Chong, C. Y., S. Mori, K. C. Chang, and W. H. Barker. 2000. Architectures and algorithms for track association and fusion. *IEEE Aerospace and Electronic Systems Magazine* 15: 5–13.

Cover, T. M. and J. A. Thomas. 2006. *Elements of Information Theory*. New York: John Wiley & Sons.

Drummond, O. E. 1996. Track fusion with feedback. *Proceedings of the SPIE Symposium on Sensor and Data Processing of Small Targets*, Vol. 2759, Orlando, FL, pp. 342–360.

Drummond, O. E. 1997a. A hybrid sensor fusion algorithm architecture and tracklets. *Proceedings of the SPIE Symposium on Signal and Data Processing of Small Targets*, Vol. 3163, San Diego, CA.

Drummond, O. E. 1997b. Tracklets and a hybrid fusion with process noise. *Proceedings of the SPIE Symposium on Signal and Data Processing of Small Targets*, Vol. 3163, San Diego, CA.

Dunham, D. T., S. S. Blackman, and R. J. Dempster. 2004. Multiple hypothesis tracking for a distributed multiple platform system. *Proceedings of the SPIE Symposium on Signal and Data Processing of Small Targets*, Orlando, FL, pp. 13–15.

Durrant-Whyte, H. F., B. S. Y. Rao, and H. Hu. 1990. Toward a fully decentralized architecture for multi-sensor data fusion. *Proceedings of the IEEE International Conference on Robotic Automation*, Cincinnati, OH, pp. 1331–1336.

Ferry, J. P. 2010. Exact association probability for data with bias and feature. *Journal of Advances in Information Fusion* 5(1): 41–66.

Gao, Y. and X. R. Li. 2010. Quasi-tracklet fusion accounting for cross-correlation. *Proceedings of the 13th International Conference on Information Fusion*, Edinburg, U.K.

Govaers, F. and W. Koch. 2010. Distributed Kalman filter fusion at arbitrary instants of time. *Proceedings of the 13th International Conference on Information Fusion*, Edinburg, U.K.

Govaers, F. and W. Koch. 2011. On the globalized likelihood function for exact track-to-track fusion at arbitrary instants of time. *Proceedings of the 14th International Conference on Information Fusion*, Chicago, IL.

Hashemipour, H. R., S. Roy, and A. J. Laub. 1988. Decentralized structures for parallel Kalman filtering. *IEEE Transactions on Automatic Control* AC 33: 88–93.

Hurley, M. 2002. An information-theoretic justification for covariance intersection and its generalization. *Proceedings of the 5th International Conference on Information Fusion*, Annapolis, MD.

Iyengar, S. S. and R. R. Brook, Eds. 2005. *Distributed Sensor Networks*. Chapman &Hall CRC Computer & Information Science Series. Boca Raton, FL: CRC Press.

Julier, S. J. 2006. An empirical study into the use of Chernoff information for robust, distributed fusion of Gaussian mixture models. *Proceedings of the 8th International Conference on Information Fusion*, Florence, Italy.

Julier, S. J., J. K. Uhlmann, J. Walters et al. 2006. The challenge of scalable and distributed fusion of disparate sources of information. *Proceedings SPIE Conference on Multisensor, Multisource Information Fusion: Architectures, Algorithms, and Applications*, Vol. 6242, Orlando, FL.

Kalman, R. E. 1960. A new approach to linear filtering and prediction problems. *Transactions of ASME—Journal of Basic Engineering, Series D* 82: 35–45.

Kalman, R. E. and R. S. Bucy. 1960. New results in linear filtering and prediction theory. *Transactions of ASME—Journal of Basic Engineering*, 83: 95–108.

Koch, W. 2008. On optimal distributed Kalman filtering and retrodiction at arbitrary communication rates for maneuvering targets. *Proceedings of the IEEE International Conference on Multisensor Fusion and Integration for Intelligence Systems (MFI 2008)*, Seoul, Korea, pp. 457–462.

Koch, W. 2009. Exact update formulae for distributed Kalman filtering and retrodiction at arbitrary communication rates. *Proceedings of the 12th International Conference on Information Fusion*, Seattle, WA, pp. 2209–2216.

Levedahl, M. 2002. An explicit pattern matching assignment algorithm. *Proceedings of SPIE Symposium on Signal and Data Processing of Small Targets*, Vol. 4728, Orlando, FL.

Li, X. R., Y. Zhu, J. Wang et al. 2003. Optimal linear estimation fusion—Part I: Unified fusion rules. *IEEE Transactions on Information Theory* 49: 9.

Liggins II, M. E. and K. C. Chang. 2009. Distributed fusion architectures, algorithms, and performance within a network-centric architecture. In *Handbook of Multisensor Data Fusion: Theory and Practice*. M. E. Liggins, D. H. Hall, and J. Llinas (Eds.), Boca Raton, FL: CRC Press.

Liggins II, M. E., C.-Y. Chong, I. Kadar et al. 1997. Distributed fusion architecture and algorithms for target tracking. *Proceedings of the IEEE* 85: 95–107.

Lobbia, R. and M. Kent. 1994. Data fusion of decentralized tracker outputs. *IEEE Transactions on Aerospace and Electronic Systems* 30: 787–799.

Miller, M. D., O. E. Drummond, and A. J. Perrella. 1998. Tracklets and covariance truncation options for theater missile tracking. *Proceedings of the 1st International Conference on Multisource-Multisensor Data Fusion*, Las Vegas, NV.

Moore, J. R. and W. D. Blaire. 2000. Practical aspects of multisensor tracking (Chapter 1). *Multitarget-Multisensor Tracking—Applications and Advances*, Vol. III. Y. Bar-Shalom and W. D. Blair (Eds.), Boston, MA: Artech House.

Mori, S., W. H. Barker, C.-Y. Chong, and K.-C. Chang. 2002. Track association and track fusion with non-deterministic target dynamics. *IEEE Transaction on Aerospace and Electronic Systems* 38(2): 659–668.

Mori, S. and C.-Y. Chong. 2003. Track-to-track association metric. *Proceedings of the 6th International Conference on Information Fusion*, Cairns, Queensland, Australia.

Mori, S. and C.-Y. Chong. 2007. Comparison of bias removal algorithms in track-to-track association. *Proceedings of SPIE Symposium on Signal and Data Processing of Small Targets*, Vol. 6699, San Diego, CA.

Mori, S., C.-Y. Chong, and K. C. Chang. 2009. Track association and fusion using Janossy measure density functions. *Proceedings of the 12th International Conference on Information Fusion*, Seattle, WA.

Nicholson, D., S. J. Julier, and J. K. Uhlmann. 2001. DDF: An evaluation of covariance intersection. *Proceedings of the 4th International Conference on Information Fusion*, Montreal, Quebec, Canada.

Nicholson, D., C. M. Lloyd, S. J. Julier et al. 2002. Scalable distributed data fusion. *Proceedings of the Fifth International Conference on Information Fusion*, Annapolis, MD, pp. 630–635.

Rao, B. S. Y., H. F. Durrant-Whyte, and J. A. Sheen. 1993. Fully decentralized multi-sensor system for tracking and surveillance. *International Journal of Robotics Research* 12(1): 20–44.

Rhodes, I. B. 1971. A tutorial introduction to estimation and filtering. *IEEE Transactions on Automatic Control* AC 16(6): 688–706.

Singer, R. A. and A. J. Kanyuck. 1971. Computer control of multiple site correlation. *Automatica* 7: 455–463.

Speyer, J. L. 1979. Computation and transmission requirements for a decentralized linear-quadratic-Gaussian control problem. *IEEE Transactions on Automatic Control* AC 24: 266–269.

Vebber, P. W. 1991. An examination of target tracking in the antisubmarine warfare system evaluation tool (ASSET), Master thesis, Naval Postgraduate School, Monterey, CA.

Washburn, A. 1969. Probability density of a moving particle. *Operations Research* 17(5): 861–871.

Wiener, N. 1949. *Extrapolation, Interpolation, and Smoothing of Stationary Time Series*. New York: Technology Press of MIT.

Zhu, Y. and X. R. Li. 1999. Best linear unbiased estimation fusion. *Proceedings of the 2nd International Conference on Information Fusion*, Sunnyvale, CA, pp. 1054–1061.

7 Decentralized Data Fusion

Formulation and Algorithms

Paul Thompson, Eric Nettleton,
and Hugh Durrant-Whyte

CONTENTS

7.1 Decentralized Data Fusion Introduction 162
7.2 Information Form Introduction 163
7.3 Decentralized Fusion and Communication 164
    7.3.1 Tree Network Topology, Channel Cache 165
    7.3.2 Related Channel Filter Approaches 167
    7.3.3 Summary 169
7.4 Dynamic Systems 170
    7.4.1 State Dynamics 172
        7.4.1.1 State Dynamic Model 172
        7.4.1.2 Trajectory Information Approach 172
        7.4.1.3 Equivalence to the Conventional Approach 173
        7.4.1.4 Multiple Trajectory States 174
    7.4.2 Dynamics in Decentralized Data Fusion 177
        7.4.2.1 Common Process Noise Problem 177
        7.4.2.2 Delayed and Asequent Observations 178
        7.4.2.3 Burst Communications 180
        7.4.2.4 Solution Using Trajectory States 180
        7.4.2.5 Filtering the Trajectory State System 181
        7.4.2.6 Filtering with Stored Filter Estimates 181
        7.4.2.7 Operation of Channel Caches with Trajectory States 183
    7.4.3 Summary 183
7.5 *k*-Tree Topologies for Redundant and Dynamic Networks 184
    7.5.1 Decentralized Data Fusion on *k*-Trees 187
    7.5.2 Data-Tagging Sets 187
    7.5.3 Separator and Neighborhood Properties 188
        7.5.3.1 Separator Property 188
        7.5.3.2 Local Neighborhood Property 189
    7.5.4 *k*-Tree Communications Algorithm 190
        7.5.4.1 Data-Tag Set Elimination 191

     7.5.5 Link and Node Failure Robustness.. 191

     7.5.6 Summary 192

7.6 Conclusion 193

7.A Appendix 193

     7.A.1 Marginalization in the Information Form 193

     7.A.2 Trajectory Information Form Equivalence 194

Acknowledgments.. 196

References.. 196

7.1 DECENTRALIZED DATA FUSION INTRODUCTION

A decentralized data fusion (DDF) system consists of a network of sensing and computing nodes that aim to cooperatively estimate a common state [12]. Fusion occurs on each node using locally obtained observations and communications from neighboring nodes, without relying on a centralized decision or fusion system. This chapter summarizes and builds on previous research in DDF, including [18,22].

DDF systems have been characterized by three constraints [9,12]:

1. There should be no single central fusion center; no single node should be central to the successful operation of the network.
2. There is no common communication facility; nodes cannot broadcast results and communication must be kept on a strictly node-to-node basis.
3. Sensor nodes do not have any global knowledge of the network topology; nodes should only know about connections in their own neighborhood.

The resulting estimates in the decentralized system can be compared to an equivalent centralized estimator operating with the same observations and modeling assumptions. The focus of this chapter is on exact solutions for DDF, which are equivalent to centralized data fusion, in the following sense:

- The use of consistent fusion with information terms which are conditionally independent given the state, as opposed to methods which double count or miss information terms or use conservative fusion methods.
- The use of direct solution methods as opposed to iterative or convergent methods.

This chapter is organized as follows. Section 7.2 introduces the information form, which is used in this chapter as the expression for fusion operations. Section 7.3 discusses the fusion update and communication aspects of DDF and discusses the operation of DDF on tree topology networks. Section 7.4 introduces the trajectory state formulation for dynamics and uses this to operate decentralized networks for dynamic systems, including handling delayed, asequent and burst communications issues. Section 7.5 extends the tree topology to k-tree topologies for redundant and dynamic decentralized topologies.

7.2   INFORMATION FORM INTRODUCTION

For the decentralized algorithms presented in this chapter, it is convenient to express the fusion operations in terms of *information* by reformulating *multiplication* of probability as *summation* of log-probability.

The main properties that motivate the use of the information form are as follows:

1. Additivity of fusion and observation updates
2. Sparsity of the information matrix

Consider a random variable $\mathbf{x}$ with prior probability density function (PDF) described by a Gaussian PDF, together with a linear observation, described by a Gaussian likelihood:

$$p(\mathbf{x}) = \frac{1}{b}\exp\left(-\frac{1}{2}(\mathbf{x}-\hat{\mathbf{x}})^T\mathbf{P}^{-1}(\mathbf{x}-\hat{\mathbf{x}})\right) \tag{7.1}$$

$$p(\mathbf{z}\mid\mathbf{x}) = \frac{1}{c}\exp\left(-\frac{1}{2}(\mathbf{H}\mathbf{x}-\mathbf{z})^T\mathbf{R}^{-1}(\mathbf{H}\mathbf{x}-\mathbf{z})\right) \tag{7.2}$$

Where the observation is modeled as

$$\mathbf{z} = \mathbf{H}\mathbf{x}+\mathbf{w} \quad E[\mathbf{w}]=\mathbf{0} \quad E[\mathbf{w}\mathbf{w}^T]=\mathbf{R} \tag{7.3}$$

Under Bayes' rule, $p(\mathbf{x}\mid\mathbf{z})=p(\mathbf{z}\mid\mathbf{x})p(\mathbf{x})/p(\mathbf{z})$, the posterior PDF given the prior and observation is

$$p(\mathbf{x}\mid\mathbf{z}) = \frac{1}{d}\exp\left(-\frac{1}{2}(\mathbf{x}-\hat{\mathbf{x}})^T\mathbf{P}^{-1}(\mathbf{x}-\hat{\mathbf{x}})-\frac{1}{2}(\mathbf{H}\mathbf{x}-\mathbf{z})^T\mathbf{R}^{-1}(\mathbf{H}\mathbf{x}-\mathbf{z})\right) \tag{7.4}$$

$$= \frac{1}{d}\exp\left(-\frac{1}{2}(\mathbf{x}-\hat{\mathbf{x}}_+)^T\mathbf{P}_+^{-1}(\mathbf{x}-\hat{\mathbf{x}}_+)\right) \tag{7.5}$$

The two expressions for the posterior must equate

$$(\mathbf{x}-\hat{\mathbf{x}}_+)^T\mathbf{P}_+^{-1}(\mathbf{x}-\hat{\mathbf{x}}_+) = \begin{array}{l} (\mathbf{x}-\hat{\mathbf{x}})^T\mathbf{P}^{-1}(\mathbf{x}-\hat{\mathbf{x}}) \\ +(\mathbf{H}\mathbf{x}-\mathbf{z})^T\mathbf{R}^{-1}(\mathbf{H}\mathbf{x}-\mathbf{z}) \end{array} \tag{7.6}$$

By matching first and second derivatives of each side with respect to $\mathbf{x}$, this results in

$$\mathbf{P}_+^{-1} = \mathbf{P}^{-1}+\mathbf{H}^T\mathbf{R}^{-1}\mathbf{H} \tag{7.7}$$

$$\mathbf{P}_+^{-1}\hat{\mathbf{x}}_+ = \mathbf{P}^{-1}\hat{\mathbf{x}}_+ +\mathbf{H}^T\mathbf{R}^{-1}\mathbf{z} \tag{7.8}$$

The information form is defined by these terms $\mathbf{P}^{-1}$ and $\mathbf{P}^{-1}\hat{\mathbf{x}}$:

$$
\begin{aligned}
\mathbf{Y} \quad &\triangleq \quad -\frac{\partial^2 \log\{p(\mathbf{x})\}}{\partial \mathbf{x}^2} \quad = \quad \mathbf{P}^{-1} \\[2mm]
\hat{\mathbf{y}} \quad &\triangleq \quad -\left.\frac{\partial \log\{p(\mathbf{x})\}}{\partial \mathbf{x}}\right|_{@\mathbf{x}=0} \quad = \quad \mathbf{P}^{-1}\,\hat{\mathbf{x}}
\end{aligned}
\tag{7.9}
$$

Consequently, given $\mathbf{Y}$ and $\hat{\mathbf{y}}$, the estimate is recovered by the solution of the linear system

$$
\mathbf{Y}\hat{\mathbf{x}} = \hat{\mathbf{y}}
\tag{7.10}
$$

The estimate $\hat{\mathbf{x}}$ will be identical to that obtained by a covariance-based Gaussian estimator such as a Kalman filter operating under identical assumptions.

So, given a prior PDF described by information matrix $\mathbf{Y}$ and information vector $\mathbf{y}$, the posterior following the observation is

$$
\mathbf{Y}^+ = \mathbf{Y} + \mathbf{H}^T \mathbf{R}^{-1} \mathbf{H}
\tag{7.11}
$$

$$
\mathbf{y}^+ = \mathbf{y} + \mathbf{H}^T \mathbf{R}^{-1} \mathbf{z}
\tag{7.12}
$$

It is convenient to label the observation as contributing *observation information* in the form

$$
\mathbf{I} = \mathbf{H}^T \mathbf{R}^{-1} \mathbf{H}
\tag{7.13}
$$

$$
\mathbf{i} = \mathbf{H}^T \mathbf{R}^{-1} \mathbf{z}
\tag{7.14}
$$

In general, the fusion of multiple, statistically independent terms is a straightforward addition:

$$
\mathbf{Y}^+ = \sum_i \mathbf{Y}_i \quad \mathbf{y}^+ = \sum_i \mathbf{y}_i
\tag{7.15}
$$

7.3   DECENTRALIZED FUSION AND COMMUNICATION

This section discusses DDF with a focus on the fusion and update steps resulting from communication and observations. These aspects are highlighted by considering DDF on a static state variable. Section 7.4 extends this discussion by considering system dynamics and temporal aspects.

The static case highlights the basic properties of the *fusion* component of the decentralized system, since the present formulation of DDF is built on the *additive* properties of the information fusion operations.

The system then consists of a common static state $\mathbf{x}$, which is to be estimated on the multiple platforms. The decentralized system is required to obtain an estimate in exact agreement with an equivalent centralized estimator using the same observations and modeling assumptions. In a static system, the posterior information is identically the sum of the individual independent observation information terms

$$\mathbf{Y} = \sum_k \mathbf{H}_k^T \mathbf{R}_k^{-1} \mathbf{H}_k \quad \hat{\mathbf{y}} = \sum_k \mathbf{H}_k^T \mathbf{R}_k^{-1} \mathbf{z}_k \qquad (7.16)$$

$$\hat{\mathbf{x}} = \mathbf{Y}/\mathbf{y} \qquad (7.17)$$

In essence, the DDF nodes communicate to obtain the $\mathbf{Y}$ and $\hat{\mathbf{y}}$ sums in Equation 7.16. The required globally agreeing estimate is then obtained through the solution of Equation 7.17 separately at each node.

Different algorithms on different topologies operate different methods for obtaining the sums in Equation 7.16:

- In a centralized estimator, each $\mathbf{H}_k^T \mathbf{R}_k^{-1} \mathbf{H}_k$ and $\mathbf{H}_k^T \mathbf{R}_k^{-1} \mathbf{z}_k$ is communicated to the central estimator, which performs the sum in Equation 7.16.
- In a fully connected decentralized topology, each node transmits $\mathbf{H}_k^T \mathbf{R}_k^{-1} \mathbf{H}_k$ and $\mathbf{H}_k^T \mathbf{R}_k^{-1} \mathbf{z}_k$ to *each other node*. Each node is then able to separately perform the sum in Equation 7.16.
- In a tree-connected decentralized topology, nodes accumulate partial sums of Equation 7.16 and communicate in a tree to obtain the global sums. This topology is discussed further later.

7.3.1 Tree Network Topology, Channel Cache

This section considers the singly connected or tree decentralized topology. Under this topology, the graph properties of the tree and the distributivity of the addition are exploited in order to perform the required summation in Equation 7.16.

A tree topology has no cycles. This means that for each node any communications to a neighbor cannot affect any other neighbor. Also, any communications from a neighbor cannot be affected by any other neighbor. This occurs because at any node, a, the neighbors of the neighbors of a, excluding a, are disjoint.

$$\text{For every node } a \text{ in a tree: } \{\mathcal{N}(\mathcal{N}(a))\}\backslash a \text{ are disjoint} \qquad (7.18)$$

where
$\mathcal{N}(a)$ is the neighbor of node a
$S\backslash a$ is the set S excluding a

This means that the sum in Equation 7.16 can be written as a hierarchy of partial sums over disjoint subsets:

$$\mathbf{Y}_a = \mathbf{I}_a + \sum_{i \in \mathcal{N}(a)} \left\{ \mathbf{I}_i + \sum_{j \in \{\mathcal{N}(i)\backslash a\}} \left\{ \mathbf{I}_j + \sum_{k \in \{\mathcal{N}(j)\backslash i\}} \left\{ \mathbf{I}_k + \cdots \right\} \right\} \right\} \qquad (7.19)$$

The tree topology guarantees that the terms inside each summation are disjoint (independent from each other) and therefore prevents double counting of observation information.

The algorithm developed in this section will be referred to as the *channel cache* algorithm. The channel cache algorithm is a variant of the well-known *channel filter* algorithm [6,9,12,18–20,22]. The channel cache algorithm is also inspired by junction tree algorithm for inference in graphical models [21] and [16].

The operation of a tree topology decentralized network is illustrated in Figure 7.1. Figure 7.1 shows a *branch* in a tree network.

Each node stores its own observation information (dark gray) in the form $\mathbf{I} = \mathbf{H}^T \mathbf{R}^{-1} \mathbf{H}$ and $\mathbf{i} = \mathbf{H}^T \mathbf{R}^{-1} \mathbf{z}$. These correspond to the $\mathbf{I}$ terms in Equation 7.19. These observation information terms are required to be statistically independent information unique to one node.

The communicated term from a node i to a is an information matrix $\mathbf{C}_{ia}$ (and its information vector counterpart). $\mathbf{C}_{ia}$ consists of the transmit node's own independent observation information plus the sum of all communicated terms received from the "upstream" part of the tree network:

$$\mathbf{C}_{ia} = \mathbf{I}_i + \sum_{j \in \{\mathcal{N}(i)\backslash a\}} \mathbf{C}_{ji} \qquad (7.20)$$

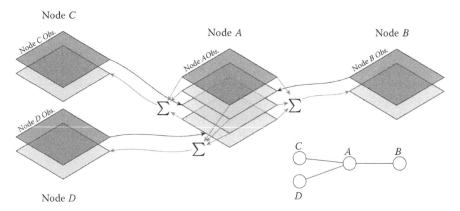

FIGURE 7.1 DDF with channel caches. Four nodes are arranged in the topology shown in the lower right. Each node stores its own fused observations (dark gray). Each node caches the received communication term from each of its neighbors (light gray). The total fused information at each node is the sum of each stack, since each layer consists of independent information. The transmitted communication term is the sum of the stack excluding the destination's cache term.

$$= \mathbf{I}_i + \sum_{j \in \{\mathcal{N}(i) \backslash a\}} \left\{ \mathbf{I}_j + \sum_{k \in \{\mathcal{N}(j) \backslash i\}} \{\mathbf{C}_{kj}\} \right\} \qquad (7.21)$$

Each node locally caches the received communication term from each of its neighbors (light gray) in a so-called channel cache. All of the channel cache, $\mathbf{C}$, and the observation, $\mathbf{I}$, information terms are statistically independent.

Transmission of a communication term has no effect at the transmitting node, so transmissions can be lost without breaking the consistency of the estimates. On reception of a communication term, the received term is simply stored in the channel cache. This means that duplicate transmissions and/or duplicate receptions are acceptable.

Each node can obtain the total network sum, $\mathbf{Y}$, by summing its own observation information together with all the locally cached communication terms, e.g., at node a:

$$\mathbf{Y}_a = \mathbf{I}_a + \sum_{i \in \mathcal{N}(a)} \mathbf{C}_{ia} \qquad (7.22)$$

The net result is that the network computes a series of partial sums, with each node obtaining the sum as in Equation 7.19. For each node, the evaluation of Equation 7.19 operates as a series of messages propagating inward on the tree toward that node.

Nodes initialize their observation and communication cache information terms to zero, $\mathbf{I} = 0$, $\mathbf{C} = 0$, such that nodes can produce estimates even before the network has finished propagating terms across the network span.

7.3.2 RELATED CHANNEL FILTER APPROACHES

The previous section presented the channel cache algorithm for tree topology networks. The channel cache is closely related to the channel filter algorithm, which has been discussed in various papers [6,9,12,18–20,22].

The approach used in a channel filter is to maintain the total information estimate at each node and maintain the *common* information between pairs of nodes on a tree network. The channel filter's use of the common information is motivated by the following equation for the fusion of a local $\mathbf{Y}_A$ and a received communication $\mathbf{Y}_B$ from a remote node:

$$\mathbf{Y}_{A \cup B} = \mathbf{Y}_A + \mathbf{Y}_B - \mathbf{Y}_{A \cap B} \qquad (7.23)$$

Each primary operation of the channel filter is described next (referring to operations at a node i, transmitting to a node a and receiving from a node j). Table 7.1 summarizes the channel algorithms. The channel filter consists of the following operations:

- *Observation*: $\mathbf{Y}_i += \mathbf{I}$. Observation information $\mathbf{I}$ adds simply into the total information $\mathbf{Y}_i$ without affecting any channel common information.
- *Transmit:* The node's current total information is transmitted, $\mathbf{C}_{ia} = \mathbf{Y}_i$. It is assumed that the destination node a will successfully receive the communication; therefore, the common information is set $\mathbf{Y}_{ia} = \mathbf{Y}_i$.

TABLE 7.1

Summary of the Primary Operations for the Channel Cache and Channel Filter Algorithms

	Channel Cache	Channel Filter
Obs Update	$I_i + = I$	$Y_i + = I$
Transmit	$C_{ia} = I_i + \sum_{j \in \{\mathcal{N}(i) \backslash a\}} C_{ji}$	$C_{ia} = Y_i$ $Y_{ia} = Y_i$
Receive	Store C_{ji}	$Y_i += C_{ji} - Y_{ji}$ $Y_{ji} = C_{ji}$
Result	$Y_i = I_i + \sum_{j \in \mathcal{N}(i)} C_{ji}$	Use Y_i

- *Receive:* The received information is Y_{ji}, so given the existing common information C_{ji}, the total information is updated: $Y_i += C_{ji} - Y_{ji}$. The nodes now have common information from the communication, so the node sets $Y_{ji} = C_{ji}$.
- *Result:* The total information is maintained in Y_i.

Both the channel filter and the channel cache algorithms are designed to exploit a tree topology network. At each node, both the channel filter and channel cache algorithms store an information matrix and vector for each neighbor, intended to ensure correct consistent interaction with that neighbor. The basic difference between the channel filter and the channel cache algorithms is as follows:

- The channel filter algorithm maintains the *common information* between the local total information and each neighbor's total information.
- The channel cache algorithm maintains the *contributed information* from each neighbor.

Using the *common information* requires both nodes to maintain identical copies of the common information, which is vulnerable to failure if the two copies differ (cases in which this can happen are discussed later). The common information maintained at both nodes on a channel is required to be identical, since the common information, $Y_{i \cap j}$, is symmetrical between two nodes, i.e., $Y_{i \cap j} = Y_{j \cap i}$ [12].

By contrast, the channel cache algorithm maintains a local record of the *contributed information* from the neighbor. This decouples the communication between nodes such that the changes only occur locally when information is received, not when it is transmitted. Table 7.2 shows a decentralized communication transaction between two nodes, showing both the contributed information and the common information, in the case of an ideal communication. The communication must update the common information at *both* nodes, which requires an *assumption* of successful communication for the sender. On the other hand, the communication

TABLE 7.2

Common Information versus Contributed Information in an Ideal Pair of Communications

| | At Node A | | | At Node B | | |
| | Contrib | Common | Common | Contrib | Local | |
Local Total	from B	with B	with A	from A	Total	Comment
a	0	0	0	0	b	Start.
a	0	a	a	a	$b+a$	$A \rightarrow B$
$a+b$	B	$a+b$	$a+b$	a	$b+a$	$A \leftarrow B$

only needs to update the *contributed* information record once (at the destination node) and only upon an actual successful communication.

Table 7.2 also shows that the two contributed information terms sum to the common information:

$$\mathbf{Y}_{A \text{ contributed to } B} + \mathbf{Y}_{B \text{ contributed to } A} = \mathbf{Y}_{A \cap B} \qquad (7.24)$$

The channel filter algorithm can fall into cases where the two common-information records can differ due to miscommunication:

- *Asynchronous operation.* If nodes send messages which "cross over," then their common-information records can become misaligned. An example is shown in Table 7.3. Consider a pair of nodes i,j which transmit almost simultaneously at times t_i, t_j and receive at times r_i, r_j. This asynchronous case arises if $r_j > t_j$ or $r_i > t_i$.
- *Lost transmissions.* The channel filter algorithm can also become misaligned in the case of transmissions which are lost. This occurs if a node completes the "transmit" update to its channel filter but the destination fails to receive the message. An example is shown in Table 7.4.

7.3.3 SUMMARY

This section described DDF with a focus on the observation update and the decentralized communication, particularly in tree topology networks.

The fusion of independent observation and/or communicated information is additive when performed in the log-likelihood or information form. Therefore, the problem of forming a decentralized estimate which is identical to a centralized equivalent reduces down to a decentralized algorithm for forming a correct sum of the observation information terms.

When applied to tree topology networks, it suffices to maintain a local node information matrix and vector and one information matrix and vector for each neighbor in the network.

TABLE 7.3
Asynchronous Operation

Channel Filter

	At Node A			At Node B		
Local Total	Common with B	Comms	Comms	Common with A	Local Total	Comment
a	0	—	—	0	b	Start.
a	a	a (out) ⟍ ⟋ (out) b		b	b	Transmit.
b	b	b (in) ⟋⟍ (in) a		a	a	Receive.

Channel Cache

	At Node A			At Node B		
Local Total	Common with B	Comms	Comms	Common with A	Local Total	Comment
a	0	—	—	0	b	Start.
a	0	a (out) ⟍ ⟋ (out) b		0	b	Transmit.
$a+b$	b	b (in) ⟋⟍ (in) a		a	$a+b$	Receive.

Note: In the channel filter algorithm, nodes that both transmit simultaneously result in incorrectly swapped estimates rather than fused estimates. The channel cache algorithm modifies the node state only on reception, not transmission, and hence is able to operate correctly.

This section presented the channel cache algorithm for handling the local node information and decentralized communications operations. The channel cache algorithm handles imperfect communications such as asynchronous transmissions, lost transmissions in a simple manner. The channel cache algorithm operates on records of *contributed information* from each neighbor. This is in addition to the capabilities previously possible with channel filters on tree networks, especially the avoidance of double counting, avoidance of conservative fusion, while achieving global agreement among nodes in a decentralized network.

The aforementioned discussion has focused on the observation update and the decentralized communication. The following section extends the discussion of DDF into dynamic systems.

7.4   DYNAMIC SYSTEMS

The observation and communication updates, as described in the previous section, were discussed with respect to a static system, i.e., a single-state vector $\mathbf{x}$. This section extends the discussion into dynamic systems and reviews the smoothing or trajectory state formulation of dynamic systems to formulate DDF for estimation of dynamic systems. This trajectory state formulation of dynamics is then applied to address the issues of delayed and asequent observations and burst communications in DDF.

TABLE 7.4

Lost Transmission

Channel Filter

At Node A				At Node B		
	Common			Common		
Local Total	with B	Comms	Comms	with A	Local Total	Comment
a	0	—	—	0	b	Start.
a	a	a (out,lost)	—	0	b	A → B (lost)
b	b	b (in)◀————(out) b		b	b	A ← B
b	b	b (out)————▶(in) b		b	b	A → B

Channel Cache

At Node A				At Node B		
	Contrib			Contrib		
Local Total	from B	Comms	Comms	from A	Local Total	Comment
a	0	—	—	0	b	Start.
a	0	a (out,lost)	—	0	b	A → B (lost)
a+b	b	b (in)	(out) b	0	b	A ← B
a+b	b	a (out)	(in) a	a	a+b	A → B

Note: In the channel filter algorithm, a lost transmission results in a loss of the new information in the lost transmission. The information is not recovered at subsequent communications, due to the operation of the algorithm subtracting (incorrect) common information. For the channel cache algorithm, a lost transmission has no effect, and hence the information is correctly gained upon the next successful communication.

When the decentralized system has observation and/or communication interruptions and delays, it becomes important to decide when and where the dynamic propagation of the estimate is to be applied. Furthermore, at each decentralized node, there are stored communication terms relating to other nodes, and so it is also necessary to consider the dynamic propagation of these.

This section presents the *trajectory state* approach to representing system dynamics. The trajectory state approach expands the state for a dynamic system into a joint state consisting of a sequence (*trajectory*) of states. The tools for manipulating joint probabilities in several dimensions and tools for manipulating probabilities and decentralization of static states then become applicable to the dynamic system.

The trajectory state approach relates to smoothing methods used in Kalman smoothing [17]. It is also known as delayed states and has been used to account for delayed decision making in estimation such as delayed associations [15]. The use of delayed states in the information form, with the resulting sparse structure, has been applied in localization and mapping [8,10]. Delayed states have more recently been applied to DDF as a tool for delayed measurements [1] and for delayed and asequent measurements and communications [4].

This section focuses on correct approaches to dealing with delayed, asequent, and burst communications with dynamic models that are known and can be applied by each node. The trajectory state approach can be extended to allow the decentralized system to distribute dynamic models which originate from one node (known as model distribution). The issue of dynamic communication topologies in DDF is discussed separately in Section 7.5.

7.4.1 STATE DYNAMICS

This section explains the state dynamics and trajectory state form, in general. Section 7.4.2 applies these to DDF specifically.

7.4.1.1 State Dynamic Model

We consider a basic, linear discrete time state dynamic model in the form

$$\mathbf{x}_{k+1} = \mathbf{F}\mathbf{x}_k + \mathbf{B}\mathbf{u}_k + \mathbf{G}\mathbf{v}_k \tag{7.25}$$

where
  $\mathbf{x}_k$ is the state vector at instant k
  $\mathbf{F}$ is the state transition matrix
  $\mathbf{v}_k$ is unknown, zero mean, white noise, $E[\mathbf{v}_k]=0,\ E[\mathbf{v}_k \mathbf{v}_k^T]=\mathbf{Q}$
  $\mathbf{u}_k$ is a known control signal, if available

The conventional treatment is to form a *prediction* by using a dynamic *transformation* of the estimate [2]

$$\hat{\mathbf{x}}_{k+1} = \mathbf{F}\hat{\mathbf{x}}_k + \mathbf{B}\mathbf{u}_k \tag{7.26}$$

$$\mathbf{P}_{k+1} = \mathbf{F}\mathbf{P}_k\mathbf{F}^T + \mathbf{G}\mathbf{Q}\mathbf{G}^T \tag{7.27}$$

where this is considered to be a *transformation* of the estimate, replacing the estimate for time k by that for time $k+1$, as a discrete operation.

7.4.1.2 Trajectory Information Approach

Equation 7.27 can actually be considered (see later) to consist of an *augmentation* of the estimate into the latter timestep $k+1$, followed immediately by a *marginalization* to remove timestep k. In this way, the prediction operation in Equation 7.27 moves the estimate forward in time but removes the state components for the past timestep. This removal of the past timestep makes it impossible (or difficult) to fuse late observations or communicated information. This transformative prediction approach thus requires observations to be fused in at the appropriate timestep.

This section describes an alternative approach known as *delayed state* or *trajectory state* approach, which is designed to address the aforementioned issues. In the trajectory state approach, instead of considering *prediction* equations to explicitly transform the estimate from time k to $k+1$, we instead consider the *trajectory* described by a joint state $\mathbf{X} = \begin{bmatrix} \mathbf{x}_k & \mathbf{x}_{k+1} \end{bmatrix}$.

The key reason why this is useful is as follows: If we operate with a joint trajectory state vector $\mathbf{X} = \begin{bmatrix} \mathbf{x}_k & \mathbf{x}_{k+1} & \cdots & \mathbf{x}_{k+n} \end{bmatrix}$, then observations and communications in any time (k to $k+n$), and the dynamic model all act *additively* on the joint state $\mathbf{X}$. Thus the methods of Section 7.3 remain applicable, since they are designed to exploit additive operations over decentralized networks.

Therefore, we rearrange Equation 7.25 to focus on the joint trajectory state $\mathbf{X} = \begin{bmatrix} \mathbf{x}_k & \mathbf{x}_{k+1} \end{bmatrix}$:

$$\mathbf{Bu} = -\mathbf{Fx}_k + \mathcal{I}\mathbf{x}_{k+1} - \mathbf{Gv} \tag{7.28}$$

$$\mathbf{Bu} = \begin{bmatrix} -\mathbf{F} & \mathcal{I} \end{bmatrix} \begin{bmatrix} \mathbf{x}_k & \mathbf{x}_{k+1} \end{bmatrix}^T - \mathbf{Gv} \tag{7.29}$$

where $\mathcal{I}$ denotes the identity matrix. We can then consider Equation 7.29 in the form of an observation, as in Equation 7.3:

$$
\begin{array}{ccccccc}
\mathbf{z} & = & \mathbf{H} & \mathbf{x} & +\mathbf{w} & E[\mathbf{ww}^T] & = & \mathbf{R} \\
\mathbf{Bu} & = & \begin{bmatrix} -\mathbf{F} & \mathcal{I} \end{bmatrix} & \begin{bmatrix} \mathbf{x}_k & \mathbf{x}_{k+1} \end{bmatrix}^T & -\mathbf{Gv} & E[\mathbf{Gvv}^T\mathbf{G}^T] & = & \mathbf{GQG}^T
\end{array}
\tag{7.30}
$$

Considering the dynamic model in the form of an observation requires the following replacements:

$$\mathbf{H} \leftarrow \begin{bmatrix} -\mathbf{F} & \mathcal{I} \end{bmatrix} \quad \mathbf{R} \leftarrow \mathbf{GQG}^T \quad \mathbf{z} \leftarrow \mathbf{Bu} \tag{7.31}$$

By analogy with Equation 7.13, the dynamic model can then be represented as an information matrix and vector in the joint trajectory state $\mathbf{X}$:

$$\mathbf{I} = \mathbf{H}^T\mathbf{R}^{-1}\mathbf{H} \quad \mathbf{i} = \mathbf{H}^T\mathbf{R}^{-1}\mathbf{z} \tag{7.32}$$

$$\mathbf{I} = \begin{pmatrix} \mathbf{F}^T\mathbf{Q}^{-1}\mathbf{F} & -\mathbf{F}^T\mathbf{Q}^{-1} \\ -\mathbf{Q}^{-1}\mathbf{F} & \mathbf{Q}^{-1} \end{pmatrix} \quad \mathbf{i} = \begin{pmatrix} -\mathbf{F}^T\mathbf{Q}^{-1}\mathbf{Bu} \\ \mathbf{Q}^{-1}\mathbf{Bu} \end{pmatrix} \tag{7.33}$$

where $\mathbf{Q} \triangleq \mathbf{GQG}^T$.

7.4.1.3 Equivalence to the Conventional Approach

We next show the equivalence of the trajectory state approach to existing *prediction* equations in the information and covariance forms. To show the equivalence, we setup the same initial conditions and steps as for a prediction:

1. Define some prior information in the earlier timestep $\mathbf{x}_k$: $\mathbf{Y}_k$ and $\mathbf{y}_k$ satisfying $\mathbf{Y}_k\hat{\mathbf{x}}_k = \mathbf{y}_k$.
2. Allow no prior information in the latter $\mathbf{x}_{k+1}$.
3. Apply the dynamic model to the joint $\mathbf{x}_k$ and $\mathbf{x}_{k+1}$.
4. Evaluate the marginal information in the latter timestep $\mathbf{x}_{k+1}$.

The posterior information, $\mathbf{Y}$ and $\mathbf{y}$, after step 3 (i.e., given the prior and the dynamic model) is

$$
\mathbf{Y} = \begin{pmatrix} \mathbf{Y}_k & \mathbf{0} \\ \mathbf{0} & \mathbf{0} \end{pmatrix} + \mathbf{H}^T \mathbf{R}^{-1} \mathbf{H} \quad \mathbf{y} = \begin{pmatrix} \mathbf{y}_k \\ \mathbf{0} \end{pmatrix} + \mathbf{H}^T \mathbf{R}^{-1} \mathbf{z} \tag{7.34}
$$

$$
\mathbf{Y} = \begin{pmatrix} \mathbf{Y}_k + \mathbf{F}^T \mathbf{Q}^{-1} \mathbf{F} & -\mathbf{F}^T \mathbf{Q}^{-1} \\ -\mathbf{Q}^{-1} \mathbf{F} & \mathbf{Q}^{-1} \end{pmatrix} \quad \mathbf{y} = \begin{pmatrix} \mathbf{y}_k - \mathbf{F}^T \mathbf{Q}^{-1} \mathbf{Bu} \\ \mathbf{Q}^{-1} \mathbf{Bu} \end{pmatrix} \tag{7.35}
$$

Equation 7.35 is equivalent to the conventional prediction in Equation 7.27 (proofs are provided in the appendix):

- The joint $\hat{\mathbf{X}}$ satisfies $\mathbf{Y}\hat{\mathbf{X}} = \mathbf{y}$, with $\hat{\mathbf{X}} = \begin{pmatrix} \hat{\mathbf{x}}_k \\ \hat{\mathbf{x}}_{k+1} \end{pmatrix} = \begin{pmatrix} \hat{\mathbf{x}}_k \\ \mathbf{F}\hat{\mathbf{x}}_k + \mathbf{Bu} \end{pmatrix}$.
- The $\mathbf{x}_k$ marginal of $\mathbf{Y}$ remains as the given $\mathbf{Y}_k$ and $\mathbf{y}_k$.
- The $\mathbf{x}_{k+1}$ marginal of $\mathbf{Y}$ yields known expressions [2,19] for the prediction in covariance and information forms:

$$
\mathbf{Y}_{k+1|k} = \{\mathbf{FP}_{k|k}\mathbf{F}^T + \mathbf{Q}\}^{-1} \tag{7.36}
$$

$$
= \mathbf{M} - \mathbf{M}\mathbf{G}\mathbf{\Sigma}^{-1}\mathbf{G}^T\mathbf{M} \tag{7.37}
$$

$$
\mathbf{y}_{k+1|k} = \left[\mathcal{I} - \mathbf{M}\mathbf{G}\mathbf{\Sigma}^{-1}\mathbf{G}^T\right]\mathbf{F}^{-T}\mathbf{y}_{k|k} + \mathbf{Y}_{k+1|k}\mathbf{Bu} \tag{7.38}
$$

$$
\left(\mathbf{\Sigma} = \mathbf{G}^T\mathbf{M}\mathbf{G} + \mathbf{Q}^{-1}\right) \tag{7.39}
$$

$$
\left(\mathbf{M} = \mathbf{F}^{-T}\mathbf{Y}_{k|k}\mathbf{F}^{-1}\right) \tag{7.40}
$$

7.4.1.4 Multiple Trajectory States

Earlier we discussed the formation of a pair of joint successive dynamic states, $\mathbf{x}_k$ and $\mathbf{x}_{k+1}$. Now consider a longer sequence of trajectory states. The original discrete time dynamic model in Equation 7.25 holds for each pair of successive dynamic states; therefore, each successive pair has the dynamic model information added, as in Equation 7.33. The information matrix and vector for the dynamic model between any successive states k and $k+1$ is

$$
\mathbf{I}_k^{\mathrm{dyn}} = \begin{pmatrix} \mathbf{F}^T\mathbf{Q}^{-1}\mathbf{F} & -\mathbf{F}^T\mathbf{Q}^{-1} \\ -\mathbf{Q}^{-1}\mathbf{F} & \mathbf{Q}^{-1} \end{pmatrix} \quad \mathbf{i}_k^{\mathrm{dyn}} = \begin{pmatrix} -\mathbf{F}^T\mathbf{Q}^{-1}\mathbf{Bu} \\ \mathbf{Q}^{-1}\mathbf{Bu} \end{pmatrix} \tag{7.41}
$$

We later re-write $\mathbf{I}_k^{\mathrm{dyn}} = \begin{pmatrix} \mathbf{A} & \mathbf{D} \\ \mathbf{D}^T & \mathbf{C} \end{pmatrix}$ and $\mathbf{i}_k^{\mathrm{dyn}} = \begin{pmatrix} \mathbf{a} & \mathbf{c} \end{pmatrix}^T$ to save space.

Over a sequence of trajectory states, these pairwise $\mathbf{I}_{\text{dyn}}$ blocks add up to form a sparse *banded* matrix:

$$\sum_{k \in [1,7]} \mathbf{I}_k^{\text{dyn}} = \begin{pmatrix} \mathbf{A} & \mathbf{D} & 0 & 0 & 0 & 0 & 0 & 0 \\ \mathbf{D}^T & \mathbf{C}+\mathbf{A} & \mathbf{D} & 0 & 0 & 0 & 0 & 0 \\ 0 & \mathbf{D}^T & \mathbf{C}+\mathbf{A} & \mathbf{D} & 0 & 0 & 0 & 0 \\ 0 & 0 & \mathbf{D}^T & \mathbf{C}+\mathbf{A} & \mathbf{D} & 0 & 0 & 0 \\ 0 & 0 & 0 & \mathbf{D}^T & \mathbf{C}+\mathbf{A} & \mathbf{D} & 0 & 0 \\ 0 & 0 & 0 & 0 & \mathbf{D}^T & \mathbf{C}+\mathbf{A} & \mathbf{D} & 0 \\ 0 & 0 & 0 & 0 & 0 & \mathbf{D}^T & \mathbf{C}+\mathbf{A} & \mathbf{D} \\ 0 & 0 & 0 & 0 & 0 & 0 & \mathbf{D}^T & \mathbf{C} \end{pmatrix}$$

(7.42)

The benefit of the trajectory state formulation is that observations of states within the trajectory appear *additively* in the information matrix. For example, the total information for the trajectory system from times 1 to 8, including a prior $\mathbf{Y}_{1|1}$ at time $k=1$, an observation $\mathbf{I}_5$ at time $k=5$, and the dynamic model information between each time is given by

$$\mathbf{Y}_{1:8} = \mathbf{Y}_{1|1} + \mathbf{I}_5 + \sum_{k \in [1,7]} \mathbf{I}_k^{\text{dyn}} =$$

$$\begin{pmatrix} \mathbf{Y}_{1|1}+\mathbf{A} & \mathbf{D} & 0 & 0 & 0 & 0 & 0 & 0 \\ \mathbf{D}^T & \mathbf{C}+\mathbf{A} & \mathbf{D} & 0 & 0 & 0 & 0 & 0 \\ 0 & \mathbf{D}^T & \mathbf{C}+\mathbf{A} & \mathbf{D} & 0 & 0 & 0 & 0 \\ 0 & 0 & \mathbf{D}^T & \mathbf{C}+\mathbf{A} & \mathbf{D} & 0 & 0 & 0 \\ 0 & 0 & 0 & \mathbf{D}^T & \mathbf{C}+\mathbf{A}+\mathbf{I}_5 & \mathbf{D} & 0 & 0 \\ 0 & 0 & 0 & 0 & \mathbf{D}^T & \mathbf{C}+\mathbf{A} & \mathbf{D} & 0 \\ 0 & 0 & 0 & 0 & 0 & \mathbf{D}^T & \mathbf{C}+\mathbf{A} & \mathbf{D} \\ 0 & 0 & 0 & 0 & 0 & 0 & \mathbf{D}^T & \mathbf{C} \end{pmatrix}$$

$$\mathbf{y}_{1:8} = \begin{pmatrix} \mathbf{y}_{1|1}+\mathbf{a} & \mathbf{c}+\mathbf{a} & \mathbf{c}+\mathbf{a} & \mathbf{c}+\mathbf{a} & \mathbf{c}+\mathbf{a}+\mathbf{i}_5 & \mathbf{c}+\mathbf{a} & \mathbf{c}+\mathbf{a} & \mathbf{c} \end{pmatrix}^T$$

(7.43)

where the observation information $\mathbf{I}_5$ appears as an addition on the diagonal of $\mathbf{Y}$ corresponding to the state at the observed time.

To propagate the trajectory state system forward in time (maintaining a fixed duration trajectory), there are two steps:

1. *Augmenting* the system with the additional timestep. This requires expanding the state vector for the new timestep $(k+1)$ and adding the dynamic model information $\mathbf{I}_k^{dyn}$ and $\mathbf{i}_k^{dyn}$.
2. *Marginalizing* away the earliest timestep.

The system following propagation by one timestep is now given by

$$
\mathbf{Y}_{2:9} = \begin{pmatrix}
\mathbf{A} + \mathbf{Y}_{2|1} & \mathbf{D} & 0 & 0 & 0 & 0 & 0 & 0 \\
\mathbf{D}^T & \mathbf{C} + \mathbf{A} & \mathbf{D} & 0 & 0 & 0 & 0 & 0 \\
0 & \mathbf{D}^T & \mathbf{C} + \mathbf{A} & \mathbf{D} & 0 & 0 & 0 & 0 \\
0 & 0 & \mathbf{D}^T & \mathbf{C} + \mathbf{A} + \mathbf{I}_5 & \mathbf{D} & 0 & 0 & 0 \\
0 & 0 & 0 & \mathbf{D}^T & \mathbf{C} + \mathbf{A} & \mathbf{D} & 0 & 0 \\
0 & 0 & 0 & 0 & \mathbf{D}^T & \mathbf{C} + \mathbf{A} & \mathbf{D} & 0 \\
0 & 0 & 0 & 0 & 0 & \mathbf{D}^T & \mathbf{C} + \mathbf{A} & \mathbf{D} \\
0 & 0 & 0 & 0 & 0 & 0 & \mathbf{D}^T & \mathbf{C}
\end{pmatrix}
$$

$$
\mathbf{y}_{2:9} = \begin{pmatrix} \mathbf{a} + \mathbf{y}_{2|1} & \mathbf{c} + \mathbf{a} & \mathbf{c} + \mathbf{a} & \mathbf{c} + \mathbf{a} + \mathbf{i}_5 & \mathbf{c} + \mathbf{a} & \mathbf{c} + \mathbf{a} & \mathbf{c} + \mathbf{a} & \mathbf{c} \end{pmatrix}^T
$$

$$(7.44)$$

where $\mathbf{Y}_{2|1}$ and $\mathbf{y}_{2|1}$ are

$$
\mathbf{Y}_{2|1} = \mathbf{C} - \mathbf{D}^T (\mathbf{Y}_{1|1} + \mathbf{A})^{-1} \mathbf{D} \tag{7.45}
$$

$$
= \mathbf{Q}^{-1} - \mathbf{Q}^{-1} \mathbf{F} (\mathbf{Y}_{1|1} + \mathbf{F}^T \mathbf{Q}^{-1} \mathbf{F})^{-1} \mathbf{F}^T \mathbf{Q}^{-1} \tag{7.46}
$$

$$
\mathbf{y}_{2|1} = \mathbf{c} - \mathbf{D}^T (\mathbf{Y}_1 + \mathbf{A})^{-1} (\mathbf{y}_1 + \mathbf{a}) \tag{7.47}
$$

$$
= \mathbf{Q}^{-1} \mathbf{B} \mathbf{u} + \mathbf{Q}^{-1} \mathbf{F} (\mathbf{Y}_{1|1} + \mathbf{F}^T \mathbf{Q}^{-1} \mathbf{F})^{-1} (\mathbf{y}_{1|1} - \mathbf{F}^T \mathbf{Q}^{-1} \mathbf{B} \mathbf{u}) \tag{7.48}
$$

$\mathbf{Y}_{2|1}$ is actually the same expression as for the predicted $\mathbf{Y}_{k+1|k}$ in Equation 7.35. This is proven in the appendix. The earlier prior information, $\mathbf{Y}_{1|1}$, clearly resides in a *nonadditive* form, in the expression $\mathbf{D}^T (\mathbf{Y}_{1|1} + \mathbf{A})^{-1} \mathbf{D}$.

In summary:

- The *augmentation* process, which extends the system to further timesteps, continues the same sparse banded pattern in the information matrix.
- The fusion of observations within the duration of the trajectory states is a straightforward *addition* in the information matrix and vector.
- Marginalization of the earliest timestep in a succession of trajectory states follows the same pattern as for information filtering prediction, leaving any observations or prior information in the removed timestep k in a *nonadditive* form.

7.4.2 DYNAMICS IN DECENTRALIZED DATA FUSION

The previous section discussed the state dynamics generally, resulting in the formation of a trajectory state system. The key advantage of using a sequence of trajectory states is that for observations of states within the trajectory states, the observation information is additive, just as for observations of a static state. This additivity of observation information applies regardless of the timing or sequence of observations, as long as the state at the observed time exists in the current set of trajectory states.

This section describes the application of the trajectory state approach for handling timing issues in DDF. In particular, we consider the following problem cases:

- *Delayed and asequent data fusion*, in which an observation from an earlier time becomes available after a prediction step, has been performed (delayed) or after other data have been fused for later times (asequent). Delayed and asequent observations usually refer to local sensor node observations.
- *Burst communication*, which occurs when decentralized communications is resumed after a period of interruption. The communications that occurs after the interruption is referred to as burst communication, since it aims to deliver a large amount of information in a short time (or single message) to re-establish agreement between the nodes. Burst communications can also be thought of as delayed/asequent fusion across multiple decentralized nodes.

The key issue behind the aforementioned difficulties is *additivity* of observation and communicated information, and the fact that for states that have been replaced by *predictions* cannot be updated additively by other predictions. This is explained in further detail next.

7.4.2.1 Common Process Noise Problem

The underlying issue of concern relates to the common process noise problem. They are so called because separately predicted terms ignore their common use of the same process noise. This can also be expressed as the problem that the fusion of predicted information is unequal to the prediction of fused information:

$$\text{Predict}(\text{Fuse}(A, B)) \neq \text{Fuse}(\text{Predict}(A), \text{Predict}(B)) \tag{7.49}$$

Consider a case where a fused estimate is predicted forward. The fused estimate at time k is obtained as the sum of two independent information terms, e.g., $\mathbf{Y}_{k,a}$ and $\mathbf{Y}_{k,b}$.

$$\mathbf{Y}_k = \mathbf{Y}_{k,a} + \mathbf{Y}_{k,b} \tag{7.50}$$

The correct expression for the predicted information for time $k+1$ requires the prediction of the sum

$$\mathbf{Y}_{k+1}^{\text{exact}} = \text{Predict}(\mathbf{Y}_k) \qquad (7.51)$$

$$= \{\mathbf{F}\mathbf{Y}_k^{-1}\mathbf{F}^T + \mathbf{Q}\}^{-1} \qquad (7.52)$$

If, however, the term $\mathbf{Y}_{k,a}$ has already been predicted forward, a common approximation to $\mathbf{Y}_{k+1}$ is to take

$$\mathbf{Y}_{k+1}^{\text{approx}} = \text{Predict}(\mathbf{Y}_{k,a}) + \text{Predict}(\mathbf{Y}_{k,b}) \qquad (7.53)$$

$$= \{\mathbf{F}\mathbf{Y}_{k,a}^{-1}\mathbf{F}^T + \mathbf{Q}\}^{-1} + \{\mathbf{F}\mathbf{Y}_{k,b}^{-1}\mathbf{F}^T + \mathbf{Q}\}^{-1} \qquad (7.54)$$

The approximate form is not generally equal to the exact form $\mathbf{Y}_{k+1}^{\text{approx}} \neq \mathbf{Y}_{k+1}^{\text{exact}}$. The approximate form ignores the fact that there is only one underlying process; hence, the two prediction instances share common process noise, $\mathbf{v}$ (of which $E[\mathbf{v}\mathbf{v}^T] = \mathbf{Q}$). To consider the approximation further, consider a simpler worst case where $\mathbf{Y}_a = \mathbf{Y}_b$:

$$\frac{\mathbf{Y}_{k+1}^{\text{approx}}}{\mathbf{Y}_{k+1}^{\text{exact}}} = \frac{\{\mathbf{F}\mathbf{Y}_a^{-1}\mathbf{F}^T + \mathbf{Q}\}^{-1} + \{\mathbf{F}\mathbf{Y}_b^{-1}\mathbf{F}^T + \mathbf{Q}\}^{-1}}{\{\mathbf{F}(\mathbf{Y}_a + \mathbf{Y}_b)^{-1}\mathbf{F}^T + \mathbf{Q}\}^{-1}} \qquad (7.55)$$

$$= \frac{\{\mathcal{I} + 2\mathbf{a}\}}{\{\mathcal{I} + \mathbf{a}\}} \qquad (7.56)$$

where

$$\mathbf{a} = \mathbf{F}^{-T}\mathbf{Y}_a\mathbf{F}^{-1}\mathbf{Q} = (\mathbf{F}\mathbf{Y}_a^{-1}\mathbf{F}^T)^{-1}\mathbf{Q} \qquad (7.57)$$

for which it can be seen that

$$\lim_{\mathbf{a}\to 0} \frac{\mathbf{Y}_{k+1}^{\text{approx}}}{\mathbf{Y}_{k+1}^{\text{exact}}} = \mathcal{I} \qquad \lim_{\mathbf{a}\to\infty} \frac{\mathbf{Y}_{k+1}^{\text{approx}}}{\mathbf{Y}_{k+1}^{\text{exact}}} = 2\mathcal{I} \qquad (7.58)$$

$$\mathbf{Y}_{k+1}^{\text{exact}} \leq \mathbf{Y}_{k+1}^{\text{approx}} \leq 2\mathbf{Y}_{k+1}^{\text{exact}} \qquad (7.59)$$

This shows that $\mathbf{Y}_{k+1}^{\text{approx}}$ is always slightly overconfident, but is close to $\mathbf{Y}_{k+1}^{\text{exact}}$ for small Q. However, for large Q, the $\mathbf{Y}_{k+1}^{\text{approx}}$ is *overconfident*, being up to $2\mathbf{Y}_{k+1}^{\text{exact}}$ in the worst case. Predicting the $\mathbf{Y}_{k,a}$ and $\mathbf{Y}_{k,b}$ independently is equivalent to claiming that there are two independent process models available.

7.4.2.2 Delayed and Asequent Observations

A delayed observation occurs when an observation from an earlier time becomes available after a prediction step has been performed [19]. The problem of delayed observations can occur in any form of estimator, not only decentralized estimators.

Delayed observations can occur as a result of processing and/or communication delays before observations are available at the estimator.

An asequent observation occurs when an observation from an earlier time becomes available after other data have been fused for later times [19]. Asequent observations may occur if multiple sensors are used locally on a single node, and these sensors have differing observation delays. The case of asequent observations occurring on distinct decentralized nodes is similar, but since it involves the communication aspect it is more similar to the burst communications case discussed later.

The problem with delayed or asequent data fusion is that once the estimator has predicted the local state forward to time $k+1$, the (late) incoming information for time k needs to be considered. If the late arriving information is predicted forward separately, the common process noise problem applies (as discussed in Section 7.4.2) and the result will be approximate and over-confident.

The problem with delayed and asequent data fusion is basically caused by the filter architecture *destructively* predicting estimates forward. That is, applying the prediction equations in a way that *replaces* a local estimate.

The proposed solution instead applies the trajectory state approach to avoid *destructively* predicting estimates until after a window of time has passed, while still obtaining correct current-time filter estimates given all available past observations.

The trajectory information matrix is constructed as in Section 7.4.1:

$$
\mathbf{Y}_{k:k+4} =
\begin{pmatrix}
\mathbf{A}+\mathbf{Y}_{k|k-1}+\mathbf{I}_k & \mathbf{D} & 0 & 0 & 0 \\
\mathbf{D}^T & \mathbf{C}+\mathbf{A}+\mathbf{I}_{k+1} & \mathbf{D} & 0 & 0 \\
0 & \mathbf{D}^T & \mathbf{C}+\mathbf{A}+\mathbf{I}_{k+2} & \mathbf{D} & 0 \\
0 & 0 & \mathbf{D}^T & \mathbf{C}+\mathbf{A}+\mathbf{I}_{k+3} & \mathbf{D} \\
0 & 0 & 0 & \mathbf{D}^T & \mathbf{C}+\mathbf{I}_{k+4}
\end{pmatrix}
$$

$$(7.60)$$

where $\mathbf{Y}_{k:k+4}$ is written with observation information on each timestep, indicating how current, delayed, and/or asequent observations can be fused additively in the trajectory information matrix and vector at their appropriate timestep, as long as that timestep is available within the trajectory state system.

Given the trajectory state system, the estimate solution for the current (latest) timestep will be equivalent to a filtered solution, correctly accounting for the late and asequent observations. Methods for obtaining the solution are discussed in Section 7.4.2.

The trajectory state approach with N timesteps of trajectory states defers the destructive prediction of the earliest state by N timesteps, allowing delayed and asequent observations in that duration. However, very late observations beyond N timesteps will still be subject to the same common process noise problem preventing their use. Very late observations beyond N timesteps are expected to occur less frequently and be less informative to the present estimate and should be discarded (which is conservative). Note that the intention of the trajectory state method is to use N timesteps such that the system can still benefit from the observations with small delay which are very likely to occur and very beneficial to the present estimate.

7.4.2.3 Burst Communications

Burst communication occurs when decentralized communications are resumed after a period of interruption. The communications that occurs after the interruption is referred to as burst communication since it aims to deliver a large amount of information in a short time (or single message) to re-establish agreement between the nodes.

The problem with burst communications occurs when the estimator predicts the local state forward during a period of interrupted communications. The problem is that other decentralized nodes will also perform the same prediction on their local estimates. When the nodes re-connect and communicate, the common process noise problem arises, since the information from each node will have been separately predicted.

The problem with delayed and asequent data fusion is again caused by the filter architecture *destructively* predicting estimates forward. That is, applying the prediction equations in a way that *replaces* a local estimate.

The proposed solution, as for asequent observations, involves using trajectory states in order to maintain a window of some duration in which communications can be late, but still fuse additively into states in the trajectory window. Estimates for the current time can still be obtained from the system, conditioned on all the available past observations.

Referring to Equation 7.60, the decentralized system can communicate the diagonal matrix consisting of the $\mathbf{I}_k$ blocks:

$$\mathbf{I}_{k:k+4} = \begin{pmatrix} \mathbf{I}_k & 0 & 0 & 0 & 0 \\ 0 & \mathbf{I}_{k+1} & 0 & 0 & 0 \\ 0 & 0 & \mathbf{I}_{k+2} & 0 & 0 \\ 0 & 0 & 0 & \mathbf{I}_{k+3} & 0 \\ 0 & 0 & 0 & 0 & \mathbf{I}_{k+4} \end{pmatrix} \tag{7.61}$$

This becomes equivalent to a sequence of static decentralized problems, one for each timestep. The band structure corresponding to the dynamic model can be applied locally at each node.

For normal operation, with frequent communications, the nodes transmit their current $\mathbf{I}_k$ block. But if the communications is blocked for an interval of time and then later resumed, the resulting burst communications will contain the diagonal blocks $\mathbf{I}_{j:k}$ for the fused observations for the blocked interval.

The methods for obtaining the solution are discussed in Section 7.4.2.

7.4.2.4 Solution Using Trajectory States

The cases of delayed and asequent observations and burst communications presented earlier can be addressed using a set of trajectory states. This section focuses on how to solve the resulting trajectory state system:

$$\mathbf{Y}_{k-4:k} = \begin{pmatrix} \mathbf{A} + \mathbf{Y}_{k-n|k-n-1} + \mathbf{I}_{k-n} & \mathbf{D} & 0 & 0 & 0 \\ \mathbf{D}^T & \mathbf{C} + \mathbf{A} + \mathbf{I}_{k-n+1} & \mathbf{D} & 0 & 0 \\ 0 & \mathbf{D}^T & \mathbf{C} + \mathbf{A} + \mathbf{I}_{...} & \mathbf{D} & 0 \\ 0 & 0 & \mathbf{D}^T & \mathbf{C} + \mathbf{A} + \mathbf{I}_{k-1} & \mathbf{D} \\ 0 & 0 & 0 & \mathbf{D}^T & \mathbf{C} + \mathbf{I}_k \end{pmatrix}$$

$$(7.62)$$

$$\mathbf{y}_{k:k+4} = \begin{pmatrix} \mathbf{a} + \mathbf{y}_{k-n|k-n-1} + \mathbf{i}_{k-n} \\ \mathbf{a} + \mathbf{c} + \mathbf{i}_{k-n+1} \\ \mathbf{a} + \mathbf{c} + \mathbf{i}_{...} \\ \mathbf{a} + \mathbf{c} + \mathbf{i}_{k-1} \\ \mathbf{c} + \mathbf{i}_k \end{pmatrix}$$

$$(7.63)$$

The system in Equation 7.62 is a block tridiagonal sparse linear system. Such a system can be solved very efficiently in $O(n)$ time, for n trajectory states [11]. It is also possible to obtain *smoothing* estimates for the duration of the trajectory states by solving the joint system fully. This basically corresponds to solving for the latest estimate as described later, together with back-substitution for the smoothed estimates of the earlier states.

7.4.2.5   Filtering the Trajectory State System

The solution process for the filtered estimate of a trajectory state system is very similar to an online filtering process. In that case, the dynamic model in the trajectory state system need only be defined implicitly, leaving only the diagonal blocks (observations and prior) to be explicitly stored. So the current filtering estimate can be obtained basically by running the information filtering prediction cycles, starting from the prior information in the start of the trajectory and using the stored fused observation information at each time. Note that we only need to run this when requiring an estimate for the present state. Multiple observations can be added into the trajectory system without requiring this solve process. This approach is less general than the next, which will be described in greater detail.

7.4.2.6   Filtering with Stored Filter Estimates

In most cases, it is likely that observations and decentralized communications will arrive with only a small delay, and thus only affect the latter part of the trajectory state system. In that case, it is inefficient to process the entire trajectory state system for its whole duration. Also, allowing re-processing only the affected portion of the trajectory may allow a longer trajectory system to be used. When an estimate of the present state is required, it is only necessary to process forward from timestep $k - n$ to the present, where timestep $k - n$ is the *earliest changed* state in the trajectory. (Changes include local observations and decentralized communications.) In this way, the cost in computation to re-process following delayed or asequent observations or newly arrived burst communications depends on how far back the observation occurs, such that the normal case of short delays can proceed forward with little overhead.

This method is similar to that described for asequent data fusion in Ref. [19]. The difference is that we store the observations I_k in the trajectory state approach, not only the filtered estimates. This allows the handling of burst communications and also simplifies the case for asequent observations.

This method corresponds to filtering, but stores in memory the filtered estimates for a few key timesteps in the trajectory duration. The system needs to store I_j for each timestep, and $Y_{j|j-1}$ and $y_{j|j-1}$ for a few key timesteps.

The I_j terms for each j and the filtered $Y_{k-n|k-n-1}$ for the earliest trajectory state timestep, $k - n$ are statistically independent of each other. These are regarded as "source" information. The stored filter $Y_{j|j-1}$ estimates for the other timesteps are *not* independent of each other, and not independent of the I_k or $Y_{k-n|k-n-1}$. These are to be regarded as a computational aid, storing partial results. These filter $Y_{j|j1}$ could instead be re-processed from the initial $Y_{k-n|k-n-1}$ and the observation information terms I_j.

This occurs as follows:

1. The forward filtering can start from time $k - n$ where timestep $k - n$ is the *earliest changed* state in the trajectory, or wherever a starting or stored prior information exists. Starting from time $k - n$, define a current information matrix and vector of the size of the state at a single time:

$$Y_c = Y_{k-n|k-n-1} \quad y_c = y_{k-n|k-n-1} \tag{7.64}$$

2. For each time $j = k - n$: k
   a. Fuse observations for time j:

$$Y_c = Y_{j|j} = Y_{j|j-1} + I_j \quad y_c = y_{j|j} = y_{j|j-1} + i_j \tag{7.65}$$

   b. Predict the current $Y_{j|j}$ and $y_{j|j}$ to time $j+1$ using Equation 7.37 (except at the last time k):

$$Y_c = Y_{j+1|j} = M - MG\Sigma^{-1}G^T M \tag{7.66}$$

$$y_c = y_{j+1|j} = \left[\mathcal{I} - MG\Sigma^{-1}G^T\right]F^{-T}y_{j|j} + Y_c Bu \tag{7.67}$$

$$\left(\Sigma = G^T MG + Q^{-1}\right) \tag{7.68}$$

$$\left(M = F^{-T}Y_{j|j}F^{-1}\right) \tag{7.69}$$

   c. The resulting $Y_c = Y_{j+1|j}$ and $y_c = y_{j+1|j}$ can be stored if desired, so processing can resume from time $j+1$ later.
3. The resulting $Y_c = Y_{k|k}$ and $y_c = y_{k|k}$ is the *filtered* information for the state at time k. $\hat{X}_k$ is obtained from

$$\hat{x}_k = Y_{k|k}^{-1} y_{k|k} \tag{7.70}$$

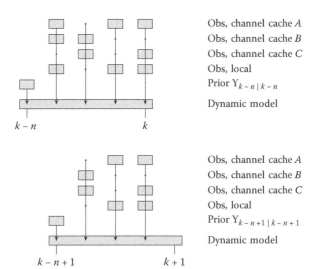

FIGURE 7.2 Illustration of combined trajectory state and channel cache–based DDF at a single node. The lower portion shows the whole system propagated forward by one timestep.

7.4.2.7 Operation of Channel Caches with Trajectory States

Figure 7.2 shows the combined trajectory state and channel cache–based DDF node. The node stores the observations $\mathbf{I}_j$ and $\mathbf{i}_j$ for each timestep $k - n$ to k in the trajectory state window for each channel. These are the channel cache *contributed information* terms from the neighbors. The node similarly stores its own observations for each timestep. The node also stores the prior $\mathbf{Y}_{k-n|k-n}$ and $\mathbf{y}_{k-n|k-n}$. The complete trajectory state system is formed by summing all the entries for each timestep, including the dynamic model information.

To shift the combined system forward by one timestep, at the *back* of the trajectory state window (earliest timestep), the prior $\mathbf{Y}_{k-n|k-n}$ is propagated forward as in Equation 7.43. The observations $\mathbf{I}_{k-n+1}$ are added into $\mathbf{Y}_{k-n+1|k-n}$, leaving the next prior $\mathbf{Y}_{k-n+1|k-n+1}$. The observations for time $k - n + 1$ are then popped out of the trajectory system, so that the guaranteed conditional independence of all information terms is maintained. The *front* end (latest and current timesteps) of the trajectory state window is extended into timestep $k + 1$, ready for new observations or channel terms.

7.4.3 Summary

This section presented the smoothing or trajectory state formulation of dynamic systems to formulate DDF for estimation of dynamic systems. This trajectory state formulation of dynamics was then applied to address the issues of delayed and asequent observations and burst communications in DDF.

The solution method for the trajectory state formulation requires nodes to store the fused observation information for each timestep for a finite duration, allowing for observation and communication delays. For efficient solving it is also useful to store the results of filter estimates.

7.5   *K*-TREE TOPOLOGIES FOR REDUNDANT AND DYNAMIC NETWORKS

The algorithms presented in Section 7.3.1 relate to tree topology networks. Exact decentralized estimation has, in the past, largely been restricted to singly connected tree networks [7,9,18,20]. The key point relating to tree networks is that the DDF problem can be reduced down to a problem of finding a global sum of information terms, which is performed in an efficient local manner on a tree network. In tree networks, there is only one path between any two nodes. This is used in the decentralized algorithm to ensure exact fusion, especially avoiding cases of double counting or rumor propagation in the network. However, the single path property of tree networks also means that tree networks are vulnerable to the failure of nodes and links, since the failure of any nonleaf node or link would leave the network in multiple disconnected pieces. Tree networks include both "star" and chain topologies as well as branching trees.

This section presents an extension beyond tree communications network topologies into so-called *k*-tree network topologies. The *k*-tree topologies are more general than tree topologies but are more specialized for scalability than arbitrary topologies. The presentation of this section is based on Ref. [24]. The *k*-tree is an extension beyond tree topologies, which keeps an overall strict tree-like pattern on a large scale (N nodes $\gg k$), as shown in Figure 7.3f, but allows redundant, looped, dynamic topologies or other subsets of full connection within groups of nodes smaller or equal to $k+1$. The costs in storage and communication grow with k but not with N, the total number of nodes in the network. The *k*-tree topologies are intended to be used with as small k as possible.

The motivation behind using *k*-tree topologies is to improve redundancy and dynamism while maintaining scalability and correctness. For redundancy and fault robustness, it is desirable to allow the network to include multiple redundant paths such that some links or nodes can fail without disconnecting the network topology. It is also desirable to improve dynamism so that some topology changes are able to allow for link failures and re-connections, especially for mobile decentralized networks. The dynamic topology capability is closely related to the link redundancy capability, because once the algorithm is capable of handling multiple paths redundantly, then the network can pick and choose among them dynamically. These capabilities are obtained while ensuring the scalability and correctness of the DDF network.

The *k*-tree topology is used to define an *allowable topology*; the allowable set of links for the decentralized network. Once this *k*-tree *allowable topology* is established, it defines which nodes can communicate on which links and establishes what each node needs to store and communicate in order to ensure correct and exact DDF, as will be described later in this section. This use of a defined restricted topology is similar to how *spanning-tree* algorithms can be used to define an allowable tree of links in an otherwise unstructured network for DDF [16,18]. A spanning-tree can then be used with tree-topology decentralized networks (as in Section 7.3.1), but these do not offer a simple, exact method for handling the data fusion aspects of changing topology or dealing with link failures.

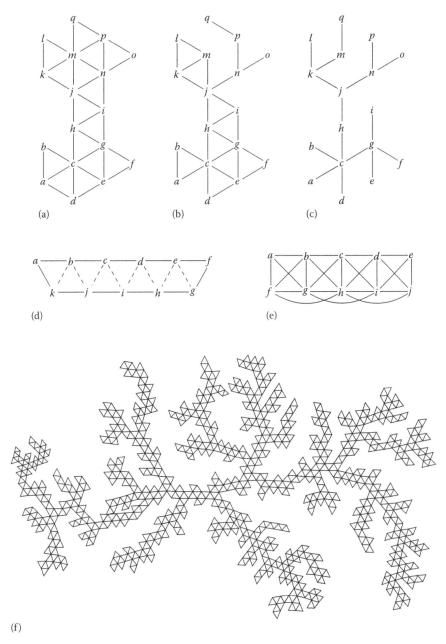

(a)   (b)   (c)

(d)   (e)

(f)

FIGURE 7.3 Example k-tree topologies. Each line is an allowable decentralized communication link, each vertex is a decentralized node. (a) A complete two-tree topology, (b) a mixed one-two-tree topology, (c) a one-tree topology over the same nodes, (d) a ring topology (black lines) is a subset of a two-tree (gray dashed lines), (e) a complete three-tree network, (f) a larger two-tree example showing the broad scale tree topology for $N \gg k$.

In a k-tree allowable topology (and using the k-tree algorithms presented here), nodes can communicate dynamically on all/any links *within* the k-tree topology, even if there are multiple redundant paths or loops and links fail and reconnect unpredictably.

A k-tree allowable topology becomes an arbitrary *unrestricted* topology for $k \geq N$, the number of nodes. This would allow completely arbitrary dynamic and redundant decentralized communications, but would however result in expensive storage and communication.

The treewidth of a graph is well known for its role in limiting the complexity of algorithms in graph theory [3,13,14], graphical models [5], and sparse linear algebra [21]. Given the strong effect of the treewidth on the complexity of the algorithms and network, we considered generalizations of one-tree topologies into k-tree topologies, focusing in particular on the next-highest k; $k=2$, since it is the simplest topology that demonstrates the novel properties of the k-tree approach. It is notable that arbitrarily large *ring* networks can be expressed as a two-tree network. Example k-tree topologies are shown in Figure 7.3.

A complete k-tree graph is made up of cliques of $k+1$ nodes [13,14]. Each adjacent pair of cliques overlaps at k nodes (a *junction* or *separator*). The overall graph of connections between the cliques is a tree. A k-tree graph has *treewidth* of k, so called because the separators are made of k nodes.

Table 7.5 shows the number of links in various k-tree topologies compared with those of a fully connected topology. This shows that the number of k-tree links grows at $O(n^2)$ up until $n = k+1$ (when the first $k+1$ clique is formed), after

TABLE 7.5

Number of Allowable Links Using Trees of Different k versus Fully Connected

N Nodes	1-Tree	2-Tree	3-Tree	4-Tree	k-Tree	Full
1	0	0	0	0		0
2	1	1	1	1		1
3	2	3	3	3		3
4	3	5	6	6	$\frac{1}{2}(n^2 - n)$	6
5	4	7	9	10	for $n \leq (k+1)$	10
6	5	9	12	14		15
7	6	11	15	18		21
8	7	13	18	22	$kn - \frac{1}{2}(k^2 + k)$	28
9	8	15	21	26	for $n > (k+1)$	36
10	9	17	24	30		45
20	19	37	54	70		190
$n > (k+1)$	$n-1$	$2n-3$	$3n-6$	$4n-10$	$kn - \frac{1}{2}(k^2 + k)$	$\frac{1}{2}(n^2 - n)$
$n \gg 1$	$O(n)$	$O(n)$	$O(n)$	$O(n)$	$O(n)$	$O(n^2)$

which each additional node only adds an extra k links. Hence, the number of k-tree links grows as $O(n)$ ultimately. By contrast, the fully connected topology always grows as $O(n^2)$.

7.5.1 Decentralized Data Fusion on k-Trees

The decentralized algorithm defines what each node needs to store and communicate such that each node can obtain the global fused information. The algorithm actively limits the data sizes communicated and stored, leading to the scalable performance of the system. The goal of the topology and message passing is to produce a set of terms $p_i(x)$, such that the fusion of these is a consistent estimate for the state x:

$$p_\cup(x) = \frac{1}{c} \prod_i p_i(x) \tag{7.71}$$

These $p_i(x)$ are probabilities which are conditionally independent of each other given x, or equivalently, that they have independent errors.

As shown earlier, it is convenient to express this as a sum of information terms:

$$\mathbf{Y}_\cup = \sum_i \mathbf{Y}_i \tag{7.72}$$

$$\mathbf{y}_\cup = \sum_i \mathbf{y}_i \tag{7.73}$$

7.5.2 Data-Tagging Sets

The approach used here guarantees against double counting of information by using explicit "data-tagging" sets. A data-tagging set is a set of separate information terms, $\mathbf{Y}_i$, each with a unique identifier. Each data-tagging set stores only conditionally independent terms, so Equation 7.72 can be used on all items in a data-tagging set to recover a consistent fused estimate. Fusion of two or more sets is performed as a *set union* followed by Bayesian fusion (Equation 7.72). The set union step identifies any terms with matching labels and ensures that these are counted only once in the Bayesian fusion. Thus data-tagging avoids double counting of information.

The approach used here ensures scalability by summarizing every stored or communicated data-tagging set into a minimal size. This summarization process exploits the global k-tree property and uses the local topology around the sending and receiving nodes. The necessary local topology properties are guaranteed by designing the global network topology as a k-tree.

The proposed approach uses an efficient, minimal form of data-tagging. This is in contrast to the inefficient *full data-tagging* approach. In the full data-tagging method, each node maintains a set of independent information terms (conditionally independent of each other, given the true state), including its own sensor observations.

In communicating out to any neighbor, the full set of information terms is sent. In receiving communication from a neighbor, the received set is merged (unioned) into the local set. The full data-tagging approach guarantees avoidance of double counting in arbitrary network topology and allows arbitrary dynamism, but is expensive for large-scale networks. Eventually every node's storage and every communicated set have the full list of the conditionally independent information terms arising from every other node. In the full data-tagging approach, the node storage and communication size is $O(n)$ for n nodes in the whole network. This increasing storage and communication size limits the scalability of the network for large n. The full data-tagging approach is equivalent to a k-tree operating with $k \geq n$ for n nodes in the network.

The proposed k-tree approach is obtained by reducing the data-tagging sets to exploit the tree nature of the communications network. The communications and storage scheme proposed achieves correct operation in k-tree networks without using full data-tagging, thus obtaining a decentralized algorithm which is scalable in the number of nodes.

The "stack" of channel cache terms, in Section 7.3.1, Figure 7.1 is actually a minimal data-tag set for the tree network. Each node has $n + 1$ entries corresponding to the n neighbors and a single entry for itself.

7.5.3 Separator and Neighborhood Properties

Before explaining the k-tree decentralized algorithm, it is necessary to discuss some properties of the k-tree.

7.5.3.1 Separator Property

An important k-tree property is the existence of tree separators, as shown in Figure 7.4. In a k-tree any k-clique is a separator. Each separator divides the network into distinct parts. Within each part, the effect of all other parts can be summarized into the separator. Separators enable efficient summarization of entire branches of the k-tree network. Separators use the k-tree separator property: in a k-tree, if any path between any two nodes i,k passes through the separator, then all paths between nodes i,k pass through the separator.

These separators are used at the *borders* of the local neighborhood $\mathcal{L}$ to summarize the fused total of the rest of the network beyond the local neighborhood. For example in Figure 7.4, the total information in each half can be expressed as

FIGURE 7.4 Illustration of the separator property. In a k-tree any k-clique is a separator. Each separator divides the network into two parts. In this figure, $b - d$ is the separator. The two parts and the intersection are shown. Within each part, the effect of the other part can be summarized into the separator.

$$\mathbf{Y}_{\text{rhs}} = \{\mathbf{Y}_{\text{interior}}\} + [\mathbf{Y}_{\text{separator}}] \tag{7.74}$$

$$= \{\mathbf{Y}_b + \mathbf{Y}_e + \mathbf{Y}_h + \mathbf{Y}_d + \mathbf{Y}_f\} + [\mathbf{Y}_{bd}] \tag{7.75}$$

$$\mathbf{Y}_{\text{lhs}} = \{\mathbf{Y}_a + \mathbf{Y}_b + \mathbf{Y}_c + \mathbf{Y}_d + \mathbf{Y}_g\} + [\mathbf{Y}_{bd}] \tag{7.76}$$

where $\mathbf{Y}_{bd}$ represents information in the *separator* $b - d$.

The identifiers in the data-tagging sets are used to identify which node or *branch* of the tree network the information originates from. This means that the identifier should be a *set of node labels*, to allow reference to one neighbor, or k neighbors on a k-tree branch separator.

7.5.3.2 Local Neighborhood Property

A consequence of the separator property is that the local neighborhood around a node becomes a sufficient representation for that node's interaction with the whole rest of the network. The k-tree networks allow an efficient decentralized and local neighborhood representation to serve as the only required topology awareness at the nodes. This is important for scalability, allowing the representation of a global network with only small local neighborhood representations. The local neighborhood is therefore an important data structure used in the algorithm proposed in this chapter.

At any node, $\mathcal{V}_i$, the *local neighborhood subgraph* consists of $\mathcal{V}_i$, the neighbors of $\mathcal{V}_i$ and the links and cliques between them, as shown in Figure 7.5.

The local neighborhood representation is motivated by the k-tree "junction path covering property": in a k-tree, if any path between any two nodes i,k passes through the local neighborhood of a node j, then all paths between nodes i,k pass through the local neighborhood of j.

This junction path covering property means that the local neighborhood around a node j has control over how any messages can pass from one side to the other. The local neighborhood encodes which neighbors to communicate with, which information terms must be maintained separately in data-tag sets (for correctness), and which terms can be fused into others (for scalability).

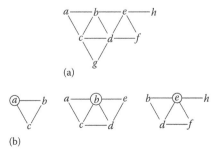

FIGURE 7.5 Illustration of the local neighborhood representation, $\mathcal{L}$. In k-tree networks, the local neighborhood $\mathcal{L}$ is an efficient local summary of the relevant parts of the global topology: (a) global network topology, (b) local neighborhood representations, $\mathcal{L}$, at a, b, e respectively.

For one-tree topologies used in prior works, the local neighborhood representation is simply the list of neighboring vertices and list of the corresponding edges to those neighbors.

7.5.4 *k*-Tree Communications Algorithm

This section explains the decentralized communications algorithm for *k*-trees. We explain the algorithm in the case that the complete *k*-tree is present. Note, however, that the full set of links is not *required*.

The algorithm will be described by referring to the sending node, $\mathcal{V}_i$ ("transmitting vertex") and the receiving node $\mathcal{V}_d$ ("destination vertex"). The transmitting node knows the topology of the allowable links within its own neighborhood of the allowable *k*-tree topology, denoted as $\mathcal{L}$. The sending node has an existing data-tag set. The objective of the algorithm is to calculate a *reduced* data-tag set to send to the destination, $\mathcal{V}_d$.

The communications algorithm is simply stated as follows:

- The data-tag set is *reduced* into the *intersection* of the local and destination neighborhoods.

The communications algorithm is given in algorithm 1 and illustrated in Figure 7.6. In step 1, the algorithm initially copies the local data-tag set to the output data-tag set. This corresponds to the full data-tagging solution. The subsequent steps erase and/or summarize some of the entries, thus ensuring scalability. For step 2, data-tag terms involving the destination vertex are redundant and can be explicitly deleted. Step 3 eliminates any data-tag terms which are not neighbors of the destination vertex. This is explained in the following section.

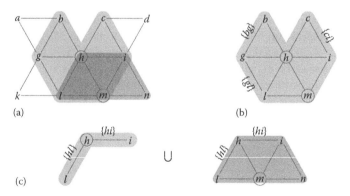

FIGURE 7.6 Summary of the communications algorithm. The local neighborhood at the source node is summarized into the neighborhood *separator* for the destination. This summarized separator set is sent to the destination and merged into the local set. (a) The full network, highlighting neighbor-hoods of V_h and V_m and their intersection. (b) At V_h the network beyond the immediate neighbors is already summarized within the neighborhood. (c) To prepare a communication set, the source V_h can summarize its local neighborhood set into the intersection with the destination neighbor V_m (resulting in the left hand set). This set is communicated to V_m. V_h keeps the union of the received set (left) with its local set (right).

TABLE 7.6
Leaf Vertex Elimination

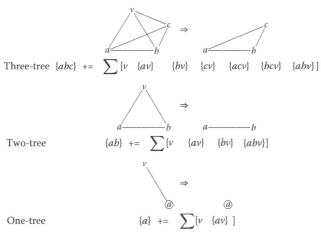

Three-tree $\{abc\} \mathrel{+}= \sum [v \quad \{av\} \quad \{bv\} \quad \{cv\} \quad \{acv\} \quad \{bcv\} \quad \{abv\}]$

Two-tree $\{ab\} \mathrel{+}= \sum [v \quad \{av\} \quad \{bv\} \quad \{abv\}]$

One-tree $\{a\} \mathrel{+}= \sum [v \quad \{av\}]$

Note: The decentralized algorithm uses leaf vertex elimination to reduce the size of the set communicated to a neighbor. In each case in the table, vertex v is to be eliminated. Vertex v has been identified as unnecessary to communicate explicitly, so is instead all terms involving v are merged into the resulting separator term.

7.5.4.1 Data-Tag Set Elimination

This section explains the marginalization process which summarizes nonlocal information, in Algorithm 1.

Elimination proceeds at each step by eliminating a so-called *leaf vertex*, which reduces the size of the data-tag set. A leaf vertex in an ordinary tree would be any vertex with exactly one edge. More generally, however, there are k-tree leaves which are defined as follows: A vertex which is part of exactly one clique of $k+1$ is a k-tree leaf.

Eliminating a k-tree leaf results in its former $k+1$ clique being reduced to a clique of k. To eliminate a vertex v in a $k+1$ clique

- The *result* data-tag term r is that with identifier containing the k node labels of the *neighbors* of v
- For each data-tag term, t, whose identifier contains v: add t into r

Examples of leaf vertex elimination for $k \le 3$ are shown in Table 7.6.

7.5.5 LINK AND NODE FAILURE ROBUSTNESS

The key properties of the proposed approach are correct fusion, scalability and robustness against node and link loss. Achieving these properties simultaneously is achieved by using the bounded treewidth network topology.

ALGORITHM 1: K-TREE DECENTRALIZED COMMUNICATIONS

Compute the communication output data-tag set to send
Input: $\mathcal{L}$: a copy of the local neighborhood graph
Input: $\mathcal{V}_t$: this transmitting node in $\mathcal{L}$
Input: $\mathcal{V}_d$: the destination neighbor in $\mathcal{L}$
Input: localTags: the local data-tag set
Result: destTags: the output data-tag set to send
1. Starting case: No summarization:
Copy destTags ← localTags
2. Delete terms involving $\mathcal{V}_d$:
Erase term $\mathcal{V}_d$ from destTags
Erase any terms for $\mathcal{V}_d$ separators from destTags
3. Summarize away parts not local to $\mathcal{V}_d$:
Determine the region to summarize out, S:
S is all vertices in $\mathcal{L}$ except $\mathcal{V}_d$ and its neighbors
while S is not empty **do**
    Find a leaf vertex $\mathcal{V}_l$ of $\mathcal{L}$ in S
    Eliminate $\mathcal{V}_l$, updating destTags
    Erase $\mathcal{V}_l$ from S

The proposed method is robust against link and node failures simply because it can send information terms on multiple paths. This still yields correct and consistent fusion since the method uses data-tagging to avoid double counting and/or the need for conservative fusion. Furthermore, the method still yields a scalable solution for large networks since the multiple-path and data-tagging is only performed within the nodes and separators of the $k+1$ cliques.

Figure 7.7 shows the pattern of communication of individual information terms in the data-tag sets. In various cases, there are multiple sources redundantly communicating the same term. The receiving node always stores the incoming information terms into a given data-tag set entry. The receive process has no effect other than storing the information, so it is acceptable to receive the same term multiple times from different paths.

7.5.6 Summary

This section presented an algorithm for scalable DDF based on k-tree topologies. The k-tree topologies are more densely connected than 1-trees, but still have an overall sparse (k-)tree topology which gives scalability for large networks. The k-tree topologies have some redundancy in the topology, which makes them more robust to node or link failures than 1-tree topologies. The k-tree topologies allow dynamic changes to the communications topology within subsets of links in the k-tree. Finally, k-tree topologies transition into the fully connected topology and fully data-tagged decentralized algorithm as k increases to N, the number of nodes in the network. Thus k-trees allow some trade-off via k between the tree-based approaches ($k \ll N$) and the unstructured approaches ($k \sim N$).

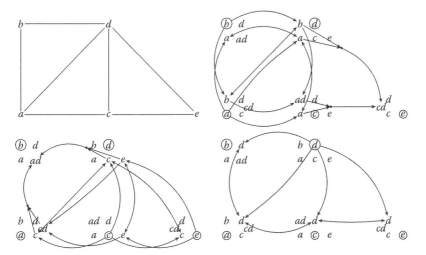

FIGURE 7.7 Diagrams showing the individual data-tag terms which would be stored and communicated in the given topology. At each node the diagram shows the stored terms at that node (cluster of labels), including its own independent information (circled labels). Each arrow indicates the communication of an individual data-tag term. Terms which originate from each node are shown in different shades. Communications which result from the fusion of multiple terms are shown in dashed lines. Communications is strictly with nearest neighbors only, but the sum of all data-tag terms at each node equals the global sum of independent information.

7.6 CONCLUSION

This chapter presented and reviewed methods for DDF. This chapter focused on the channel cache algorithm for DDF in tree topologies and robustness to imperfect communications. In the second part, this chapter reviewed the trajectory state formulation of dynamic systems to formulate DDF for the estimation of dynamic systems. This trajectory state formulation of DDF was applied to address the issues of delayed and asequent observations and burst communications in DDF. In the final part, this chapter extended the operation of DDF on tree topology networks into so-called k-tree topologies. The k-tree topologies are tree-like on the broad scale, which gives good scalability for large networks of nodes. The k-tree topologies allow loops, dense connections, and hence redundancy and dynamic changes among groups of up to $k+1$ nodes.

Taken together, these algorithms contribute significantly toward achieving DDF that is robust to communications latencies and failures, but still yield centralized equivalent estimator performance and are scalable for larger networks.

7.A APPENDIX

7.A.1 MARGINALIZATION IN THE INFORMATION FORM

This appendix states the expressions required for *marginalization* in the information form. Consider an information matrix partitioned into state variables $\mathbf{x}_a$ and $\mathbf{x}_c$:

$$\mathbf{Y} = \begin{pmatrix} \mathbf{A} & \mathbf{B} \\ \mathbf{B}^T & \mathbf{C} \end{pmatrix} \quad \hat{\mathbf{x}} = \begin{pmatrix} \hat{\mathbf{x}}_a \\ \hat{\mathbf{x}}_c \end{pmatrix} \quad \hat{\mathbf{y}} = \begin{pmatrix} \mathbf{a} \\ \mathbf{c} \end{pmatrix} \tag{7.77}$$

where these satisfy $\mathbf{Y}\hat{\mathbf{X}} = \hat{\mathbf{y}}$

Then the *marginal* information matrix $\mathbf{Y}_a$ and marginal information vector $\mathbf{y}_a$ which satisfy $\mathbf{Y}_a\hat{\mathbf{X}}_a = \hat{\mathbf{y}}_a$, and similarly for $\mathbf{Y}_c$ are

$$\mathbf{Y}_a = \mathbf{A} - \mathbf{B}\mathbf{C}^{-1}\mathbf{B}^T \quad \mathbf{y}_a = \mathbf{a} - \mathbf{B}\mathbf{C}^{-1}\mathbf{c} \quad \mathbf{Y}_a\hat{\mathbf{x}}_a = \hat{\mathbf{y}}_a \tag{7.78}$$

$$\mathbf{Y}_c = \mathbf{C} - \mathbf{B}^T\mathbf{A}^{-1}\mathbf{B} \quad \mathbf{y}_c = \mathbf{c} - \mathbf{B}^T\mathbf{A}^{-1}\mathbf{a} \quad \mathbf{Y}_c\hat{\mathbf{x}}_c = \hat{\mathbf{y}}_c \tag{7.79}$$

7.A.2 TRAJECTORY INFORMATION FORM EQUIVALENCE

As stated earlier, the following are both equivalent:

$$\mathbf{Y} = \begin{pmatrix} \mathbf{Y}_k + \mathbf{F}^T\mathbf{Q}^{-1}\mathbf{F} & -\mathbf{F}^T\mathbf{Q}^{-1} \\ -\mathbf{Q}^{-1}\mathbf{F} & \mathbf{Q}^{-1} \end{pmatrix} \quad \mathbf{y} = \begin{pmatrix} \mathbf{y}_k - \mathbf{F}^T\mathbf{Q}^{-1}\mathbf{B}\mathbf{u} \\ \mathbf{Q}^{-1}\mathbf{B}\mathbf{u} \end{pmatrix}$$

$$\mathbf{P}_{k+1} = \mathbf{F}\mathbf{P}_k\mathbf{F}^T + \mathbf{G}\mathbf{Q}\mathbf{G}^T \quad \hat{\mathbf{x}}_{k+1} = \mathbf{F}\hat{\mathbf{x}}_k + \mathbf{B}\mathbf{u}_k$$

This can be shown in a few ways:

- The joint $\hat{\mathbf{X}}$ satisfies $\mathbf{Y}\hat{\mathbf{X}} = \mathbf{y}$, with $\hat{\mathbf{X}} = \begin{pmatrix} \hat{\mathbf{x}}_k \\ \hat{\mathbf{x}}_{k+1} \end{pmatrix} = \begin{pmatrix} \hat{\mathbf{x}}_k \\ \mathbf{F}\hat{\mathbf{x}}_k + \mathbf{B}\mathbf{u} \end{pmatrix}$:

$$\mathbf{Y}\hat{\mathbf{X}} = \begin{pmatrix} \mathbf{Y}_k + \mathbf{F}^T\mathbf{Q}^{-1}\mathbf{F} & -\mathbf{F}^T\mathbf{Q}^{-1} \\ -\mathbf{Q}^{-1}\mathbf{F} & \mathbf{Q}^{-1} \end{pmatrix}\begin{pmatrix} \hat{\mathbf{x}}_k \\ \mathbf{F}\hat{\mathbf{x}}_k + \mathbf{B}\mathbf{u} \end{pmatrix} \tag{7.80}$$

$$= \begin{pmatrix} \mathbf{Y}_k\hat{\mathbf{x}} + \mathbf{F}^T\mathbf{Q}^{-1}\mathbf{F}\hat{\mathbf{x}} - \mathbf{F}^T\mathbf{Q}^{-1}\{\mathbf{F}\hat{\mathbf{x}}_k + \mathbf{B}\mathbf{u}\} \\ -\mathbf{Q}^{-1}\mathbf{F}\hat{\mathbf{x}} + \mathbf{Q}^{-1}\{\mathbf{F}\hat{\mathbf{x}}_k + \mathbf{B}\mathbf{u}\} \end{pmatrix} \tag{7.81}$$

$$= \begin{pmatrix} \mathbf{y}_k - \mathbf{F}^T\mathbf{Q}^{-1}\mathbf{B}\mathbf{u} \\ \mathbf{Q}^{-1}\mathbf{B}\mathbf{u} \end{pmatrix} \tag{7.82}$$

$$= \mathbf{y} \tag{7.83}$$

- The $\mathbf{x}_k$ marginal of $\mathbf{Y}$ is equal to the prior $\mathbf{Y}_k$ and $\mathbf{y}_k$. This means that augmenting a predicted state $\mathbf{x}_{k+1}$ onto a given $\mathbf{x}_k$ system (including the addition of the dynamic model information) *does not* alter the marginal PDF for $\mathbf{x}_k$. This is shown as follows.

- We write the $\mathbf{x}_k$ marginal of $\mathbf{Y}$ as $\mathbf{Y}_k^{\mathrm{marg}}$, leaving $\mathbf{Y}_k$ to mean the prior information matrix of timestep k

$$\mathbf{Y}_k^{\mathrm{marg}} = \{\mathbf{Y}_k + \mathbf{F}^T\mathbf{Q}^{-1}\mathbf{F}\} - \mathbf{F}^T\mathbf{Q}^{-1}\{\mathbf{Q}^{-1}\}^{-1}\mathbf{Q}^{-1}\mathbf{F} \tag{7.84}$$

$$= \mathbf{Y}_k \tag{7.85}$$

- The $\mathbf{x}_{k+1}$ marginal of $\mathbf{Y}$ yields known expressions [2,19] for the prediction in covariance and information forms:

$$\mathbf{Y}_{k+1} = \{\mathbf{F}\mathbf{P}_k\mathbf{F}^T + \mathbf{Q}\}^{-1} \tag{7.86}$$

$$= \mathbf{M} - \mathbf{M}\mathbf{G}(\mathbf{G}^T\mathbf{M}\mathbf{G} + \mathbf{Q}^{-1})^{-1}\mathbf{G}^T\mathbf{M} \tag{7.87}$$

$$\left(\mathbf{M} = \mathbf{F}^{-T}\mathbf{Y}_k\mathbf{F}^{-1}\right) \tag{7.88}$$

The $\mathbf{x}_{k+1}$ marginal of $\mathbf{Y}$, using Equation 7.79, is

$$\mathbf{Y}_{k+1} = \mathbf{Q}^{-1} - \mathbf{Q}^{-1}\mathbf{F}\left[\mathbf{Y}_k + \mathbf{F}^T\mathbf{Q}^{-1}\mathbf{F}\right]^{-1}\mathbf{F}^T\mathbf{Q}^{-1} \tag{7.89}$$

Using the matrix inversion lemma:

$$[\mathbf{B}\mathbf{C}\mathbf{D} + \mathbf{A}]^{-1} = \mathbf{A}^{-1} - \mathbf{A}^{-1}\mathbf{B}\left[\mathbf{C}^{-1} + \mathbf{D}\mathbf{A}^{-1}\mathbf{B}\right]^{-1}\mathbf{D}\mathbf{A}^{-1} \tag{7.90}$$

With $\mathbf{A} \to \mathbf{Q} \quad \mathbf{B} \to \mathbf{F} \quad \mathbf{C} \to \mathbf{Y}_k^{-1} \quad \mathbf{D} \to \mathbf{F}'$

$$\mathbf{Y}_{k+1} = [\mathbf{F}\mathbf{Y}_k^{-1}\mathbf{F}^T + \mathbf{Q}]^{-1} \tag{7.91}$$

$$= [\mathbf{F}\mathbf{P}_k\mathbf{F}^T + \mathbf{G}\mathbf{Q}\mathbf{G}^T]^{-1} \tag{7.92}$$

which is the covariance form prediction equation.

The information form prediction equation is obtained by a different use of the matrix inversion lemma, using

$$\mathbf{A} \to \mathbf{F}\mathbf{Y}_k^{-1}\mathbf{F}^T \quad \mathbf{B} \to \mathbf{G} \quad \mathbf{C} \to \mathbf{Q} \quad \mathbf{D} \to \mathbf{G}^T$$

$$\mathbf{Y}_{k+1} = [\mathbf{F}\mathbf{Y}_k^{-1}\mathbf{F}^T + \mathbf{G}\mathbf{Q}\mathbf{G}^T]^{-1} \tag{7.93}$$

$$= \mathbf{M} - \mathbf{M}\mathbf{G}(\mathbf{Q}^{-1} + \mathbf{G}^T\mathbf{M}\mathbf{G})^{-1}\mathbf{G}^T\mathbf{M} \tag{7.94}$$

where

$$\mathbf{M} = [\mathbf{F}\mathbf{Y}_k^{-1}\mathbf{F}^T]^{-1} = \mathbf{F}^{-T}\mathbf{Y}_k\mathbf{F}^{-1}$$

Equations 7.89, 7.92, and 7.94 can also be found systematically from the following *augmented system* [23]:

$$\begin{pmatrix} \mathbf{Y}_k & 0 & \mathbf{F} & 0 \\ 0 & 0 & -\mathbf{I} & 0 \\ \mathbf{F} & -\mathbf{I} & 0 & \mathbf{G} \\ 0 & 0 & \mathbf{G}^T & \mathbf{Q}^{-1} \end{pmatrix} \begin{pmatrix} \mathbf{x}_k \\ \mathbf{x}_{k+1} \\ \nu \\ \mathbf{v}_k \end{pmatrix} = \begin{pmatrix} \mathbf{y}_k \\ 0 \\ -\mathbf{Bu}_k \\ 0 \end{pmatrix} \tag{7.95}$$

where ν is a vector of *Lagrange multipliers* [23] and $\mathbf{v}_k$ is the (unknown) process noise, as in Equation 7.25.

- Marginalizing (7.95) in the ordering $(\mathbf{v}, \nu, \text{then } \mathbf{x}_k)$ results in the predicted information marginal:

$$\mathbf{Y}_{k+1} = \mathbf{Q}^{-1} - \mathbf{Q}^{-1}\mathbf{F}(\mathbf{Y} + \mathbf{F}^T\mathbf{Q}^{-1}\mathbf{F})^{-1}\mathbf{F}^T\mathbf{Q}^{-1} \tag{7.96}$$

- Marginalizing (7.95) in the ordering $(\mathbf{x}_k, \nu \text{ then } \mathbf{v})$ results in the same predicted information marginal, in the form conventionally used in information filtering:

$$\mathbf{Y}_{k+1} = \mathbf{M} - \mathbf{MG}(\mathbf{Q}^{-1} + \mathbf{G}^T\mathbf{MG})^{-1}\mathbf{G}^T\mathbf{M} \tag{7.97}$$

$$\mathbf{M} = \mathbf{F}^{-T}\mathbf{YF}^{-1} \tag{7.98}$$

- Marginalizing (7.95) in the ordering $(\mathbf{x}_k, \mathbf{v} \text{ then } \nu)$ results in the inverse of the expression used in the covariance form Kalman filtering:

$$\mathbf{Y}_{k+1} = (\mathbf{FPF}^T + \mathbf{GQG}^T)^{-1} \tag{7.99}$$

ACKNOWLEDGMENTS

The work in this chapter was supported by the Australian Centre for Field Robotics. The authors would also like to thank Tim Bailey and Chris Lloyd for related technical discussions.

REFERENCES

1. T. Bailey and H. Durrant-Whyte. Decentralised data fusion with delayed states for consistent inference in mobile ad hoc networks. Technical report, 2007, http://www. personal.acfr.usyd.edu.au/tbailey/papers/delayedstated.pdf
2. Y. Bar-Shalom, X. Rong Li, and T. Kirubarajan. *Estimation with Applications to Tracking and Navigation.* Wiley, Hoboken, NJ, 2001.
3. B. Bollobás. *Modern Graph Theory.* Springer, New York, 1998.
4. J. Capitan, L. Merino, F. Caballero, and A. Ollero. Decentralized delayed-state information filter (DDSIF): A new approach for cooperative decentralized tracking. *Robotics and Autonomous Systems,* 59(6):376–388, 2011.
5. V. Chandrasekaran, N. Srebro, and P. Harsha. Complexity of inference in graphical models. In *24th Conference on Uncertainty in Artificial Intelligence and Statistics,* Helsinki, Finland, pp. 70–78, 2008.

6. K.C. Chang, C. Chong, and S. Mori. On scalable distributed sensor fusion. International Fusion, 2008 11th International conference Cologne, Germany, pp. 1–8, 2008.

7. K. Chang, C.-Y. Chong, and S. Mori. Analytical and computational evaluation of scalable distributed fusion algorithms. *IEEE Transactions on Aerospace and Electronic Systems*, 46(4):2022–2034, October 2010.

8. F. Dellaert and M. Kaess. Square root SAM: Simultaneous localization and mapping via square root information smoothing. *International Journal of Robotics Research*, 25(12):1181–1203, 2006.

9. H. Durrant-Whyte, M. Stevens, and E. Nettleton. Data fusion in decentralised sensing networks. In *4th International Conference on Information Fusion*, Montreal, Quebec, Canada, pp. 302–307, 2001.

10. R. Eustice, H. Singh, and J. Leonard. Exactly sparse delayed-state filters. In *Proceedings of the 2005 IEEE International Conference on Robotics and Automation*, Barcelona, Spain, pp. 2417–2424, 2005.

11. G.H. Golub and C.F. Van Loan. *Matrix Computations*, 3rd edn. The Johns Hopkins University Press, Baltimore, MD, 1996.

12. S. Grime and H.F. Durrant-Whyte. Data fusion in decentralized sensor networks. *Control Engineering Practice*, 2(5):849–863, 1994.

13. T. Kloks. Treewidth, computations and approximations. Lecture Notes in Computer Science. 1994.

14. E. Korach and N. Solel. Tree-width, path-width and cut-width. *Discrete Applied Mathematics*, 43(1):97–101, 1993.

15. J.J. Leonard and R J. Rikoski. Incorporation of delayed decision making into stochastic mapping. In *International Symposium on Experimental Robotics*, Montreal, Quebec, Canada, pp. 533–542, 2000.

16. A. Makarenko, A. Brooks, T. Kaupp, H. Durrant-Whyte, and F. Dellaert. Decentralised data fusion: A graphical model approach. In *Proceedings of the 12th International Conference on Information Fusion*, Seattle, WA, pp. 545–554, July 2009.

17. P.S. Maybeck. Stochastic models estimation and control. *Mathematics in Science and Engineering*, 1:423, 1979.

18. E. Nettleton. Decentralised architectures for tracking and navigation with multiple flight vehicles. PhD thesis, Australian Centre for Field Robotics, Department of Aerospace, Mechanical and Mechatronic Engineering, The University of Sydney, Sydney, Australia, 2003.

19. E.W. Nettleton and H.F. Durrant-Whyte. Delayed and asequent data in decentralised sensing networks. *Proceedings of SPIE—The International Society for Optical Engineering*, 4571:1–9, 2001.

20. D. Nicholson, C.M. Lloyd, S.J. Julier, and J.K. Uhlmann. Scalable distributed data fusion. In *Information Fusion, 2002. Proceedings of the Fifth International Conference*, Annapolis, MD, Vol. 1, pp. 630–635, 2002.

21. M.A. Paskin and G.D. Lawrence. Junction tree algorithms for solving sparse linear systems. Technical Report UCB/CSD-03-1271, University of California, Berkeley, CA, 2003.

22. S. Sukkarieh, E. Nettleton, J.-H. Kim, M. Ridley, A. Goktogan, and H. Durrant-Whyte. The ANSER project: Data fusion across multiple uninhabited air vehicles. *International Journal of Robotics Research*, 22(7–8):505–539, 2003.

23. P. Thompson. A novel augmented graph approach for estimation in localisation and mapping. PhD thesis, The University of Sydney, Sydney, Australia, March 2009.

24. P. Thompson and H. Durrant-Whyte. Decentralised data fusion in 2-tree sensor networks. In *Proceedings of the 13th International Conference on Information Fusion*, Edinburgh, U.K., pp. 1–8, July 2010.

8 Toward a Theoretical Foundation for Distributed Fusion

Ronald Mahler

CONTENTS

8.1 Introduction 200
8.2 Single-Target Distributed Fusion: Review 202
    8.2.1 Single-Target Bayes Filter..202
    8.2.2 T^2F with Independent Sources..204
    8.2.3 T^2F with Known Double-Counting 206
    8.2.4 Covariance Intersection 207
    8.2.5 Exponential Mixture Fusion 209
8.3 Finite-Set Statistics: Review..212
    8.3.1 Multitarget Recursive Bayes Filter 212
    8.3.2 Multitarget Calculus 214
    8.3.3 PHD Filter..216
    8.3.4 CPHD Filter..218
    8.3.5 Significant Recent Developments 220
8.4 General Multitarget Distributed Fusion..222
    8.4.1 Multitarget T^2F of Independent Sources....................................222
    8.4.2 Multitarget T^2F with Known Double-Counting 223
    8.4.3 Multitarget XM Fusion 223
8.5 CPHD/PHD Filter Distributed Fusion..224
    8.5.1 CPHD Filter T^2F of Independent Sources 224
    8.5.2 PHD Filter T^2F of Independent Sources 226
    8.5.3 CPHD Filter T^2F with Known Double-Counting........................227
    8.5.4 PHD Filter T^2F with Known Double-Counting...........................227
    8.5.5 CPHD Filter XM Fusion..228
    8.5.6 PHD Filter XM Fusion 229
8.6 Computational Issues...230
    8.6.1 Implementation: Exact T^2F Formulas...230
        8.6.1.1 Case 1: GM-PHD Tracks 231
        8.6.1.2 Case 2: Particle-PHD Tracks 232
    8.6.2 Implementation: XM T^2F Formulas 232
        8.6.2.1 Case 1: GM-PHD Tracks 232
        8.6.2.2 Case 2: Particle-PHD Tracks 234

8.7   Mathematical Derivations...234
        8.7.1   Proof: CPHD T^{2}F Fusion—Independent Sources 234
        8.7.2   Proof: PHD T^{2}F Fusion—Independent Sources..............................236
        8.7.3   Proof: CPHD Filter with Double-Counting...................................236
        8.7.4   Proof: CPHD Filter XM Fusion..237
        8.7.5   Proof: PHD Filter XM Fusion 238
        8.7.6   Proof: PHD Filter Chernoff Information..239
        8.7.7   Proof: XM Implementation 240
8.8   Conclusions...241
References...242

8.1   INTRODUCTION

Measurement-to-track fusion (MTF) refers to the process of collecting measurement data and then using it to improve the accuracy of the most recent estimates of the numbers and states of targets. Over the last two decades, both the theory and the practice of MTF have become increasingly mature. But, in parallel, another development has occurred: the increasing prevalence of physically dispersed sensors connected by communications networks, ad hoc or otherwise. One response to this development might be to try to apply MTF techniques to such situations. But because transmission links are often bandwidth-limited, it is often not possible to transmit raw measurements in a timely fashion, if at all. Consequently, emphasis has shifted to the transmission of *track data* and to *track-to-track fusion*, hereafter abbreviated as "T^{2}F." Most commonly, the term "track data" refers to target state estimates and their associated error-covariance matrices—as supplied, for example, by a radar equipped with an extended Kalman filter (EKF). T^{2}F refers to the process of merging single- or multi-target track data from multiple sensor sources, with the aim of achieving more accurate localization, increased track continuity, and fewer false tracks.

T^{2}F is fundamentally different than MTF. In particular, it cannot be addressed by processing tracks in the same way as measurements. Both MTF theory and practice are commonly based on two independence assumptions. First, measurements are statistically independent from time-step to time-step. Second, measurements generated by different sensor sources are statistically independent.

However, single-target track data is the consequence of some recursive filtering process, such as an EKF, and consequently is inherently time-correlated. If it is processed in the same way as measurements, spuriously optimistic target localization estimates will be the result. "Tracklet" approaches [1], such as inverse Kalman filters, decorrelate tracks so that they can be processed in the same way as measurements. However, such techniques cannot be effectively applied when targets are rapidly maneuvering, since decorrelation must be performed over some extended time-window.

Furthermore, multisource track data (like multisource measurement data) in distributed networks can be corrupted by "double counting" [2]. A simple example: data from node A is passed to nodes X and Y, which then pass it to node B. If node B processes this data as though it were independent, then spuriously optimistic target localization will again be the result. Many T^2 fusion solutions have been devised for networks with pre-specified topologies. But such methods will not be applicable to ad hoc networks. "Pedigree" techniques have been proposed to address this challenge, by having every node "stamp" the tracks with suitable metadata before passing them on. In a large network, however, accumulated metadata can eventually greatly exceed the size of the track data that it documents. This problem can be sidestepped through node-to-node querying methods—but at the cost of increased bandwidth requirements. (A more practical difficulty: the large number of legacy networks makes it unlikely that any pedigree convention is likely to be accepted, standardized, and implemented across all or even some of them.)

In part because of such issues, T^2F theory is probably as underdeveloped now as MTF theory was two or three decades ago. The goal of this chapter is to try to remedy this situation by proposing the elements of a general theoretical foundation for T^2F, building on ideas that I first suggested in 2000 [3]. These ideas have recently been greatly refined, especially by Daniel Clark and his associates [4–6].

The methodology will be the same as that which I have previously applied to MTF and which has been described in *Statistical Multisource-Multitarget Information Fusion* [7]:

1. Model an entire multisensor-multitarget system as a single, evolving stochastic process using the theory of random finite sets.
2. Formulate an optimal solution to the problem at hand—typically in the form of some kind of multisource-multitarget recursive Bayes filter.
3. Recognize that one way to accomplish this is to find an optimal solution to the corresponding single-sensor, single-target problem and then generalize it to the multisensor-multitarget case.
4. Recognize that this optimal solution will almost always be computationally intractable, and thus that principled statistical approximations of it must be formulated.

The principled approximation methods that I have most frequently advocated are as follows:

1. Probability hypothesis density (PHD) filters, in which the multitarget process is approximated as an evolving *Poisson process* [7, chapter 16].
2. Cardinalized PHD (CPHD) filters, in which it is approximated as an evolving *identically, independently distributed cluster* (i.i.d.c.) *process* [7, chapter 16].
3. Multi-Bernoulli filters, in which it is approximated as an evolving multi-Bernoulli process [7, chapter 17].

In what follows I will consider only the first two approximation methods, which will be applied to three successively more difficult multisource-multitarget track fusion challenges:

1. Exact T^2F of independent track sources.
2. Exact T^2F of track sources with known double-counting.
3. Approximate T^2F of track sources having unknown correlations, using multitarget generalizations of Uhlmann and Julier's covariance intersection (CI) approach.

In each of these cases I proceed by formulating a general approach to multisource-multitarget T^2F and then by deriving more computationally tractable approximations using CPHD and PHD filters in the manner proposed by Clark et al.

The chapter is organized as follows:

1. Section 8.2: Review of single-target T^2F theory.
2. Section 8.3: Review of those aspects of finite-set statistics (FISST) required to understand the chapter.
3. Section 8.4: Direct generalization of single-target T^2F to multitarget T^2F.
4. Section 8.5: Approximation of this general approach using CPHD and PHD filters.
5. Section 8.6: A discussion of possible implementation approaches.
6. Section 8.7: Mathematical derivations.
7. Section 8.8: Summary and conclusions.

8.2   SINGLE-TARGET DISTRIBUTED FUSION: REVIEW

In this section, I summarize some major aspects of single-target T^2F that will be required for what follows:

1. Section 8.2.1: The single-target recursive Bayes filter is the foundation of the material in this section. I summarize the basic elements of this filter and define the concept of a "track" in general.
2. Section 8.2.2: Single-target T^2F when the track sources are independent. Approach: the track-merging formula of Chong et al. and its special case, Bayes parallel combination.
3. Section 8.2.3: Single-target T^2F when the track sources are dependent because of known double-counting. Approach: the generalized track-merging formula of Chong et al.
4. Section 8.2.4: Single-target T^2F when the track sources are linear-Gaussian but their correlations are completely unknown. Approach: the CI method of Uhlmann and Julier.
5. Section 8.2.5: Single-target T^2F when the track sources are arbitrary and their correlations are completely unknown. Approach: Mahler's generalized CI method, rechristened by Julier and Uhlmann as "exponential mixture" (XM) fusion.

8.2.1   SINGLE-TARGET BAYES FILTER

The approach in this section is based on the Bayesian theoretical foundation for single-target tracking, the *single-target Bayes nonlinear filter* (see Chapter 2 of [7]). This filter propagates a Bayes posterior distribution $f_{k|k}(\mathbf{x}|Z^k)$ through time

$$\cdots \to f_{k|k}(\mathbf{x}\,|\,Z^k) \overset{\text{predictor}}{\to} f_{k+1|k}(\mathbf{x}\,|\,Z^k) \overset{\text{corrector}}{\to} f_{k+1|k+1}(\mathbf{x}\,|\,Z^{k+1}) \to \cdots \qquad (8.1)$$

where

    $\mathbf{x}$ is the single-target state-vector

    Z^k: $\mathbf{z}_1,\ldots,\mathbf{z}_k$ is a time-sequence of measurements collected by the sensor at times $t_1,\ldots,t_k$

The Bayes filter presumes the existence of models for the sensor and for the presumed interim target motion, for example the additive models

$$\mathbf{X}_{k+1|k} = \varphi_k(\mathbf{x}) + \mathbf{W}_k, \quad \mathbf{Z}_{k+1} = \eta_{k+1}(\mathbf{x}) + \mathbf{V}_{k+1}, \tag{8.2}$$

where (1) $\mathbf{x}$ is the target state, (2) the deterministic motion model $\varphi_k(\mathbf{x})$ is a nonlinear function of $\mathbf{x}$, (3) $\mathbf{W}_k$ is a zero-mean random vector (the "plant noise"), (4) the deterministic measurement model $\eta(\mathbf{x})$ is a nonlinear function of $\mathbf{x}$, and (5) $\mathbf{V}_k$ is a zero-mean random vector (the sensor measurement noise). Given these models one can construct a Markov transition density and likelihood function. For the additive models, for example, these have the form

$$f_{k+1|k}(\mathbf{x} \mid \mathbf{x}') = f_{\mathbf{W}_k}(\mathbf{x} - \varphi_k(\mathbf{x}')), \quad f_{k+1}(\mathbf{z}|\mathbf{x}) = f_{\mathbf{V}_{k+1}}(\mathbf{z} - \eta_{k+1}(\mathbf{x})). \tag{8.3}$$

The single-target recursive Bayes filter is defined by the time-update and measurement-update equations

$$f_{k+1|k}(\mathbf{x} \mid Z^k) = \int f_{k+1|k}(\mathbf{x} \mid \mathbf{x}') \cdot f_{k|k}(\mathbf{x}' \mid Z^k) d\mathbf{x}' \tag{8.4}$$

$$f_{k+1|k+1}(\mathbf{x} \mid Z^{k+1}) = \frac{f_{k+1}(\mathbf{z}_{k+1} \mid \mathbf{x}) \cdot f_{k+1|k}(\mathbf{x} \mid Z^k)}{f_{k+1}(\mathbf{z}_{k+1} \mid Z^k)} \tag{8.5}$$

where the Bayes normalization factor is

$$f_{k+1}(\mathbf{z}_{k+1} \mid Z^k) = \int f_{k+1}(\mathbf{z}_{k+1} \mid \mathbf{x}) \cdot f_{k+1|k}(\mathbf{x} \mid Z^k) d\mathbf{x}. \tag{8.6}$$

Information of interest—target position, velocity, type, etc.—can be extracted from $f_{k|k}(\mathbf{x}|Z^k)$ using a Bayes-optimal multitarget state estimator. The maximum a posteriori (MAP) estimator, for example, determines the most probable target state:

$$\mathbf{x}_{k+1|k+1}^{\text{MAP}} = \arg\sup_{\mathbf{x}} f_{k+1|k+1}(\mathbf{x} \mid Z^{k+1}). \tag{8.7}$$

Multisensor, single-target MTF with *independent* sensors is accomplished by applying Equation 8.5 successively for each sensor. Suppose, for example, that there are s sensors. Their respective, simultaneously collected measurements $\overset{1}{\mathbf{z}},\ldots,\overset{s}{\mathbf{z}}$ are mediated by likelihood functions $\overset{1}{f}_{k+1}(\overset{1}{\mathbf{z}} \mid \mathbf{x}),\ldots,\overset{s}{f}_{k+1}(\overset{s}{\mathbf{z}} \mid \mathbf{x})$. By applying Equation 8.5 first using $\overset{1}{f}_{k+1}(\overset{1}{\mathbf{z}} \mid \mathbf{x})$ and then using $\overset{2}{f}_{k+1}(\overset{2}{\mathbf{z}} \mid \mathbf{x})$ and so on, the measurements $\overset{1}{\mathbf{z}},\ldots,\overset{s}{\mathbf{z}}$ are not only fused, but differences in sensor noise, sensor geometry, sensor obscurations, etc., are taken into account. Equivalently, one can apply Equation 8.5 to the joint likelihood function

$$\overset{1}{f}_{k+1}(Z \mid \mathbf{x}) = \overset{1}{f}_{k+1}(\overset{1}{\mathbf{z}} \mid \mathbf{x}) \cdots \overset{s}{f}_{k+1}(\overset{s}{\mathbf{z}} \mid \mathbf{x}) \tag{8.8}$$

where $Z = \{\overset{1}{\mathbf{z}},\ldots,\overset{s}{\mathbf{z}}\}$ denotes the set of multisensor measurements.

When motion and measurement models are linear-Gaussian, the Bayes filter reduces to the Kalman filter. Likewise, the multisensor Bayes filter (for independent sensors) reduces to the multisensor Kalman filter. In either case, a "track" can mean any of the following: (1) an instantaneous state-estimate $\mathbf{x}_{k+1|k+1}$, (2) $\mathbf{x}_{k+1|k+1}$ together with its error covariance matrix $P_{k+1|k+1}$, (3) a labeled time-sequence of state-estimates, or (4) a labeled time-sequence of state-estimates and error covariance matrices.

Remark 1: Since my goal is to develop a more general T²F theory, in what follows a "track" at a particular time-step k will refer to the entire distribution $f_{k|k}(\mathbf{x}|Z^k)$, rather than to the estimates $\mathbf{x}_{k|k}$ or $(\mathbf{x}_{k|k}, P_{k|k})$ extracted from it. Also, for the sake of notational simplicity, I will typically suppress measurement-dependence and employ the abbreviation

$$f_{k|k}(\mathbf{x}) \overset{\text{abbr.}}{=} f_{k|k}(\mathbf{x} \mid Z^k). \tag{8.9}$$

8.2.2   T²F WITH INDEPENDENT SOURCES

Suppose that a single target is being tracked and that s independent sources, relying on their own dedicated local sensors, provide track data about this target to a T²F site. The jth sensor suite collects a time-sequence $\overset{j}{Z}{}^k : \overset{j}{Z}_1,\ldots,\overset{j}{Z}_k$, where $\overset{j}{Z}_l$ denotes the set of measurements supplied by the jth source's sensors at time t_l. The source does not pass its measurements directly to the fusion site. Rather, it passes the following information:

- Measurement-updated, single-target track data, in the form of posterior distributions $\overset{j}{f}_{k|k}(\mathbf{x}) \overset{\text{abbr.}}{=} \overset{j}{f}_{k|k}(\mathbf{x} \mid \overset{j}{Z}{}^k)$
- Time-updated, single-target track data, in the form of distributions $\overset{j}{f}_{k+1|k}(\mathbf{x}) \overset{\text{abbr.}}{=} \overset{j}{f}_{k+1|k}(\mathbf{x} \mid \overset{j}{Z}{}^k)$

Let $f_{k|k}(\mathbf{x}) \overset{\text{abbr.}}{=} f_{k|k}(\mathbf{x} \mid Z^k)$ be the fusion node's determination of the target state, given the accumulated track data Z^k supplied by all of the sensor sources. Then Chong et al. [2] noted that the fused data at time-step $k+1$ is exactly specified by the following track-merging formula:

$$f_{k+1|k+1}(\mathbf{x}) \propto \frac{\overset{1}{f}_{k+1|k+1}(\mathbf{x})}{\overset{1}{f}_{k+1|k}(\mathbf{x})} \cdots \frac{\overset{s}{f}_{k+1|k+1}(\mathbf{x})}{\overset{s}{f}_{k+1|k}(\mathbf{x})} \cdot f_{k+1|k}(\mathbf{x}) \tag{8.10}$$

where the constant of proportionality is

$$K = \int \frac{\overset{1}{f}_{k+1|k+1}(\mathbf{x})}{\overset{1}{f}_{k+1|k}(\mathbf{x})} \cdots \frac{\overset{s}{f}_{k+1|k+1}(\mathbf{x})}{\overset{s}{f}_{k+1|k}(\mathbf{x})} \cdot f_{k+1|k}(\mathbf{x}) d\mathbf{x}. \tag{8.11}$$

This formula also applies to the asynchronous-sensor case. If each source has its own data rate, then the measurement-collection times $t_1,\dots,t_k$ can be taken to refer to the arrival times of data from all of the sources, taken collectively. If at time t_l only s_l of the sources provide data, then Equation 8.10 is replaced by the corresponding formula for those sources only.

Equation 8.10 is an immediate consequence of Bayes' rule. Let $f_{k+1}(\overset{j}{Z} \mid \mathbf{x})$ be the joint likelihood function for the jth source's local sensors. Then

$$f_{k+1|k+1}(\mathbf{x}) \propto f_{k+1}(\overset{1}{Z}_{k+1} \mid \mathbf{x}) \cdots f_{k+1}(\overset{s}{Z}_{k+1} \mid \mathbf{x}) \cdot f_{k+1|k}(\mathbf{x}) \tag{8.12}$$

and thus Equation 8.10 follows from the fact that $\overset{j}{f}_{k+1|k+1}(\mathbf{x}) \propto f_{k+1}(\overset{j}{Z}_{k+1} \mid \mathbf{x}) \cdot \overset{j}{f}_{k+1|k}(\mathbf{x})$ for all $j = 1,\dots,s$.

Suppose, now, that the sources do not pass on their time-updated track data $\overset{j}{f}_{k+1|k}(\mathbf{x})$ but, rather, only their measurement-updated track data $\overset{j}{f}_{k+1|k+1}(\mathbf{x})$. (This is what happens with radars equipped with EKFs, for example.) In this case, Equation 8.10 can no longer be constructed, and some approximation must be devised.

One approach is to presume that all of the sources employ identical target motion models. That is, the sources' Markov densities $\overset{j}{f}_{k+1|k}(\mathbf{x} \mid \mathbf{x}')$ are identical to the fusion site's Markov density: $\overset{j}{f}_{k+1|k}(\mathbf{x} \mid \mathbf{x}') = f_{k+1|k}(\mathbf{x} \mid \mathbf{x}')$ for all $j = 1,\dots,s$. Under this assumption, the fusion site can itself construct time-updated track data for the sources, using the prediction integral

$$\overset{j}{f}_{k+1|k}(\mathbf{x}) = \int f_{k+1|k}(\mathbf{x} \mid \mathbf{x}') \cdot \overset{j}{f}_{k|k}(\mathbf{x}) d\mathbf{x}, \tag{8.13}$$

and then apply Equation 8.10.

A second but more restrictive approximation is also possible. It is based on the presumption that the sources' time-updated track data is identical to the fusion site's: $\overset{j}{f}_{k+1|k}(\mathbf{x}) = f_{k+1|k}(\mathbf{x})$ for all $j = 1,\dots,s$. In this case, Equation 8.10 reduces to

$$f_{k+1|k+1}(\mathbf{x}) \propto \overset{1}{f}_{k+1|k+1}(\mathbf{x}) \cdots \overset{s}{f}_{k+1|k+1}(\mathbf{x}) \cdot f_{k+1|k}(\mathbf{x})^{1-s}. \tag{8.14}$$

This formula is known as "Bayes parallel combination" [7, p. 137].

8.2.3 T²F WITH KNOWN DOUBLE-COUNTING

In the previous section, it was assumed that each data source is equipped with its own suite of *dedicated* sensors—that is, the sources share no sensors in common. That is, expressed with greater mathematical precision, let $\overset{i}{Z}_1,\ldots,\overset{i}{Z}_k$ be the time-sequence of measurement-sets for the ith source and let $\overset{j}{Z}_1,\ldots,\overset{j}{Z}_k$ be the time-sequence of measurement-sets for the jth source. Then $\overset{i}{Z}_l \cap \overset{j}{Z}_l\, \phi$ whenever $i \neq j$, for all $l = 1,\ldots,k$.

If on the other hand $\overset{i}{Z}_l \cap \overset{j}{Z}_l\, \phi$, then the sources are sharing at least some sensors and *double-counting* of measurements occurs. Chong et al. [2] generalized Equation 8.10 to this case—assuming that one knows, a priori, which sensors are being shared by which sources. Define $Z_{k+1} = \overset{1}{Z}_{k+1} \cup \cdots \cup \overset{s}{Z}_{k+1}$. Let

- $\overset{12}{Z}_{k+1}$ be the measurements supplied to the second source that are not in $\overset{1}{Z}_{k+1}$
- $\overset{13}{Z}_{k+1}$ the measurements supplied to the third source that are not in $\overset{1}{Z}_{k+1} \cup \overset{12}{Z}_{k+1}$
- $\overset{14}{Z}_{k+1}$ the measurements supplied to the fourth source that are not in $\overset{1}{Z}_{k+1} \cup \overset{12}{Z}_{k+1} \cup \overset{13}{Z}_{k+1}$

and so on. Define

$$\overset{(j)}{Z}_{k+1} = \overset{j}{Z}_{k+1} - \overset{1j}{Z}_{k+1}. \tag{8.15}$$

Then Equation 8.10 generalizes to

$$f_{k+1|k+1}(\mathbf{x}) \propto \frac{\overset{1}{f}_{k+1|k+1}(\mathbf{x})}{\overset{1}{f}_{k+1|k}(\mathbf{x})} \cdot \frac{\overset{2}{f}_{k+1|k+1}(\mathbf{x})}{\overset{2}{f}_{k+1|k}(\mathbf{x}\,|\,\overset{(2)}{Z})} \cdots \frac{\overset{s}{f}_{k+1|k+1}(\mathbf{x})}{\overset{s}{f}_{k+1|k}(\mathbf{x}\,|\,\overset{(s)}{Z})} \cdot f_{k+1|k}(\mathbf{x}). \tag{8.16}$$

If Equation 8.16 is to be applied, the jth source must know which sensors it shares with each of sources $1,\ldots,j-1$, and must pass on $\overset{j}{f}_{k+1|k}(\mathbf{x}\,|\,\overset{(j)}{Z})$ in addition to $\overset{j}{f}_{k+1|k}(\mathbf{x})$. Clearly, as the number of sensors increases, the problem becomes more complex, in terms of both computational cost and communications requirements.

Equation 8.16 is, once again, an immediate consequence of Bayes' rule:

$$f_{k+1|k+1}(\mathbf{x}) \propto f_{k+1}(\overset{1}{Z}_{k+1}\,|\,\mathbf{x}) \cdot f_{k+1}(\overset{12}{Z}_{k+1}\,|\,\mathbf{x}) \cdots f_{k+1}(\overset{1s}{Z}_{k+1}\,|\,\mathbf{x}) \cdot f_{k+1|k}(\mathbf{x}) \tag{8.17}$$

$$= f_{k+1}(\overset{1}{Z}_{k+1}\,|\,\mathbf{x}) \cdot \frac{f_{k+1}(\overset{2}{Z}_{k+1}\,|\,\mathbf{x})}{f_{k+1}(\overset{(2)}{Z}_{k+1}\,|\,\mathbf{x})} \cdots \frac{f_{k+1}(\overset{s}{Z}_{k+1}\,|\,\mathbf{x})}{f_{k+1}(\overset{(s)}{Z}_{k+1}\,|\,\mathbf{x})} \cdot f_{k+1|k}(\mathbf{x}) \tag{8.18}$$

$$\propto \frac{\overset{1}{f}_{k+1|k+1}(\mathbf{x})}{\overset{1}{f}_{k+1|k}(\mathbf{x})} \cdot \frac{\overset{2}{f}_{k+1|k+1}(\mathbf{x})}{\overset{2}{f}_{k+1|k}(\mathbf{x} \mid \overset{(2)}{Z})} \cdots \frac{\overset{s}{f}_{k+1|k+1}(\mathbf{x})}{\overset{s}{f}_{k+1|k}(\mathbf{x} \mid \overset{(s)}{Z})} \cdot f_{k+1|k}(\mathbf{x}). \tag{8.19}$$

As an example, set $s = 2$ and suppose that $f_{k+1|k}(\mathbf{x}) = \overset{1}{f}_{k+1|k}(\mathbf{x})$. Then Equation 8.16 reduces to the following formula of Chong et al. [2]:

$$f_{k+1|k+1}(\mathbf{x}) \propto \frac{\overset{1}{f}_{k+1|k+1}(\mathbf{x}) \cdot \overset{2}{f}_{k+1|k+1}(\mathbf{x})}{\overset{2}{f}_{k+1|k}(\mathbf{x} \mid \overset{1}{Z}_{k+1} \cap \overset{2}{Z}_{k+1})}. \tag{8.20}$$

For in this case, $\overset{12}{Z}_{k+1} = \overset{2}{Z}_{k+1} - (\overset{1}{Z}_{k+1} \cap \overset{2}{Z}_{k+1})$ and so $\overset{(2)}{Z}_{k+1} = \overset{2}{Z}_{k+1} - \overset{12}{Z}_{k+1} = \overset{1}{Z}_{k+1} \cap \overset{2}{Z}_{k+1}$. Thus

$$f_{k+1|k+1}(\mathbf{x}) \propto \frac{\overset{1}{f}_{k+1|k+1}(\mathbf{x})}{\overset{1}{f}_{k+1|k}(\mathbf{x})} \cdot \frac{\overset{2}{f}_{k+1|k+1}(\mathbf{x})}{\overset{2}{f}_{k+1|k}(\mathbf{x} \mid \overset{(2)}{Z})} \cdot f_{k+1|k}(\mathbf{x}) \tag{8.21}$$

$$= \frac{\overset{1}{f}_{k+1|k+1}(\mathbf{x})}{\overset{1}{f}_{k+1|k}(\mathbf{x})} \cdot \frac{\overset{2}{f}_{k+1|k+1}(\mathbf{x})}{\overset{2}{f}_{k+1|k}(\mathbf{x} \mid \overset{1}{Z}_{k+1} \cap \overset{2}{Z}_{k+1})} \cdot f_{k+1|k}(\mathbf{x}) \tag{8.22}$$

$$= \frac{\overset{1}{f}_{k+1|k+1}(\mathbf{x}) \cdot \overset{2}{f}_{k+1|k+1}(\mathbf{x})}{\overset{2}{f}_{k+1|k}(\mathbf{x} \mid \overset{1}{Z}_{k+1} \cap \overset{2}{Z}_{k+1})}. \tag{8.23}$$

8.2.4 COVARIANCE INTERSECTION

Sections 8.2.2 and 8.2.3 address situations in which enough a priori knowledge is available to make exact track merging possible. In general, however, this will not be possible. This is because not enough a priori information is available, or because even if available it cannot be effectively utilized. This situation is, in part, what the CI method of Uhlmann and Julier [8–10] is intended to address.

Suppose that a single target is being observed by two track sources. At time-step k, the first source provides a track $(\overset{0}{\mathbf{x}}_{k|k}, \overset{0}{P}_{k|k})$ and the second source provides a track $(\overset{1}{\mathbf{x}}_{k|k}, \overset{1}{P}_{k|k})$. CI is a method for merging $(\overset{0}{\mathbf{x}}_{k|k}, \overset{0}{P}_{k|k})$ and $(\overset{1}{\mathbf{x}}_{k|k}, \overset{1}{P}_{k|k})$ into a single $(\mathbf{x}_{k|k}, P_{k|k})$ that is robust with respect to ambiguity. This means, in particular, that the uncertainty $P_{k|k}$ in $\mathbf{x}_{k|k}$ is neither too small (over-confidence) nor too large (under-confidence). Let $0 \leq \omega \leq 1$ and define $(\overset{\omega}{\mathbf{x}}_{k|k}, \overset{\omega}{P}_{k|k})$ by

$$\overset{\omega}{P}_{k|k}^{-1} = (1 - \omega) \overset{0}{P}_{k|k}^{-1} + \omega \overset{1}{P}_{k|k}^{-1} \tag{8.24}$$

$$\overset{\omega}{P}{}^{-1}_{k|k} \overset{\omega}{\mathbf{x}}_{k|k} = (1 - \omega)\, \overset{0}{P}{}^{-1}_{k|k} \overset{0}{\mathbf{x}}_{k|k} + \omega\, \overset{1}{P}{}^{-1}_{k|k} \overset{1}{\mathbf{x}}_{k|k}. \tag{8.25}$$

The matrix $\overset{\omega}{P}_{k|k}$ is positive-definite regardless of the value of ω, and $(\overset{\omega}{\mathbf{x}}_{k|k}, \overset{\omega}{P}_{k|k})$ instantiates to $(\overset{0}{\mathbf{x}}_{k|k}, \overset{0}{P}_{k|k})$ resp. $(\overset{1}{\mathbf{x}}_{k|k}, \overset{1}{P}_{k|k})$ when $\omega = 0$ resp. $\omega = 1$.

Suppose that

$$(\mathbf{x} - \overset{0}{\mathbf{x}}_{k|k})^T\, \overset{0}{P}{}^{-1}_{k|k} (\mathbf{x} - \overset{0}{\mathbf{x}}_{k|k}) \le \sigma^2 \tag{8.26}$$

$$(\mathbf{x} - \overset{1}{\mathbf{x}}_{k|k})^T\, \overset{1}{P}{}^{-1}_{k|k} (\mathbf{x} - \overset{1}{\mathbf{x}}_{k|k}) \le \sigma^2 \tag{8.27}$$

are the error hyper-ellipsoids of size σ associated with the tracks $(\overset{0}{\mathbf{x}}_{k|k}, \overset{0}{P}_{k|k})$ and $(\overset{1}{\mathbf{x}}_{k|k}, \overset{1}{P}_{k|k})$. Then it can be shown that, for any $0 \le \omega \le 1$ and any $\sigma > 0$,

$$(\mathbf{x} - \overset{\omega}{\mathbf{x}}_{k|k})^T\, \overset{\omega}{P}{}^{-1}_{k|k} (\mathbf{x} - \overset{\omega}{\mathbf{x}}_{k|k}) \le \sigma^2. \tag{8.28}$$

That is, the error hyper-ellipsoid of the merged track always contains the intersection of the interiors of the error hyper-ellipsoids of the original tracks.

Intuitively speaking, ω should be chosen so that the hypervolume of the hyper-ellipsoid $(\mathbf{x} - \overset{\omega}{\mathbf{x}}_{k|k})^T\, \overset{\omega}{P}{}^{-1}_{k|k} (\mathbf{x} - \overset{\omega}{\mathbf{x}}_{k|k}) = \sigma^2$ is as *small* as possible. That is, the merged hyper-ellipsoid should have the best possible fit to the intersection-region of the two original hyper-ellipsoids. Uhlmann and Julier proposed choosing $\omega = \hat{\omega}$ so that it minimizes either the trace tr $\overset{\omega}{P}_{k|k}$ or the determinant det $\overset{\omega}{P}_{k|k}$. They demonstrated that this approach yields an approximation of the exact merged track that is unbiased and whose degree of uncertainty is not overstated.

Fränken and Hüpper [11] subsequently proposed a more computationally tractable "fast CI" approximation. Here, ω is chosen according to the formula

$$1 - \omega = \frac{\det(\overset{0}{P}{}^{-1}_{k|k} + \overset{1}{P}{}^{-1}_{k|k}) - \det \overset{1}{P}{}^{-1}_{k|k} + \det \overset{0}{P}{}^{-1}_{k|k}}{2 \cdot \det(\overset{0}{P}{}^{-1}_{k|k} + \overset{1}{P}{}^{-1}_{k|k})}. \tag{8.29}$$

These authors also proposed the following generalization. Consider the multisource CI problem defined by

$$\overset{\omega}{P}{}^{-1}_{k|k} = \omega_1 \overset{1}{P}{}^{-1}_{k|k} + \cdots + \omega_n \overset{n}{P}{}^{-1}_{k|k} \tag{8.30}$$

$$\overset{\omega}{P}{}^{-1}_{k|k} \overset{\omega}{\mathbf{x}}_{k|k} = \omega_1 \overset{1}{P}{}^{-1}_{k|k} \overset{0}{\mathbf{x}}_{k|k} + \cdots + \omega_n \overset{n}{P}{}^{-1}_{k|k} \overset{n}{\mathbf{x}}_{k|k} \tag{8.31}$$

with $\omega_1 + \cdots + \omega_n = 1$. Then their proposed approximation is

$$\omega_i = \frac{\det P_{kk}^{-1} - \det(P_{kk}^{-1} - \overset{i}{P_{k|k}^{-1}}) + \det \overset{i}{P_{k|k}^{-1}}}{n \cdot \det P_{kk}^{-1} + \sum_{j=1}^{n} \left[\det \overset{j}{P_{k|k}^{-1}} - \det(P_{kk}^{-1} - \overset{j}{P_{k|k}^{-1}}) \right]} \tag{8.32}$$

where

$$P_{kk}^{-1} = \sum_{i=1}^{n} \overset{i}{P_{k|k}^{-1}}.$$

Much research has been devoted to determining the effectiveness of CI. The emerging consensus seems to be that CI tends to produce estimates of the fused track that are pessimistic. That is, the fused target-localizations are significantly worse than what one would get from an exact fused solution. This behavior is exactly what one would expect, given that, by design, CI must address worst-case situations in which to-be-fused tracks could be highly correlated.

8.2.5 EXPONENTIAL MIXTURE FUSION

The CI method addresses the merging of only linear-Gaussian track sources. How might it be generalized to more general sources? In 2000 [3], I observed that the following identity is true:

$$\frac{N_{\underset{P_{k|k}}{0}}(\mathbf{x} - \mathbf{x}_{k|k})^{1-\omega} \cdot N_{\underset{P_{k|k}}{1}}(\mathbf{x} - \mathbf{x}_{k|k})^{\omega}}{\int N_{\underset{P_{k|k}}{0}}(\mathbf{y} - \mathbf{x}_{k|k})^{1-\omega} \cdot N_{\underset{P_{k|k}}{1}}(\mathbf{y} - \mathbf{x}_{k|k})^{\omega} d\mathbf{y}} = N_{\underset{P_{k|k}}{\omega}}(\mathbf{x} - \mathbf{x}_{k|k}) \tag{8.33}$$

where, in general, $N_{P_0}(\mathbf{x} - \mathbf{x}_0)$ denotes a multidimensional Gaussian distribution with mean $\mathbf{x}_0$ and covariance matrix P_0. That is, CI can be expressed entirely in terms of density functions rather than covariance matrices. I proposed, therefore, that the following definition be taken as the obvious generalization of the CI merging formula to arbitrary track sources:

$$\overset{\omega}{f}_{k+1|k+1}(\mathbf{x}) = \frac{\overset{0}{f}_{k+1|k+1}(\mathbf{x})^{1-\omega} \cdot \overset{1}{f}_{k+1|k+1}(\mathbf{x})^{\omega}}{\int \overset{0}{f}_{k+1|k+1}(\mathbf{y})^{1-\omega} \cdot \overset{1}{f}_{k+1|k+1}(\mathbf{y})^{\omega} d\mathbf{y}}. \tag{8.34}$$

Hurley independently proposed Equation 8.34 in 2002 [12]. He also justified its theoretical reasonableness on the basis of its similarity to Chernoff information, which is defined as follows:

$$C(\overset{1}{f}_{k+1|k+1}; \overset{0}{f}_{k+1|k+1}) = \sup_{0 \le \omega \le 1} \left(-\log \int \overset{0}{f}_{k+1|k+1}(\mathbf{x})^{1-\omega} \cdot \overset{1}{f}_{k+1|k+1}(\mathbf{x})^{\omega} d\mathbf{x} \right). \tag{8.35}$$

As it turns out, Equation 8.34 is a special case of "logarithmic opinion pooling," when the opinions of only two experts are being pooled [13]. This means that CI is itself a special case of logarithmic opinion pooling, given that the opinions of two linear-Gaussian experts are being pooled. Julier and Uhlmann have described Equation 8.34 as an "XM model" for track fusion [14,15]. (It has also been given the name "Chernoff fusion" [16].) I will adopt their terminology in what follows, abbreviating it as "XM fusion." (Julier has also suggested approximations for computing the XM fusion formula when the original distributions $\overset{0}{f}_{k+1|k+1}(\mathbf{x})$ and $\overset{1}{f}_{k+1|k+1}(\mathbf{x})$ are Gaussian mixtures [14].)

The XM fusion density has several appealing properties. First, and perhaps most importantly, Julier has shown that it is invariant with respect to double counting [17]. That is, suppose that the distributions $\overset{0}{f}_{k+1|k+1}(\mathbf{x})$ and $\overset{1}{f}_{k+1|k+1}(\mathbf{x})$ have double-counted information in the sense of Section 8.2.3. Then $\overset{\omega}{f}_{k+1|k+1}(\mathbf{x})$ incorporates the double-counted information only once, in the same sense as does Equation 8.20.

Second, for all $0 \le \omega \le 1$ [9]:

$$\min\{\overset{0}{f}_{k+1|k+1}(\mathbf{x}), \overset{1}{f}_{k+1|k+1}(\mathbf{x})\} \le \overset{\omega}{f}_{k+1|k+1}(\mathbf{x}) \quad \text{(all } \mathbf{x}) \tag{8.36}$$

$$\max\{\overset{0}{f}_{k+1|k+1}(\mathbf{x}_0), \overset{1}{f}_{k+1|k+1}(\mathbf{x}_0)\} \le \overset{\omega}{f}_{k+1|k+1}(\mathbf{x}_0) \quad \text{(there exists } \mathbf{x}_0). \tag{8.37}$$

The first inequality indicates that $\overset{\omega}{f}_{k+1|k+1}(\mathbf{x})$ does not reduce information (as compared to the original distributions), whereas the second one indicates that it can also increase it.

In Ref. [3], I proposed the following as the most theoretically reasonable procedure for optimizing ω

$$\hat{\omega} = \arg \sup_{\omega} \sup_{\mathbf{x}} \overset{\omega}{f}_{k+1|k+1}(\mathbf{x}), \tag{8.38}$$

in which case $\overset{\omega}{f}_{k+1|k+1}(\mathbf{x})$ with $\omega = \hat{\omega}$ results in the best choice of track merging. That is, the optimal value of ω is the one that results in the largest MAP estimate. (Note that Equation 8.38 can be approximated by computing the covariance matrix $\overset{\omega}{P}_{k|k}$ of $\overset{\omega}{f}_{k+1|k+1}(\mathbf{x})$ and minimizing its determinant or trace, as originally proposed by Uhlmann and Julier [8–10].)

Julier has proposed [14] that, rather than Equation 8.38, a more theoretically principled optimization procedure would be to choose ω as the maximizing value in Equation 8.35:

$$\breve{\omega} = \arg\inf_{\omega} \int \overset{0}{f}_{k+1|k+1}(\mathbf{x})^{1-\omega} \cdot \overset{1}{f}_{k+1|k+1}(\mathbf{x})^{\omega} d\mathbf{x}. \tag{8.39}$$

This has the effect of minimizing the degree of overlap between the distributions $\overset{0}{f}_{k+1|k+1}(\mathbf{x})^{1-\omega}$ and $\overset{1}{f}_{k+1|k+1}(\mathbf{x})^{\omega}$. His reasoning is as follows. First, $\breve{\omega}$ reflects the information contained in the distribution $\overset{\omega}{f}_{k+1|k+1}(\mathbf{x})$ as an entirety—rather than just the information contained at a single point, the MAP estimate. Second, $\overset{\breve{\omega}}{f}_{k+1|k+1}(\mathbf{x})$ can be shown to be equally distant from $\overset{0}{f}_{k+1|k+1}(\mathbf{x})$ and $\overset{0}{f}_{k+1|k+1}(\mathbf{x})$ in a Kullback–Leibler information-theoretic sense.

Nevertheless, I argue that Equation 8.38 is a more justifiable theoretical choice, for two reasons:

1. In target tracking, a track distribution $f_{k|k}(\mathbf{x})$ is of little interest unless one can extract from it an accurate estimate of target state. *Using the entire distribution $f_{k|k}(\mathbf{x})$ for this purpose is typically a bad idea.* For example, in practical application, most of the modes of $f_{k|k}(\mathbf{x})$ will be minor modes caused by clutter returns, along with (if SNR is large enough) a single larger target-associated mode. Thus an estimator that employs all of $f_{k|k}(\mathbf{x})$—the expected value $\bar{\mathbf{x}}_{k|k}$ of $f_{k|k}(\mathbf{x})$ for example—can produce unstable and very unaccurate estimates. The MAP estimator, Equation 8.7, is usually more appropriate for practical application, since it tends to produce more stable and accurate state estimates.

2. Abstract information-theoretic distances should be treated with caution when isolated from physical intuition. There is a literal infinitude of information-based distance concepts—most obviously, the Csiszár-divergence family [18,19]

$$K_c(\overset{1}{f}_{k+1|k+1}; \overset{0}{f}_{k+1|k+1}) = \int \overset{0}{f}_{k+1|k+1}(\mathbf{x}) \cdot c\left(\frac{\overset{1}{f}_{k+1|k+1}(\mathbf{x})}{\overset{0}{f}_{k+1|k+1}(\mathbf{x})}\right) d\mathbf{x} \tag{8.40}$$

and its multitarget generalizations [20], where $c(x)$ is some nonnegative convex function. For example, choose the convex kernel $c(x)$ to be $c_{\omega}(x) = (1 - \omega)x + \omega - x^{\omega}$. Then Chernoff information can be expressed in terms of $K_{c_{\omega}}$, which is

$$K_{c_{\omega}}(\overset{0}{f}_{k+1|k+1}; \overset{1}{f}_{k+1|k+1}) = 1 - \int \overset{0}{f}_{k+1|k+1}(\mathbf{x})^{1-\omega} \cdot \overset{1}{f}_{k+1|k+1}(\mathbf{x})^{\omega} d\mathbf{x}. \tag{8.41}$$

In addition to these, there are many distance metrics on probability distributions, such as Wasserstein distance. Which of these is "best," why is it best, and what might its physical interpretation be?

The reasoning behind Equation 8.38, by way of contrast, inherently arises from the practical goal of trying to achieve the most accurate and stable state estimates possible. For each ω, consider the following statements about $\overset{\omega}{f}_{k+1|k+1}(\mathbf{x})$:

1. It is the distribution of the merged track.
2. The MAP estimate for this track is $\overset{\omega}{\mathbf{x}}_{k+1|k+1} = \arg\sup_{\mathbf{x}} \overset{\omega}{f}_{k+1|k+1}(\mathbf{x})$.
3. The larger the value of $\sup_{\mathbf{x}} \overset{\omega}{f}_{k+1|k+1}(\mathbf{x})$, the more probable—and therefore the more sharply localized—$\overset{\omega}{\mathbf{x}}_{k+1|k+1}$ will be.
4. Thus one should choose that value $\hat{\omega}$ of ω which corresponds to the most-probable (best localized) MAP estimate.

The necessity of this line of reasoning will become apparent when I propose multi-target generalizations of XM fusion later in the chapter. In this situation, concepts such as covariance or trace can no longer even be defined. Concepts such as Chernoff information and Csiszár discrimination can still be defined, but their physical meaning is even less evident than in the single-target case. The primary difficulty is a practical one, namely that in multitarget problems the computability of Equation 8.38 will be questionable. Thus computational tractability will usually be the primary motivation for choosing information-theoretic or other optimization approaches in preference to Equation 8.38.

8.3 FINITE-SET STATISTICS: REVIEW

In this section, I briefly review basic elements of finite-set statistics (FISST) [7,21,22] that are required for the material that follows:

1. Section 8.3.1: The multisensor-multitarget recursive Bayes filter. This is the foundation for the approach to T²F that will be introduced shortly.
2. Section 8.3.2: A brief summary of the basic elements of the FISST differential and integral multitarget calculus, including Poisson processes and i.i.d.c. processes.
3. Section 8.3.3: The PHD filter. This is the first computational approximation of the multitarget Bayes filter.
4. Section 8.3.4: The CPHD filter. This is the second computational approximation of the multitarget Bayes filter.
5. Section 8.3.5: A brief summary of significant recent advances involving PHD and CPHD filters.

8.3.1 MULTITARGET RECURSIVE BAYES FILTER

My approach to multisource-multitarget T²F is based on the *multisensor-multitarget recursive Bayes filter* [7, chapter 14]. Let $Z^{(k)}$: $Z_1,...,Z_k$ be a time-sequence of multisensor-multitarget measurement-sets Z_i collected at times $t_1,...,t_k$. That is, each Z_i consists of the measurements collected by all available sensors at or near time-step i. They can have the form $Z_i = $; (no measurements collected); $Z_i = \{z_1\}$ (one measurement z_1 collected); $Z_i = \{z_1, z_2\}$ (two measurements z_1, z_2 collected); and so on. Given this, the multitarget Bayes filter propagates a multitarget posterior distribution $f_{k|k}(X|Z^k)$ through time:

$$\cdots \rightarrow f_{k|k}(X \mid Z^{(k)}) \overset{\text{predictor}}{\rightarrow} f_{k+1|k}(X \mid Z^{(k)}) \overset{\text{corrector}}{\rightarrow} f_{k+1|k+1}(X \mid Z^{(k+1)}) \rightarrow \cdots \quad (8.42)$$

Here, X is the single-target state-set—i.e., $X=$ if no targets are present, $X = \{x_1\}$ if a single target with state x_1 is present, $X = \{x_1, x_2\}$ if two targets with states x_1, x_2 are present, etc. The "cardinality distribution"

$$p_{k+1|k+1}(n \mid Z^{(k+1)}) = \int_{|X|=n} f_{k+1|k+1}(X \mid Z^{(k+1)})\delta X \quad (8.43)$$

defines the posterior probability that the multitarget scene contains n targets, where $\int \cdot \delta X$ indicates a multitarget "set integral" as defined in Section 8.3.2.

The multitarget Bayes filter presumes the existence of multitarget motion and measurement models, for example:

$$\Xi_{k+1|k} = S_k(X) \cup B_k, \quad \Sigma_{k+1|k} = T_{k+1}(X) \cup C_{k+1} \quad (8.44)$$

where
  $S_k(X)$ is the random finite subset (RFS) of persisting targets
  B_k is the RFS of appearing targets
  $T_{k+1}(X)$ is the RFS of target-generated measurements
  C_{k+1} is the RFS of clutter measurements

Given these models, using multitarget calculus (Section 8.3.2) one can construct a multitarget Markov transition density and a multitarget likelihood function

$$f_{k+1|k}(X \mid X'), \quad f_{k+1}(Z \mid X) \quad (8.45)$$

(see Chapters 12 and 13 of [7]). Because of this systematic specification of models, at any given time-step the distribution $f_{k|k}(X|Z^{(k)})$ systematically encapsulates all relevant information regarding the presumed strengths and weaknesses of the targets, and the known strengths and weaknesses of the sensors.

The multitarget Bayes filter is defined by the predictor and corrector equations

$$f_{k+1|k}(X \mid Z^{(k)}) = \int f_{k+1|k}(X \mid X') \cdot f_{k|k}(X' \mid Z^{(k)})\delta X' \quad (8.46)$$

$$f_{k+1|k+1}(X \mid Z^{(k+1)}) = \frac{f_{k+1}(Z_{k+1} \mid X) \cdot f_{k+1|k}(X \mid Z^{(k)})}{f_{k+1}(Z_{k+1} \mid Z^{(k)})} \quad (8.47)$$

where

$$f_{k+1}(Z_{k+1} \mid Z^{(k)}) = \int f_{k+1}(Z_{k+1} \mid X) \cdot f_{k+1|k}(X \mid Z^{(k)}) \delta X. \tag{8.48}$$

In what follows, I will abbreviate, for all $k \geq 0$,

$$f_{k|k}(X) \overset{\text{abbr.}}{=} f_{k|k}(X \mid Z^{(k)}) \tag{8.49}$$

$$f_{k+1|k}(X) \overset{\text{abbr.}}{=} f_{k+1|k}(X \mid Z^{(k)}). \tag{8.50}$$

Information of interest—number of targets, the positions, velocities, and types of the targets, etc.—can be jointly extracted from $f_{k|k}(X|Z^{(k)})$ using a Bayes-optimal *multitarget state estimator* (see Section 14.5 of [7]). For example, the *joint multitarget* (JoM) estimator is defined by

$$X_{k+1|k+1}^{\text{JoM}} = \arg \sup_{X} f_{k+1|k+1}(X \mid Z^{(k+1)}) \cdot \frac{c^{|X|}}{|X|!} \tag{8.51}$$

where c is a fixed constant which has the same units of measurement as the single-target state $\mathbf{x}$.

Remark 2: Generally speaking, c should be approximately equal to the accuracy to which the state is to be estimated, as long as the following inequality is satisfied [7, p. 500]: $f_{k+1|k+1}(X|Z^{(k+1)}) \cdot c^{\hat{n}} \leq 1$ for all X, where $\hat{n}$ is the MAP estimate derived from the cardinality distribution.

8.3.2 MULTITARGET CALCULUS

The finite-set statistics *multitarget integral-differential calculus* is central to the approach that I advocate. Functional derivatives and set derivatives [7, chapter 11] are key to the construction of "true" multitarget Markov densities and multitarget likelihood functions. They are also key to the construction of principled approximations of the multitarget Bayes filter, such as the PHD and CPHD filters.

A *set integral* accounts for random variability in target number as well as in target state. Let $f_{k|k}(X)$ be a multitarget probability distribution. Then it has the form

$$\int f(X) \delta X = f(\phi) + \sum_{n=1}^{\infty} \frac{1}{n!} \int f_{k|k}(\{\mathbf{x}_1, \dots, \mathbf{x}_n\}) d\mathbf{x}_1 \cdots d\mathbf{x}_n. \tag{8.52}$$

Let $F[h]$ be any functional—i.e., a scalar-valued function whose argument h is a function $h(\mathbf{x})$. Then the *functional derivative* of F with respect to any finite set $X = \{\mathbf{x}_1, \dots, \mathbf{x}_n\}$ with $|X| = n \geq 0$ is given by

$$\frac{\delta F}{\delta X}[h] = \frac{\delta}{\delta \mathbf{x}_1} \cdots \frac{\delta}{\delta \mathbf{x}_n} F[h]$$ (8.53)

$$\frac{\delta}{\delta \mathbf{x}} F[h] = \lim_{\varepsilon \searrow 0} \frac{F[h + \varepsilon \delta_{\mathbf{x}}] - F[h]}{\varepsilon}$$ (8.54)

where $\delta_{\mathbf{x}}(\mathbf{x}')$ denotes the Dirac delta function concentrated at $\mathbf{x}$. Functional derivatives and set integrals are inverse operations, in the sense that

$$F[h] = \int h^X \cdot \frac{\delta F}{\delta X}[0] \delta X$$ (8.55)

$$\left[\frac{\delta}{\delta X} \int h^Y \cdot f(X) \delta X \right]_{h=0} = f(X).$$ (8.56)

Here, for any function $h(\mathbf{x})$,

$$h^X = \begin{cases} 1 & \text{if} & X = \phi \\ \prod_{\mathbf{x} \in X} h(\mathbf{x}) & \text{if} & \text{otherwise} \end{cases}.$$ (8.57)

In this chapter, we will require frequent use of two special multitarget processes. Suppose that $f(X)$ is a multitarget probability distribution. Then it is the distribution of

- A *Poisson process* (Poisson RFS) if

$$f(X) = e^{-N} \cdot D^X$$ (8.58)

where

$$N = \int D(\mathbf{x}) d\mathbf{x}$$

$D(\mathbf{x})$ is the PHD, or "intensity function," of the process

- An independent, identically distributed cluster (i.i.d.c.) process (i.i.d.c. RFS) if

$$f(X) = |X|! \cdot p(|X|) \cdot s^X$$ (8.59)

where
$s(\mathbf{x})$ is the *spatial density*
$p(n)$ is the *cardinality distribution* of the process

Equation 8.58 is a special case of Equation 8.59 with $p(n) = e^{-N} \cdot N^n/n!$
As an example, one can verify that Equation 8.58 defines a multitarget probability distribution:

$$\int f(X)\delta X = e^{-N} \cdot D^{\phi} + e^{-N} \sum_{n=1}^{\infty} \frac{1}{n!} \int D(\mathbf{x}_1) \cdots D(\mathbf{x}_n) d\mathbf{x}_1 \cdots d\mathbf{x}_n \qquad (8.60)$$

$$= e^{-N} + e^{-N} \sum_{n=1}^{\infty} \frac{1}{n!} N^n = e^{-N} \cdot e^N = 1. \qquad (8.61)$$

Likewise, Equation 8.59 defines a multitarget probability distribution:

$$\int f(X)\delta X = 0! \cdot p(0) \cdot s^{\phi} + \sum_{n=1}^{\infty} \frac{n! \cdot p(n)}{n!} \int s(\mathbf{x}_1) \cdots s(\mathbf{x}_n) d\mathbf{x}_1 \cdots d\mathbf{x}_n \qquad (8.62)$$

$$= p(0) + \sum_{n=1}^{\infty} p(n) = 1. \qquad (8.63)$$

8.3.3   PHD FILTER

Constant-gain Kalman filters—the alpha-beta filter, for example—provide the most computationally tractable approximation of the single-sensor Bayes filter. A constant-gain Kalman filter propagates the first statistical moment (posterior expectation) $\hat{\mathbf{x}}_{k|k}$ in place of $f_{k|k}(\mathbf{x}|Z^k)$, using alternating predictor steps $\hat{\mathbf{x}}_{k|k} \to \hat{\mathbf{x}}_{k+1|k}$ and corrector steps $\hat{\mathbf{x}}_{k+1|k} \to \hat{\mathbf{x}}_{k+1|k+1}$.

The PHD filter mimics this basic idea, but at a more abstract, *statistical* level [7, Chapter 16] [23]. It propagates *a first-order multitarget moment* of the multitarget posterior $f_{k|k}(X|Z^{(k)})$ instead of $f_{k|k}(X|Z^{(k)})$ itself:

$$\cdots \to D_{k|k}(\mathbf{x} \mid Z^{(k)}) \overset{\text{predictor}}{\to} D_{k+1|k}(\mathbf{x} \mid Z^{(k)}) \overset{\text{corrector}}{\to} D_{k+1|k+1}(\mathbf{x} \mid Z^{(k+1)}) \to \cdots \qquad (8.64)$$

This moment, the PHD, is the density function on single-target states $\mathbf{x}$ defined by

$$D_{k|k}(\mathbf{x}) \overset{\text{abbr.}}{=} D_{k|k}(\mathbf{x} \mid Z^{(k)}) = \int f_{k|k}(X \cup \{\mathbf{x}\} \mid Z^{(k)})\delta X. \qquad (8.65)$$

It is not a probability density, since its integral is in general not 1. Rather, $N_{k|k} = \int D_{k|k}(\mathbf{x})d\mathbf{x}$ is the total expected number of targets in the scenario. Intuitively speaking, $D_{k|k}(\mathbf{x})$ is the *track density* at $\mathbf{x}$. The peaks of $D_{k|k}(\mathbf{x})$ are approximately at the locations of the most likely target states. So, one way of estimating the number $\hat{n}$ and states $\hat{\mathbf{x}}_1, \ldots, \hat{\mathbf{x}}_{\hat{n}}$ of the predicted tracks is to take $\hat{n}$ to be the nearest integer $\hat{n}$ in $N_{k+1|k}$ and then determine the $\hat{n}$ highest peaks of $D_{k|k}(\mathbf{x})$.

The PHD can be propagated through time using the following predictor (time-update) and corrector (data-update) equations. Neglecting the spawning of targets by other targets, these are

$$D_{k+1|k}(\mathbf{x}) = N_{k+1|k}^B s_{k+1|k}^B(\mathbf{x}) + \int p_S(\mathbf{x}') \cdot f_{k+1|k}(\mathbf{x} \mid \mathbf{x}') \cdot D_{k|k}(\mathbf{x}') d\mathbf{x}' \qquad (8.66)$$

$$\frac{D_{k+1|k+1}(\mathbf{x})}{D_{k+1|k}(\mathbf{x})} = 1 - p_D(\mathbf{x}) + \sum_{\mathbf{z} \in Z_{k+1}} \frac{p_D(\mathbf{x}) \cdot L_{\mathbf{z}}(\mathbf{x})}{\lambda_{k+1} c_{k+1}(\mathbf{z}) + \tau_{k+1}(\mathbf{z})}. \qquad (8.67)$$

Here,

- $N_{k+1|k}^B$ is the expected number, and $s_{k+1|k}^B(\mathbf{x})$ the spatial distribution, of newly appearing targets.
- $p_S(\mathbf{x}') \overset{\text{abbr.}}{=} p_{S,k+1|k}(\mathbf{x}')$ is the probability that a target with state $\mathbf{x}'$ at time-step k will survive into time-step $k+1$.
- $f_{k+1|k}(\mathbf{x} \mid \mathbf{x}')$ is the single-target Markov transition density.
- $p_D(\mathbf{x}) \overset{\text{abbr.}}{=} p_{D,k+1}(\mathbf{x})$ is the probability that a target with state $\mathbf{x}$ at time-step $k+1$ will generate a measurement.
- $L_{\mathbf{z}}(\mathbf{x}) \overset{\text{abbr.}}{=} f_{k+1}(\mathbf{z} \mid \mathbf{x})$ is the single-target likelihood function.
- λ_{k+1} is the clutter rate and $c_{k+1}(\mathbf{z})$ is the spatial distribution of the Poisson clutter process, where

$$\tau_{k+1}(\mathbf{z}) = \int p_D(\mathbf{x}) \cdot L_{\mathbf{z}}(\mathbf{x}) \cdot D_{k+1|k}(\mathbf{x}) d\mathbf{x}. \qquad (8.68)$$

One can get an intuitive understanding of how the PHD filter works by noticing that the measurement-updated expected number of targets is

$$N_{k+1|k+1} = \int D_{k+1|k+1}(\mathbf{x}) d\mathbf{x} = \overset{\text{ND}}{N}_{k+1|k+1} + \sum_{\mathbf{z} \in Z_{k+1}} \overset{\text{D}}{N}_{k+1|k+1}(\mathbf{z}) \qquad (8.69)$$

where

$$\overset{\text{ND}}{N}_{k+1|k+1} = \int (1 - p_D(\mathbf{x})) \cdot D_{k+1|k}(\mathbf{x}) d\mathbf{x} \qquad (8.70)$$

$$\overset{\text{D}}{N}_{k+1|k+1}(\mathbf{z}) = \frac{\tau_{k+1}(\mathbf{z})}{\lambda_{k+1} c_{k+1}(\mathbf{z}) + \tau_{k+1}(\mathbf{z})} \leq 1. \qquad (8.71)$$

The nondetection term $\overset{\text{ND}}{N}_{k+1|k+1}$ is an estimate of the number of targets that have not been detected. The detection ratio $\overset{\text{D}}{N}_{k+1|k+1}(\mathbf{z})$ assesses whether or not $\mathbf{z}$ originated with clutter or with a target. If $\overset{\text{D}}{N}_{k+1|k+1}(\mathbf{z}) > 1/2$—that is, if $\tau_{k+1}(\mathbf{z}) > \lambda_{k+1} c_{k+1}(\mathbf{z})$—then $\mathbf{z}$ is "target-like." If $\overset{\text{D}}{N}_{k+1|k+1}(\mathbf{z}) < 1/2$ then it is "clutter-like."

The derivation of Equation 8.67 requires the following simplifying assumption: *the predicted target process is approximately Poisson.* As is evident from Equation 8.67, the PHD filter does not require explicit measurement-to-track association. It has computational order $O(mn)$, where m is the current number of measurements and n is the current number of targets. It tends to produce inaccurate (high variance) instantaneous estimates $N_{k|k}$ of target number. Thus it is typically necessary to average $N_{k|k}$ over some time window.

The PHD filter can be implemented using both sequential Monte Carlo (SMC, a.k.a. particle-system) approximation, or Gaussian-mixture approximation. In the first case, it is called a "particle-PHD filter" and in the second case a "GM-PHD filter" (see Chapter 16 of [7] and [45–47]).

8.3.4   CPHD FILTER

The CPHD *filter* generalizes the PHD filter [7, chapter 16] [23]. It admits more general false alarm models (called "independent, identically distributed cluster" [i.i.d.c.] models) than the Poisson models assumed in the PHD filter. It propagates two things: a *spatial distribution* $s_{k|k}(\mathbf{x})$ and a *cardinality distribution* $p_{k|k}(n) \overset{\text{abbr.}}{=} p_{k|k}(n \mid Z^{(k)})$ on target number n:

$$\cdots \rightarrow \begin{cases} s_{k|k}(\mathbf{x} \mid Z^{(k)}) \\ p_{k|k}(n \mid Z^{(k)}) \end{cases} \overset{\text{predictor}}{\rightarrow} \begin{cases} s_{k+1|k}(\mathbf{x} \mid Z^{(k)}) \\ p_{k+1|k}(n \mid Z^{(k)}) \end{cases} \overset{\text{corrector}}{\rightarrow} \begin{cases} s_{k+1|k+1}(\mathbf{x} \mid Z^{(k+1)}) \\ p_{k+1|k+1}(n \mid Z^{(k+1)}) \end{cases} \rightarrow \cdots \quad (8.72)$$

If $N_{k|k} = \sum_{n \geq 0} n \cdot p_{k|k}(n \mid Z^{(k)})$ is the expected number of targets, then $D_{k|k}(\mathbf{x}|Z^{(k)}) = N_{k|k} \cdot s_{k|k}(\mathbf{x}|Z^{(k)})$ is the corresponding PHD. Or, equivalently, $s_{k|k}(\mathbf{x} \mid Z^{(k)}) = N_{k|k}^{-1} D_{k|k}(\mathbf{x} \mid Z^{(k)})$.

CPHD Filter Time-Update Equations. The predictor equations for the CPHD filter are

$$D_{k+1|k}(\mathbf{x}) = b_{k+1|k}(\mathbf{x}) + \int p_S(\mathbf{x}') \cdot f_{k+1|k}(\mathbf{x} \mid \mathbf{x}') \cdot D_{k|k}(\mathbf{x}')d\mathbf{x}' \qquad (8.73)$$

$$p_{k+1|k}(n) = \sum_{n' \geq 0} p_{k+1|k}(n \mid n') \cdot p_{k|k}(n') \qquad (8.74)$$

where $p_{k+1|k}^B(n-j)$ is the cardinality distribution of the birth process and where

$$p_{k+1|k}(n \mid n') = \sum_{j=0}^{n} p_{k+1|k}^B(n-j) \cdot C_{n',j} \cdot \psi_k^j \left(1 - \psi_k\right)^{n'-j} \qquad (8.75)$$

$$\psi_k = \int p_S(\mathbf{x}) \cdot s_{k+1|k}(\mathbf{x})d\mathbf{x} \qquad (8.76)$$

$$N_{k+1|k} = N_{k+1|k}^B + \int p_S(\mathbf{x}') \cdot D_{k|k}(\mathbf{x}')d\mathbf{x}' \qquad (8.77)$$

$$N_{k+1|k}^B = \int b_{k+1|k}(\mathbf{x})\,d\mathbf{x} \tag{8.78}$$

$$C_{n',j} = \begin{cases} \dfrac{n'!}{j!\,(n'-j)!} & \text{if} \quad 0 \le j \le n' \\[2mm] 0 & \text{if} \quad \text{otherwise} \end{cases}. \tag{8.79}$$

CPHD Filter Measurement-Update Equations. If $m = |Z_{k+1}|$ where $Z_{k+1} = \{\mathbf{z}_1,\dots,\mathbf{z}_m\}$ is the newly collected measurement-set, then the corrector equations for the CPHD filter are

$$\frac{D_{k+1|k+1}(\mathbf{x})}{s_{k+1|k}(\mathbf{x})} = \left(1 - p_D(\mathbf{x})\right) \cdot \overset{\text{ND}}{E}_{k+1} + \sum_{\mathbf{z} \in Z_{k+1}} p_D(\mathbf{x}) \cdot L_\mathbf{z}(\mathbf{x}) \cdot \overset{\text{D}}{E}_{k+1}(\mathbf{z}) \tag{8.80}$$

$$\frac{p_{k+1|k+1}(n)}{p_{k+1|k}(n)} = \frac{\displaystyle\sum_{j=0}^{\min\{m,n\}} (m-j)!\, p_{k+1}^\kappa(m-j) \cdot P_{n,j} \cdot \phi_k^{n-j} \cdot \sigma_j(Z_{k+1})}{\displaystyle\sum_{l=0}^{m} (m-l)!\, p_{k+1}^\kappa(m-l) \cdot \sigma_l(Z_{k+1}) \cdot G_{k+1|k}^{(l)}(\phi_k)} \tag{8.81}$$

where

$$\overset{\text{ND}}{E}_{k+1} = \frac{\displaystyle\sum_{j=0}^{m} (m-j)!\, p_{k+1}^\kappa(m-j) \cdot \sigma_j(Z_{k+1}) \cdot G_{k+1|k}^{(j+1)}(\phi_k)}{\displaystyle\sum_{l=0}^{m} (m-l)!\, p_{k+1}^\kappa(m-l) \cdot \sigma_l(Z_{k+1}) \cdot G_{k+1|k}^{(l)}(\phi_k)} \tag{8.82}$$

$$\overset{\text{D}}{E}_{k+1}(\mathbf{z}) = \frac{1}{c_{k+1}(\mathbf{z})} \cdot \frac{\displaystyle\sum_{j=0}^{m-1} (m-j-1)!\, p_{k+1}^\kappa(m-j-1) \cdot \sigma_j(Z_{k+1}-\{\mathbf{z}_j\}) \cdot G_{k+1|k}^{(j+1)}(\phi_k)}{\displaystyle\sum_{l=0}^{m} (m-l)!\, p_{k+1}^\kappa(m-l) \cdot \sigma_l(Z_{k+1}) \cdot G_{k+1|k}^{(l)}(\phi_k)} \tag{8.83}$$

and where

$$\sigma_j(Z_{k+1}) = \sigma_{m,j}\left(\frac{\tau_{k+1}(\mathbf{z}_1)}{c_{k+1}(\mathbf{z}_1)},\dots,\frac{\tau_{k+1}(\mathbf{z}_m)}{c_{k+1}(\mathbf{z}_m)}\right) \tag{8.84}$$

$$G_{k+1|k}^{(l)}(\phi_k) = \sum_{n'\ge l} P_{n',l} \cdot p_{k+1|k}(n') \cdot \phi_k^{n'-l} \tag{8.85}$$

$$G_{k+1|k}^{(j+1)}(\phi_k) = \sum_{n'\ge j+1} P_{n',j+1} \cdot p_{k+1|k}(n') \cdot \phi_k^{n'-j-1} \tag{8.86}$$

$$\phi_k = \int (1 - p_D(\mathbf{x})) \cdot s_{k+1|k+1}(\mathbf{x}) d\mathbf{x} \qquad (8.87)$$

$$\tau_{k+1}(\mathbf{z}) = \int p_D(\mathbf{x}) \cdot L_{\mathbf{z}}(\mathbf{x}) \cdot s_{k+1|k}(\mathbf{x}) d\mathbf{x} \qquad (8.88)$$

where $P_{n,i} = n!/(n-i)!$ is the permutation coefficient.

The corrector equations for the CPHD filter require the following simplifying assumption: that the predicted target process is approximately an i.i.d.c. process. The CPHD filter has computational order $O(m^3n)$, though this can be reduced to $O(m^2n)$ using special numerical techniques.

The CPHD filter can be implemented using both particle approximation and Gaussian-mixture approximation. In the first case, it is called a "particle-CPHD filter" and in the second case a "GM-CPHD filter."

8.3.5 SIGNIFICANT RECENT DEVELOPMENTS

The theory and practice of random set filters has developed rapidly in recent years. In this section, I briefly summarize a few of the most recent advances:

1. *Track-before-detect filtering in pixelized images without preprocessing.* Most multitarget tracking algorithms using pixelized image data rely on some kind of image preprocessing step to extract detection-type features: threshold detectors, edge detectors, blob detectors, etc. In Ref. [24], Vo, Vo, and Pham have demonstrated a computationally tractable multitarget detection and tracking algorithm that does not require such preprocessing. It is based on a suitable modification of the "multi-Bernoulli filter" introduced in Ref. [7, chapter 17] and then corrected and implemented in Ref. [25].

2. *Simultaneous localization and mapping (SLAM).* When neither GPS nor terrain maps are available, a robotic platform must detect landmarks, use them to construct a terrain map on the fly, and simultaneously orient the platform with respect to that map. The current state-of-the-art in SLAM is the FastSLAM approach, which employs measurement-to-track association, in conjunction with heuristic procedures for clutter rejection and initiation and termination of landmarks. Mullane, Vo, Adams, and Vo have shown that a PHD filter-based SLAM filter significantly outperforms FastSLAM in regard to the accuracy of both platform trajectory estimation and landmark detection and localization [26,27]. Clark has devised an even faster and more accurate SLAM-PHD filter based on a cluster-process formulation [28].

3. *"Background agnostic" (BAG) CPHD filters.* The "classical" CPHD filter relies on an a priori model $\lambda_{k+1}, c_{k+1}(\mathbf{z}), p_{k+1}^{\kappa}(m)$ of the clutter process and on an a priori model $p_D(\mathbf{x})$ of the state-dependent probability of detection. In 2009, I initiated a study of PHD and CPHD filters that do not require a

priori clutter models but, rather, are capable of estimating them, on the fly, directly from the measurements. In Refs. [29,30], the clutter process was assumed to be a finite superposition of Poisson clutter processes, each with an intensity function of the form $\kappa(\mathbf{z}) = \lambda \cdot \theta_c(\mathbf{z})$ with clutter rate $0 \leq \lambda \leq 1$ and spatial distribution $\theta_c(\mathbf{z})$ parameterized by $\mathbf{c}$. Unfortunately, the resulting PHD/CPHD filters are combinatorially complex. Subsequently, in Ref. [31], I derived computationally tractable version CPHD filters. In this case, the clutter process is assumed to be an *infinite* superposition of *Bernoulli* clutter processes, each with an intensity function of the form $\kappa(\mathbf{z}) = \lambda \cdot \theta_c(\mathbf{z})$ with $0 \leq \lambda \leq 1$. Then, in Ref. [32], I showed how to further extend filters when both the clutter process and $p_D(\mathbf{x})$ are unknown. This filter has been implemented in certain special cases and shown to perform reasonably well under simulated conditions [33,34].

4. *"Background agnostic" multi-Bernoulli filters.* Vo, Vo, Hoseinnezhad, and Mahler have generalized the just-mentioned approach to nonlinear situations, via a particle-filter implementation of a background-agnostic multi-Bernoulli filter [35–37].

5. *Principled, tractable multisensor CPHD/PHD filters.* The PHD/CPHD filter measurement-update steps described in Sections 8.3.3 and 8.3.4 are inherently *single-sensor* formulas. What of the multisensor case? In practical application, the de facto approach has been to employ the "iterated corrector" approximation. That is, apply the measurement-update equations successively, once for each sensor. It is well known that this approach is not invariant to changes in the order of the sensors. Moreover, for the PHD filter (but apparently not for the CPHD filter) it turns out that the iterated-corrector approach leads to performance degradation when the probabilities of detection for the sensors are significantly different [38]. In Ref. [39], I introduced a new approximation that leads to principled, order-invariant, computationally tractable multisensor PHD and CPHD filters. Nagappa et al. have shown that this approximation outperforms the interated-corrector approach and, for the PHD filter, is also a good approximation of the theoretically correct two-sensor PHD filter [40].

6. *Joint multisensor-multitarget tracking and sensor-bias estimation.* Current multitarget detection and tracking algorithms presume that all sensors are spatially registered—i.e., that all sensor states are precisely specified with respect to some common coordinate system. In actuality, any particular sensor's observations may be contaminated by *spatial misregistration biases* that may take translational, rotational, and other forms. In Ref. [41], I proposed an approach that leverages any unknown targets that may be in the scene, if there are enough of them present, to estimate the spatial biases of the sensors while simultaneously detecting and tracking the targets. Ristić and Clark have implemented a cluster-process variant of this approach for a specific kind of spatial misregistration, and found that it performs well [42].

8.4   GENERAL MULTITARGET DISTRIBUTED FUSION

In this section, I show how to directly generalize the single-target T^2F theory of Section 8.2 to the multitarget situation. The section is organized as follows:

1. Section 8.4.1: Multitarget T^2F when the track sources are independent. Approach: multitarget generalization of Equation 8.10.
2. Section 8.4.2: Multitarget T^2F when the track sources are dependent because of known double-counting. Approach: multitarget generalization of Equation 8.16.
3. Section 8.4.3: Multitarget T^2F when the track sources are arbitrary and their correlations are completely unknown. Approach: multitarget generalization of XM fusion, Equations 8.34 through 8.38.

8.4.1   MULTITARGET T²F OF INDEPENDENT SOURCES

Suppose that multiple targets are being tracked, and that s independent sources, relying on their own dedicated local sensors, provide track data about these targets to a T^2F site. The jth sensor suite collects a time-sequence $Z^k : \overset{j}{Z}_1,...,\overset{j}{Z}_k$, where $\overset{j}{Z}_l$ denotes the set of measurements supplied by the jth source's sensors at time t_l. The source does not pass this information directly to the fusion site. Rather, it passes the following information:

- Measurement-updated multitarget track data, in the form of multitarget probability distributions $\overset{j}{f}_{k|k}(X) \overset{abbr}{=} \overset{j}{f}_{k|k}(X \mid Z^{(k)})$
- Time-updated multitarget track data, in the form of multitarget distributions $\overset{j}{f}_{k+1|k}(X) \overset{abbr}{=} \overset{j}{f}_{k+1|k}(X \mid Z^{(k)})$

Let $f_{k|k}(X) \overset{abbr.}{=} f_{k|k}(X \mid Z^{(k)})$ be the fusion node's determination of the multitarget state, given all the accumulated track data supplied by the sensor sources. Then the exact multitarget generalization of Equation 8.10 is the following multitarget track-merging formula, first introduced in Ref. [3]:

$$f_{k+1|k+1}(X) \propto \frac{\overset{1}{f}_{k+1|k+1}(X)}{\overset{1}{f}_{k+1|k}(X)} \cdots \frac{\overset{s}{f}_{k+1|k+1}(X)}{\overset{s}{f}_{k+1|k}(X)} \cdot f_{k+1|k}(X). \qquad (8.89)$$

This is the fundamental formula for multitarget T^2F with independent sources. As in Section 8.2.2, it is being assumed here that the sources provide their data in lock-step, simultaneously at every time-step. Once again, however, it also applies to the asynchronous case. If each source has its own data rate, then the measurement-collection times $t_1,...,t_k$ can be taken to refer to the arrival times of data from all of the track sources, taken collectively. If at time t_l only s_l of the sources provide data, then Equation 8.10 is replaced by the corresponding formula for only those sources.

The approximations described in Section 8.2.2 apply equally well here. Suppose that the sources do not pass on their time-update track data but, rather, only their measurement-update track data. Presume that all of the sources employ identical target motion models. Then (in principle) the fusion site can construct time-update track data for the sources, using the multitarget prediction integral

$$\overset{j}{f}_{k+1|k}(X) = \int f_{k+1|k}(X \mid X') \cdot \overset{j}{f}_{k|k}(X)\delta X, \tag{8.90}$$

and then apply Equation 8.89.

Alternatively, assume that the sources' time-updated track data is identical to the fusion site's: $\overset{j}{f}_{k+1|k}(X) = f_{k+1|k}(X)$ for all $j = 1, \ldots, s$. Then Equation 8.89 reduces to

$$f_{k+1|k+1}(X) \propto \overset{1}{f}_{k+1|k+1}(X) \cdots \overset{s}{f}_{k+1|k+1}(X) \cdot f_{k+1|k}(X)^{1-s}, \tag{8.91}$$

which is the multitarget version of Bayes parallel combination, Equation 8.14.

Equations 8.89 and 8.91 are computationally intractable in general. The task of devising more tractable approximations of them will be taken up in Sections 8.5.1 and 8.5.2.

8.4.2 Multitarget T²F with Known Double-Counting

Suppose now that the data sources share sensors, but that it is known which sensors are being shared by which sources. As in Section 8.2.3, define $Z_{k+1} = \overset{1}{Z}_{k+1} \cup \ldots \cup \overset{s}{Z}_{k+1}$. Let $\overset{12}{Z}_{k+1}$ be the measurements supplied to the second source that are not in $\overset{1}{Z}_{k+1}$, $\overset{13}{Z}_{k+1}$ the measurements supplied to the third source that are not in $\overset{1}{Z}_{k+1} \cup \overset{12}{Z}_{k+1}$, $\overset{14}{Z}_{k+1}$ the measurements supplied to the fourth source that are not in $\overset{1}{Z}_{k+1} \cup \overset{12}{Z}_{k+1} \cup \overset{13}{Z}_{k+1}$, and so on. Let $\overset{(j)}{Z}_{k+1} = \overset{j}{Z}_{k+1} - \overset{1j}{Z}_{k+1}$. Then the multitarget version of Equation 8.16 is

$$f_{k+1|k+1}(X) \propto \frac{\overset{1}{f}_{k+1|k+1}(X)}{\overset{1}{f}_{k+1|k}(X)} \cdot \frac{\overset{2}{f}_{k+1|k+1}(X)}{\overset{2}{f}_{k+1|k}(X \mid \overset{(2)}{Z})} \cdots \frac{\overset{s}{f}_{k+1|k+1}(X)}{\overset{s}{f}_{k+1|k}(X \mid \overset{(s)}{Z})} \cdot f_{k+1|k}(X). \tag{8.92}$$

This is the fundamental formula for multitarget T²F with known double-counting. As in Section 8.2.3, the jth source must know which sensors it shares with each of sources $1, \ldots, j-1$, and must pass on $\overset{j}{f}_{k+1|k}(\mathbf{x} \mid \overset{(j)}{Z})$ in addition to $\overset{j}{f}_{k+1|k}(\mathbf{x})$.

Equation 8.92 is computationally intractable in general. More tractable approximations of it will be taken up in Sections 8.5.3 and 8.5.4.

8.4.3 Multitarget XM Fusion

Suppose that multiple targets are being observed by two track sources. At time-step k, the first source provides a multitarget distribution $\overset{1}{f}_{k+1|k+1}(X)$ and the second source provides a multitarget distribution $\overset{2}{f}_{k+1|k+1}(X)$. Then the multitarget version of the single-target XM fusion formula, Equation 8.34, is [3]

$$\overset{\omega}{f}_{k+1|k+1}(X) = \frac{\overset{1}{f}_{k+1|k+1}(X)^{1-\omega} \cdot \overset{2}{f}_{k+1|k+1}(X)^{\omega}}{\int \overset{1}{f}_{k+1|k+1}(Y)^{1-\omega} \cdot \overset{2}{f}_{k+1|k+1}(Y)^{\omega} \delta Y}. \tag{8.93}$$

This is the general formula for the XM fusion of multitarget track sources with completely unknown correlations. As I did in Ref. [3] and at the end of Section 8.2.5, I argue that the most theoretically reasonable optimal XM fusion procedure is as follows:

$$\overset{XM}{f}_{k+1|k+1}(X) = \overset{\hat{\omega}}{f}_{k+1|k+1}(X) \tag{8.94}$$

where

$$\hat{\omega} = \arg \sup_{\omega} \sup_{X} \overset{\omega}{f}_{k+1|k+1}(X) \cdot \frac{c^{|X|}}{|X|!} \tag{8.95}$$

where c is as defined in Equation 8.51: a fixed constant which has the same units of measurement as the single-target state $\mathbf{x}$.

However this may be, Equations 8.93 through 8.95 are computationally intractable in general. More tractable approximations of these equations will be taken up in Sections 8.5.5 and 8.5.6.

8.5 CPHD/PHD FILTER DISTRIBUTED FUSION

In this section, I derive CPHD filter-based approximations of the multitarget T²F approaches described in Section 8.4. The section is organized as follows:

1. Sections 8.5.1 and 8.5.2: Multitarget T²F when the track sources are independent. Approach: CPHD and PHD filter approximations of Equation 8.89.
2. Section 8.5.3: Multitarget T²F when the track sources are dependent because of known double-counting. Approach: CPHD and PHD filter approximations of Equation 8.92.
3. Sections 8.5.5 and 8.5.6: Multitarget T²F when the track sources are arbitrary and their correlations are completely unknown. Approach: CPHD and PHD filter approximations of Equation 8.93, as proposed by Clark et al.

8.5.1 CPHD FILTER T²F OF INDEPENDENT SOURCES

Suppose as in Section 8.4.1 that multiple targets are being tracked, and that s independent sources, relying on their own dedicated local sensors, provide track data about these targets to a T²F site. The jth sensor suite collects its measurements, processes them using a CPHD filter, and then passes on the following to a central T²F site:

- Measurement-update multitarget track data, in the form of spatial distributions $\overset{j}{s}_{k|k}(\mathbf{x}) \overset{\text{abbr}}{=} \overset{j}{s}_{k|k}(\mathbf{x} \mid Z^{(k)})$ and cardinality distributions $\overset{j}{p}_{k|k}(n) \overset{\text{abbr}}{=} \overset{j}{p}_{k|k}(n \mid Z^{(k)})$

- Time-update multitarget track data, in the form of spatial distributions $\overset{j}{s}_{k+1|k}(\mathbf{x}) \overset{\text{abbr}}{=} \overset{j}{s}_{k+1|k}(\mathbf{x} \mid Z^{(k)})$ and cardinality distributions $\overset{j}{p}_{k+1|k}(n) \overset{\text{abbr}}{=} \overset{j}{p}_{k+1|k}(n \mid Z^{(k)})$

Then the multitarget track merging formula—i.e., a CPHD filter approximation of Equation 8.89—is as follows (see Section 8.7.1):

$$p_{k+1|k+1}(n) = \frac{1}{\mu_{k+1|k+1}} \cdot \frac{\overset{1}{p}_{k+1|k+1}(n)}{\overset{1}{p}_{k+1|k}(n)} \cdots \frac{\overset{s}{p}_{k+1|k+1}(n)}{\overset{s}{p}_{k+1|k}(n)} \cdot \sigma_{k+1|k+1}^n \cdot p_{k+1|k}(n) \qquad (8.96)$$

$$s_{k+1|k+1}(\mathbf{x}) = \frac{1}{\sigma_{k+1|k+1}} \cdot \frac{\overset{1}{s}_{k+1|k+1}(\mathbf{x})}{\overset{1}{s}_{k+1|k}(\mathbf{x})} \cdots \frac{\overset{s}{s}_{k+1|k+1}(\mathbf{x})}{\overset{s}{s}_{k+1|k}(\mathbf{x})} \cdot s_{k+1|k}(\mathbf{x}) \qquad (8.97)$$

where

$$\mu_{k+1|k+1} = \sum_{n \geq 0} \frac{\overset{1}{p}_{k+1|k+1}(n)}{\overset{1}{p}_{k+1|k}(n)} \cdots \frac{\overset{s}{p}_{k+1|k+1}(n)}{\overset{s}{p}_{k+1|k}(n)} \cdot \sigma_{k+1|k+1}^n \cdot p_{k+1|k}(n) \qquad (8.98)$$

$$\sigma_{k+1|k+1} = \int \frac{\overset{1}{s}_{k+1|k+1}(\mathbf{x})}{\overset{1}{s}_{k+1|k}(\mathbf{x})} \cdots \frac{\overset{s}{s}_{k+1|k+1}(\mathbf{x})}{\overset{s}{s}_{k+1|k}(\mathbf{x})} \cdot s_{k+1|k}(\mathbf{x}) d\mathbf{x}. \qquad (8.99)$$

Suppose that we use the approximations $p_{k+1|k}(n) = \overset{1}{p}_{k+1|k}(n) = \cdots = \overset{s}{p}_{k+1|k}(n)$ and $s_{k+1|k}(\mathbf{x}) = \overset{1}{s}_{k+1|k}(\mathbf{x}) = \cdots = \overset{s}{s}_{k+1|k}(\mathbf{x})$. Then we get the CPHD filter analog of the Bayes parallel combination formula, Equation 8.14:

$$p_{k+1|k+1}(n) = \frac{1}{\mu_{k+1|k+1}} \cdot \overset{1}{p}_{k+1|k+1}(n) \cdots \overset{s}{p}_{k+1|k+1}(n) \cdot \sigma_{k+1|k+1}^n \cdot p_{k+1|k}(n)^{1-s} \qquad (8.100)$$

$$s_{k+1|k+1}(\mathbf{x}) = \frac{1}{\sigma_{k+1|k+1}} \cdot \overset{1}{s}_{k+1|k+1}(\mathbf{x}) \cdots \overset{s}{s}_{k+1|k+1}(\mathbf{x}) \cdot s_{k+1|k}(\mathbf{x})^{1-s} \qquad (8.101)$$

where

$$\mu_{k+1|k+1} = \sum_{n \geq 0} \overset{1}{p}_{k+1|k+1}(n) \cdots \overset{s}{p}_{k+1|k+1}(n) \cdot \sigma_{k+1|k+1}^n \qquad (8.102)$$

$$\sigma_{k+1|k+1} = \int \overset{1}{s}_{k+1|k+1}(\mathbf{x}) \cdots \overset{s}{s}_{k+1|k+1}(\mathbf{x}) \cdot s_{k+1|k}(\mathbf{x})^{1-s} d\mathbf{x}. \qquad (8.103)$$

Remark 3: Equations 8.100 and 8.101 have been employed as the basis for the principled approximate multisensor CPHD and PHD filters [39] mentioned in Section 8.3.5.

8.5.2 PHD Filter T²F of Independent Sources

What is the analog of Equation 8.97 for PHD filters? That is, suppose that the track sources use PHD filters rather than CPHD filters, and thus pass on PHDs rather than spatial distributions and cardinality distributions. Then what is the formula for the merged PHD? This turns out to be (see Section 8.7.2)

$$D_{k+1|k+1}(\mathbf{x}) = \frac{\overset{1}{D}_{k+1|k+1}(\mathbf{x})}{\overset{1}{D}_{k+1|k}(\mathbf{x})} \cdots \frac{\overset{s}{D}_{k+1|k+1}(\mathbf{x})}{\overset{s}{D}_{k+1|k}(\mathbf{x})} \cdot D_{k+1|k}(\mathbf{x}). \qquad (8.104)$$

This merging formula is potentially problematic because, in the single-target case, it does not reduce to the correct single-target formula. For example, suppose that $D_{k+1|k}(\mathbf{x}) = \overset{1}{D}_{k+1|k}(\mathbf{x}) = \cdots = \overset{s}{D}_{k+1|k}(\mathbf{x})$, in which case

$$D_{k+1|k+1}(\mathbf{x}) = \overset{1}{D}_{k+1|k+1}(\mathbf{x}) \cdots \overset{s}{D}_{k+1|k+1}(\mathbf{x}) \cdot D_{k+1|k}(\mathbf{x})^{1-s}. \qquad (8.105)$$

Equation 8.105 should reduce to Bayes parallel combination, Equation 8.14. However, it does not. To see this, note that with the single-target Bayes recursive filter, there are (1) no missed detections or false alarms; (2) the integrals of $D_{k+1|k}(\mathbf{x}) = f_{k+1|k}(\mathbf{x})$ and $\overset{i}{D}_{k+1|k+1}(\mathbf{x}) = \overset{i}{f}_{k+1|k+1}(\mathbf{x})$ for all i should equal 1; and (3) the integral of $D_{k+1|k}+1(\mathbf{x})$ should equal 1.

By way of contrast, Equations 8.96 and 8.97 *do* reduce to the correct single-target formula in the single-target case. Thus one must conclude:

- Equation 8.104 is unlikely to provide an accurate approximate track-merging formula when the number of targets in the scenario is small.

Remark 4: Let $\overset{i}{D}_{k+1|k+1}(\mathbf{x}) = \overset{i}{L}_{Z_{k+1}}(\mathbf{x}) \cdot \overset{i}{D}_{k+1|k}(\mathbf{x})$ be the PHD filter measurement-update formula for the ith source, as defined in Equation 8.67. Then Equation 8.105 becomes

$$D_{k+1|k+1}(\mathbf{x}) = \overset{1}{L}_{Z_{k+1}}^{1}(\mathbf{x}) \cdots \overset{s}{L}_{Z_{k+1}}^{s}(\mathbf{x}) \cdot D_{k+1|k}(\mathbf{x}). \qquad (8.106)$$

This is the multisensor PHD measurement-update formula as described in Equation 8.106 of reference [43]. It follows that this update formula is likely to be inaccurate when the number of targets is small.

8.5.3 CPHD Filter T²F with Known Double-Counting

Suppose that the data sources share sensors, but that it is known which sensors are being shared by which sources. The sources use CPHD filters to process these measurements: As in Section 8.4.2, define $Z_{k+1} = \overset{1}{Z}_{k+1} \cup \cdots \cup \overset{s}{Z}_{k+1}$. Let $\overset{12}{Z}_{k+1}$ be the measurements supplied to the second source that are not in $\overset{1}{Z}_{k+1}$, $\overset{13}{Z}_{k+1}$ the measurements supplied to the third source that are not in $\overset{1}{Z}_{k+1} \cup \overset{12}{Z}_{k+1}$, $\overset{14}{Z}_{k+1}$ the measurements supplied to the fourth source that are not in $\overset{1}{Z}_{k+1} \cup \overset{12}{Z}_{k+1} \cup \overset{13}{Z}_{k+1}$, and so on. Let $\overset{(j)}{Z}_{k+1} = \overset{j}{Z}_{k+1} - \overset{1j}{Z}_{k+1}$.

At time-step k, the jth source provides spatial distributions $\overset{j}{s}_{k|k}(\mathbf{x})$ and $\overset{j}{s}_{k+1|k}(\mathbf{x} \mid \overset{(j)}{Z})$, and cardinality distributions $\overset{j}{p}_{k|k}(n)$ and $\overset{j}{p}_{k+1|k}(n \mid \overset{(j)}{Z})$. Then the CPHD filter version of Equation 8.92 is

$$p_{k+1|k+1}(n) = \frac{1}{\mu_{k+1|k+1}} \cdot \frac{\overset{1}{p}_{k+1|k+1}(n)}{\overset{1}{p}_{k+1|k}(n)} \cdot \frac{\overset{s}{p}_{k+1|k+1}(n)}{\overset{2}{p}_{k+1|k}(n \mid \overset{(2)}{Z})} \cdots \frac{\overset{s}{p}_{k+1|k+1}(n)}{\overset{s}{p}_{k+1|k}(n \mid \overset{(s)}{Z})} \cdot \sigma^n_{k+1|k+1} \cdot p_{k+1|k}(n) \quad (8.107)$$

$$s_{k+1|k+1}(\mathbf{x}) = \frac{1}{\sigma_{k+1|k+1}} \cdot \frac{\overset{1}{s}_{k+1|k+1}(\mathbf{x})}{\overset{1}{s}_{k+1|k}(\mathbf{x})} \cdot \frac{\overset{2}{s}_{k+1|k+1}(\mathbf{x})}{\overset{2}{s}_{k+1|k}(\mathbf{x} \mid \overset{(2)}{Z})} \cdots \frac{\overset{s}{s}_{k+1|k+1}(\mathbf{x})}{\overset{s}{s}_{k+1|k}(\mathbf{x} \mid \overset{(s)}{Z})} \cdot s_{k+1|k}(\mathbf{x}) \quad (8.108)$$

where

$$\mu_{k+1|k+1} = \sum_{n \geq 0} \frac{\overset{1}{p}_{k+1|k+1}(n)}{\overset{1}{p}_{k+1|k}(n)} \cdot \frac{\overset{s}{p}_{k+1|k+1}(n)}{\overset{2}{p}_{k+1|k}(n \mid \overset{(2)}{Z})} \cdots \frac{\overset{s}{p}_{k+1|k+1}(n)}{\overset{s}{p}_{k+1|k}(n \mid \overset{(s)}{Z})} \cdot \sigma^n_{k+1|k+1} \cdot p_{k+1|k}(n) \quad (8.109)$$

$$\sigma_{k+1|k+1} = \int \frac{\overset{1}{s}_{k+1|k+1}(\mathbf{x})}{\overset{1}{s}_{k+1|k}(\mathbf{x})} \cdot \frac{\overset{2}{s}_{k+1|k+1}(\mathbf{x})}{\overset{2}{s}_{k+1|k}(\mathbf{x} \mid \overset{(2)}{Z})} \cdots \frac{\overset{s}{s}_{k+1|k+1}(\mathbf{x})}{\overset{s}{s}_{k+1|k}(\mathbf{x} \mid \overset{(s)}{Z})} \cdot s_{k+1|k}(\mathbf{x}) d\mathbf{x}. \quad (8.110)$$

8.5.4 PHD Filter T²F with Known Double-Counting

The PHD filter version of these equations is

$$D_{k+1|k+1}(\mathbf{x}) = \frac{\overset{1}{D}_{k+1|k+1}(\mathbf{x})}{\overset{1}{D}_{k+1|k}(\mathbf{x})} \cdot \frac{\overset{2}{D}_{k+1|k+1}(\mathbf{x})}{\overset{2}{D}_{k+1|k}(\mathbf{x} \mid \overset{(2)}{Z})} \cdots \frac{\overset{s}{D}_{k+1|k+1}(\mathbf{x})}{\overset{s}{D}_{k+1|k}(\mathbf{x} \mid \overset{(s)}{Z})} \cdot D_{k+1|k}(\mathbf{x}). \quad (8.111)$$

This update formula is unlikely to offer good performance when the number of targets is small.

8.5.5   CPHD FILTER XM FUSION

Clark et al. have considered the special case of Equation 8.93 when the distributions are i.i.d. cluster processes—that is, when track fusion is based on CPHD or PHD filters [4–6]. Suppose that multiple targets are being observed by two track sources equipped with CPHD filters. At time-step k, the first source provides a spatial distribution $\overset{0}{s}_{k|k}(\mathbf{x})$ and cardinality distribution $\overset{0}{p}_{k|k}(n)$; and the second source provides a spatial distribution $\overset{1}{s}_{k|k}(\mathbf{x})$ and cardinality distribution $\overset{1}{p}_{k|k}(n)$.

Given this, the CPHD filter approximation of the multitarget XM fusion formula, Equation 8.93, is (see Section 8.7.4)

$$\overset{\omega}{p}_{k+1|k+1}(n) = \frac{1}{\overset{\omega}{\mu}_{k+1|k+1}} \cdot \overset{0}{p}_{k+1|k+1}(n)^{1-\omega} \cdot \overset{1}{p}_{k+1|k+1}(n)^{\omega} \cdot \overset{\omega}{\sigma}{}^{n}_{k+1|k+1} \tag{8.112}$$

$$\overset{\omega}{s}_{k+1|k+1}(\mathbf{x}) = \frac{1}{\overset{\omega}{\sigma}_{k+1|k+1}} \cdot \overset{0}{s}_{k+1|k+1}(\mathbf{x})^{1-\omega} \cdot \overset{1}{s}_{k+1|k+1}(\mathbf{x})^{\omega} \tag{8.113}$$

where

$$\overset{\omega}{\mu}_{k+1|k+1} = \sum_{n \geq 0} \overset{0}{p}_{k+1|k+1}(n)^{1-\omega} \cdot \overset{1}{p}_{k+1|k+1}(n)^{\omega} \cdot \overset{\omega}{\sigma}{}^{n}_{k+1|k+1} \tag{8.114}$$

$$\overset{\omega}{\sigma}_{k+1|k+1} = \int \overset{0}{s}_{k+1|k+1}(\mathbf{x})^{1-\omega} \cdot \overset{1}{s}_{k+1|k+1}(\mathbf{x})^{\omega} d\mathbf{x}. \tag{8.115}$$

From a theoretical point of view, optimization of Equations 8.112 and 8.113 would be obtained via Equation 8.95:

$$\hat{\omega} = \arg \sup_{\omega} \sup_{X} \overset{\omega}{f}_{k+1|k+1}(X) \cdot \frac{c^{|X|}}{|X|!} = \arg \sup_{\omega} \sup_{X} p_{k|k}(|X|) \cdot \left(c s_{k+1|k+1} \right)^{X}. \tag{8.116}$$

However, this formula will usually be computationally problematic, as will be the multitarget version of the Chernoff information, Equation 8.35. A very approximate approach would be to first apply Chernoff optimization to the spatial distributions:

$$\breve{\omega}_1 = \arg \inf_{\omega} \int \overset{0}{s}_{k+1|k+1}(\mathbf{x})^{1-\omega} \cdot \overset{1}{s}_{k+1|k+1}(\mathbf{x})^{\omega} d\mathbf{x}. \tag{8.117}$$

Then setting

$$\breve{\sigma} = \int \overset{0}{s}_{k+1|k+1}(\mathbf{x})^{1-\breve{\omega}_1} \cdot \overset{1}{s}_{k+1|k+1}(\mathbf{x})^{\breve{\omega}_1} d\mathbf{x}, \tag{8.118}$$

one could apply Chernoff optimization once again to the cardinality distributions:

$$\check{\omega}_2 = \arg\inf_{\omega} \sum_{n \geq 0} \overset{0}{p}_{k+1|k+1}(n)^{1-\omega} \cdot \overset{1}{p}_{k+1|k+1}(n)^{\omega} \cdot \check{\sigma}^n \tag{8.119}$$

The final distributions would then be

$$\overset{XM}{p}_{k+1|k+1}(n) = \overset{\check{\omega}_2}{p}_{k+1|k+1}(n), \quad \overset{XM}{s}_{k+1|k+1}(\mathbf{x}) = \overset{\check{\omega}_1}{s}_{k+1|k+1}(\mathbf{x}). \tag{8.120}$$

Clark et al. have implemented Equations 8.112 through 8.115 and have considered a broad range of optimization procedures [5,6,44]. They have further demonstrated that these equations lead to good distributed-fusion performance.

8.5.6 PHD FILTER XM FUSION

The PHD filter approximation of the multitarget XM fusion formula, Equation 8.93, can be shown to be (see Section 8.7.5)

$$\overset{\omega}{D}_{k+1|k+1}(\mathbf{x}) = \overset{0}{D}_{k+1|k+1}(\mathbf{x})^{1-\omega} \cdot \overset{1}{D}_{k+1|k+1}(\mathbf{x})^{\omega}. \tag{8.121}$$

This formula is an equality, not a proportionality. Thus it does not reduce to the single-target XM fusion formula, Equation 8.34, in the single-target case. Thus it is unlikely to perform well when the number of targets is small.

How might we optimize Equation 8.121? Once again, Equation 8.116 will be computationally problematic. One alternative is as follows. It can be shown (see Section 8.7.6) that Chernoff information can be defined for PHDs and has the form

$$C(\overset{0}{D}_{k+1|k+1}, \overset{1}{D}_{k+1|k+1}) = \sup_{0 \leq \omega \leq 1} (K_{\omega} - \overset{\omega}{N}_{k+1|k+1}) \tag{8.122}$$

where

$$\overset{\omega}{N}_{k+1|k+1} = \int \overset{\omega}{D}_{k+1|k+1}(\mathbf{x})d\mathbf{x} = \int \overset{0}{D}_{k+1|k+1}(\mathbf{x})^{1-\omega} \cdot \overset{1}{D}_{k+1|k+1}(\mathbf{x})^{\omega} d\mathbf{x} \tag{8.123}$$

is the expected number of targets corresponding to ω, and where

$$K_{\omega} = (1-\omega)\overset{0}{N}_{k+1|k+1} + \omega\overset{1}{N}_{k+1|k+1} \tag{8.124}$$

is the weighted expected number of targets. Equation 8.122 is, at least in principle, potentially computationally tractable. Thus one would chose

$$\overset{\text{XM}}{D}_{k+1|k+1}(\mathbf{x}) = \overset{\check{\omega}}{D}_{k+1|k+1}(\mathbf{x}) \tag{8.125}$$

where

$$\check{\omega} = \arg\sup_{\omega}(K_\omega - \overset{\omega}{N}_{k+1|k+1}). \tag{8.126}$$

As an example, suppose that expected target numbers are identical: $\overset{0}{N}_{k+1|k+1} = \overset{1}{N}_{k+1|k+1} = N_{k+1|k+1}$. Then $K_\omega = N_{k+1|k+1}$ is constant and it is easily shown that

$$C(\overset{0}{D}_{k+1|k+1}, \overset{1}{D}_{k+1|k+1}) = N_{k+1|k+1} \cdot \sup_{0 \le \omega \le 1}\left(1 - \int \overset{0}{s}_{k+1|k+1}(\mathbf{x})^{1-\omega} \cdot \overset{1}{s}_{k+1|k+1}(\mathbf{x})^{\omega} d\mathbf{x}\right). \tag{8.127}$$

As another example, let the spatial distributions be identical: $\overset{0}{s}_{k+1|k+1}(\mathbf{x}) = \overset{1}{s}_{k+1|k+1}(\mathbf{x}) = s_{k+1|k+1}(\mathbf{x})$. Then it is easily shown that

$$C(\overset{0}{D}_{k+1|k+1}, \overset{1}{D}_{k+1|k+1}) = \sup_{0 \le \omega \le 1}\left((1-\omega)\overset{0}{N}_{k+1|k+1} + \omega\overset{1}{N}_{k+1|k+1} - N_{k+1|k+1}^{1-\omega}\, N_{k+1|k+1}^{\omega}\right). \tag{8.128}$$

8.6　COMPUTATIONAL ISSUES

In this section, I address the practical computability of the T²F CPHD/PHD filter formulas derived in the previous sections. There are two general fusion architectures that can be envisioned. In the first architecture, the track sources use GM-CPHD or GM-PHD filters, and transmit their Gaussian-mixture PHDs to the T²F site. In the second architecture, the track sources use particle-CPHD or particle-PHD filters, and transmit their particle-PHDs to the T²F site. In either case, the most serious obstacle to practical implementation is the following:

- The exact fusion formulas in Sections 8.5.1 through 8.5.3 involve division by PHDs.
- The XM fusion formulas in Sections 8.5.5 through 8.5.6 involve fractional powers of PHDs.

I deal with these two situations in the two sections that follow.

8.6.1　Implementation: Exact T²F Formulas

In what follows, I consider implementation of the exact fusion formulas in Sections 8.5.1 through 8.5.3. I consider two cases: the track sources employ GM-CPHD or GM-PHD filters; or the track sources employ particle-CPHD or particle-PHD filters.

8.6.1.1 Case 1: GM-PHD Tracks

Suppose that the track sources send their PHDs to the T²F site in the form of Gaussian mixtures—or, more precisely, as finite sets of the form $\{(w_1,\mathbf{x}_1,P_1),\ldots,(w_n,\mathbf{x}_n,P_n)\}$ where each triple $(w_i,\mathbf{x}_i,P_i)$ is a Gaussian component. Here I sketch the outlines of a possible implementation approach, using a hybridization of particle and Gaussian-mixture techniques.

For the sake of clarity, consider the simplest CPHD/PHD filter track-merging formula, Equation 8.105:

$$D_{k+1|k+1}(\mathbf{x}) = \overset{1}{D}_{k+1|k+1}(\mathbf{x})\cdots \overset{s}{D}_{k+1|k+1}(\mathbf{x}) \cdot D_{k+1|k}(\mathbf{x})^{1-s}. \tag{8.129}$$

If we can devise an implementation solution in this case, it should be possible to devise solutions for the more complex track-merging formulas in Sections 8.5.1 through 8.5.3. Rewrite Equation 8.129 as

$$D_{k+1|k+1}(\mathbf{x}) = \overset{1}{D}_{k+1|k+1}(\mathbf{x})\cdots \overset{s}{D}_{k+1|k+1}(\mathbf{x}) \cdot D_{k+1|k}(\mathbf{x})^{-s} \cdot D_{k+1|k}(\mathbf{x}). \tag{8.130}$$

where $D_{k+1|k}(\mathbf{x})$ and each $\overset{i}{D}_{k+1|k+1}(\mathbf{x})$ is a Gaussian mixture. Then:

1. Use standard GM-PHD filter merging and pruning techniques to reduce the product $\overset{1}{D}_{k+1|k+1}(\mathbf{x})\cdots \overset{s}{D}_{k+1|k+1}(\mathbf{x})$ to a new Gaussian-mixture PHD $\overset{1\ldots s}{D}_{k+1|k+1}(\mathbf{x})$.

2. Draw a statistical sample from the normalized predicted PHD:

$$\mathbf{x}^1_{k+1|k},\ldots,\mathbf{x}^v_{k+1|k} \sim \frac{D_{k+1|k}(\mathbf{x})}{N_{k+1|k}}. \tag{8.131}$$

3. Approximate $D_{k+1|k}(\mathbf{x})$ as the Dirac mixture

$$D_{k+1|k}(\mathbf{x}) \cong \frac{N_{k+1|k}}{v}\sum_{i=1}^{v}\delta_{\mathbf{x}^i_{k+1|k}}(\mathbf{x}). \tag{8.132}$$

4. Determine the corresponding particle approximation of $D_{k+1|k+1}(\mathbf{x})$:

$$\tilde{D}_{k+1|k+1}(\mathbf{x}) = \frac{N_{k+1|k}}{v}\sum_{i=1}^{v}\overset{1\ldots s}{D}_{k+1|k+1}(\mathbf{x}^i_{k+1|k}) \cdot D_{k+1|k}(\mathbf{x}^i_{k+1|k})^{-s}\cdot\delta_{\mathbf{x}^i_{k+1|k}}(\mathbf{x}). \tag{8.133}$$

For this formula to be effective, there have to be enough particles nearby the means of the Gaussian components of $\overset{1\ldots s}{D}_{k+1|k+1}(\mathbf{x})$.

5. Employ some particle-resampling technique to convert Equation 8.133 into distribution-sample form:

$$\breve{D}_{k+1|k+1}(\mathbf{x}) = \frac{\tilde{N}_{k+1|k+1}}{v} \sum_{i=1}^{v} \delta_{\mathbf{y}_{k+1|k}^{i}}(\mathbf{x}) \qquad (8.134)$$

where $\mathbf{y}_{k+1|k}^{1}, \ldots, \mathbf{y}_{k+1|k}^{v}$ are the resampled particles and where

$$\tilde{N}_{k+1|k+1} = N_{k+1|k} \sum_{i=1}^{v} {}^{1\ldots s} D_{k+1|k+1}(\mathbf{x}_{k+1|k}^{i}) \cdot D_{k+1|k}(\mathbf{x}_{k+1|k}^{i})^{-s}. \qquad (8.135)$$

6. Use the EM algorithm, or some other particle-regularization procedure, to approximate $\breve{D}_{k+1|k+1}(\mathbf{x})$ as a Gaussian mixture.
7. Iterate.

8.6.1.2   Case 2: Particle-PHD Tracks

This approach can be modified to address the case in which the track sources send their PHDs to the T^{2}F site in the form of Dirac mixtures:

1. The $\overset{i}{D}_{k+1|k+1}(\mathbf{x})$ are Dirac mixtures. Use the EM algorithm, or some other particle-regularization procedure, to approximate each of them as a Gaussian mixture.
2. Apply steps 1, 4, and 5 from the previous implementation approach.

8.6.2   IMPLEMENTATION: XM T^{2}F FORMULAS

In what follows, I consider implementation of the XM fusion formulas in Sections 8.5.5 through 8.5.6. I consider two cases: the track sources employ GM-CPHD or GM-PHD filters; or the track sources employ particle-CPHD or particle-PHD filters.

8.6.2.1   Case 1: GM-PHD Tracks

Suppose that the track sources send their PHDs to the T^{2}F site in the form of Gaussian mixtures. For the sake of conceptual clarity, consider the simplest of the XM fusion formulas, Equation 8.121:

$$\overset{\omega}{D}_{k+1|k+1}(\mathbf{x}) = \overset{0}{D}_{k+1|k+1}(\mathbf{x})^{1-\omega} \cdot \overset{1}{D}_{k+1|k+1}(\mathbf{x})^{\omega}. \qquad (8.136)$$

If we can devise an implementation solution in this case, it should be possible to devise solutions for the more complex track-merging formulas in Section 8.5.5.

One approach is to adapt the approximation suggested by Julier, one that appears to be surprisingly effective [15]:

$$\left(\sum_{i=1}^{n} x_i\right)^{\omega} \cong \sum_{i=1}^{n} x_i^{\omega}, \quad \left(\sum_{i=1}^{n} x_i\right)^{1-\omega} \cong \sum_{i=1}^{n} x_i^{1-\omega} \qquad (8.137)$$

Thus, if

$$\overset{0}{D}_{k+1|k+1}(\mathbf{x}) = \sum_{i=1}^{\overset{0}{\nu}} \overset{0}{w}_i \cdot N_{\overset{0}{P}_i}(\mathbf{x} - \overset{0}{\mathbf{x}}_i) \tag{8.138}$$

$$\overset{1}{D}_{k+1|k+1}(\mathbf{x}) = \sum_{j=1}^{\overset{1}{\nu}} \overset{1}{w}_j \cdot N_{\overset{1}{P}_j}(\mathbf{x} - \overset{1}{\mathbf{x}}_j), \tag{8.139}$$

it can be shown (see Section 8.7.7) that the corresponding XM fusion formula is

$$\left\{ \left(\overset{0}{w}_i, \overset{0}{\mathbf{x}}_i, \overset{0}{P}_i \right) \right\}_i, \left\{ \left(\overset{1}{w}_j, \overset{1}{\mathbf{x}}_j, \overset{1}{P}_j \right) \right\}_j \rightarrow \left\{ \left(\overset{\omega}{w}_{i,j}, \overset{\omega}{\mathbf{x}}_{i,j}, \overset{\omega}{P}_{i,j} \right) \right\}_{i,j} \tag{8.140}$$

where for $i = 1, \ldots, \overset{0}{\nu}$ and $j = 1, \ldots, \overset{1}{\nu}$,

$$\overset{\omega}{P}_{i,j}^{-1} = (1 - \omega) \overset{0}{P}_i^{-1} + \omega \overset{1}{P}_j^{-1} \tag{8.141}$$

$$\overset{\omega}{P}_{i,j}^{-1} \overset{\omega}{\mathbf{x}}_{i,j} = (1 - \omega) \overset{0}{P}_i^{-1} \overset{0}{\mathbf{x}}_i + \omega \overset{1}{P}_j^{-1} \overset{1}{\mathbf{x}}_j \tag{8.142}$$

$$\overset{\omega}{w}_{i,j} = \frac{\overset{0}{w}_i^{1-\omega} \cdot \overset{1}{w}_j^{\omega}}{\sqrt{\omega^N (1-\omega)^N}} \cdot N_{\overset{1}{P}_i/(1-\omega)+\overset{0}{P}_j/\omega}(\overset{1}{\mathbf{x}}_j - \overset{0}{\mathbf{x}}_i) \tag{8.143}$$

and where N is the dimension of the underlying Euclidean space.

As for optimization of ω, it follows that

$$\overset{\omega}{N}_{k+1|k+1} = \int \overset{\omega}{D}_{k+1|k+1}(\mathbf{x}) d\mathbf{x} \tag{8.144}$$

$$= \sum_{i=1}^{\overset{0}{\nu}} \sum_{j=1}^{\overset{1}{\nu}} \frac{\overset{0}{w}_i^{1-\omega} \cdot \overset{1}{w}_j^{\omega}}{\sqrt{\omega^N (1-\omega)^N}} \cdot N_{\overset{1}{P}_i/(1-\omega)+\overset{0}{P}_j/\omega}(\overset{1}{\mathbf{x}}_j - \overset{0}{\mathbf{x}}_i) \tag{8.145}$$

and thus from Equation 8.122 that the Chernoff information is, approximately, the supremum with respect to ω of the quantity

$$K_\omega - \overset{\omega}{N}_{k+1|k+1} \tag{8.146}$$

$$= \sum_{i=1}^{\overset{0}{\nu}} \sum_{j=1}^{\overset{1}{\nu}} \left(\frac{(1-\omega)\cdot \overset{0}{w_i}}{\overset{1}{\nu}} + \frac{\omega\cdot \overset{1}{w_j}}{\overset{0}{\nu}} - \frac{\overset{0}{w_i}^{1-\omega}\cdot \overset{1}{w_j}^{\omega}}{\sqrt{\omega^N (1-\omega)^N}} \cdot N_{0,\; \overset{1}{P_i}/(1-\omega)+\overset{0}{P_j}/\omega} \left(\overset{1}{\mathbf{x}}_j - \overset{0}{\mathbf{x}}_i \right) \right). \tag{8.147}$$

8.6.2.2   Case 2: Particle-PHD Tracks

This approach can be modified to address the case in which the track sources send their PHDs to the T²F site in the form of Dirac mixtures. Use the EM algorithm, or another particle-regularization procedure, to approximate the particle-PHDs $\overset{0}{D}_{k+1|k+1}(\mathbf{x})$ and $\overset{1}{D}_{k+1|k+1}(\mathbf{x})$ as Gaussian mixtures. Then proceed as before.

8.7   MATHEMATICAL DERIVATIONS

8.7.1   PROOF: CPHD T²F FUSION—INDEPENDENT SOURCES

Suppose that the track distributions of the sources are i.i.d.c. processes, i.e.,

$$f_{k+1|k}(X) = |X|!\, p_{k+1|k}(|X|)\cdot s_{k+1|k}^X \tag{8.148}$$

$$\overset{j}{f}_{k+1|k+1}(X) = |X|!\, \overset{j}{p}_{k+1|k+1}(|X|)\cdot s_{k+1|k+1}^{X\,\,\overset{j}{}} \tag{8.149}$$

$$\overset{j}{f}_{k+1|k}(X) = |X|!\, \overset{j}{p}_{k+1|k}(|X|)\cdot s_{k+1|k}^{X\,\,\overset{j}{}} \tag{8.150}$$

Then I show that the multitarget track-merging distribution of Equation 8.89 is also that of an i.i.d.c. process:

$$f_{k+1|k+1}(X) = |X|!\, p_{k+1|k+1}(|X|)\cdot s_{k+1|k+1}^X \tag{8.151}$$

where

$$p_{k+1|k+1}(n) = \frac{1}{\mu_{k+1|k+1}} \cdot \frac{\overset{1}{p}_{k+1|k+1}(n)}{\overset{1}{p}_{k+1|k}(n)} \cdots \frac{\overset{s}{p}_{k+1|k+1}(n)}{\overset{s}{p}_{k+1|k}(n)} \cdot \sigma_{k+1|k+1}^n \cdot p_{k+1|k}(n) \tag{8.152}$$

$$s_{k+1|k+1}(\mathbf{x}) = \frac{1}{\sigma_{k+1|k+1}} \frac{\overset{1}{s}_{k+1|k+1}(\mathbf{x})}{\overset{1}{s}_{k+1|k}(\mathbf{x})} \cdots \frac{\overset{s}{s}_{k+1|k+1}(\mathbf{x})}{\overset{s}{s}_{k+1|k}(\mathbf{x})} \cdot s_{k+1|k}(\mathbf{x}) \tag{8.153}$$

and where

$$\mu_{k+1|k+1} = \sum_{n\geq 0} \frac{\overset{1}{p}_{k+1|k+1}(n)}{\overset{1}{p}_{k+1|k}(n)} \cdots \frac{\overset{s}{p}_{k+1|k+1}(n)}{\overset{s}{p}_{k+1|k}(n)} \cdot \sigma_{k+1|k+1}^n \cdot p_{k+1|k}(n) \tag{8.154}$$

$$\sigma_{k+1|k+1} = \int \frac{\overset{1}{s}_{k+1|k+1}(\mathbf{x})}{\overset{1}{s}_{k+1|k}(\mathbf{x})} \cdots \frac{\overset{s}{s}_{k+1|k+1}(\mathbf{x})}{\overset{s}{s}_{k+1|k}(\mathbf{x})} \cdot s_{k+1|k}(\mathbf{x}) d\mathbf{x} \tag{8.155}$$

For, from Equation 8.89,

$$f_{k+1|k+1}(X) \propto \frac{\overset{1}{f}_{k+1|k+1}(X)}{\overset{1}{f}_{k+1|k}(X)} \cdots \frac{\overset{s}{f}_{k+1|k+1}(X)}{\overset{s}{f}_{k+1|k}(X)} \cdot f_{k+1|k}(X) \tag{8.156}$$

$$= \frac{|X|! \, \overset{1}{p}_{k+1|k+1}(|X|) \cdot \overset{X}{s}_{k+1|k+1}^{1}}{|X|! \, \overset{1}{p}_{k+1|k}(|X|) \cdot \overset{X}{s}_{k+1|k}^{1}} \cdots \frac{|X|! \, \overset{s}{p}_{k+1|k+1}(|X|) \cdot \overset{X}{s}_{k+1|k+1}^{s}}{|X|! \, \overset{s}{p}_{k+1|k}(|X|) \cdot \overset{X}{s}_{k+1|k}^{s}} \tag{8.157}$$

$$\cdot |X|! \, p_{k+1|k}(|X|) \cdot \overset{X}{s}_{k+1|k}$$

$$= |X|! \frac{\overset{1}{p}_{k+1|k+1}(|X|)}{\overset{1}{p}_{k+1|k}(|X|)} \cdots \frac{\overset{s}{p}_{k+1|k+1}(|X|)}{\overset{s}{p}_{k+1|k}(|X|)} \cdot p_{k+1|k}(|X|) \cdot \left(\frac{\overset{1}{s}_{k+1|k+1}}{\overset{1}{s}_{k+1|k}} \cdots \frac{\overset{s}{s}_{k+1|k+1}}{\overset{s}{s}_{k+1|k}} \cdot s_{k+1|k} \right)^{X} \tag{8.158}$$

$$\propto |X|! \, p_{k+1|k+1}(|X|) \cdot \overset{X}{s}_{k+1|k+1} \tag{8.159}$$

where the probability distributions $p_{k+1|k+1}(n)$ and $s_{k+1|k+1}(\mathbf{x})$ are defined by

$$p_{k+1|k+1}(n) \propto \frac{\overset{1}{p}_{k+1|k+1}(n)}{\overset{1}{p}_{k+1|k}(n)} \cdots \frac{\overset{s}{p}_{k+1|k+1}(n)}{\overset{s}{p}_{k+1|k}(n)} \cdot \sigma_{k+1|k+1}^{n} \cdot p_{k+1|k}(n) \tag{8.160}$$

$$s_{k+1|k+1}(\mathbf{x}) \propto \frac{\overset{1}{s}_{k+1|k+1}(\mathbf{x})}{\overset{1}{s}_{k+1|k}(\mathbf{x})} \cdots \frac{\overset{s}{s}_{k+1|k+1}(\mathbf{x})}{\overset{s}{s}_{k+1|k}(\mathbf{x})} \cdot s_{k+1|k}(\mathbf{x}). \tag{8.161}$$

Thus

$$f_{k+1|k+1}(X) = K \cdot |X|! \, p_{k+1|k+1}(|X|) \cdot \overset{X}{s}_{k+1|k+1} \tag{8.162}$$

for some K which is independent of X. Integrating both sides, and making note of Equation 8.63, we get

$$1 = \int f_{k+1|k+1}(X) \delta X = K \cdot \int |X|! \, p_{k+1|k+1}(|X|) \cdot \overset{X}{s}_{k+1|k+1} \delta X = K \cdot 1 = K \tag{8.163}$$

which completes the derivation.

8.7.2   PROOF: PHD T²F FUSION—INDEPENDENT SOURCES

Suppose that the track distributions of the sources are Poisson processes, i.e.,

$$f_{k+1|k}(X) = e^{-N_{k+1|k}} \cdot D_{k+1|k}^X \tag{8.164}$$

$$\overset{j}{f}_{k+1|k+1}(X) = e^{-\overset{j}{N}_{k+1|k+1}} \cdot \overset{j}{D}_{k+1|k+1}^X \tag{8.165}$$

$$\overset{j}{f}_{k+1|k}(X) = e^{-\overset{j}{N}_{k+1|k}} \cdot \overset{j}{D}_{k+1|k}^X . \tag{8.166}$$

Then I show that the XM fusion of the sources is also an i.i.d.c. process, where

$$N_{k+1|k+1} = \int \frac{\overset{1}{D}_{k+1|k+1}(\mathbf{x})}{\overset{1}{D}_{k+1|k}(\mathbf{x})} \cdots \frac{\overset{s}{D}_{k+1|k+1}(\mathbf{x})}{\overset{s}{D}_{k+1|k}(\mathbf{x})} \cdot D_{k+1|k}(\mathbf{x}) d\mathbf{x} \tag{8.167}$$

$$D_{k+1|k+1}(\mathbf{x}) = \frac{\overset{1}{D}_{k+1|k+1}(\mathbf{x})}{\overset{1}{D}_{k+1|k}(\mathbf{x})} \cdots \frac{\overset{s}{D}_{k+1|k+1}(\mathbf{x})}{\overset{s}{D}_{k+1|k}(\mathbf{x})} \cdot D_{k+1|k}(\mathbf{x}). \tag{8.168}$$

For, in this case

$$f_{k+1|k+1}(X) \propto \frac{\overset{1}{f}_{k+1|k+1}(X)}{\overset{1}{f}_{k+1|k}(X)} \cdots \frac{\overset{s}{f}_{k+1|k+1}(X)}{\overset{s}{f}_{k+1|k}(X)} \cdot f_{k+1|k}(X) \tag{8.169}$$

$$= \frac{e^{-\overset{1}{N}_{k+1|k+1}} \cdot \overset{1}{D}_{k+1|k+1}^X}{e^{-\overset{1}{N}_{k+1|k}} \cdot \overset{1}{D}_{k+1|k}^X} \cdots \frac{e^{-\overset{s}{N}_{k+1|k+1}} \cdot \overset{s}{D}_{k+1|k+1}^X}{e^{-\overset{s}{N}_{k+1|k}} \cdot \overset{s}{D}_{k+1|k}^X} \cdot e^{-N_{k+1|k}} \cdot D_{k+1|k}^X \tag{8.170}$$

$$\propto \left(\frac{\overset{1}{D}_{k+1|k+1}}{\overset{1}{D}_{k+1|k}} \cdots \frac{\overset{s}{D}_{k+1|k+1}}{\overset{s}{D}_{k+1|k}} \cdot D_{k+1|k} \right)^X \tag{8.171}$$

$$= e^{-N_{k+1|k+1}} \cdot D_{k+1|k+1}^X \tag{8.172}$$

as claimed.

8.7.3   PROOF: CPHD FILTER WITH DOUBLE-COUNTING

Suppose that the track distributions of the s track sources are i.i.d.c. processes, i.e.,

$$f_{k+1|k}(X) = |X|! \, p_{k+1|k}(|X|) \cdot s_{k+1|k}^X \tag{8.173}$$

$$\overset{j}{f}_{k+1|k+1}(X) = |X|! \cdot \overset{j}{p}_{k+1|k+1}(|X|) \cdot s^{\overset{j}{X}}_{k+1|k+1} \tag{8.174}$$

$$\overset{j}{f}_{k+1|k}(X|Z) = |X|! \cdot \overset{j}{p}_{k+1|k}(|X|) \cdot s^{\overset{j}{X}}_{k+1|k} \tag{8.175}$$

where

$$\overset{j}{p}_{k+1|k}(n) \overset{abbr.}{=} \overset{j}{p}_{k+1|k}(n|\overset{(j)}{Z}) \tag{8.176}$$

$$\overset{j}{s}_{k+1|k}(\mathbf{x}) \overset{abbr.}{=} \overset{j}{s}_{k+1|k}(\mathbf{x}|\overset{(j)}{Z}) \tag{8.177}$$

Then I show that the multitarget track-merging distribution of Equation 8.92 is also that of an i.i.d.c. process:

$$f_{k+1|k+1}(X) = |X|! \cdot p_{k+1|k+1}(|X|) \cdot s^{X}_{k+1|k+1} \tag{8.178}$$

where

$$p_{k+1|k+1}(n) = \frac{1}{\mu_{k+1|k+1}} \cdot \frac{\overset{1}{p}_{k+1|k+1}(n)}{\overset{1}{p}_{k+1|k}(n)} \cdot \frac{\overset{s}{p}_{k+1|k+1}(n)}{\overset{2}{p}_{k+1|k}(n|\overset{(2)}{Z})} \cdots \frac{\overset{s}{p}_{k+1|k+1}(n)}{\overset{s}{p}_{k+1|k}(n|\overset{(s)}{Z})} \cdot \sigma^{n}_{k+1|k+1} \cdot p_{k+1|k}(n) \tag{8.179}$$

$$s_{k+1|k+1}(\mathbf{x}) = \frac{1}{\sigma_{k+1|k+1}} \cdot \frac{\overset{1}{s}_{k+1|k+1}(\mathbf{x})}{\overset{1}{s}_{k+1|k}(\mathbf{x})} \cdot \frac{\overset{2}{s}_{k+1|k+1}(\mathbf{x})}{\overset{2}{s}_{k+1|k}(\mathbf{x}|\overset{(2)}{Z})} \cdots \frac{\overset{s}{s}_{k+1|k+1}(\mathbf{x})}{\overset{s}{s}_{k+1|k}(\mathbf{x}|\overset{(s)}{Z})} \cdot s_{k+1|k}(\mathbf{x}) \tag{8.180}$$

and where

$$\mu_{k+1|k+1} = \sum_{n\geq 0} \frac{\overset{1}{p}_{k+1|k+1}(n)}{\overset{1}{p}_{k+1|k}(n)} \cdot \frac{\overset{s}{p}_{k+1|k+1}(n)}{\overset{2}{p}_{k+1|k}(n|\overset{(2)}{Z})} \cdots \frac{\overset{s}{p}_{k+1|k+1}(n)}{\overset{s}{p}_{k+1|k}(n|\overset{(s)}{Z})} \cdot \sigma^{n}_{k+1|k+1} \cdot p_{k+1|k}(n) \tag{8.181}$$

$$\sigma_{k+1|k+1} = \int \frac{\overset{1}{s}_{k+1|k+1}(\mathbf{x})}{\overset{1}{s}_{k+1|k}(\mathbf{x})} \cdot \frac{\overset{2}{s}_{k+1|k+1}(\mathbf{x})}{\overset{2}{s}_{k+1|k}(\mathbf{x}|\overset{(2)}{Z})} \cdots \frac{\overset{s}{s}_{k+1|k+1}(\mathbf{x})}{\overset{s}{s}_{k+1|k}(\mathbf{x}|\overset{(s)}{Z})} \cdot s_{k+1|k}(\mathbf{x})d\mathbf{x} \tag{8.182}$$

The derivation is essentially identical to that in Section 8.7.1.

8.7.4 PROOF: CPHD FILTER XM FUSION

We are to prove the CPHD filter version of XM fusion as defined in Equations 8.112 and 8.113. Thus assume that the original multitarget distributions redistributions of i.i.d.c. processes:

$$\overset{0}{f}_{k+1|k+1}(X) = |X|! \cdot \overset{0}{p}_{k+1|k+1}(|X|) \cdot s_{k+1|k+1}^{X\,0} \tag{8.183}$$

$$\overset{0}{f}_{k+1|k}(X) = |X|! \cdot \overset{0}{p}_{k+1|k}(|X|) \cdot s_{k+1|k}^{X\,0} \tag{8.184}$$

$$\overset{1}{f}_{k+1|k+1}(X) = |X|! \cdot \overset{1}{p}_{k+1|k+1}(|X|) \cdot s_{k+1|k+1}^{X\,1} \tag{8.185}$$

$$\overset{1}{f}_{k+1|k+1}(X) = |X|! \cdot \overset{1}{p}_{k+1|k}(|X|) \cdot s_{k+1|k}^{X\,1} \tag{8.186}$$

$$f_{k+1|k}(X) = |X|! \cdot p_{k+1|k+1}(|X|) \cdot s_{k+1|k}^{X} \tag{8.187}$$

Then Equation 8.93 becomes

$$\overset{\omega}{f}_{k+1|k+1}(X) \propto \overset{0}{f}_{k+1|k+1}(X)^{1-\omega} \cdot \overset{1}{f}_{k+1|k+1}(X)^{\omega}$$

$$\propto \left(|X|! \cdot \overset{0}{p}_{k+1|k+1}(|X|) \cdot s_{k+1|k+1}^{X\,0} \right)^{1-\omega} \left(|X|! \cdot \overset{1}{p}_{k+1|k+1}(|X|) \cdot s_{k+1|k+1}^{X\,1} \right)^{\omega} \tag{8.188}$$

$$= |X|! \cdot \overset{0}{p}_{k+1|k+1}(|X|)^{1-\omega} \cdot \overset{1}{p}_{k+1|k+1}(|X|)^{\omega} \cdot \left(s_{k+1|k+1}^{1-\omega\,0} \cdot s_{k+1|k+1}^{\omega\,1} \right)^{X}$$

$$\propto |X|! \cdot \overset{\omega}{p}_{k+1|k+1}(|X|) \cdot s_{k+1|k+1}^{X\,\omega} \tag{8.189}$$

where

$$\overset{\omega}{p}_{k+1|k+1}(n) \propto \overset{0}{p}_{k+1|k+1}(n)^{1-\omega} \cdot \overset{1}{p}_{k+1|k+1}(n)^{\omega} \cdot \sigma_{k+1}^{n\,\omega} \tag{8.190}$$

$$\overset{\omega}{s}_{k+1|k+1}(\mathbf{x}) \propto \overset{0}{s}_{k+1|k+1}(\mathbf{x})^{1-\omega} \cdot \overset{1}{s}_{k+1|k+1}(\mathbf{x})^{\omega} \tag{8.191}$$

and where

$$\overset{\omega}{\sigma}_{k+1} = \int \overset{0}{s}_{k+1|k+1}(\mathbf{x})^{1-\omega} \cdot \overset{1}{s}_{k+1|k+1}(\mathbf{x})^{\omega} d\mathbf{x}. \tag{8.192}$$

This concludes the proof.

8.7.5 PROOF: PHD FILTER XM FUSION

We are to prove the PHD filter version of XM fusion as defined in Equation 8.121. Thus assume that the original multitarget distributions are distributions of Poisson processes:

$$\overset{0}{f}_{k+1|k+1}(X) = e^{-\overset{0}{N}_{k+1|k+1}} \cdot \overset{0}{D}_{k+1|k+1}^{X} \tag{8.193}$$

$$\overset{0}{f}_{k+1|k}(X) = e^{-\overset{0}{N}_{k+1|k}} \cdot \overset{0}{D}_{k+1|k}^{X} \tag{8.194}$$

$$\overset{1}{f}_{k+1|k+1}(X) = e^{-\overset{1}{N}_{k+1|k+1}} \overset{1}{D}_{k+1|k+1}^{X} \tag{8.195}$$

$$\overset{1}{f}_{k+1|k+1}(X) = e^{-\overset{1}{N}_{k+1|k}} \cdot \overset{1}{D}_{k+1|k}^{X} \tag{8.196}$$

$$f_{k+1|k}(X) = e^{-N_{k+1|k}} \cdot D_{k+1|k}^{X} \tag{8.197}$$

Then from Equation 8.93, we get

$$\overset{\omega}{f}_{k+1|k+1}(X) \propto \left(e^{-\overset{0}{N}_{k+1|k+1}} \cdot \overset{0}{D}_{k+1|k+1}^{X} \right)^{1-\omega} \left(e^{-\overset{1}{N}_{k+1|k+1}} \cdot \overset{1}{D}_{k+1|k+1}^{X} \right)^{\omega} \tag{8.198}$$

$$= e^{-(1-\omega)\overset{0}{N}_{k+1|k+1} - \omega\overset{1}{N}_{k+1|k+1}} \cdot \left(\overset{0}{D}_{k+1|k+1}^{1-\omega} \cdot \overset{1}{D}_{k+1|k+1}^{\omega} \right)^{X} \tag{8.199}$$

$$\propto \left(\overset{0}{D}_{k+1|k+1}^{1-\omega} \cdot \overset{1}{D}_{k+1|k+1}^{\omega} \right)^{X} \tag{8.200}$$

$$\propto e^{-\overset{\omega}{N}_{k+1|k+1}} \cdot \overset{\omega}{D}_{k+1|k+1}^{X} \tag{8.201}$$

where

$$\overset{\omega}{D}_{k+1|k+1}(\mathbf{x}) = \overset{0}{D}_{k+1|k+1}(\mathbf{x})^{1-\omega} \cdot \overset{1}{D}_{k+1|k+1}(\mathbf{x})^{\omega} \tag{8.202}$$

$$\overset{\omega}{N}_{k+1|k+1} = \int \overset{0}{D}_{k+1|k+1}(\mathbf{x})^{1-\omega} \cdot \overset{1}{D}_{k+1|k+1}(\mathbf{x})^{\omega} d\mathbf{x}. \tag{8.203}$$

8.7.6 PROOF: PHD FILTER CHERNOFF INFORMATION

We are to show that Chernoff information directly generalizes as follows:

$$C(\overset{0}{D}_{k+1|k+1}, \overset{1}{D}_{k+1|k+1}) = \sup_{0 \le \omega \le 1} (K_{\omega} - \overset{\omega}{N}_{k+1|k+1}) \tag{8.204}$$

where

$$K_\omega = (1-\omega)\,\overset{0}{N}_{k+1|k+1} + \omega\,\overset{1}{N}_{k+1|k+1}. \tag{8.205}$$

Applying Equation 8.35 to the multitarget distributions $\overset{0}{f}_{k+1|k+1}(X)$ and $\overset{1}{f}_{k+1|k+1}(X)^\omega$, we get

$$C(\overset{0}{D}_{k+1|k+1}, \overset{1}{D}_{k+1|k+1}) = \sup_{0\le\omega\le1}\left(-\log\int \overset{0}{f}_{k+1|k+1}(X)^{1-\omega}\cdot\overset{1}{f}_{k+1|k+1}(X)^\omega\,\delta X\right) \tag{8.206}$$

$$= \sup_{0\le\omega\le1}\left(-\log\int\left(e^{-\overset{0}{N}_{k+1|k+1}}\,\overset{0}{D}\,^X_{k+1|k+1}\right)^{1-\omega}\left(e^{-\overset{1}{N}_{k+1|k+1}}\,\overset{1}{D}\,^X_{k+1|k+1}\right)^\omega\,\delta X\right) \tag{8.207}$$

$$= \sup_{0\le\omega\le1}\left(K_\omega-\log\int\left(\overset{0}{D}\,^{1-\omega}_{k+1|k+1}\cdot\overset{1}{D}\,^\omega_{k+1|k+1}\right)^X\,\delta X\right). \tag{8.208}$$

According to Equation 8.61,

$$\int\left(\overset{0}{D}\,^{1-\omega}_{k+1|k+1}\cdot\overset{1}{D}\,^\omega_{k+1|k+1}\right)^X\,\delta X = e^{\overset{\omega}{N}_{k+1|k+1}} \tag{8.209}$$

where

$$\overset{\omega}{N}_{k+1|k+1} = \int \overset{0}{D}_{k+1|k+1}(\mathbf{x})^{1-\omega}\cdot\overset{1}{D}_{k+1|k+1}(\mathbf{x})^\omega\,d\mathbf{x}. \tag{8.210}$$

Thus

$$C(\overset{0}{D}_{k+1|k+1}, \overset{1}{D}_{k+1|k+1}) = \sup_{0\le\omega\le1}\left(K_\omega-\log e^{\overset{\omega}{N}_{k+1|k+1}}\right) \tag{8.211}$$

$$= \sup_{0\le\omega\le1}\left(K_\omega-\overset{\omega}{N}_{k+1|k+1}\right). \tag{8.212}$$

Thus we can define Equation 8.212 to be the "Chernoff information" of the PHDs $\overset{0}{D}_{k+1|k+1}(\mathbf{x})$ and $\overset{1}{D}_{k+1|k+1}(\mathbf{x})$.

8.7.7 PROOF: XM IMPLEMENTATION

We are to establish Equations 8.141 through 8.143. Substituting Equations 8.138 and 8.139 into Equation 8.136 we get

$$\overset{\omega}{D}_{k+1|k+1}(\mathbf{x}) \cong \left(\sum_{i=1}^{\overset{0}{\nu}} w_i^{1-\omega} \cdot \underset{\overset{0}{P_i}}{N_0} (\mathbf{x} - \overset{0}{\mathbf{x}}_i)^{1-\omega} \right) \left(\sum_{j=1}^{\overset{1}{\nu}} w_j^{\omega} \cdot \underset{\overset{1}{P_j}}{N_1} (\mathbf{x} - \overset{1}{\mathbf{x}}_j)^{\omega} \right) \quad (8.213)$$

$$= \sum_{i=1}^{\overset{0}{\nu}} \sum_{j=1}^{\overset{1}{\nu}} w_i^{1-\omega} \cdot w_j^{\omega} \cdot \underset{\overset{0}{P_i}}{N_0} (\mathbf{x} - \overset{0}{\mathbf{x}}_i)^{1-\omega} \cdot \underset{\overset{1}{P_j}}{N_1} (\mathbf{x} - \overset{1}{\mathbf{x}}_j)^{\omega} \quad (8.214)$$

$$= \sum_{i=1}^{\overset{0}{\nu}} \sum_{j=1}^{\overset{1}{\nu}} w_i^{1-\omega} \cdot w_j^{\omega} \cdot \sqrt{\frac{\det 2\pi \overset{0}{P_i} / (1-\omega)}{\det 2\pi \overset{0}{P_i}}} \cdot \sqrt{\frac{\det 2\pi \overset{1}{P_j} / \omega}{\det 2\pi \overset{1}{P_j}}}$$

$$\cdot \underset{\overset{0}{P_i}/(1-\omega)}{N_0} (\mathbf{x} - \overset{0}{\mathbf{x}}_i) \cdot \underset{\overset{1}{P_j}/\omega}{N_1} (\mathbf{x} - \overset{1}{\mathbf{x}}_j) \quad (8.215)$$

$$= \sum_{i=1}^{\overset{0}{\nu}} \sum_{j=1}^{\overset{1}{\nu}} \frac{w_i^{1-\omega} \cdot w_j^{\omega}}{\sqrt{\omega^N (1-\omega)^N}} \cdot \underset{\overset{1}{P_i}/(1-\omega)+\overset{1}{P_j}/\omega}{N_0} (\overset{1}{\mathbf{x}}_j - \overset{0}{\mathbf{x}}_i) \cdot \underset{\overset{\omega}{P_{i,j}}}{N_\omega} (\mathbf{x} - \overset{\omega}{\mathbf{x}}_{i,j}) \quad (8.216)$$

where

$$\overset{\omega}{P}_{i,j}^{-1} = (1-\omega) \overset{0}{P}_i^{-1} + \omega \overset{1}{P}_j^{-1} \quad (8.217)$$

$$\overset{\omega}{P}_{i,j}^{-1} \overset{\omega}{\mathbf{x}}_{i,j} = (1-\omega) \overset{0}{P}_i^{-1} \overset{0}{\mathbf{x}}_i + \omega \overset{1}{P}_j^{-1} \overset{1}{\mathbf{x}}_j \quad (8.218)$$

This establishes the result.

8.8 CONCLUSIONS

In this chapter, I have proposed the elements of a general theoretical foundation for multisource-multitarget track-to-track fusion (T²F). After summarizing three major single-target T²F situations and approaches—exact T²F without double-counting, exact T²F with double-counting, and approximate T²F in the manner of Clark et al.—I showed how to directly generalize them to the multisource-multitarget case. Since the resulting algorithms are computationally intractable in general, I showed how to derive approximate versions of them using CPHD and PHD filter-based approaches. I also suggested notional implementation techniques for these approaches.

The ideas proposed in this chapter are, of course, just a beginning. Further research into practical implementation is necessary. Also, since the XM fusion approach is a generalization of the CI approach, it is necessarily also an overly conservative approach. More work needs to be conducted on modifications of single-target XM fusion that model different assumptions about the degree of correlations between the original distributions. If such modifications can be devised, then it should be possible to generalize them to the multitarget case using the methodology advocated in this chapter.

REFERENCES

1. O.E. Drummond, On track and tracklet fusion filtering, in O.E. Drummond (Ed.), *Signal and Data Processing of Small Targets 2002*, *SPIE Proceedings*, Vol. 4728, pp. 176–195, Bellingham, WA, 2002.
2. C.-Y. Chong, S. Mori, and K.-C. Chang, Distributed multitarget multisensor tracking, in Y. Bar-Shalom (Ed.), *Multitarget-Multisensor Tracking: Advanced Applications*, Chapter 8, Artech House, London, U.K., 1990; Re-published as *Multitarget-Multisensor Tracking: Advances and Applications*, Vol. I, YBS, Storrs, CT, 1996.
3. R. Mahler, Optimal/robust distributed data fusion: A unified approach, in I. Kadar (Ed.), *Signal Processing, Sensor Fusion, and Target Recognition IX*, *SPIE Proceedings*, Vol. 4052, pp. 128–138, Orlando, FL, 2000.
4. D. Clark, S. Julier, R. Mahler, and B. Ristić, Robust multi-object sensor fusion with unknown correlations, *Proceedings of the Conference on Sensor Signal Processing for Defence 2010 (SSPD2010)*, Imperial College, London, U.K., September 29–30, 2010.
5. M. Uney, D. Clark, and S. Julier, Information measures in distributed multitarget tracking, *Proceedings of the 14th International Conference on Information Fusion*, Chicago, IL, July 5–8, 2011.
6. M. Uney, S. Julier, D. Clark, and B. Ristić, Monte Carlo realisation of a distributed multi-object fusion algorithm, *Proceedings of the Conference on Sensor Signal Processing for Defence 2010 (SSPD2010)*, Imperial College, London, U.K., September 29–30, 2010.
7. R. Mahler, *Statistical Multisource-Multitarget Information Fusion*, Artech House, Norwood, MA, 2007.
8. J.K. Uhlmann, General data fusion for estimates with unknown cross covariances, *SPIE Proceedings*, 2755, 536–547, 1996.
9. S. Julier and J. Uhlmann, A non-divergent estimation algorithm in the presence of unknown correlations, *Proceedings of the IEEE American Control Conference*, Vol. 4, pp. 2369–2373, Albuquerque, NM, June 4–6, 1997.
10. S. Julier and J. Uhlmann, General decentralized data fusion with covariance intersection, in D.L. Hall and J. Llinas (Eds.), *Handbook of Multisensor Data Fusion*, 2nd edn., Chapter 14, pp. 319–343, CRC Press, Boca Raton, FL, 2008.
11. D. Fränken and A. Hüpper, Improved fast covariance intersection for distributed data fusion, *Proceedings of the 8th International Conference on Information Fusion*, pp. 154–160, Philadelphia, PA, July 25–28, 2005.
12. M. Hurley, An information-theoretic justification for covariance intersection and its generalization, *Proceedings of the 5th International Conference on Information Fusion*, Annapolis, MD, July 7–11, 2002.
13. T. Heskes, Selecting weighting factors in logarithm opinion pools, *Advances in Neural Information Processing Systems*, 10, 266–272, 1998.
14. S. Julier, T. Bailey, and J. Uhlmann, Using exponential mixture models for suboptimal distributed data fusion, *Proceedings of the 2006 IEEE Nonlinear Signal Processing Workshop*, pp. 13–15, Birmingham, U.K., September 13–15, 2006.
15. J. Julier, An empirical study into the use of Chernoff information for robust, distributed fusion of Gaussian mixture models, *Proceedings of the 9th International Conference on Information Fusion*, Florence, Italy, July 10–13, 2006.

16. W. Farrell III and C. Ganesh, Generalized Chernoff fusion approximation for practical distributed data fusion, *Proceedings of the 12th International Conference on Information Fusion*, pp. 555–562, Seattle, WA, July 6–9, 2009.

17. S. Julier, Fusion without independence, *Proceedings of the 2008 IET Seminar on Tracking and Data Fusion: Algorithms and Applications*, pp. 1–5, Birmingham, U.K., April 15–16, 2008.

18. I. Csiszár, I-divergence geometry of probability distributions and minimization problems, *Annals of Probability*, 3(1), 146–158, 1975.

19. I. Csiszár, Information-type measures of difference of probability distributions and indirect observations, *Studia Scientiarum Mathematicarum Hungarica*, 2, 299–318, 1967.

20. T. Zajic and R. Mahler, Practical information-based data fusion performance evaluation, *SPIE Proceedings*, 3720, 92–103, 1999.

21. R. Mahler, Random set theory for target tracking and identification, in D.L. Hall and J. Llinas (Eds.), *Handbook of Multisensor Data Fusion*, 2nd edn., Chapter 16, pp. 369–410, CRC Press, Boca Raton, FL, 2008.

22. R. Mahler, 'Statistics 101' for multisensor, multitarget data fusion, *IEEE Aerospace and Electronics Systems Magazine, Part 2: Tutorials*, 19(1), 53–64, 2004.

23. R. Mahler, PHD filters of higher order in target number, *IEEE Transactions on Aerospace and Electronic Systems*, 43(4), 1523–1543, 2007.

24. B.-N. Vo, B.-T. Vo, and N. Pham, Bayesian multi-object estimation from image observations, *Proceedings of the 12th International Conference on Information Fusion*, Seattle, WA, July 6–9, 2009.

25. B.-T. Vo, B.-N. Vo, and A. Cantoni, The cardinality balanced multi-target multi-Bernoulli filter and its implementations, *IEEE Transactions on Aerospace and Electronic Systems*, 57(2), 409–423, 2009.

26. J. Mullane, B.-N. Vo, M. Adams, and B.-T. Vo, A random set approach to Bayesian SLAM, Accepted for publication in *IEEE Trans. Robotics and Automation,* 27(2), 268–282, 2011.

27 J. Mullane, B.-N. Vo, M. Adams, and B.-T. Vo, *Random Finite Sets in Robotic Map Building and SLAM,* Springer, 2011.

28. C.S. Lee, D. Clark, and J. Salvi, SLAM with single cluster PHD filters, *Proceedings of the 2012 IEEE International Conference on Robotics and Automation (ICRA2012)*, St. Paul, MN, 2012.

29. R. Mahler, CPHD and PHD filters for unknown backgrounds, I: Dynamic data clustering, in J. Cox and P. Motaghedi (Eds.), *Sensors and Systems for Space Applications III*, *SPIE Proceedings*, Vol. 7330, 2009.

30. R. Mahler, CPHD and PHD filters for unknown backgrounds, II: Multitarget filtering in dynamic clutter, in J. Cox and P. Motaghedi (Eds.), *Sensors and Systems for Space Applications III*, *SPIE Proceedings*, Vol. 7330, 2009.

31. R. Mahler and A. El-Fallah, CPHD and PHD filters for unknown backgrounds, III: Tractable multitarget filtering in dynamic clutter, in O. Drummond (Ed.), *Signals and Data Processing of Small Targets 2010*, *SPIE Proceedings*, Vol. 7698, 2010.

32. R. Mahler and A. El-Fallah, CPHD filtering with unknown probability of detection, in I. Kadar (Ed.), *Signal Processing, Sensor Fusion, and Target Recognition XIX*, *SPIE Proceedings*, Vol. 7697, 2010.

33. R. Mahler, B.-T. Vo, and B.-N. Vo, CPHD filtering with unknown clutter rate and detection profile, *Proceedings of the 14th International Conference on Information Fusion*, Chicago, IL, July 5–8, 2011.

34. R. Mahler, B.-T. Vo, and B.-N. Vo, CPHD filtering with unknown clutter rate and detection profile, *IEEE Transactions on Signal Processing*, 59(6), 3497–3513, 2011.

35. B.-T. Vo, B.-N. Vo, R. Hoseinnezhad, and R. Mahler, Multi-Bernoulli filtering with unknown clutter intensity and sensor field-of-view, *Proceedings of the 2011 IEEE Conference on Information Sciences and Systems*, Baltimore, MD, March 23–25, 2011.

36. B.-T. Vo, B.-N. Vo, R. Hoseinnezhad, and R. Mahler, Multi-Bernoulli filtering with unknown clutter intensity and sensor field-of-view, *Proceedings of the 45th Annual Conference on Information Sciences and Systems (CICS2011)*, Johns Hopkins University, Baltimore, MD, March 23–25, 2011.

37. B.-T. Vo, B.-N. Vo, R. Hoseinnezhad, and R. Mahler, Multi-Bernoulli filtering with unknown clutter intensity and sensor field-of-view, Submitted to *IEEE Transactions on Aerospace and Electronic Systems*, 2011.

38. S. Nagappa and D. Clark, On the ordering of the sensors in the iterated-corrector probability hypothesis density (PHD) filter, in I. Kadar (Ed.), *Signal Processing, Sensor Fusion, and Target Recognition XX*, *SPIE Proc.*, Vol. 8050, Orlando, FL, April 26–28, 2011.

39. R. Mahler, Approximate multisensor CPHD and PHD filters, *Proceedings of the 13th International Conference on Information Fusion*, Edinburgh, Scotland, July 26–29, 2010.

40. S. Nagappa, D. Clark, and R. Mahler, Incorporating track uncertainty into the OSPA metric, *Proceedings of the 14th International Conference on Information Fusion*, Chicago, IL, July 5–8, 2011.

41. R. Mahler and A. El-Fallah, Unified Bayesian registration and tracking, in I. Kadar (Ed.), *Signal Processing, Sensor Fusion, and Target Recognition XX*, *SPIE Proceedings*, Vol. 8050, Orlando, FL, April 26–28, 2011.

42. R. Ristić and D. Clark, Particle filter for joint estimation of multi-object dynamic state and multi-sensor bias, *Proceedings of the 37th IEEE International Conference on Acoustics, Speech and Signal Processing (ICASSP2012)*, Kyoto, Japan, March 25–30, 2012.

43. R. Mahler, Multitarget filtering via first-order multitarget moments, *IEEE Transactions on Aerospace and Electronics Systems*, 39(4), 1152–1178, 2003.

44. R. Hoseinnezhad, B.-N. Vo, D. Suter, and B.-T. Vo, Multi-object filtering from image sequence without detection, *Proceedings of the 2010 IEEE International Conference on Acoustics, Speech and Signal Processing (ICASSP)*, Dallas, TX, March 14–19, 2010.

45. D. Clark and B.-N. Vo, Convergence analysis of the Gaussian mixture PHD filter, *IEEE Transactions on Signal Processing*, 55(4), 1204–1212, 2007.

46. K. Panta, D. Clark, and B.-N. Vo, Data association and track management for the Gaussian mixture probability hypothesis density filter, *IEEE Transactions on Aerospace and Electronic Systems*, 45(3), 1003–1016, 2009.

47. B.-N. Vo and W.-K. Ma, The Gaussian mixture probability hypothesis density filter, *IEEE Transactions on Signal Processing*, 54(11), 4091–4104, 2006.

James Llinas and Chee-Yee Chong

CONTENTS

9.1 Introduction 245
9.2 Overview of Object Classification Approaches 246
9.3 Architectural Options for Object Classification 248
9.4 Issues in Distributed Object Classification 250
    9.4.1 Explicit Double-Counting 251
    9.4.2 Implicit Double-Counting (Statistically Dependent) 251
    9.4.3 Legacy Systems with Hard Declarations 252
    9.4.4 Mixed Uncertainty Representations 253
9.5 Classifier Fusion 253
    9.5.1 Taxonomy of Classifier Fusion Methods 254
    9.5.2 Combining Classifiers 254
9.6 Optimal Distributed Bayesian Object Classification 256
    9.6.1 Centralized Object Classification Algorithm 258
    9.6.2 Distributed Object Classification 258
        9.6.2.1 Bayesian Distributed Fusion Algorithm 259
        9.6.2.2 Optimal Distributed Object Classification 260
    9.6.3 Communication Strategies 262
    9.6.4 Performance Evaluation 263
        9.6.4.1 Simulation Scenario and Data Generation 263
        9.6.4.2 Performance Evaluation Approach 263
        9.6.4.3 Comparison of Fusion Algorithms 264
9.7 Conclusions 268
References 269

9.1 INTRODUCTION

Object classification is the process of providing some level of identification of an object, whether it is at a very specific level or at a general level. In a defense or security environment, such classes can range from distinguishing people from vehicles, a type of vehicle, or in more specific cases a particular object or entity, even to the "serial number." Of course, another fundamental classification for such purposes is the traditional friend-foe-neutral-unknown class categories that are used in developing action-taking policies and decisions. The concept of identity is an interesting one and is nonmetric in that there is no generalized or unified scale of units by which we "measure" the identity of an object. The nonmetric nature of the identity of an object has motivated the formation of several alternative approaches to the representation, quantification, and manipulation of uncertainty in identity. Statistical methods, fuzzy set (possibilistic), and neural network methods are some of the popular methods that have been applied to this problem but two statistically based techniques remain the most popular: Bayesian methods and Dempster–Shafer methods. For decentralized/distributed applications, these two approaches have the attractive features that they are both commutative and associative; so results are independent of the order in which data are received and processed.

Dempster–Shafer methods (Al-Ani and Deriche 2002) were largely developed to overcome the apparent inability of the Bayesian formalism to adequately deal with "unknown" classes, as opposed to an estimated degree of ignorance attempted in Dempster–Shafer. However, achieving this feature is usually at the expense of computationally more intensive algorithms. The distributed classification problem raises a number of architectural problems, and two key challenges are (1) how to connect all the sensors and processing nodes together (these are organizational and architecture issues) and (2) how to fuse the data for estimates generated by the nodal operations (this is a data fusion algorithm design issue).

By and large, an object's class is discerned by an examination of its features or attributes where we distinguish a feature as an observed characteristic of an object and an attribute as an inherent characteristic. As regards information fusion processing, automated classification techniques are placed in the "Level 1" category of the JDL data fusion process model where, in conjunction with tracking and kinematic estimation operations, they help to answer two fundamental questions "Where is it?" and "What is it?" One can quickly get philosophical about the concept of class and identity, and there are various works the reader can examine to explore these viewpoints, such as Zalta (2012) and Hartshorne et al. (1931–1958). As just mentioned, identity and/or class membership is determined from features and attributes (F&A henceforth) by some type of pattern recognition process that is able to assign a label to the object according to the F&A estimates and perhaps some ontology or taxonomy of object types and classes. Assigning an identity label is not an absolute process; classification of an object may, for example, be context-dependent (e.g., whether an object is in the class "large" or not may be dependent on its surroundings).

The structure of this chapter is as follows: in Section 9.2 we provide a limited overview of object classification approaches, in Section 9.3 a summary of classification architectures is provided. Section 9.4 addresses distributed object classification issues, Section 9.5 discusses classifier fusion, and Section 9.6 provides an extended discussion on distributed Bayesian classification and performance evaluation, as the Bayes formalism remains a central methodology for many classification problems.

9.2 OVERVIEW OF OBJECT CLASSIFICATION APPROACHES

The data and information fusion community, as well as the remote sensing community, have a long history of research in exploring and developing automated methods for estimating an object's class from available observational data on F&A, which is the fundamental goal of object/entity classification. In a broad sense, the basic processing steps for classification involve sensor-dependent preprocessing that in the multisensor fusion case includes common referencing or alignment (for imaging sensors often used for classification operations, this is typically called co-registration), F&A extraction, F&A association, and class-estimation. There are many complexities in attempting to automate this process since the mapping between F&A and the ability to assign an object label is a very complex relationship, which is complicated by the usual observation errors and, in the military domain, by efforts of the adversary to incorporate decoys and deception techniques to create ambiguous F&A characteristics. In the military domain, these works have usually come under the labels of "automated" or "assisted" or "aided" target recognition or "ATR" methods, also non-cooperative target recognition (NCTR) methods, and there have been many conferences and published works on these topics. We find no very recent open-source publications explicitly directed to state-of-the-art assessments of ATR/NCTR technology but some state-of-the-art and survey literature does exist, such as those by Bhanu (1986), Roth (1990), Cohen (1991), Novak (2000), and Murphy and Taylor (2002), but nothing very recent in the timeframe of this book. The most central source for publications on ATR and the broader topic of object classification are the SPIE defense conferences held annually for a number of years, and the interested reader is referred to this source for additional information. On the other hand, there are some more recent works that review the broader domain of classification methods as a general problem. Still, no overarching state-of-the-art work has been done, but there are various survey papers directed to certain classes of methods.

One way to summarize alternative approaches to classification is provided by bisecting the methods into the "discriminative" and "generative" types; a number of recent references address this topic, such as Ng and Jordan (2002). Generative classification methods assume some functional form for the conditional relation between the F&A data and the class labeling, and estimate the parameters of these functions directly from truthed training data. An example is Naïve Bayes where the generative approach takes the form of the relation as P(Features|Class Label), and the model of the prior existence of a class label as P(Class Label); in typical notation these are $P(X|Y)$ and $P(Y)$. Additional generative techniques include model-based recognition, Gaussian mixture densities, and Bayesian networks. Discriminative classifiers are methods constructed to directly learn the relation P(Class Label|F&A Data), or directly learn $P(Y|X)$. An example often cited in the literature is logistic regression but other methods include neural networks, various discriminant classifiers, and support vector machines.

Montillo (2010) asserts the advantages and disadvantages of the generative approach in Table 9.1. And similarly, the trade-offs are also asserted by Montillo for the discriminative approaches in Table 9.2.

The interested reader is also directed to the various citations for both the ATR/ NCTR literature that, while somewhat dated, still gives some valuable insight, and into the generative, discriminative comparative literature for a somewhat more modern view.

TABLE 9.1

Features of Generative Approaches to Classification

Advantages	Disadvantages
Ability to introduce prior knowledge	Marred by generic optimization criteria
Do not require large training sets	Potentially wasteful modeling
Generation of synthetic inputs	Reliant on domain expertise
	Do not scale well to large number of classes

Source: Adapted from Montillo, A., Generative versus discriminative modeling frameworks, Lecture at Temple University, http://www.albertmontillo.com/lectures,presentations/Generative%20vs%20Discriminative.pdf, 2010.

TABLE 9.2

Features of Discriminative Approaches to Classification

Advantages	Disadvantages
Fast prediction speed	Task specific
Potentially more accurate prediction	Long training time
	Do not easily handle compositionality

Source: Adapted from Montillo, A., Generative versus discriminative modeling frameworks, Lecture at Temple University, http://www.albertmontillo.com/lectures,presentations/Generative%20vs%20Discriminative.pdf, 2010.

It is beyond the scope of this chapter to provide further detail on these state-of-the-art and survey remarks, nor a tutorial-type overview of this extensive field; for the most up-to-date open-source information, we again suggest the SPIE conference literature, the *IEEE Pattern Analysis and Machine Intelligence Transactions*, and for tutorials the interested reader is referred to such books as those by Tait (2005), Sadjadi and Javidi (2007), or Nebabin (1994).

9.3 ARCHITECTURAL OPTIONS FOR OBJECT CLASSIFICATION

To better understand the content of this chapter, we offer some perspectives on the structural design of an object classification process as determined by multiple observational data. A notional set of architectural configurations for object classification processing is shown in Figure 9.1.

In a distributed context, it could be for example that each sensor-specific processing thread is a node in a network, and the Declaration Fusion process is occurring at a receiving node that collects each sensor-specific declaration of class. There are many possible variations of the different ways a distributed classification process may operate but this figure should be adequate for our purpose. We characterize the process as involving multiple sensor systems (S1, S2) that each generates (imperfect)

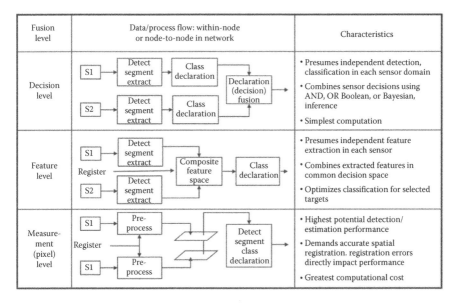

FIGURE 9.1 Notional multisensor object classification process options.

observational data. Three alternative processing schemes are shown: a measurement-based approach, a feature-based approach, and a decision-based approach. As mentioned earlier, the processes shown can happen within a node in a network (local/single-node-multisensor fusion) or they can represent internodal, distributed operations where sensor-specific operations at one node are then sent, via the network, to another (receiving node) where the multisensor fusion would take place. In either case, these are the following characteristics of each operation:

- The measurement-based approach ideally would use combined raw sensor data at the measurement (pre-feature) level to form information-rich features and attributes that would then provide the evidential foundation for a classification/recognition/identification algorithm. Note that in the distributed case this involves sending raw measurement data across the network, often considered prohibitive due to bandwidth limitations.
- The feature-based approach involves sensor-specific computation of available features, then a combining/fusing of the features as the evidential basis for classification. In the distributed case, the features would be communicated to other nodes.
- The decision-based approach has the sensor-specific processing proceeding to the declaration/decision level, after which the decisions are fused. Decisions can be sent over the network using lower bandwidth links or by consuming small amounts of any given bandwidth.

Note too that for measurement- and feature-based approaches, there is a requirement to co-register or align the data (in the "common referencing" function of a fusion process) before the combining operations take place. Registration requirements are most stringent for the most detailed type of data, so that the highest precision in registration is needed for the measurement-based operations.

In the case of distributed object classification, clearly what happens regarding fusion operations at any receiving node is dependent on what has been transmitted; of course, it is possible that mixtures of types of data may also arrive at the receiving node. For any distributed environment, it should be noted that any node can only fuse two things: those data that it "owns" (typically called "organic" data), and data that come to it somehow over the network. Thus, in the distributed case, a crucial design issue is the specification of what we call here "information-sharing strategies" that govern who sends what to whom, how often, in what format, etc.; this issue is crucial but outside the scope of this chapter.

9.4   ISSUES IN DISTRIBUTED OBJECT CLASSIFICATION

Distributed target tracking and classification have become a popular area of research because of the proliferation of sensors (Brooks et al. 2003, Kotecha et al. 2005). The issues related to distributed object classification are analogous to the issues associated with distributed tracking (Liggins et al. 1997), which include the issue of correlated data and decisions and the associated effects on the fused result. As in any fusion process, consideration of the common referencing, data association, and state estimation functions and how they might be affected by the peculiarities of the distributed environment needs to be carried out. In large networks, it is typical that any given node has limited knowledge of the functions and capabilities of the other nodes in the network; it is often assumed that any given node only knows specific information about its neighboring nodes but not distant nodes. This situation imputes the requirement for sending/transmitting nodes to append metadata to their messages to provide adequate information to receiving nodes such as to allow proper processing of the information in the received messages. This metadata has come to be called "pedigree" information by the fusion community, and one definition of pedigree is that it is information needed by a receiving node such as to assure the mathematical or formal integrity of whatever operation it applies to the received data, although the extent of the pedigree content may include yet other information.

Later we provide a brief description and a simple diagrammatic summary of some typical issues in distributed classification operations. Four cases will be described: "explicit" double-counting, "implicit" double-counting (statistically dependent case), hard declarations resulting from legacy systems, and mixed uncertainty representations; many other issues are possible in the distributed object ID case but these are some typical examples. It should also be noted that combinations of these cases could arise in any practical environment. Note also that none of these case models depict the further ramifications of feedback from fusion nodes to sender/contributor nodes; such feedback adds another layer of complexity to the pedigree operations. The later sections of this chapter address some of the issues raised herein in some detail.

9.4.1 EXPLICIT DOUBLE-COUNTING

One concern in such environments, if the connectivity structure and message-handling protocols are not specified in a very detailed way, is about the problem of "double-counting," wherein information in sent-messages appear to return to the original sending node as newly received messages with seemingly new, complementary information. If such messages had a "lineage" or "pedigree" tag describing their processing history, the recipient could understand that the message was indeed the same as what it had sent out earlier. This type of double-counting (also known as "data incest," "rumor propagation," and "self-intoxication") can be called "explicit," in that it is literally the result of reception of a previously sent and processed message, simply returning to the sender. This process is shown in Figure 9.2.

9.4.2 IMPLICIT DOUBLE-COUNTING (STATISTICALLY DEPENDENT)

A fundamental concern, due to the nature of the statistical mathematics often involved in classifier or ID fusion operations, is the issue of statistical independence (or dependence) of the communicated or shared information. The major issue here revolves about the complexity of the required modeling and mathematics if the information to be fused is statistically dependent, and the resulting errors in processing if independence assumptions are violated. This problem can be called one of "implicit" double-counting, in that the redundant portion of the fused information is an idiosyncrasy of the sensing and classification operations involved and are hidden, in effect. Figure 9.3 shows the idea.

In this case, the local/sending node uses two sensor-specific classifiers, and the performance of the individual classifiers is quantified using confusion matrices, resulting in the statistical quantification as shown (Prob(ID|Features)), which is sent to the fusion node along with the declarations. Note that the sensors provide a common feature/attribute F1 that is used by both sensor-specific classifiers, resulting in the individual classifier outputs being correlated. The receiving fusion node operates on the probabilities in a Bayesian way to develop the fused estimate but may erroneously assume independence of the received soft declarations, and thereby double-count the effects of Feature F1, upon which both individual declarations are based. Pedigree tags would signify the features upon which the decisions are based.

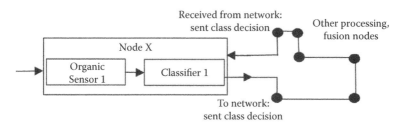

FIGURE 9.2 Explicit double-counting.

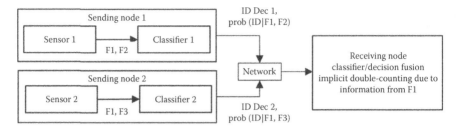

FIGURE 9.3   Implicit double-counting.

9.4.3   LEGACY SYSTEMS WITH HARD DECLARATIONS

Further, if the distributed network includes nodes that are "legacy" systems (i.e., previously built components constructed without forethought of embedding them in a fusion network), it is likely that either (1) ID classifiers will be "hard" classifiers, producing ID declarations without confidence measures ("votes" in essence) or (2) the confidence measures employed will not be consistent with other confidence measures employed by other network classifiers. In the first case, fusion can only occur with a type of voting strategy and likely without formal consideration of statistical independence aspects and any pedigree-related processing regarding this issue. Improvements can be made to this fusion approach if the rank-order of the ID declarations is also available (see Figure 9.4). In the second case, some type of uncertainty-measure-transformations need to be made to normalize the uncertainty representation into some common framework for a fusion operation.

In this representative situation, existing legacy sensor-classifier systems produce "hard" ID declarations, i.e., without qualification—these are "votes" in effect. Fusion methods can include simple voting-based strategies (majority, plurality, etc.), which may overtly ignore possible correlations as shown (common ID feature F1 as in the previous example). However, it is possible to develop voting schemes that derive from trained data, which, if the training data are truly representative, may empirically account for the effects of correlated parameters to some degree. Here again the pedigree tag would include the parameters upon which the local-node/sent IDs are based.

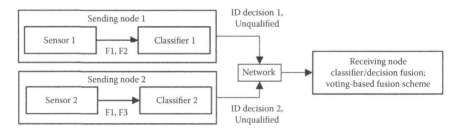

FIGURE 9.4   Legacy systems with hard decisions.

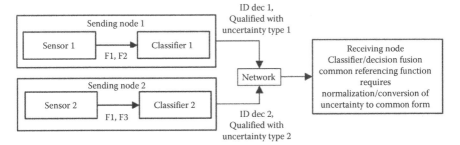

FIGURE 9.5   Classifiers with mixed uncertainty representations.

9.4.4   MIXED UNCERTAINTY REPRESENTATIONS

In this case, we have classifiers whose performance is quantified using different uncertainty types, e.g., probabilistic and evidential (Figure 9.5 just shows uncertainty types 1 and 2). The confounding effects of correlated observations may still exist. The fusion process is dependent on which uncertainty form is normalized, e.g., if the evidential form is transformed to the probabilistic form, then a Bayesian fusion process could be employed. Here, the pedigree tag must also include the type of uncertainty employed by each classifier evaluation scheme. As will be discussed later, it is also very important that the nature of such statistical or other transformations is well-understood and properly included in the fusion scheme.

As noted earlier, it is possible that real-world systems and networks may employ several sensor-classifier systems as input or in some shared/distributed framework, and combinations of the aforementioned cases could result, adding to the overall complexity of the fusion operations at some receiving node. Depending on how the overall system is architected, it may be necessary to have the pedigree tags include all possible effects, even if some particular tag elements are null.

9.5   CLASSIFIER FUSION

Here, we provide a brief overview of classifier combining methods drawn from some often-cited works that contain most of the details (e.g., Ho et al. 1994, Kittler et al. 1998, Kuncheva 2002). We begin with a taxonomic categorization as asserted in Ruta and Gabrys (2000), and then provide a point-by-point commentary on the most traditional classifier combining techniques. There are two main motivations for combining classifier results: efficiency and accuracy. Efficiencies can be achieved in multiclassifier systems by tuning the system to use the most appropriate, least computationally intensive classifier at any given moment in an application. Accuracy improvements can be realized in multiclassifier systems by combining methods that employ different sensor data and/or features, or have nuanced algorithmic features that offer some special benefit; the broad idea here is exploitation of diversity although diversity in classification can be a controversial topic. In a distributed object classification system, the classifier results from multiple nodes also have to be combined.

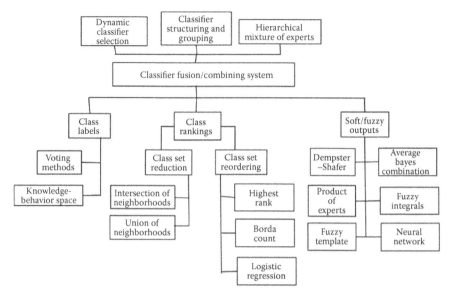

FIGURE 9.6 A taxonomy of classifier fusion techniques. (Adapted from Ruta, D. and Gabrys, B., *Comput. Inf. Syst.*, 7, 1, 2000.)

9.5.1 Taxonomy of Classifier Fusion Methods

One taxonomy of classifier fusion methods is offered in Ruta and Gabrys (2000) and is shown in Figure 9.6.

As noted earlier, the choice of fusion approach (also often called "classifier combining" in the literature) is dependent on the type of information being shared. Here, when only hard (unqualified) labels or decisions are provided, a typical approach to fusion is via voting methods. (It should be noted that there are many different strategies for voting, well beyond the usual plurality, majority-based techniques [see Llinas et al. 1998].) Much of the fusion literature has addressed the cases involving soft or fuzzy outputs as shown in Figure 9.6, where the class labels provided by any classifier are quantitatively qualified, for example with the conditional probabilities that come from the testing/calibration of that classifier, as just described. When such information is available, the fusion or combining processes can exploit the qualifier (probabilistic) information in statistical-type operations.

9.5.2 Combining Classifiers

The options available for classifier combining are influenced by various factors. If, for example, only "labels" are available,* various voting techniques can be employed; see Llinas et al. (1998) for an overview of voting-based combining

---

* The term "labels" is generally used in the pattern recognition community to mean an object-class or class-type declaration; it could be considered a "vote" by a classifier in the context of an object-class label. If that vote or label is declared without an accompanying level of uncertainty, then such declaration is also called a "hard" declaration.

methods for either label declarations or hard declarations. If continuous outputs like posterior probabilities are supplied (these are called "soft" declarations), then a variety of combining strategies are possible, again assuming a consistent, prob-abilistically based uncertainty representational form is used by all contributing classifiers (recall Section 9.4.4). For other than probabilistic but consistent repre-sentational forms of uncertainty such as fuzzy membership values, possibilistic forms, and evidential forms, the algebra of each method can be used in the classi-fier combining process (e.g., the Dempster Rule of Combination in the evidential case). Of course, all of these comments presume that the class-context of each contributing classifier is semantically consistent, otherwise some type of semantic, and associated mathematical transforms would need to be made. For example, it makes no sense to combine posterior probabilities if they are probabilities about inconsistent classes or class-types.

Kittler et al. (1998) describes what he says are two basic classifier-combining scenarios: one in which all classifiers are provided a common data set (one could call this an "algorithm-combining" approach, since the effects on the combined result are the consequence of distinctions in the various classifier algorithms), and an approach where each classifier is receiving unique data from disparate sources (as in a distributed system); Kittler's often-cited paper focuses on the latter. This paper describes the commonly used classifier combination schemes such as the product rule, sum rule, min rule, max rule, median rule, and majority voting, and conducts comparative experiments that show the sum rule to outperform the others in these particular tests.

To show the flavor of these combining rules, we follow Kittler's development for the product rule as one example; interested readers should refer to the original paper and yet other references for additional detail (e.g., Duda et al. 2001). Consider sev-eral classifier algorithms that are each provided with specific sensor measurement data and feature/attribute set. Using Kittler's notation, let the measurement vector used by the ith classifier denoted as x_i, and each class or class-type denoted as ω_i, with prior probability given by $P(\omega_i)$. The Bayesian (the only paradigm addressed in this chapter) classification problem is to determine the class, given a measurement vector Z, which is the concatenation of the x_i, as follows:

$$\text{assign} \quad Z \to \omega_f \quad \text{if}$$

$$P(\omega_f \mid x_1,\ldots,x_R) = \max_{\omega_k} P(\omega_k \mid x_1,\ldots,x_R) \tag{9.1}$$

That is, the fused class label ω_f is the maximum posterior probability (MAP) solu-tion. It should be noted that the inherent calculation of the posterior does not, in and of itself, yield a class declaration; deciding a class estimate from the posteriors is a separate, decision-making step. Strictly speaking, these calculations ideally require knowledge about the conditional probability $P(x_1,\ldots,x_R \mid \omega_k)$ which would reveal the insights about inter-source dependence in $(x_1,\ldots,x_R)$. If we allow, as is very often done, the assumption that the sources are conditionally independent given each class, i.e.,

$$P(x_1, \ldots, x_R \mid \omega_k) = \prod_{i=1}^{R} P(x_i \mid \omega_k) \tag{9.2}$$

then

$$\text{assign} \quad Z \to \omega_f \quad \text{if}$$

$$P(\omega_f) \prod_{i=1}^{R} P(x_i \mid \omega_f) = \max_{\omega_k} P(\omega_k) \prod_{i=1}^{R} P(x_i \mid \omega_k) \tag{9.3}$$

Or, in terms of the individual posteriors declared by each contributing classifier,

$$\text{assign} \quad Z \to \omega_f \quad \text{if}$$

$$P^{-(R-1)}(\omega_f) \prod_{i=1}^{R} P(\omega_f \mid x_i) = \max_{\omega_k} P^{-(R-1)}(\omega_k) \prod_{i=1}^{R} P(\omega_k \mid x_i) \tag{9.4}$$

Equation 9.4 is called the "Product Rule" for combining classifiers. It removes double-counting due to the common prior $P(\omega_i)$ but assumes conditional independence of the supporting evidence for each of the contributing Bayes classifiers. This rule is called "naïve" Bayes combination rule because in many cases there may be unmodeled statistical dependence in $(x_1, \ldots, x_R)$, as was discussed in Section 9.4. While strictly correct in a Bayesian sense, this rule has the vulnerability that a single bad classifier declaration having a very small posterior can inhibit an otherwise correct or appealing declaration. Other rules such as the sum rule follow similar developments but involve different assumptions (e.g., equal priors). Each of these strategies has trade-off factors regarding performance and accuracy; Kittler et al. (1998) developed an interesting analysis that explains why the sum rule seems to perform best in their experiments. Similar research is shown in the work of Kuncheva (2002) in which some six different classifier fusion strategies were examined for a simplified, two-class problem, and for which the individual-classifier error distributions were defined as either normal or uniform, and for which the individual classifiers were assumed independent. The particular fusion strategies examined were similar to some of those asserted in Kittler et al. (1998): minimum, maximum, average, median, majority vote.

9.6 OPTIMAL DISTRIBUTED BAYESIAN OBJECT CLASSIFICATION

The naïve classifier fusion approach described in Section 9.5 is easy to implement. However, it does not address the "double-counting" issues discussed in Section 9.4. This section presents an optimal distributed Bayesian object classification approach and compares its performance to the naïve fusion approach. Most of this section is from Chong and Mori (2005).

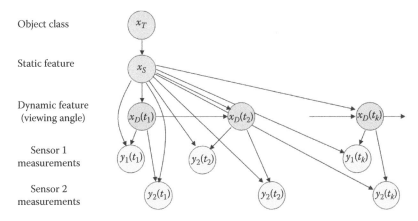

Object class

Static feature

Dynamic feature
(viewing angle)

Sensor 1
measurements

Sensor 2
measurements

FIGURE 9.7   Bayesian network model for object classification.

Bayesian object classification uses the generative approach discussed in
Section 9.2. We use the Bayesian network formalism to represent the model relating
the object class to the sensor measurements. A Bayesian network (Jensen 2001, Pearl
1988) is a graphical representation of a probabilistic model. It is a directed graph
where the nodes represent random variables or random vectors of interest and the
directed links represent probabilistic relationships between the variables. A dynamic
Bayesian network is one in which some of the nodes are indexed by time. A key feature of Bayesian networks is the explicit representation of conditional independence
needed for optimal distributed object classification.

Figure 9.7 is the dynamic Bayesian network for object classification using two
sensors. The variables in this network are the object class x_T, static feature x_S such
as object size, dynamic feature $x_D(t)$ (e.g., viewing angle) that varies with time, and
sensor measurement $y_i(t)$ for each sensor i.

The graph of the Bayesian network displays structural information such as the
dependence among the random variables. The quantitative information for the model is
represented by conditional probabilities of the random variables at the nodes given their
predecessor nodes. For the network of Figure 9.7, these probabilities are as follows:

- $P(x_T)$: prior probability of the object class.
- $P(x_S|x_T)$: conditional probability of the static feature given the object class.
- $P(x_D(t_k)|x_D(t_{k-1}), x_S)$: conditional probability of the dynamic feature $x_D(t_k)$ at
  time t_k given the dynamic feature $x_D(t_{k-1})$ at time t_{k-1} and the static feature
  x_S. If the evolution of the dynamic state does not depend on the static fea-
  ture, then the conditional probability becomes $P(x_D(t_k)|x_D(t_{k-1}))$.
- $P(y_i(t_k)|x_D(t_k), x_S)$: conditional probability of the measurement $y_i(t_k)$ of
  sensor i at time t_k given the dynamic feature $x_D(t_k)$ and the static feature
  x_S. This probability captures the effect of the dynamic feature on the measurement, e.g., measured size as a function of viewing angle as well as
  measurement error.

9.6.1 Centralized Object Classification Algorithm

Suppose $Y(t_k) = (y(t_1), y(t_2), ..., y(t_k))$ are the cumulative measurements from one or more sensors. The objective of object classification is to compute $P(x_T|Y(t_k))$, the probability of the object class given $Y(t_k)$. It can be shown (see Chong and Mori [2005] for details) that this probability can be computed by the following steps:

1. Estimating dynamic and static features $(x_S, x_D(t_k))$
   Prediction:

$$P(x_S, x_D(t_k) | Y(t_{k-1}))$$

$$= \int P(x_D(t_k) | x_D(t_{k-1}), x_S) P(x_S, x_D(t_{k-1}) | Y(t_{k-1})) dx_D(t_{k-1}) \qquad (9.5)$$

   Update:

$$P(x_S, x_D(t_k) | Y(t_k)) = C^{-1} P(y(t_k) | x_D(t_k), x_S) P(x_S, x_D(t_k) | Y(t_{k-1})) \qquad (9.6)$$

2. Estimating static feature x_S

$$P(x_S | Y(t_k)) = \int P(x_S, x_D(t_k) | Y(t_k)) dx_D(t_k) \qquad (9.7)$$

3. Estimating object class x_T

$$P(x_T | Y(t_k)) = \int P(x_S | Y(t_k)) P(x_T | x_S) dx_S \qquad (9.8)$$

Note that the joint estimation of the dynamic and state features is crucial since the sensor measurements depend on both features. The posterior probability of these features is updated whenever a new measurement is received. Object classification is basically a decoupled problem, with the posterior-type probability updated only when this information is needed.

Equations 9.5 through 9.8 can be used by a local fusion agent processing the measurements from a single sensor or a central fusion agent processing all sensor measurements by incorporating the appropriate sensor models. The prediction Equation 9.5 is independent of the sensor, except when different sensors observe different static features. In that case, Equation 9.5 predicts only the part of the static feature x_{Si} that is relevant to that sensor i.

9.6.2 Distributed Object Classification

The distributed object classification problem is combining estimates from the local classification agents to obtain an improved estimate of the object class. There are two tightly coupled parts of this problem. This first is determining the information sent from the local agent to the fusion agent. The second is combining the received information.

It is well known that combining the local object declarations, i.e., tank or truck, is suboptimal since the confidence in the declaration is not used. A common and better approach is to communicate the posterior-type probabilities $P(x_T|Y_i)$ given the measurements for each sensor i and then combine the probabilities by the product rule

$$P(x_T \mid Y_1, Y_2) = C^{-1} P(x_T \mid Y_1) P(x_T \mid Y_2) \tag{9.9}$$

However, this rule is still suboptimal because the measurement sets Y_1 and Y_2 are in general not conditionally independent given the object class. This section applies the general approach for distributed estimation (Chong et al. 2012) to the distributed object classification algorithm.

9.6.2.1 Bayesian Distributed Fusion Algorithm

Let x be the state of interest, which may include the object class, as well as the static or dynamic features. Let Y_1 and Y_2 be the measurements collected by two sensors 1 and 2. Suppose the local measurement sets are conditionally independent given the state, i.e., $P(Y_1, Y_2|x) = P(Y_1|x)P(Y_2|x)$. Then the global posterior probability can be reconstructed from the local posterior probabilities by (Chong et al. 1990)

$$P(x \mid Y_1, Y_2) = C^{-1} \frac{P(x \mid Y_1) P(x \mid Y_2)}{P(x)} \tag{9.10}$$

where C is a normalization constant. Note that the prior probability $P(x)$ appears in the denominator since this common information is used to compute each local posterior probability. Equation 9.10 can also written as

$$P(x \mid Y_1, Y_2) = C^{-1} P(Y_1 \mid x) P(x \mid Y_2) = C^{-1} P(Y_2 \mid x) P(x \mid Y_1) \tag{9.11}$$

or

$$P(x \mid Y_1, Y_2) = C^{-1} P(Y_2 \mid x) P(x \mid Y_1)$$

by recognizing that $P(x \mid Y_i) = C_i^{-1} P(Y_i \mid x) P(x)$ where C_i is a normalization constant.

For the hierarchical classification architecture, let Y_F be the set of cumulative measurements at the fusion agent F, and Y_i be the new measurements from the agent i sending the estimate. The measurement sets of the two agents are disjoint, i.e., $Y_F \cap Y_i = \phi$. Suppose the state x is chosen such that Y_F and Y_i are conditionally independent given x. Then Equations 9.10 and 9.11 imply the fusion equations

$$P(x \mid Y_i, Y_F) = C^{-1} \frac{P(x \mid Y_i) P(x \mid Y_F)}{P(x)} \tag{9.12}$$

or

$$P(x \mid Y_i, Y_F) = C^{-1} P(Y_i \mid x) P(x \mid Y_F) \tag{9.13}$$

Equations 9.12 and 9.13 represent two different but equivalent approaches of communication and fusion. The first is for the local agent to send the local posterior probability $P(x|Y_i)$ given the new measurements Y_i. The fusion agent then combines it with its posterior probability $P(x|Y_F)$ using Equation 9.12. The second is for the local agent to send the local likelihood $P(Y_i|x)$, which is then combined with the probability $P(x|Y_F)$ using Equation 9.13.

A necessary condition for Equations 9.10 through 9.13 to be optimal is the conditional independence assumption of Y_1 and Y_2. From Figure 9.7, it can be seen that choosing object class as the state will not satisfy the conditional independence condition. Thus, fusing the object class probability may be suboptimal.

9.6.2.2 Optimal Distributed Object Classification

Consider the model of Figure 9.7 and the hierarchical without feedback fusion architecture. Each local classification agent processes its sensor measurements to generate the posterior probability of a "state." Periodically, it sends the local probability based upon the measurements received since the last communication to the high-level fusion agent. The high-level fusion agent then combines the received probabilities with its current posterior probability.

Figure 9.8 is the network of Figure 9.7 redrawn to show the measurement sets without displaying the high-level nodes x_T and x_S. At the current fusion time t_{F+k}, the local fusion agent i sends the conditional probability given $Y_i(t_{F+1}, t_{F+k}) \triangleq (y_i(t_{F+1}),..., y_i(t_{F+k}))$, the measurements collected since the last fusion time t_F. The fusion agent then fuses this probability with its probability based upon the measurements $Y_F(t_1, t_F) = ((y_i(t_1), ..., y_i(t_F)); i = 1, 2)$.

A key problem is determining the random variable or state that satisfies the conditional independence assumption for the measurement sets so that the fusion equation can be used. The Bayesian network (Chong and Mori 2004) provides an easy way of determining this conditional independence. From Figures 9.7 and 9.8, the measurement sets $Y_i(t_{F+1}, t_{F+k})$ and $Y_F(t_1, t_F)$ are conditionally independent given x_S and the dynamic features $x_D(t_{F+1}, t_{F+k}) \triangleq (x_D(t_{F+1}), x_D(t_{F+2}), ..., x_D(t_{F+k}))$. However, they are not conditionally independent given x_S and the dynamic feature $x_D(t_k)$ at a single time. Thus the state to be used for fusion is $(x_S, x_D(t_{F+1}, t_{F+k}))$ and the fusion process consists of the following steps:

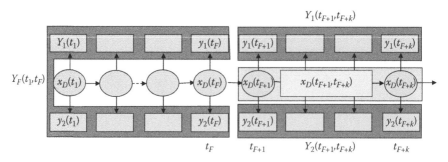

FIGURE 9.8 Conditionally independent measurements.

1. *Communication by local agent.* At time t_{F+k}, the local agent i sends to the high-level fusion agent the probability distribution $P(x_S, x_D(t_{F+1}, t_{F+k})|Y_i(t_{F+1}, t_{F+k}))$, which can be computed by the following recursive algorithm. Let $m=F+1$, $n=F+1, \ldots, F+k-1$, $x_D(t_{F+1}, t_{F+1})=x_D(t_{F+1})$, and $Y_i(t_{F+1}, t_{F+1})=y_i(t_{F+1})$. Then

$$P(x_S, x_D(t_m, t_{n+1}) \mid Y_i(t_m, t_{n+1}))$$

$$= C^{-1} P(y_i(t_{n+1}) \mid x_S, x_D(t_{n+1})) P(x_D(t_{n+1}) \mid x_S, x_D(t_n))$$

$$P(x_S, x_D(t_m, t_n) \mid Y_i(t_m, t_n)) \qquad (9.14)$$

Alternatively, the local agent sends the likelihood $P(Y_i(t_{F+1}, t_{F+k})|x_S, x_D(t_{F+1}, t_{F+k}))$ to the high-level fusion agent. This likelihood is computed directly from the probability model

$$P(Y_i(t_{F+1}, t_{F+k}) \mid x_S, x_D(t_{F+1}, t_{F+k})) = \prod_{m=F+1}^{F+k} P(Y_i(t_m) \mid x_S, x_D(t_m)) \qquad (9.15)$$

The prior probability $P(x_S, x_D(t_{F+1}, t_{F+k}))$ is given by the recursion

$$P(x_S, x_D(t_m, t_{n+1})) = P(x_S, x_D(t_m, t_n), x_D(t_{n+1}))$$

$$= P(x_D(t_{n+1}) \mid x_S, x_D(t_n)) P(x_S, x_D(t_m, t_n)) \qquad (9.16)$$

2. *Extrapolation by high-level agent.* The high-level fusion agent computes the probability $P(x_S, x_D(t_{F+1}, t_{F+k})|Y_F(t_1, t_F))$ by extrapolating $P(x_S, x_D(t_F)|Y_F(t_1, t_F))$ to obtain

$$P(x_S, x_D(t_{F+1}) \mid Y_F(t_1, t_F))$$

$$= \int P(x_S, x_D(t_{F+1}), x_D(t_F) \mid Y_F(t_1, t_F)) \, dx_D(t_F)$$

$$= \int P(x_D(t_{F+1}) \mid x_D(t_F), x_S) P(x_S, x_D(t_F) \mid Y_F(t_1, t_F)) \, dx_D(t_F) \qquad (9.17)$$

and then using the recursion

$$P(x_S, x_D(t_{F+1}, t_{F+i+1}) \mid Y_F(t_1, t_F))$$

$$= P(x_S, x_D(t_{F+1}, t_{F+i}), x_D(t_{F+i+1}) \mid Y_F(t_1, t_F))$$

$$= P(x_D(t_{F+i+1}) \mid x_S, x_D(t_{F+i})) P(x_S, x_D(t_{F+1}, t_{F+i}) \mid Y_F(t_1, t_F)) \qquad (9.18)$$

If the high-level fusion agent has fused the probability from another sensor j, then its current posterior probability $P(x_S, x_D(t_{F+1}, t_{F+k})|Y_F(t_{F+1}, t_{F+k}))$ is already computed for $x_D(t_{F+1}, t_{F+k})$. In this case, extrapolation is not needed.

3. *Fusion by high-level agent*. The high-level fusion agent computes the fused posterior probability by Equation 9.12

$$P(x_S, x_D(t_{F+1}, t_{F+k}) \mid Y_F(t_1, t_F), Y_i(t_{F+1}, t_{F+k}))$$

$$= C^{-1} \frac{P(x_S, x_D(t_{F+1}, t_{F+k}) \mid Y_F(t_1, t_F)) P(x_S, x_D(t_{F+1}, t_{F+k}) \mid Y_i(t_{F+1}, t_{F+k}))}{P(x_S, x_D(t_{F+1}, t_{F+k}))}$$

(9.19)

or Equation 9.13

$$P(x_S, x_D(t_{F+1}, t_{F+k}) \mid Y_F(t_1, t_F), Y_i(t_{F+1}, t_{F+k}))$$

$$= C^{-1} P(Y_i(t_{F+1}, t_{F+k}) \mid x_S, x_D(t_{F+1}, t_{F+k})) P(x_S, x_D(t_{F+1}, t_{F+k}) \mid Y_F(t_1, t_F))$$

(9.20)

These three steps are repeated when information is received from another local fusion agent. The high-level fusion agent does not have to extrapolate the probability if the received probability corresponds to observation times already included in the current fused probability.

9.6.3 COMMUNICATION STRATEGIES

One advantage of distributed object classification over the centralized architecture is reduced communication bandwidth if the local agents do not have to communicate the measurements to a central location. This advantage may not be realized if the local agents have to communicate their sufficient statistics at each sensor observation time since sensor measurements are vectors while sufficient statistics are probability distributions. However, a local agent does not have to communicate its fusion results after receiving every new measurement since the new measurement may not contain enough new information. Thus, communication should take place when it is needed, and not because a sensor receives new measurements. In particular, whether a local agent should communicate or not should depend on the increase in information resulting from communication.

With an information push strategy, a local agent monitors the local fusion results and determines whether enough new information has been acquired since the last communication. This determination is based on its local information and does not require knowledge (although it will be useful) of the fusion agent.

Let t_F be the last time that the local agent i communicates to the fusion agent and $t_k > t_F$ be the current time. Let $P(x|Y_i(t_F))$ be the probability distribution of the sufficient state x at the last fusion time, and $P(x|Y_i(t_k))$ be that at the current time given the cumulative measurements $Y_i(t_k)$. This sufficient state x may be exact in the sense that it allows optimal distributed fusion or it may be approximate to reduce communication bandwidth. It may also be extrapolated from the fusion time t_F to the current time t_k. With the information push strategy, the local agent determines whether there is enough difference between the two probability distributions, i.e., $D(P(x|Y_i(t_F)), P(x|Y_i(t_k))) > d$, where $D(\cdot,\cdot)$ is an appropriate distance measure and d is a threshold on the distance.

A number of probability distance measures are available with different properties and degrees of complexity. A popular one is the Kullback–Leibler (KL) divergence (Kullback and Leibler 1951), which is $D(P_2, P_1) = \int_x P_2(x) \log(P_2(x)/P_1(x)) dx$ for two continuous distributions P_1 and P_2, and $D(P_2, P_1) = \sum_x P_2(x) \log(P_2(x)/P_1(x))$ for two discrete distributions.

We will use the KL divergence as the measure to develop communication management strategies because of its information theoretic-interpretation.

9.6.4 Performance Evaluation

This section presents simulation results to evaluate the performance of various distributed object classification approaches.

9.6.4.1 Simulation Scenario and Data Generation

The scenario assumes two sensors with the object and sensor models given by Figure 9.7. There are three object classes with uniform prior probabilities equal to 1/3. The single static feature x_S is observed by both sensors and is a Gaussian random variable with unit variance $\sigma_i^2 = 1$ and nominal mean $m_j = 3$, 0, and 3 for object class $j = 1, 2, 3$. The separation between the feature distributions for different object classes represents the difficulty of classification and is captured by $D_{ij} \triangleq (m_j - m_k)^2 / \sigma_i^2$. The measurements are generated by two sensors viewing the object with complementary geometry,

$$y_1(t_k) = x_S \sin x_D(t_k) + v_1(t_k) \tag{9.21}$$

$$y_2(t_k) = x_S \cos x_D(t_k) + v_2(t_k) \tag{9.22}$$

where the measurement noise variance is $\sigma_v^2 = 0.04$.

The dynamic feature $x_D(t_k)$ represents the viewing angle with values between 0 and π, and evolves according to Markov transition probability $P(x_D(t_{k+1})|x_D(t_k))$ with nominal transition value 0.2 from one cell to an adjacent cell.

9.6.4.2 Performance Evaluation Approach

We use the classification algorithms described in Section 9.6.2. Since the equations cannot be evaluated analytically, we discretize the static and dynamic features x_S and x_D, and convert Equation 9.5 from an integral to a sum. We retain the continuous values of the measurements in evaluating Equation 9.6.

We consider two communication strategies: fixed and adaptive. For the fixed schedule strategy, communication takes place at fixed intervals that are multiples of the observation intervals. The adaptive communication strategy uses the information push algorithm discussed in Section 9.6.3. At each communication time, the local agent sends the likelihood $P(Y_i(t_{Fi+1}, t_k)|x_S, x_D(t_k))$ for the static feature based upon the information received since the last fusion time t_{Fi}. Using only the dynamic feature at a single time is an approximation and may reduce fusion performance unless communication takes place after every measurement. However, communicating a high-dimensional dynamic feature vector is not practical.

In addition to fusing the probabilities of features and then computing the object class probability, we also fuse the object class probabilities from each sensor directly using the product rule. This approach is suboptimal since the measurements from the two sensors are not conditionally independent given the object class. However, it is used as a reference to compare with the other algorithms.

Object classification performance is evaluated by two metrics. The first is the expected posterior probability (EPP) of correct object classification defined as

$$EPP(T_j \mid T_j) = E\left[P(x_T = T_j \mid Y) \mid x_T = T_j \right] \tag{9.23}$$

where
   T_j is the true object class
   Y is the data used in computing the posterior probability

The second performance metric is the root mean square (RMS) object classification probability error defined as

$$RMS(T_j) = \sqrt{\sum_{x_T} (P(x_T \mid Y) - \delta(x_T;T_j))^2} \tag{9.24}$$

where $\delta(x_T;T_j)$ is the delta function that is 1 at $x_T = T_j$ and 0 elsewhere.

9.6.4.3   Comparison of Fusion Algorithms

Monte Carlo simulations are used to compare the performance of five algorithms: sensor 1 only, sensor 2 only, centralized fusion, fusing object class probabilities by product rule, and (approximate) distributed object classification using features. All the results assume that the true object class is 2 since it is more difficult to discriminate due to possible confusion with class 1 and class 3.

Figures 9.9 and 9.10 compare the object classification performance as measured by EPP of true object class and RMS error. The results are as expected. Centralized object classification performs best, followed by distributed classification. Even though an approximate algorithm is used to reduce bandwidth, distributed object classification actually approaches the performance of centralized fusion with increasing number of measurements. Fusing object class probabilities seems to perform well if we consider only the EPP, but the RMS error shows that it is inferior to distributed classification. The classification performance using single sensors is always the worst as expected.

Figures 9.11 and 9.12 show classification performance as a function of the separation between the object feature probability distributions. As expected, classification performance increases with larger separations. The baseline separation D_{ij} used in the simulations is 3.

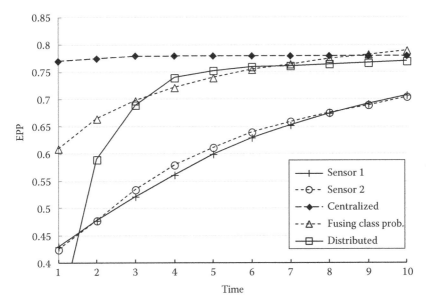

FIGURE 9.9   EPP of correct classification versus time.

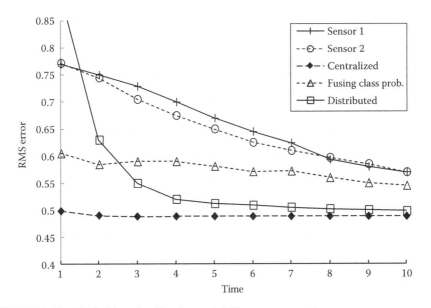

FIGURE 9.10   RMS object classification probability error versus time.

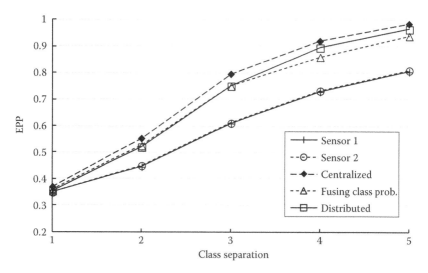

FIGURE 9.11   EPP of correct classification versus object class separation.

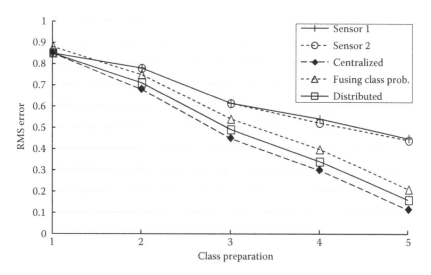

FIGURE 9.12   RMS object classification probability error versus object class separation.

Figures 9.13 and 9.14 show how the transition probability of the dynamic feature affects classification performance. Note that single sensor performance benefits more from increasing the transition probability than those of two sensors. This is probably due to better observability of the static feature from varying the dynamic feature. Larger transition probability implies more movement in viewing angles and better average observability.

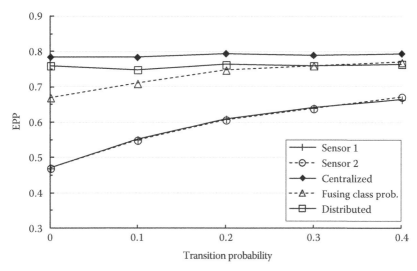

FIGURE 9.13 EPP of correct classification versus transition probability.

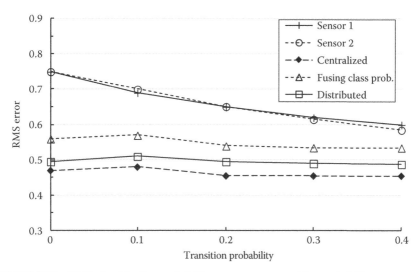

FIGURE 9.14 RMS classification probability error versus transition probability.

Figures 9.15 and 9.16 compare the performance of distributed classification using the adaptive and fixed schedule communication strategies. The results of centralized fusion, fusing object class probabilities, and single sensor classification are plotted for reference.

The adaptive communication strategy performs much better than the fixed schedule strategy, and achieves the performance of centralized fusion when the average

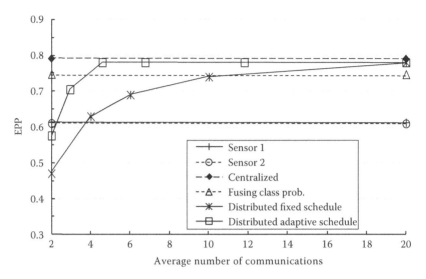

FIGURE 9.15   EPP of correct classification versus number of communications.

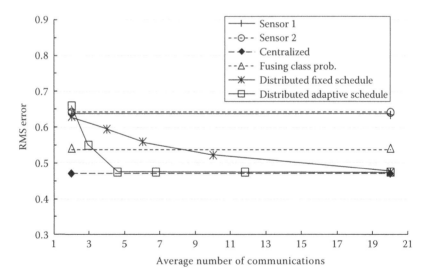

FIGURE 9.16   RMS classification probability error versus average number of communications.

number of communications is about 5. Fixed schedule communication does not achieve this level of performance with twice that number.

9.7   CONCLUSIONS

Distributed object classification is an important problem in distributed data fusion and has been extensively studied by the fusion community. Technical issues are similar to those in distributed tracking and include choice of appropriate architecture, addressing dependent information in the data to be fused, design of optimal algorithms, etc. For the important cases involving imaging techniques, there are a number of additional issues not covered in this chapter that should be considered if imaging sensors are employed; there is an extensive body of literature on these problems. Here, we have discussed a range of foundational, architectural, and algorithmic issues that in fact can also be employed with imaging sensors but have not been explicitly framed in a sensor-specific way. We also present an approach for the important case of optimal distributed Bayesian classification and compare its performance with simpler fusion rules.

REFERENCES

Al-Ani, A. and M. Deriche. 2002. A new technique for combining multiple classifiers using the Dempster–Shafer theory of evidence. *Journal of Artificial Intelligence Research*, 17:333–361.

Bhanu, B. 1986. Automatic target recognition: State of the art survey. *IEEE Transactions on Aerospace and Electronic Systems*, 22:364–379.

Brooks, R. R., P. Ramanthan, and A. M. Sayed. 2003. Distributed target classification and tracking in sensor networks. *Proceedings of IEEE*, 91:1163–1171.

Chong, C. Y., K. C. Chang, and S. Mori. 2012. Fundamentals of distributed estimation. In *Distributed Fusion for Net-Centric Operations*, D. Hall, M. Liggins, C. Y. Chong, and J. Llinas (eds.). Boca Raton, FL: CRC Press.

Chong, C. Y. and S. Mori. 2004. Graphical models for nonlinear distributed estimation. *Proceedings of the 7th International Conference on Information Fusion (FUSION 2004)*, Stockholm, Sweden, pp. 614–621.

Chong, C. Y. and S. Mori. 2005. Distributed fusion and communication management for target identification. *Proceedings of the 8th International Conference on Information Fusion*, Philadelphia, PA.

Chong, C. Y., S. Mori, and K. C. Chang. 1990. Distributed multitarget multisensor tracking. In *Multitarget-Multisensor Tracking: Advanced Applications*, Y. Bar-Shalom (ed.), chapter 8, pp. 247–295. Norwood, MA: Artech House.

Cohen, M. N. 1991. An overview of radar-based, automatic, noncooperative target recognition techniques. *Proceedings of IEEE International Conference on Systems Engineering*, Dayton, OH, pp. 29–34.

Duda, R. O., P. E. Hart, and D. G. Stork. 2001. *Pattern Classification*. New York: Wiley Interscience.

Hartshorne, C., P. Weiss, and A. Burks, eds. 1931–1958. *The Collected Papers of Charles Sanders Peirce*. Cambridge, MA: Harvard University Press.

Ho, T. K., J. H. Hull, and S. N. Srihari. 1994. Decision combination in multiple classifier systems. *IEEE Transactions on Pattern Recognition and Machine Intelligence*, 16:66–75.

Jensen, F. Y. 2001. *Bayesian Networks and Decision Graphs*. New York: Springer-Verlag.

Kittler, J., M. Hatef, R. P. W. Duin, and J. Matas. 1998. On combining classifiers. *IEEE Transactions on Pattern Analysis and Machine Intelligence*, 20:226–239.

Kotecha, J. H., V. Ramachandran, and A. M. Sayeed. 2005. Distributed multitarget classification in wireless sensor networks. *IEEE Journal in Selected Areas in Communications*, 23:703–713.

Kullback, S. and R. A. Leibler. 1951. On information and sufficiency. *Annals of Mathematical Statistics*, 22:79–86.

Kuncheva, L. I. 2002. A theoretical study on six classifier fusion strategies. *IEEE Transactions on Pattern Analysis and Machine Intelligence*, 24:281–286.

Liggins II, M., C. Y. Chong, I. Kadar, M. G. Alford, V. Vannicola, and S. Thomopoulos. 1997. Distributed fusion architectures and algorithms for target tracking. *Proceedings IEEE*, 85:95–107.

Llinas, J., R. Acharya, and C. Ke. 1998. Fusion-based methods in the absence of quantitative classifier confidence. Report CMIF 6–98, Center for Multisource Information Fusion, University at Buffalo, Buffalo, New York.

Montillo, A. 2010. Generative versus discriminative modeling frameworks. Lecture at Temple University http://www.albertmontillo.com/lectures,presentations/Generative%20vs%20 Discriminative.pdf

Murphy, R. and B. Taylor. 2002. A survey of machine learning techniques for automatic target recognition, submitted to *IEEE Transactions on Pattern Analysis and Machine Intelligence,* www.citeseer.ist.psu.edu/viewdoc/summarydoi=10.1.1.53.9244

Nebabin, V. G. 1994. *Methods and Techniques of Radar Recognition*. Norwood, MA: Artech House.

Ng, A. Y. and M. I. Jordan. 2002. On discriminative vs. generative classifiers: A comparison of logistic regression and naive Bayes. *Advances in Neural Information Processing Systems*, 2:841–848.

Novak, L. M. 2000. State of the art of SAR automatic target recognition. *Proceedings of the IEEE 2000 International Radar Conference*, Alexandria, VA, pp. 836–843.

Pearl, J. 1988. *Probabilistic Reasoning in Intelligent Systems: Networks of Plausible Inference*. San Francisco, CA: Morgan Kaufman.

Roth, M. W. 1990. Survey of neural network technology for automatic target recognition. *IEEE Transactions on Neural Networks*, 1:28–43.

Ruta, D. and B. Gabrys. 2000. An overview of classifier fusion methods. *Computing and Information Systems*, 7:1–10.

Sadjadi, F. A. and B. Javidi. 2007. *Physics of Automatic Target Recognition*. New York: Springer.

Tait, P. 2005. *Introduction to Radar Target Recognition (Radar, Sonar & Navigation)*. London, U.K.: The Institution of Engineering and Technology.

Zalta, E. N. ed. 2012. Stanford Encyclopedia of Philosophy. http://plato.stanford.edu/

10 A Framework for Distributed High-Level Fusion

Subrata Das

CONTENTS

10.1 Introduction 271
10.2 Concept and Approach 272
10.3 Distributed Fusion Environments 274
10.4 Algorithm for Distributed Situation Assessment 277
    10.4.1 Junction Tree Construction and Inference 280
10.5 Distributed Kalman Filter 282
10.6 Relevance to Network-Centric Warfare 285
10.7 Role of Intelligent Agents 286
    10.7.1 What Is an Agent? 287
    10.7.2 Use of Agents in Distributed Fusion 288
10.8 Conclusions and Further Reading 290
References 291

10.1 INTRODUCTION

This chapter presents distributed fusion from the situation assessment (SA) (a.k.a. Level 2 fusion [Das 2008b, Steinberg 2008]) perspective. It also describes the relevance of distributed fusion to network-centric warfare (NCW) environments and the role of intelligent agents in that context.

For SA in a distributed NCW environment, each node represents a sensor, software program, machine, human operator, warfighter, or a unit. A fusion node maintains the joint state of the set of variables modeling a local SA task at hand. Informally, the set of variables maintained by a fusion node is a *clique* (maximal sets of variables that are all pairwise linked), and the set of cliques in the environment together form a clique network to be transformed into a *junction tree*, where the nodes are the cliques. Thus, the cliques of a junction tree are maintained by local fusion nodes within the environment. Local fusion nodes communicate and coordinate with each other to improve their local estimates of the situation, avoiding the repeated use of identical information.

A junction tree can also be obtained by transforming (Jensen 2001) a Bayesian belief network (BN) (Pearl 1988, Jensen 2001, Das 2008b) model representing a global SA model in the context of a mission, thereby contributing to the development of a common tactical picture (CTP) of the mission via shared awareness. Each clique is maintained by a local fusion node. Inference on such a BN model for SA relies on evidence from individual local fusion nodes. We make use of the message-passing inference algorithm for junction trees that naturally fits within distributed NCW environments. A BN structure with nodes and links is a natural fit for distributing tasks in an NCW environment at various levels of abstraction and hierarchy. BNs have been applied extensively for centralized fusion (e.g., Jones et al. 1998, Das et al. 2002a, Wright et al. 2002, Mirmoeini and Krishnamurthy 2005, Su et al. 2011) where domain variables are represented by nodes.

A major source of evidence for SA is target tracks. We briefly describe distributed target tracking in an NCW environment for the sake of completeness of the chapter. We produce an overall estimate of a target at a fusion center that combines estimates from distributed sensors located at different fusion nodes which are all tracking the same target. We make use of the Kalman filter (KF) algorithm for estimating targets at local fusion nodes from sensor observations. Individual estimates from local fusion nodes are then combined at a fusion center, thereby generating evidence to be propagated into a BN model for SA.

The objective of the survey-type penultimate section on intelligent agents is to provide readers with a background on the capabilities of intelligent agents in the context of fusion, and how these intelligent agents are being exploited by fusion researchers and developers. It is clear that the agent technology is ideal for SA in a distributed NCW environment. In fact, each node in the environment can be implemented as or represented by an intelligent agent.

Readers who need supplementary background information on BN models, algorithms, and distributed fusion are recommended to consult the concluding section's list of relevant references.

10.2 CONCEPT AND APPROACH

Some sensor networks consist of a large number of nodes of sensing devices, densely distributed over the operational environment of interest. Nodes have wired or wireless connectivity tied into one or more backbone networks, such as the Internet, SIPRNET (secret Internet protocol router network), or NIPRNET (nonclassified [unclassified but sensitive] Internet protocol router network). Each sensor node has its own measurements-collection and processing facility to estimate and understand its environment. A sensor is thus "situationally aware" in terms of position and movement of targets and the threats they pose. This awareness must be shared among all other nodes to generate an assessment of the environmental situation as a whole (sometimes called the common tactical picture [CTP]) for effective coordinated action.

Sensor nodes can be conceptualized as intelligent autonomous agents that communicate, coordinate, and cooperate with each other in order to improve their local situational awareness and to assess the situation of the operational environment as a whole. The concept of distributed fusion refers to decentralized processing environments, consisting of autonomous sensor nodes, and additional processing nodes without sensors, if necessary, to facilitate message communication, data

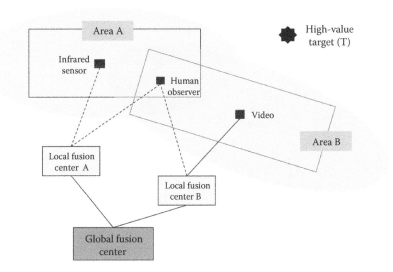

FIGURE 10.1   An example of distributed fusion environment.

storage, relaying, information aggregation, and assets scheduling. Some of the advantages of distributed fusion are reduced communication bandwidth, distribution of processing load, and improved system survivability from a single point failure. The distributed fusion concept naturally fits within the upcoming NCW paradigm and its backbone command network, the global information grid (GIG).

As a concrete example of distributed fusion, consider the decentralized processing environment as shown in Figure 10.1.

In this example, we assume there is a high-value target (top right of the figure) within a region of interest and that the designated areas A and B surrounding the target are considered to be the most vulnerable. These two areas must be under surveillance in order to detect any probing activities that indicate a possible attack threat. The sensor coverage in areas A and B, shown in grey, is by an infrared sensor (MASINT) and a video camera (IMINT), respectively. In addition, a human observer (HUMINT) is watching the area in common between A and B. There are two local fusion centers for the two areas to detect any probing activity. The infrared sensor has wireless connectivity with the local fusion center for area A, whereas the video camera has wired connectivity with the local fusion center for area B for streaming video. Moreover, the human observer communicates wirelessly with both local fusion centers. Each of the two local centers fuses the sensor data it receives in order to identify any possible probing activity. The centers then pass their assessments (i.e., higher-level abstraction, rather than raw sensor information, thus saving bandwidth) to another fusion center that assesses the overall threat level, based on the reports of probing activities and other relevant prior contextual information.

In a centralized fusion environment, where observations from IMINT, HUMINT, and MASINT are gathered in one place and fused, a BN model, such as the one in Figure 10.2, can be used for an overall SA. This model handles dependence among sensors and fusion centers via their representation in nodes and interrelationships.

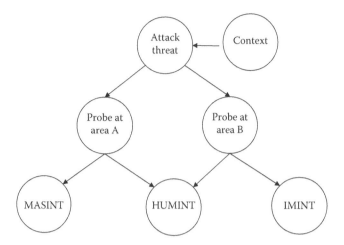

FIGURE 10.2   A centralized BN model for situation assessment.

A probing activity at an area will be observed by those sensors covering the area, and the lower half of the BN models this. For example, MASINT and HUMINT reports will be generated due to a probing activity at area A. Similarly, IMINT and HUMINT reports will be generated due to a probing activity at area B. The upper half of the BN models the threat of an attack based on the probing activities at areas A and B, together with other contextual information.

In a decentralized environment, as illustrated in Figure 10.1, each of the three fusion centers contains only a fragment of the above BN model as shown in Figure 10.3.

Local fusion centers A and B assess probing activities based on their local model fragments, and send their assessments to the global fusion center via messages. The global fusion center then uses its own models to determine the overall attack threat. If the same HUMINT report is received by both local fusion centers, the process has to ensure that this common information is used only once; otherwise, there will be a higher-than-actual level of support for a threat to be determined by the global fusion model. This is called the *data incest problem* in a distributed fusion environment, which is the result of repeated use of identical information. *Pedigree* needs to be traced, not only to identify common information but also to assign appropriate trust and confidence to data sources. An information graph (Liggins et al. 1997), for example, allows common prior information to be found.

10.3   DISTRIBUTED FUSION ENVIRONMENTS

As shown in Figure 10.4, a typical distributed fusion environment is likely to contain a variety of fusion nodes that do a variety of tasks:

- Process observations generated from a cluster of heterogeneous sensors (e.g., the local fusion centers A and B in Figure 10.1, and nodes labeled 5 and 9 in Figure 10.4).
- Process observations generated from a single sensor (e.g., nodes labeled 11, 12, and 13 in Figure 10.4).

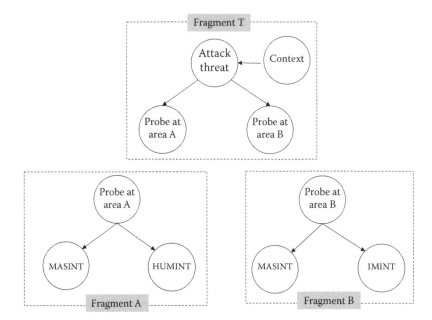

FIGURE 10.3 Distributed parts of the BN models.

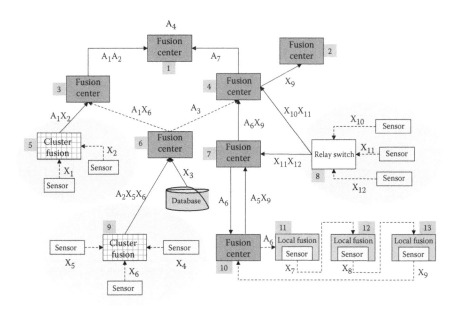

FIGURE 10.4 A typical distributed fusion environment.

- Perform a task (e.g., situation assessment [SA] and threat assessment [TA], course of action [COA] generation, planning and scheduling, CTP generation, collection management) based on information received from other sensors in the environment and from other information stored in databases (e.g., nodes labeled 1, 2, 3, 4, 6, 7, and 10 in Figure 10.4).
- Relay observations generated from sensors to other nodes (e.g., the node labeled 8 in Figure 10.4).

As shown in Figure 10.4, a fusion node receives values of some variables obtained either from sensor observations (X variables) or via information aggregation by other nodes (A variables). Such values can also be obtained from databases. For example, the fusion center labeled 6 receives values of the variables A2, X5, and X6 from the cluster fusion node labeled 9 and values of the variable X3 from a database. Note that an arrow between two nodes indicates the flow of information in the direction of the arrow as opposed to a communication link. The existence of an arrow indicates the presence of at least a one-way communication link, though not necessarily a direct link, via some communication network route. For example, there is a one-way communication link from node 3 to node 1. A reverse communication link between these two nodes will be necessary in implementing our message-passing distributed fusion algorithm to be presented later.

Each node (fusion center, cluster fusion, relay switch, or local fusion) in a distributed fusion environment has knowledge of the states of some variables, called *local variables*, as shown in Figure 10.5. For example, the fusion node labeled 6 has knowledge of the X variables X3, X5, and X6, and A variables A2 and A3. The node receives values of the variables A2, X5, and X6 from the node labeled 9 and the variable X3 from a database. The node generates values of the variable A3 via some information aggregation operation. On the other hand, fusion node 9 receives measurements X4, X5, and X6 from a cluster of sensors and generates A2; fusion

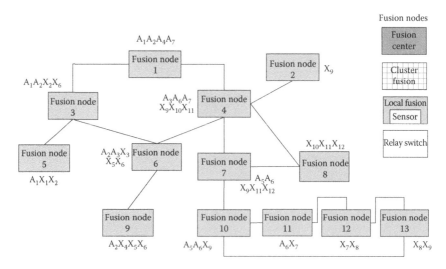

FIGURE 10.5 Network of distributed fusion nodes.

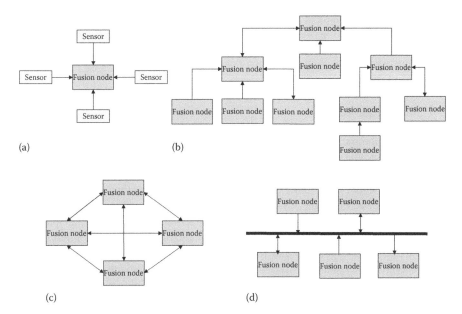

FIGURE 10.6 Possible distributed fusion environments: (a) centralized; (b) hierarchical; (c) peer-to-peer; and (d) grid-based.

node 8 relays values of the variables X10, X1, and X12 to other nodes; and fusion node 12 obtains measurements of X8 from a single sensor.

Figure 10.6 shows four possible distributed fusion environments: centralized, hierarchical, peer-to-peer, and grid-based. Note that the direction of an arrow indicates both information flow and the existence of a communication link along the direction of an arrow.

In a *centralized* environment, only the sensors are distributed, sending their observations to a centralized fusion node. The centralized node combines the sensor information to perform tracking or SA. In a *hierarchical* environment, the fusion nodes are arranged in a hierarchy, with the higher-level nodes processing results from the lower-level nodes and possibly providing some feedback. The hierarchical architecture will be natural for applications where situations are assessed with an increasing level of abstraction along a command hierarchy, starting with the tracking of targets at the bottom level. Considerable savings in communication effort can be achieved in a hierarchical fusion environment. In both *peer-to-peer* and *grid-based* distributed environments, every node is capable of communicating with every other node. This internode communication is direct in the case of a peer-to-peer environment, but some form of "publish and subscribe" communication mechanism is required in a grid-based environment.

10.4 ALGORITHM FOR DISTRIBUTED SITUATION ASSESSMENT

As mentioned in Section 10.1, there are two ways in which we can accomplish SA in a distributed environment: (1) each local fusion node maintains the state of a set of variables and (2) there is a BN model for global SA.

In the first case, we start with a distributed fusion environment such as the one shown in Figure 10.4. Our distributed SA framework in this case has four steps:

- Network formation
- Spanning tree formation
- Junction tree formation
- Message passing

The nodes of the sensor network first organize themselves into a network of fusion nodes, similar to the one shown in Figure 10.5. Each fusion node has partial knowledge of the whole environment. This network is then transformed into a *spanning tree* (a spanning tree of a connected, undirected graph, such as the one in Figure 10.5, is a tree composed of all the vertices and some or all of the edges of the graph), so that neighbor nodes establish high-quality connections. In addition, the spanning tree formation algorithm optimizes the communication required by inference in junction trees. The algorithm can recover from communication and node failures by regenerating the spanning tree. Figure 10.7 describes a spanning tree obtained from the network in Figure 10.5. The decision to sever the link between nodes 4 and 6, as opposed to between nodes 3 and 6, can be mitigated using the communication bandwidth and reliability information in the cycle of nodes 1, 3, 6, and 4.

Using pairwise communication-link information sent between neighbors in a spanning tree, the nodes compute the information necessary to transform the spanning tree into a junction tree for the inference problem. Finally, the inference problem is solved via message-passing on the junction tree.

During the formation of a spanning tree, each node chooses a set of neighbors, so that the nodes form a spanning tree where adjacent nodes have high-quality communication links. Each node's clique is then determined as follows. If i is a

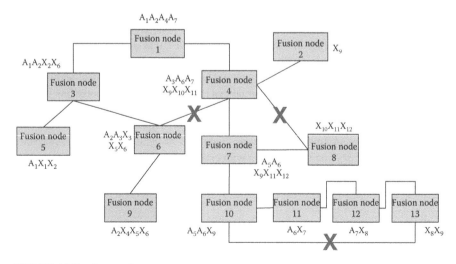

FIGURE 10.7   A spanning tree.

node and j is a neighbor of i, then the variables reachable to j from i, R_{ij}, are defined recursively as

$$R_{ij} = D_i \bigcup_{k \in nbr(i)-\{j\}} R_{ki} \tag{10.1}$$

where D_i is the set of local variables of node i. A base case corresponds to a leaf node, which is simply a collection of a node's local variables. If a node has two sets of reachable variables to two of its neighbors that both include some variable V, then the node must also carry V to satisfy the running intersection property of a junction tree. Formally, node i computes its clique C_i as

$$C_i = D_i \bigcup_{\substack{j,k \in nbr(i) \\ j \neq k}} R_{ji} \cap R_{ki} \tag{10.2}$$

A node i can also compute its *separator set* $S_{ij} = C_i \cap C_j$ with its neighbor j using reachable variables as

$$S_{ij} = C_i \cap R_{ji} \tag{10.3}$$

Figure 10.8 shows the junction tree obtained from the spanning tree in Figure 10.7.

The variables reachable to a leaf node, for example, fusion node 9, are its local variables A_2, X_4, X_5, X_6. The variables reachable to an intermediate node, for example, fusion node 1, from its neighboring nodes 3 and 4 are

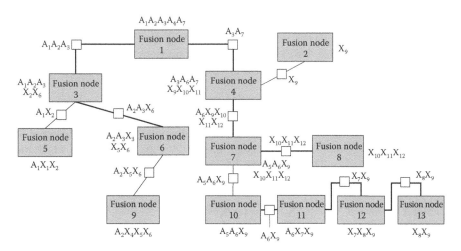

FIGURE 10.8   A junction tree from the distributed fusion environment.

$$R_{31} = \left\{A_1, A_2, A_3, X_1, X_2, X_3, X_4, X_5, X_6\right\}$$

$$R_{41} = \left\{A_3, A_5, A_6, A_7, X_7, X_8, X_9, X_{10}, X_{11}, X_{12}\right\}$$

The local variable of the fusion node 1 is $D_1 = \{A_1, A_2, A_4, A_7\}$. Therefore, its clique is

$$C_1 = \left\{A_1, A_2, A_3, A_4, A_7\right\}$$

The formation of a suitable junction tree from a BN model for SA is the only part of our distributed fusion approach that is global in nature.

Now we focus on the second case of a distributed SA where we have a BN model for global SA such as the one shown in Figure 10.2. We first apply an algorithm (Jensen 2001) that systematically transforms a BN to a junction tree in four steps: moralization, triangulation, clique identification, and junction tree formation.

10.4.1 Junction Tree Construction and Inference

The *moral graph* of a BN is obtained by adding a link between any pair of variables with a common "child" and dropping the directions of the original links in the BN. An undirected graph is *triangulated* if any cycle of length greater than 3 has a chord, that is, an edge joining two nonconsecutive nodes along the cycle. The nodes of a junction tree for a graph are the *cliques* in the graph (maximal sets of variables that are all pairwise linked).

Once we have formed a junction tree from either of the aforementioned two cases, such as the one in Figure 10.8, a message-passing algorithm then computes prior beliefs of the variables in the network via an initialization of the junction tree structure, followed by evidence propagation and marginalization. The algorithm can be run asynchronously on each node responding to changes in other nodes' states. Each time a node i receives a new separator variables message from a neighbor j, it recomputes its own clique and separator variables messages to all neighbors except j and transmits them if they have changed from their previous values. Here we briefly discuss the algorithm and how to handle evidence by computing the posterior beliefs of the variables in the network.

A junction tree maintains a joint probability distribution at each node, cluster, or separator set in terms of a belief *potential*, which is a function that maps each instantiation of the set of variables in the node into a real number. The belief potential of a set of variables $\mathbf{X}$ will be denoted as $\varphi_{\mathbf{X}}$, and $\varphi_{\mathbf{X}}(x)$ is the number onto which the belief potential maps x. The probability distribution of a set of variables $\mathbf{X}$ is just the special case of a potential whose elements add up to 1. In other words,

$$\sum_{x \in \mathbf{X}} \varphi_X(x) = \sum_{x \in \mathbf{X}} p(x) = 1 \qquad (10.4)$$

The marginalization and multiplication operations on potentials are defined in a manner similar to the same operations on probability distributions.

Belief potentials encode the joint distribution $p(\mathbf{X})$ of the BN according to the following:

$$p(\mathbf{X}) = \frac{\prod_i \phi_{C_i}}{\prod_j \phi_{S_j}} \tag{10.5}$$

where ϕ_{C_i} and ϕ_{S_j} are the cluster and separator set potentials, respectively. We have the following joint distribution for the junction tree in Figure 10.8:

$$p(A_1, \ldots, A_9, X_1, \ldots, X_{12}) = \frac{\phi_{C_1} \phi_{C_2} \cdots \phi_{C_{13}}}{\phi_{S_{13}} \phi_{S_{14}} \phi_{S_{24}} \phi_{S_{35}} \cdots \phi_{S_{12\ 13}}} \tag{10.6}$$

where

C_i represents the variable in clique i

$S_{ij} = C_i \cup C_j$ represents the separator set between nodes i and j

It is imperative that a cluster potential agrees with its neighboring separator sets on the variables in common, up to marginalization. This imperative is formalized by the concept of local consistency. A junction tree is *locally consistent* if, for each cluster C and neighboring separator set S, the following holds:

$$\sum_{C \backslash S} \phi_C = \phi_S \tag{10.7}$$

To start initialization, for each cluster C and separator set S, set the following:

$$\phi_C \leftarrow 1, \quad \phi_S \leftarrow 1 \tag{10.8}$$

Then assign each variable X to a cluster C that contains X and its parents $pa(X)$. Then set the following:

$$\phi_C \leftarrow \phi_C p(X \mid pa(X)) \tag{10.9}$$

When new evidence on a variable is entered into the tree, it becomes inconsistent and requires a global propagation to make it consistent. The posterior probabilities can be computed via marginalization and normalization from the global propagation. If evidence on a variable is updated, the tree requires re-initialization. Next, we present initialization, normalization, and marginalization procedures for handling evidence.

As before, to start initialization, for each cluster C and separator set S, set the following:

$$\phi_C \leftarrow 1, \quad \phi_S \leftarrow 1 \tag{10.10}$$

Then assign each variable X to a cluster C that contains X and its parents $pa(X)$, and then set the following:

$$\phi_C \leftarrow \phi_C p\left(X \mid pa\left(X\right)\right)$$

$$\lambda_X \leftarrow 1 \tag{10.11}$$

where λ_X is the likelihood vector for the variable X. Now, perform the following steps for each piece of evidence on a variable X:

- Encode the evidence on the variable as a likelihood λ_X^{new}.
- Identify a cluster C that contains X (e.g., one containing the variable and its parents).
- Update as follows:

$$\phi_C \leftarrow \phi_C \frac{\lambda_X^{new}}{\lambda_X}$$

$$\lambda_X \leftarrow \lambda_X^{new} \tag{10.12}$$

Now perform a global propagation using the two recursive procedures: *collect evidence* and *distribute evidence*. Note that if the belief potential of one cluster C is modified, then it is sufficient to unmark all clusters and call only *distribute evidence* (C).

The potential φ_C for each cluster C is now $p(C, e)$, where e denotes evidence incorporated into the tree. Now marginalize C into the variable as

$$p\left(X,e\right) = \sum_{C\backslash\{X\}} \phi_C \tag{10.13}$$

Compute posterior $p(X|e)$ as follows:

$$p\left(X \mid e\right) = \frac{p\left(X,e\right)}{p\left(e\right)} = \frac{p\left(X,e\right)}{\sum_X p\left(X,e\right)} \tag{10.14}$$

To update evidence for each variable X on which evidence has been obtained, first update its likelihood vector. Then initialize the junction tree by incorporating the observations. Finally, perform global propagation, marginalization, etc.

10.5   DISTRIBUTED KALMAN FILTER

As shown in Figure 10.9, we assume that a distributed KF environment consists of N local fusion nodes, producing track estimates based on a single sensor or multiple local sensors, and a fusion center, combining these local estimates into a global one. An example distributed environment for tracking a ground vehicle on a road consists of (1) a group of ground acoustic sensors laid on the road, which

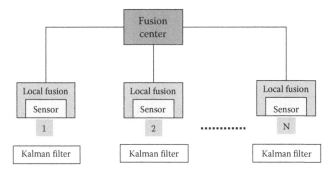

FIGURE 10.9   Distributed Kalman filter.

are coordinated by a local fusion node, (2) an aerial video feed provided by an unmanned aerial vehicle (UAV), and (3) ground moving target indicator (GMTI) data generated by joint surveillance target attack radar system (JSTARS). A local fusion node requires feedback from the global estimate to achieve the best performance. Moreover, the global estimation has to cope with sensors running at different observation rates.

The target's dynamic is modeled in a transition model as

$$X_k = FX_{k-1} + W_{k-1} \tag{10.15}$$

where
   the state vector X_k is the estimate of the target at time-instant k
   F is the transition mode matrix invariant of k
   W_k is the "white and Gaussian process" noise with zero-mean

The measurement models are given by

$$Z_{ik} = H_i X_k + V_{ik} \tag{10.16}$$

where
   Z_{ik} is the measurement or observed output state at time step k from the ith sensor
      $(i = 1,2,\ldots, N)$
   H_i is the corresponding observation matrix invariant of k
   V_{ik} is the corresponding white and Gaussian noise with zero-mean

The centralized KF algorithm for estimating the target's state and error covariance matrix has the following two recursive steps:

Prediction:

$$P_{k|k-1} = FP_{k-1|k-1}F^T + Q_{k-1}$$
$$\hat{X}_{k|k-1} = F\hat{X}_{k-1|k-1} \tag{10.17}$$

Update:

$$P_{k|k}^{-1} = P_{k|k-1}^{-1} + \sum_{i=1}^{N} H_i^T R_{ik}^{-1} H_i$$

(10.18)

$$\hat{X}_{k|k} = P_{k|k}\left[P_{k|k-1}^{-1} \hat{X}_{k|k-1} + \sum_{i=1}^{N} H_i^T R_{ik}^{-1} Z_{ik} \right]$$

where
$\hat{X}_{k|k}$ is the state estimate at time step k
$P_{k|k}$ is the error covariance matrix
Q_k and R_{ik} are covariance matrices of the process and measurement noises, respectively

The inverse P^{-1} of the covariance matrix P is a measure of the information contained in the corresponding state estimate.

In a distributed KF environment, each local fusion node i produces its own estimate $\hat{X}_{i(k|k)}$ based on the information available from its sensors, using the standard KF technique. These individual estimates are then fused together at the fusion center to produce the overall estimate $\hat{X}_{k|k}$.

As shown in Figure 10.10, there are two ways to carry out distributed KF (Liggins et al. 1997):

- Without feedback, meaning an individual fusion node performs target tracking based on its own local sensor measurements and sends its estimation of the target state and error covariance matrix to the fusion center at every time step.
- With feedback, meaning an individual fusion node sends its estimation to the fusion center as before but obtains feedback from the fusion center in terms of the center's overall estimation, using combined results from individual local fusion nodes.

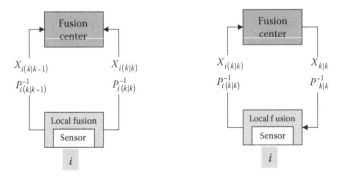

FIGURE 10.10   Distributed target tracking with and without feedback.

Without feedback:

$$P_{k|k}^{-1} = P_{k|k-1}^{-1} + \sum_{i=1}^{N} \left[P_{i(k|k)}^{-1} - P_{i(k|k-1)}^{-1} \right]$$

(10.19)

$$\hat{X}_{k|k} = P_{k|k} \left[P_{k|k-1}^{-1} \hat{X}_{k|k-1} + \sum_{i=1}^{N} \left[P_{i(k|k)}^{-1} \hat{X}_{i(k|k)} - P_{i(k|k-1)}^{-1} \hat{X}_{i(k|k-1)} \right] \right]$$

With feedback:

$$P_{k|k}^{-1} = \sum_{i=1}^{N} P_{i(k|k)}^{-1} - (N-1) P_{k|k-1}^{-1}$$

(10.20)

$$\hat{X}_{k|k} = P_{k|k} \left[\sum_{i=1}^{N} P_{i(k|k)}^{-1} \hat{X}_{i(k|k)} - (N-1) P_{k|k-1}^{-1} \hat{X}_{k|k-1} \right]$$

Note in the above two cases of estimation that the fusion center fuses only the incremental information when there is no feedback. The new information is the difference between the current and previous estimates from the local fusion nodes. When there is feedback, the fusion node must remove its own previously sent information before combining the local estimates. In other words, the new information to be sent to the fusion center is the difference between the new estimate and the last feedback from the fusion center. The process of removing an estimate is to make sure that the local estimates that are combined are independent.

10.6   RELEVANCE TO NETWORK-CENTRIC WARFARE

The NCW concept (Cebrowski and Garstka 1998, Cebrowski 2001) is a part of the DoD's effort to create a twenty-first-century military by transforming its primarily platform-centric force to a network-centric force through the use of modern information technologies. NCW is predicated upon dramatically improved capabilities for information sharing via an Internet-like infrastructure. When paired with enhanced capabilities for sensing, information sharing can enable a force to realize the full potential of dominant maneuver, precision engagement, full-dimensional protection, and focused logistics.

As shown in Figure 10.11, NCW involves working in the intersection of three interconnected domains, namely, physical, information, and cognitive.

The *physical domain* is where the situation the military seeks to influence exists. It is the domain where strikes, protections, and maneuverings take place across the environments of ground, sea, air, and space. It is the domain where physical platforms and the communications networks that connect them reside.

The *information domain* is where information is created, manipulated, and shared. It is the domain that facilitates the communication of information among

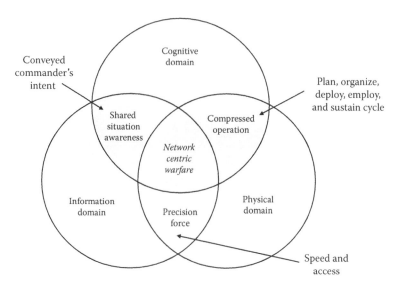

FIGURE 10.11 Conceptual vision of NCW. (From NCW, Network Centric Warfare, Department of Defense, Report to Congress, 2001.)

warfighters. It is the domain where the command and control of modern military forces is communicated and where the commander's intent is conveyed.

The *cognitive domain* is in the minds of the participants. It is the domain where perceptions, awareness, understanding, beliefs, and values reside and where, as a result of sense-making, decisions are made.

From the perspective of distributed fusion presented in the earlier sections, it is the cognitive domain that provides warfighters with the capability to develop and share high-quality situational awareness. Fusion nodes representing warfighters communicate their assessments of situations via appropriate coordination and negotiation.

The GIG is a globally interconnected, end-to-end set of information capabilities, associated processes, and personnel for collecting, processing, storing, disseminating, and managing information-on-demand for warfighters, defense policymakers, and support personnel. The GIG will operate within the information domain to enable the creation of a fusion network consisting of fusion nodes and any needed interconnections.

10.7   ROLE OF INTELLIGENT AGENTS

Recent advances in intelligent agent research (AGENTS 1997–2001, AAMAS 2002–2010) have culminated in various agent-based applications that autonomously perform a range of tasks on behalf of human operators. Examples of the kinds of tasks these applications perform include information filtering and retrieval, situation assessment and decision support, and interface personalization. Each of these tasks requires some form of human-like intelligence that must be simulated and embedded within the implemented agent-based application.

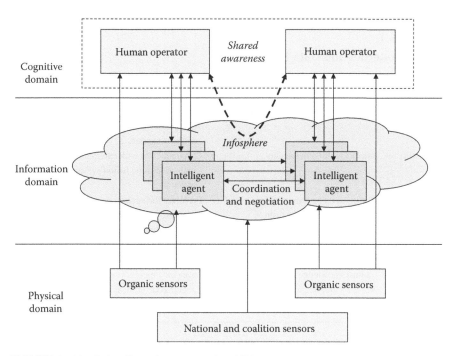

FIGURE 10.12 Role of intelligent agents for NCW.

Expectations are high that agent technologies can provide insights into and solutions for complex problems in the industrial, military, and business fusion communities. Such expectations are due to the agents' inherent capability for operating autonomously while communicating and coordinating with other agents in the environment. This makes them suitable for embedding in entities operating in hazardous and high-risk operational environments, including robots, UAVs, unattended ground sensors, etc. A recent DoD-wide thrust on NCW is by definition distributed in nature, where agents can play a vital role in the areas of cooperation, coordination, brokering, negotiation, and filtering. As shown in Figure 10.12, autonomous intelligent agents can act on behalf of warfighters (Lichtblau 2004) within an NCW environment to reduce their cognitive workload.

10.7.1 What Is an Agent?

An agent is a computational entity with intentionality that performs user-delegated tasks autonomously (Guilfoyle and Warner 1994, Caglayan and Harrison 1997). Some of the most important properties of software agents (Wooldridge and Jennings 1995), namely, autonomy, monitoring, and communication skills, are desirable features for building an ideal information fusion system. Table 10.1 summarizes the key properties of agents and agent-based fusion systems.

Intelligent agents can also be viewed as traditional artificial intelligence (AI) systems simulating human behavior. Rich AI technologies can therefore be

TABLE 10.1

Agent Properties and Data Fusion

Property	Definition	Agent-Based Fusion System
Autonomy	Operates without the direct intervention of humans or others	Autonomously executes tasks—target tracking and identification, situation and threat assessment, sensor management and decision support
Sociability	Interacts with other agents	Communicates with external environment such as sensors, fusion systems, and human operators
Reactivity	Perceives its environment and responds in a timely fashion	Perceives the environment and adjusts response accordingly
Pro-activity	Exhibits goal-directed behavior by taking the initiative	Goal of delivery of situation and threat assessment in time
Learnability	Learns from the environment to adjust knowledge and beliefs	Dynamic capabilities and behavior learned over time by observing areas of operations
Mobility	Moves with code to a node where data resides	Execute code at local sensor nodes accumulating observations

leveraged to build intelligent agents that learn from the environment, make plans and decisions, react to the environment with appropriate actions, express emotions, or revise beliefs. Intelligent agents typically represent human cognitive states using underlying knowledge and beliefs modeled in a knowledge representation language. The term "epistemic state" is used to refer to an actual or possible cognitive state that drives human behavior at a given point in time; the accurate determination (or estimation) of these epistemic states is crucial to an agent's ability to correctly simulate human behavior (Das 2008a).

10.7.2 Use of Agents in Distributed Fusion

Using agent technologies, experiments in building fusion systems (Das 2010) are being conducted at all levels (in the sense of JDL [Liggins et al. 2008, Steinberg et al. 1998, White 1988]). The communication ability of agents naturally lends itself to performing fusion tasks in a decentralized manner, where the cooperation among a set of spatially distributed agents is vital. Many important fusion problems (e.g., target tracking) are inherently decentralized.

A decentralized data fusion system, according to Durrant-Whyte and Stevens (2006),

> consists of a network of sensor nodes, each with its own processing facility, which together do not require any central fusion or central communication facility. In such a system, fusion occurs locally at each node on the basis of local observations and the information communicated from neighboring nodes.

Such decentralized systems rely on communication among nearby platforms, and therefore the number of messages that each platform sends or receives is independent of the total number of platforms in the system. This property ensures scalability to distributed systems with (almost) any number of platforms (Rosencrantz et al. 2003).

In contrast to a decentralized system, all platforms in centralized architectures communicate all sensor data to a single special agent, which processes it centrally and broadcasts the resulting state estimate back to the individual platforms. Such an approach suffers from single-point-of-failure and communication and computational bottlenecks.

There is an abundance of work in the area of distributed agent–based target tracking and in the area of distributed fusion. In general, a distributed processing architecture for estimation and fusion consists of multiple processing agents. Here we mention only some of them.

Horling et al. (2001) developed an approach to real-time distributed tracking, wherein the environment is partitioned into sectors to reduce the level of potential interaction between agents. Within each sector, agents dynamically specialize to address scanning, tracking, or other goals by taking into account resource and communication constraints. See also (Waldock and Micholson, 2007) for an approach to distributed agent-based tracking.

Hughes and Lewis (2009) investigated the track-before-detect (a method that identifies tracks before applying thresholds) problem using multiple intelligent software agents. The developed system is based on a hierarchical population of agents, with each agent representing an individual radar cell that is allowed to self-organize into target tracks.

Martin and Chang (2005) developed a tree-based distributed data fusion method for ad hoc networks, where a collection of agents share and fuse data in an ad hoc manner for estimation and decision making.

Chong and Mori (2004) highlighted the advantage of distributed estimation over centralized estimation, due to reduced communication, computation, and vulnerability to system failure, but expressed the need to address the dependence in the information. The authors developed an information graph approach to systematically represent this dependence due to communication among processing agents.

Graphical Bayesian belief networks have been applied extensively by the fusion community to perform situation assessment (Das 2008b). A network structure with nodes and links, modeling a situation assessment problem, is a natural fit for distributing tasks at various levels of abstraction and hierarchy, where nodes represent agents and messages flow between agents along the links. An approach along these lines has been adopted by Pavlin et al. (2006).

Mastrogiovanni et al. (2007) developed a framework for collaborating agents for distributed knowledge representation and data fusion based on the idea of an ecosystem of interacting artificial entities.

Rosencrantz et al. (2003) developed a decentralized technique for state estimation from multiple platforms in dynamic environments. The approach utilizes particle filters and deploys a selective communication scheme that enables individual platforms to communicate only the most informative pieces of information to other entities, thus avoiding communication overhead.

Mobile agents have also been employed for distributed fusion. Mobile agents are able to travel between the nodes of a network in order to make use of resources that are not locally available. Mobile agents enable the execution code to be moved to the data sites, thus saving network bandwidth and providing an effective way to overcome network latency.

Qi et al. (2001) developed an infrastructure for mobile agent–based distributed sensor networks (MADSNs) for multisensor data fusion. Bai et al. (2005) developed a mobile agent–based distributed fusion (MADFUSION) system for decision making in Level 2 fusion. The system environment consists of a peer-to-peer ad hoc network in which information may be dynamically distributed and collected via a publish/subscribe functionality. The software agents travel deterministically from node to node, carrying a data payload of information, which may be subscribed to by users within the network.

Focusing on fusion tasks beyond tracking and situation assessment, Nunnink and Pavlin (2005) proposed an algorithm that, based on the expected change in entropy, determines the optimal sensing resource to devote to fusion task assignment.

Jameson's (2001) Grapevine architecture for data fusion integrates intelligent agent technology, where an agent generates the information needs of the peer platform it represents.

Gerken et al. (2003) embedded intelligent agents into the mobile commander's associate (MCA) decision-aiding system to improve the situational awareness of the commander by monitoring and alerting based on the information gathered.

Das (2008a) provided three fundamental and generic approaches (logical, probabilistic, and modal) for representing and reasoning with agent epistemic states, specifically in the context of decision making. In addition, an introduction is given to the formal integration of these three approaches into a single unified approach called P3 (propositional, probabilistic, and possible world), which combines the advantages of the other approaches. The P3 approach is useful in implementing a knowledge-based intelligent agent specifically designed to perform situation assessment and decision-making tasks. Modeling an agent epistemic state (Das et al. 1997, Das and Grecu 2000, Das 2007) in such a way is analogous to modeling via the BDI (belief, desire, and intention) architecture (Rao and Georgeff 1991).

10.8 CONCLUSIONS AND FURTHER READING

In this chapter, we have presented distributed fusion from the SA and target tracking perspectives and its relevance to NCW environments. The approach to distributed fusion via message passing is a natural fit to distributed NCW environments, as it maintains the autonomy and privacy of individual agents and data sources. There are approaches along these lines, namely, distributed perception networks (DPN) (Pavlin et al. 2006) and multiply section Bayesian networks (MSBN) (Xiang et al. 1993), but the proposed approach leverages existing algorithms and reduces the overall message flow to save bandwidth.

Readers are recommended to consult Liggins et al. (1997) and Durrant-Whyte (2000) for an overall discussion on distributed fusion from the target tracking perspective. Liggins et al. (1997) also discuss an approach to address the data incest problem via information graphs. There are alternative approaches to a distributed KF algorithm, for example, that presented by Rao and Durrant-Whyte (1991). Schlosser and Kroschel (2004) present some experimental results from their study of the effect of communication rate among fusion nodes on the performance of a decentralized KF algorithm.

The book by Pearl (1988) is still the most comprehensive account of BNs, and more generally on using probabilistic reasoning to handle uncertainty. The junction tree algorithm of Lauritzen and Spiegelhalter (1988), as refined by Jensen et al. (1990) in HUGIN, is the most popular inference algorithm for general BNs. A good comprehensive procedural account of the algorithm can be found in Huang and Darwiche (1996). Jensen's books (1996, 2001) are also useful guides in this field. (See Das et al. [2002] for an application of BNs for conventional battlefield SA.)

The reader can also refer to Paskin and Guestrin (2004) for a more detailed account of a junction tree–based distributed fusion algorithm along the lines of the one presented here. The algorithm in the paper, in addition, optimizes the choice of junction tree to minimize the communication and computation required by inference. (See Das et al. [2002] for distributing components of a BN for battlefield SA across a set of networked computers to enhance inferencing efficiency and to allow computation at various levels of abstraction suitable for military hierarchical organizations.)

There is an abundance of open source literature on NCW. A "must read" on NCW is Cebrowski and Garstka (1998), and also NCW (2001) and Cebrowski (2001). Further reading of topics related to NCW are on effect-based operations (EBO) (Smith 2002) and sense and respond logistics (S&RL) (OFT 2003). The NCW vision is being realized within the DoD branches, including in the Army via its FCS (future combat systems) program, in the Navy (Antanitus 2003), and in the Air Force (Sweet 2004).

Section 10.7 on intelligent agents is nontechnical in nature; a multitude of references on intelligent agents in the fusion context are embedded within the section itself.

REFERENCES

AAMAS (2002–2010). *Proceedings of the International Joint Conferences on Autonomous Agents and Multi-Agent Systems* (1st—Bologna, Italy; 2nd—Melbourne, Victoria, Australia; 3rd—New York; 4th—the Netherlands; 5th—Hakodate, Japan; 6th—Honolulu, HI; 7th—Estoril, Portugal; 8th—Budapest, Hungary; 9th—Toronto, Ontario, Canada), ACM Press, New York.

AGENTS (1997–2001). *Proceedings of the International Conferences on Autonomous Agents* (1st—Marina del Rey, CA; 2nd—Minneapolis, MN; 3rd—Seattle, WA; 4th—Barcelona, Spain; 5th—Montreal, Quebec, Canada), ACM Press, New York.

Antanitus, D. 2003. FORCEnet architecture. *Briefing to National Defense Industrial Association (NDIA) Conference*, San Diego, CA.

Bai, L., J. Landis, J. Salerno, M. Hinman, and D. Boulware. 2005. Mobile agent-based distributed fusion (MADFUSION) system. *Proceeding of the 8th International Conference on Information Fusion*, Philadelphia, PA.

Caglayan, A. K. and C. Harrison. 1997. *Agent Sourcebook*. New York: John Wiley & Sons, Inc.

Cebrowski, A. K. 2001. The Implementation of Network-Centric Warfare. Office of the Force Transformation, Washington, DC.

Cebrowski, A. K. and J. J. Garstka. 1998. Network centric warfare: Its origin and future. *Proceedings of the Naval Institute*, 124(1): 28–35.

Chong, C.-Y. and S. Mori. 2004. Graphical models for nonlinear distributed estimation. *Proceedings of the Conference on Information Fusion*, Stockholm, Sweden, Vol. I, pp. 614–621.

Das, S. 2007. Envelope of human cognition for battlefield information processing agents. *Proceedings of the 10th International Conference on Information Fusion*, Quebec City, Quebec, Canada.

Das, S. 2008a. *Foundations of Decision-Making Agents: Logic, Probability, and Modality*. Singapore: World Scientific.

Das, S. 2008b. *High-Level Data Fusion*. Norwood, MA: Artech House.

Das, S. 2010. Agent-based information fusion, Guest Editorial, *Information Fusion (Elsevier Science)*, 11: 216–219.

Das, S., J. Fox, D. Elsdon, and P. Hammond. 1997. A flexible architecture for autonomous agents. *Journal of Experimental and Theoretical Artificial Intelligence*, 9(4): 407–440.

Das, S. and D. Grecu. 2000. COGENT: Cognitive agent to amplify human perception and cognition. *Proceedings of the 4th International Conference on Autonomous Agents*, June 2000, Barcelona, Spain, pp. 443–450.

Das, S., R. Grey, and P. Gonsalves. 2002a. Situation assessment via Bayesian belief networks. *Proceedings of the 5th International Conference on Information Fusion*, Annapolis, MD, pp. 664–671.

Das, S., K. Shuster, and C. Wu. 2002b. ACQUIRE: Agent-based complex query and information retrieval engine. *Proceedings of the 1st International Joint Conference on Autonomous Agents and Multi-Agent Systems*, Bologna, Italy, pp. 631–638.

Durrant-Whyte, H. F. 2000. *A Beginners Guide to Decentralised Data Fusion*. Sydney, New South Wales, Australia: Australian Centre for Field Robotics, The University of Sydney.

Durrant-Whyte, H. and M. Stevens. 2006. Data fusion in decentralised sensing networks. Australian Centre for Field Robotics, The University of Sydney, Sydney, New South Wales, Australia, http://www.acfr.usyd.edu.au

Gerken, P., S. Jameson, B. Sidharta, and J. Barton. 2003. Improving army aviation situational awareness with agent-based data discovery. *American Helicopter Society 59th Annual Forum*, Phoenix, AZ, pp. 602–608.

Guilfoyle, C. and E. Warner. 1994. Intelligent agents: The new revolution in software. Ovum Report.

Horling, B. et al. 2001. Distributed sensor network for real time tracking. *Proceedings of the 5th International Conference on Autonomous Agents*, Montreal, Quebec, Canada, pp. 417–424.

Huang, C. and A. Darwiche. 1996. Inference in belief networks: A procedural guide. *International Journal of Approximate Reasoning*, 15(3): 225–263.

Hughes, E. and M. Lewis. 2009. An intelligent agent based track-before-detect system applied to a range and velocity ambiguous radar. *Electromagnetic Remote Sensing Defence Technology Centre (EMRS DTC) Technical Conference*, Edinburgh, U.K.

Jameson, S. 2001. Architectures for distributed information fusion to support situation awareness on the digital battlefield. *Proceedings of the 4th International Conference on Data Fusion*, Montreal, Quebec, Canada, pp. 7–10.

Jensen, F. V. 1996. An intoduction to Bayesian Networks. New York: Springer.

Jensen, F. V. 2001. *Bayesian Networks and Decision Graphs*. New York: Springer-Verlag.

Jensen, F. V., S. L. Lauritzen, and K. G. Olesen. 1990. Bayesian updating in causal probabilistic networks by local computations. *Computational Statistics Quarterly.*, 4: 269–282.

Jones, P. et al. 1998. CoRAVEN: Modeling and design of a multimedia intelligent infrastructure for collaborative intelligence analysis. *Proceedings of the IEEE International Conference on Systems, Man, and Cybernetics (SMC'98)*, San Diego, CA, pp. 914–919.

Lauritzen, S. and D. Spiegelhalter. 1988. Local computations with probabilities on graphical structures and their applications to experts systems. *Journal of Royal Statistical Society B*, 50(2): 154–227.

Lichtblau, D. E. 2004. The critical role of intelligent software agents in enabling net-centric command and control. *Command and Control Research and Technology Symposium, The Power of Information Age Concepts and Technologies*, San Diego, CA, pp. 1–16, http://www.dtic.mil/dtic/tr/fulltext/u2/9466040.pdf

Liggins, M. E., C.-Y. Chong, I. Kadar, M. G. Alford, V. Vannicola, and S. Thomopoulos. 1997. Distributed fusion architectures and algorithms for target tracking. *Proceedings of the IEEE*, 85(1): 95–107.

Liggins, M., D. Hall, and J. Llinas (eds.). 2008. *Handbook of Multisensor Data Fusion: Theory and Practice*, 2nd edn. Boca Raton, FL: CRC Press.

Martin, T. and K. Chang. 2005. A distributed data fusion approach for mobile ad hoc networks. *Proceedings of the 8th International Conference on Information Fusion*, Philadelphia, PA, pp. 25–28.

Mastrogiovanni, F., A. Sgorbissa, and R. Zaccaria. 2007. A distributed architecture for symbolic data fusion. *Proceedings of the 20th International Joint Conference on Artificial Intelligence* (IJCAI), Hyderabad, India, pp. 2153–2158.

Mirmoeini, F. and V. Krishnamurthy. 2005. Reconfigurable Bayesian networks for adaptive situation assessment in battlespace. *Proceedings of the IEEE Conference on Networking, Sensing and Control*, Tucson, AZ, pp. 810–815.

NCW. 2001. Network Centric Warfare, Department of Defense, Report to Congress.

Nunnink, J. and G. Pavlin. 2005. A probabilistic approach to resource allocation in distributed fusion systems. *Proceedings of the Conference on Autonomous Agents and Multi-Agent Systems, AAMAS 2005*, Utrecht, the Netherlands, pp. 846–852.

OFT. 2003. Operational Sense and Respond Logistics: Co-Evolution of an Adaptive Enterprise, Concept Document, Office of Force Transformation, Washington, DC.

Paskin, M. and C. Guestrin. 2004. Robust probabilistic inference in distributed systems. *Proceedings of the 20th Conference on Uncertainty in Artificial Intelligence (UAI)*, Banff, Alberta, Canada, pp. 436–445.

Pavlin, G., P. de Oude, M. Maris, and T. Hood. 2006. Distributed perception networks: An architecture for information fusion systems based on causal probabilistic models. *Proceedings of the International Conference on Multisensor Fusion and Integration for Intelligent Systems*, Heidelberg, Germany.

Pearl, J. 1988. *Probabilistic Reasoning in Intelligent Systems: Networks of Plausible Inference*. San Mateo, CA: Morgan Kaufmann.

Qi, H., X. Wang, S. Iyengar, and K. Chakrabarty. 2001. Multisensor data fusion in distributed sensor networks using mobile agents. *Proceedings of 5th International Conference on Information Fusion*, Annapolis, MD, pp. 11–16.

Rao, B. S. and H. F. Durrant-Whyte. 1991. Fully decentralised algorithm for multisensor Kalman filtering. *IEE Proceedings-Control Theory and Applications*, 138(5): 413–420.

Rao, A. S. and M. P. Georgeff. 1991. Modeling rational agents within a BDI architecture. *Proceedings of the 2nd International Conference on Knowledge Representation and Reasoning*, Cambridge, MA, pp. 473–484.

Rosencrantz, M., G. Gordon, and S. Thrun. 2003. Decentralized sensor fusion with distributed particle filters. *Proceedings of the Conference on Uncertainty in Artificial Intelligence*, Acapulco, Mexico.

Schlosser, M. S. and K. Kroschel. 2004. Communication issues in decentralized Kalman filters. *Proceedings of the 7th International Conference on Information Fusion*, Stockholm, Sweden, pp. 731–738.

Smith, E. A. 2002. Effects based operations—Applying network centric warfare in peace, crisis, and war, DoD Command and Control Research Program, CCRP, Washington, DC.

Steinberg, A. 2008. Foundations of situation and threat assessment. In Liggins, M., D. Hall, and J. Llinas (eds.). *Handbook of Multisensor Data Fusion: Theory and Practice*, 2nd edn. Boca Raton, FL: CRC Press, pp. 437–501.

Steinberg, A. N., C. L. Bowman, and F. E. White, Jr. 1998. Revisions to the JDL data fusions models. *Proceedings of the 3rd NATO/IRIS Conference*, Quebec City, Quebec, Canada.

Su, X., P. Bai, F. Du, and Y. Feng. 2011. Application of Bayesian networks in situation assessment. In *Intelligent Computing and Information Science, Communications in Computer and Information Science*, Vol. 134. Berlin, Germany: Springer.

Sweet, N. 2004. The C2 constellation: A US air force network centric warfare program. *Command and Control Research and Technology Symposium*, San Diego CA , pp. 1–30, http://www.dodccrp.org/events/2004-CCRTS/CD/papers/164.pdf

Waldock, A. and D. Nicholson. 2007. Cooperative decentralised data fusion using probability collectives. *Proceedings of the 1st International Workshop on Agent Technology for Sensor Networks*, Honolulu, HI, pp. 47–57.

White, Jr., F. E. 1988. A model for data fusion. *Proceedings of the 1st National Symposium on Sensor Fusion*, Vol. 2, Orlando, FL.

Wooldridge, M. and N. R. Jennings. 1995. Intelligent agents: Theory and practice. *The Knowledge Engineering Review*, 10: 1–38.

Wright, E., S. Mahoney, K. Laskey, M. Takikawa, and T. Levitt. 2002. Multi-entity Bayesian networks for situation assessment. *Proceedings of the 5th International Conference on Information Fusion*, Annapolis, MD, pp. 804–811.

Xiang, Y., D. Poole, and M. Beddoes. 1993. Multiply sectioned Bayesian networks and junction forests for large knowledge based systems. *Computational Intelligence*, 9(2): 171–220.

11 Threat Analysis in Distributed Environments

Hengameh Irandoust, Abder Benaskeur,
Jean Roy, and Froduald Kabanza

CONTENTS

11.1 Introduction 296
11.2 Some Definitions 297
11.3 Threat Analysis: Primary Concepts 298
    11.3.1 Action, Event, and Reference Point 298
    11.3.2 Intentionality..299
    11.3.3 Impacts and Consequences...300
11.4 Threat Analysis as an Interference Assessment Problem............................300
    11.4.1 Intent–Capability–Opportunity Triad 301
        11.4.1.1 Intent Indicators 302
        11.4.1.2 Capability Indicators...302
        11.4.1.3 Opportunity Indicators 302
        11.4.1.4 Dual Perspective 303
    11.4.2 Threat Analysis in the Data Fusion Model.......................................303
11.5 Goal and Plan Recognition..304
11.6 Threat Analysis as a Plan Recognition Problem 306
    11.6.1 Plan Recognition...307
    11.6.2 Plan Recognition Approaches...309
        11.6.2.1 Symbolic Approaches 309
        11.6.2.2 Nontemporal Probabilistic Approaches............................309
        11.6.2.3 Probabilistic Approaches with a Temporal Dimension 311
        11.6.2.4 Mental State Modeling... 314
    11.6.3 Issues in Threat Analysis... 314
11.7 Threat Analysis in Military Operations 315
    11.7.1 Task Complexity 316
    11.7.2 Contextual Factors 316
        11.7.2.1 Uncertainty 316
        11.7.2.2 Time 316
        11.7.2.3 Nature of the Threat... 317
        11.7.2.4 Operational Environment... 318
11.8 Threat Analysis in Distributed Environments... 318
    11.8.1 Centralized and Decentralized Control.. 319
    11.8.2 Advantages of Distribution 320

    11.8.3 Operational Challenges 320
    11.8.4 Analytical Challenges... 321
    11.8.5 Collaboration Challenges.. 322
    11.8.6 Threat Analysis and Network-Centric Operations 322
11.9 Discussion 324
References... 324

11.1 INTRODUCTION

A threat to a given subject (a "subject" being a person, a vehicle, a building, a psychological state, a nation, the economy, the peace, etc.) is an individual, entity, or event that can potentially hurt, damage, kill, harm, disrupt, etc., this subject itself, or some other subjects (assets) for which this particular subject has concern. Recent years have seen a wide range of threats requiring surveillance, mitigation, and reaction in domains as diverse as military operations, cyberspace, and public security. Although each context has its own particularities, in all cases, one attempts to determine the nature of the threat and its potential to cause some form of damage.

Threat analysis consists in establishing the intent, capabilities, and opportunities of individuals, and entities that can potentially put a subject or a subject's assets in danger. Based on a priori knowledge and dynamically inferred or acquired information, threat analysis takes place in situations where there is indication of the occurrence of an event that can possibly harm a given (or set of) subject(s)/asset(s) of value.

Threat analysis involves the integration of numerous variables and calls upon several reasoning processes such as data fusion, intent, capability, opportunity estimation, goal and/or plan recognition, active observation, etc. Often performed in time-constrained and stressful conditions, the cognitive complexity of the threat analysis task can seriously challenge an individual, hence the automation of certain aspects of threat analysis in the operational domains.

The process of threat analysis can be performed by a single agent (human or software), but it can also be carried out by a team of agents, distributed over a geographic area, observing a situation from different perspectives, and attempting to merge their interpretations. This situation, while enabling information superiority, introduces a new set of challenges related to interoperability and inter-agent information sharing and collaboration. Similarly, the threat may also be comprised of multiple agents acting in coordination, which can significantly increase the difficulty of recognizing their common intent or plan. Thus, the challenges of threat analysis are multiplied significantly, as one moves from a one-on-one to a many-on-many configuration.

In the following sections, the problem of threat analysis is discussed from a theoretical perspective, while illustrating the observations by examples from the military domain.

The primary concepts of threat analysis, such as actions, goals, intentionality, consequences, reference point, are first discussed. Threat analysis is then addressed as an interference management problem where agents have to assess situations considering the intent, capabilities, and opportunities of their adversaries. By extending the scope of threat analysis to goal and then to plan recognition, it is shown that threat analysis can be viewed as an abduction problem where the observing agent is engaged in an evidence gathering—best-explaining hypotheses formulation cycle. This conceptual characterization of threat analysis enables the reader to measure the inherent difficulty of threat analysis regardless of the context of operations. The modeling frameworks and algorithmic techniques relative to different approaches, and plan recognition in particular, are extensively discussed, which shows the challenges for the automation of the threat analysis task. Next, threat analysis is presented in the context of military operations. The tasks to be performed and their complexity are discussed relatively to time, uncertainty, nature of threat, and other contextual factors.

After having grounded the problem in a military operational setting, the complexity of threat analysis in distributed environments is described, introducing the challenges of multi-threat environments and collaborative threat evaluation. The latter are analyzed both from situation analysis and collaboration perspectives. Finally, the operational challenges of threat analysis in network-centric operations are evaluated. Thus, through the document, the threat analysis problem is described in multiple contexts and at an increasing level of complexity.

11.2 SOME DEFINITIONS

A threat can be an individual, a physical entity, or an event that can potentially harm some asset of value, which is of concern to one or several agents.

It is generally accepted by the community working on the threat analysis problem (Paradis et al. 2005, Roy 2012, Steinberg 2005) that three concepts are central to the notion of threat. To constitute a threat, an entity must possess the intent or be intended to cause harm, as well as the capability and opportunity to achieve this intent.

- Intent is defined as the goal of the threat. Intent assessment determines (using all available pieces of evidence) whether the threatening entity intends to cause harm.
- Capability is defined as the ability of the threatening entity to achieve its goal and/or plan (or part thereof) as determined by the intent.
- Opportunity is defined as the existence in the environment of the required preconditions for the threat's goal/plan to succeed.

It is our contention that a threat can be defined along five dimensions, which capture these key concepts. These are

1. Negativity: the notion of a threat evokes and involves only negative connotations such as danger, harm, evil, injury, damage, hazard, destruction, loss, fear, dread, etc.
2. Intentionality: a threat can only be considered as such if it is intended so by a given goal-oriented and rational agent. Otherwise, there is danger and not threat.

3. Potential: to be a threat, an agent or entity must have the capability and opportunity to inflict the negative effect it intends to.
4. Imminence: a threat is always perceived as being in progress to achieve its goal by the agent expecting or observing it. Once the harmful event has occurred, it is no longer a threat.
5. Relativity to a point of reference: a threat is always considered as such relatively to its target(s), and the level of harm it can inflict can only be measured relatively to that point of reference and not in absolute terms. Threats are modeled in terms of potential and actualized relationships between threatening entities and threatened entities, or targets (Steinberg 2005).

Concerning the concepts 1 and 5, it must be added that causing harm includes causing distraction or negatively interfering with the goal or the objectives of an agent. Yet, negative interference must always be measured relatively to the value given by a given agent to its goal. In a situation of threat, what is threatened is a *crucial* goal of some agent, whether that goal is to change or preserve a certain state of affairs. Indeed, one cannot talk of threat for negative events that do not destroy, harm, or damage assets that are of utmost importance to one or more agents. Therefore, the expression negative impact must be interpreted relatively to a crucial goal.

Threat analysis has been defined in Roy (2012) as

> The analysis of the past, present and expected actions of external agents, covering the overall behaviour process of these agents from desires to effects/consequences, to identify menacing situations and quantitatively establish the degree of negativeness of their impact on the state and/or behaviour process of some agent of concern, and/or on some valuable human/material assets to be protected, taking into account the defensive actions that could be performed to reduce, avoid or eliminate the identified menace.

In operational environments, threat analysis is defined as the problem of determining the level of threat and the level of priority associated to it in a given situation. The level of threat indicates to what extent an entity is threatening. The level of priority indicates how much attention an observer should devote to that entity.

One should note that there are also the debatable notions of "inherent threat value" and "actual threat value" (also called *actual risk value* in Roy et al. [2002], Roy [2012]). The former is determined without consideration of a countermeasure/ defensive action, whereas the latter is established with consideration of defensive actions. In the latter case, one could also talk of "residual" threat value to refer to the threat (or risk) that remains even after a defensive action.

11.3   THREAT ANALYSIS: PRIMARY CONCEPTS

This section revisits some of the basic notions that underlie the threat analysis problem.

11.3.1   ACTION, EVENT, AND REFERENCE POINT

The process of threat analysis may implicate the observation of an action or event, which in turn involves one or several state changes in the environment.

The consequences of such action/event impact in different ways the entities and individuals concerned by the event. Thus, a given action may constitute a threat for one agent, be indifferent to another, and be an opportunity for still another. Consequences of actions are therefore positive or negative depending on the *reference point* (agent/entity) being considered. Moreover, the agents and entities concerned may be impacted by the change at different points in time. One agent may be threatened by an action instantaneously, as it occurs, while another may be affected by it only after a more or less long period of time. The effect of a negative action on different reference points varies in time and space. Often, the nature and magnitude of some state changes and/or consequences depend on the geometry (e.g., the proximity) between the "effector" (e.g., an agent performing some action) and the "affected" (e.g., a particular asset). As an example, consider the degree of severity of the explosion of a bomb as a function of the target proximity. Finally, actions can be viewed at different levels of granularity. An action can be perceived as being part of a more global action or event, which could be considered as a plan, or it can be a punctual and bounded occurrence.

11.3.2 INTENTIONALITY

Actions or events can be intentional or unintentional (e.g., potential natural disasters, accidents, or human errors). While unintentional actions can pose a *danger* to an agent or entity, only intentional actions can be considered as threats. Thus, threat analysis is concerned with characterizing, recognizing, and predicting situations in which a willful agent intends to do harm to some subject. However, actions intended for one reference point can also impact other agents and entities. Collateral damages and fratricides are examples of such unintended and unfortunate side effects.

The "belief–desire–intent" (BDI) model (Bratman 1987) used by a part of the intelligent agent community is a useful paradigm/model for a practical approach to the problem of intentionality or intent assessment. Roughly speaking, *beliefs* represent an agent's knowledge. *Desires* express what the agent views as an ideal state of the environment. These provide the agent with motivations to act. *Intention* lends deliberation to the agent's desires. Thus, intentions are viewed as something the agent has dedicated itself to trying to fulfill. They are those desires to which the agent has committed itself.

Bratman (1987) argues that unlike mere desires, intentions play the following three functional roles. Intentions normally pose problems for the agent; the agent needs to determine a way to achieve them. Intentions also provide a "screen of admissibility" for adopting other intentions. Whereas desires can be inconsistent, agents do not normally adopt intentions that they believe conflict with their present and future-directed intentions. Agents "track" the success of their attempts to achieve their intentions. Not only do agents care whether their attempts succeed but they are disposed to replan to achieve the intended effects if earlier attempts fail.

Castelfranchi (1998) defines goal-oriented or intentional agents or systems along the same lines. A goal is a mental representation of a world state or process that is candidate for (1) controlling and guiding action by means of repeated tests of the action's expected or actual results against the representation itself, (2) determining the action search and selection, and (3) qualifying its success or failure.

Intentionality is one of the core concepts used to analyze the notion of cooperation (Bratman 1987). An agent cannot be considered as cooperative if it is not intended to be, even if its actions incidentally further the goals of another agent. Likewise, a threatening agent cannot be considered as such without intention, even if its actions compromise the goals of another agent.

11.3.3 IMPACTS AND CONSEQUENCES

All actions performed by an agent perturb the environment, i.e., they produce some alterations of the state of the environment (including the state of other agents). Actions and the state changes resulting from their execution play a key role in any discussion on impact assessment and threat analysis.

The impact of an action is always relative to the perspective from which it is viewed. However, all targets do not have the same value for the adversary and the latter's intent, capability, and opportunity depend on the type of target being considered. Actions can be planned and executed to produce an overall broad effect, e.g., to demoralize the enemy, or they can be designed to produce a very specific effect, e.g., a high precision lethal attack. As the "vulnerability" of the target decreases, the required capabilities to affect it increase. As the "importance" of the target increases, so does the adversarial intent to affect it. Opportunities may also be dependent on the nature of the target. In an adversarial context, the subjects of threat will primarily protect their high-value assets. For example, in a naval task force, the oil tanker would be such an asset. All the other platforms would consider this particular platform to be the asset to protect when acting as an operational unit. However, it becomes very difficult to determine such assets or vulnerabilities in more complex systems. As a matter of fact, the impact of a threat instantiated as an attack can be physical (destruction, injury, death, etc.), psychological (instability, fear, distress, etc.), social (chaos), economical (crash, cost, etc.) etc., with one level affecting the other through complex interdependencies, leading to unpredictable consequences.

11.4 THREAT ANALYSIS AS AN INTERFERENCE ASSESSMENT PROBLEM

Let us consider two agents R and B, by reference to Red (enemy) and Blue (own or friendly) forces in the military domain. The term "agent" is used to refer to active, autonomous, goal-oriented entities.

In a world where agents co-exist, their actions, driven by inner or contingent goals, can accidentally or purposefully interfere with the actions of other agents. Therefore, in any situation, an agent needs to monitor its environment, and to assess and manage interferences with the surrounding agents and entities.

The reasoning an Agent B performs in a situation of possible negative impact is dependent on its knowledge of the type of situation and the time available for reasoning. In a situation of immediate danger, where Agent B observes Agent R's capability and opportunity to harm, Agent B's priority will be to avoid that situation.

Whether it is a car, out of control, heading toward us or a missile launched at our own ship, the first reaction would be to respond to that situation as to avoid it. The problem of the intent or the goal of the source of danger becomes irrelevant during that time frame.

In a less time-pressured environment and generally more complex situation, the negative interference of Agent R's action has to be evaluated in the light of a higher-level goal, so that Agent B can assess the scope of that action and possibly anticipate its consequences.

Similarly, in cases of positive interference, depending on the situation, an agent could simply enjoy the fortunate circumstances or in order to benefit on a larger scale or in the long term, attempt to establish the goal of the other agent(s) and assess the possibility of cooperation (mutual benefit), or exploitation (unilateral benefit).

Within this larger context, threat analysis, concerned with purposeful actions endangering crucial goals, involves reasoning on Agent R's capability/opportunity, intent/goal, and/or plan.

11.4.1 INTENT–CAPABILITY–OPPORTUNITY TRIAD

Essentially, it is sufficient to determine the intent, capability, and opportunity of Agent R to deliver damage to Agent B to establish its level of threat, if in fact a threat exists. Intent is an element of the agent's will to act. Capability is the availability of resources (e.g., physical and informational means) sufficient to undertake an action of interest. Opportunity is the presence of an operating environment in which potential targets of an action are present and are susceptible to being acted upon.

A threat can be viewed as an integral whole constituted of intent, capability, and opportunity, the disruption of any of the constituents involving the disruption of the whole (Little and Rogova 2006). From this perspective, viable threats exist when all three essential threat elements (intent, capability, opportunity) are present and form a tri-partite whole via relations of foundational dependence. Potential threats exist when at least one essential part (intent, capability, or opportunity) exists, but its corresponding relations are not established. In this sense, potential threats are threats that are not in a state of being, rather they are in a state of becoming, where portions of the item are constantly unfolding and are yet to be actualized at a given place or time.

In a military operational setting, collated information from all available sources is interpreted as part of the overall analysis of threat information in an attempt to discern patterns which may provide evidence as to the hostile entity's intent, capability, and opportunity. The threat will only have the opportunity to deliver its damage provided that it can both detect and track its target and can reach it (Paradis et al. 2005). Here, evidence consists in indicators of the presence of intent, of capability, and/or of opportunity of Agent R to harm Agent B.

Let us take the example of an air threat. Note that some of the threat indicators can be directly observed (e.g., bearing, range, speed, etc.), while others need simple calculations (e.g., Closest Point of Approach—CPA, flight profile) or advanced calculations (e.g., third party targeting), and still others could require knowledge-based inference. Indicators are derived from track characteristics, tactical data, background geopolitical situation, geography, intelligence reports, and other data. The indicators observed may be relative to any of the three key ingredients.

11.4.1.1 Intent Indicators

Within the intent–capability–opportunity triad, intent and its relationship to actions is the most complex one to assess. Intent assessment can involve both data-driven methods (i.e., explaining the purpose of observed activity) and goal-driven methods (i.e., seeking means to assumed ends) (Laird et al. 1991).

Intent can be derived based on the observation of current behavior, but also on the basis of information provided by other sources (e.g., intelligence). Also, it can be directed by a priori knowledge gained on an agent, or by own experience of its past actions.

To assess intent, one generally verifies if a certain number of criteria are satisfied, i.e., if a number of indicators are available. To do this, one has to create predictive models of behaviors that a purposeful agent might exhibit and determine the distinctive observable indicators of those behaviors (Steinberg 2007). In the military domain, while some indicators such as the track position, speed, identity, or responses (or the absence thereof) to Identification Friend or Foe (IFF) interrogations are readily available from the tactical picture, a priori databases, or from communications with other units in the force, others such as complex behaviors, e.g., threat maneuvers, tactics, group composition, and deception, can be very hard to analyze.

11.4.1.2 Capability Indicators

Capability determines the possibility for a given agent to carry out its intent. This concept refers to inherent or structural capability and can be measured independently of any particular situation, such as the lethality of a missile. Opportunity, on the other hand, refers to situational contingencies.

Capability indicators are generally available from a priori data (e.g., intelligence, database, etc.). Observations made during operations come to confirm the a priori information on the threat capability (e.g., characteristics of platforms, sensors, and weapons). It must be noted that one of the challenges with capability evidence gathering and exploitation is when dealing with asymmetric threats (see Section 11.7.2.3.3).

11.4.1.3 Opportunity Indicators

Opportunity, which we also refer to as the "situational capability," is the presence of favorable factors for actions to occur (Roy et al. 2002), and thus depends on the dynamics of the situation. Assuming that Agent R has the intent and the (structural) capability to inflict harm to Agent B, several conditions may be required in the environment in order for Agent R to make the delivery of this harm possible.

Like intent indicators, some of opportunity indicators are readily available from the tactical picture or a priori data, some are easily calculated, and others are much more difficult to determine. Indeed, some instances of opportunity assessment require a more elaborate predictive analysis of the agent behavior, e.g., analyzing a trajectory taking into account the engagement geometry, dynamic models of the entities in the volume of interest, and potential obstructions.

11.4.1.4 Dual Perspective

Both the capability and the opportunity of Agent R are directly related to the vulnerability of Agent B. This vulnerability can also be either structural or situational. Structural vulnerability is a function of the very nature of Agent B and directly determines, and is impacted by, the capability of Agent R. The situational vulnerability of Agent B offers opportunity to Agent R. It can be expressed in terms of its observability, i.e., the extent to which it can be seen/sensed by Agent R, and its reachability, which is the likelihood that it will be reached and affected by the action of Agent R. If these two conditions do not hold simultaneously, then Agent R will not be considered as having the opportunity to deliver harm.

Opportunities for action can be characterized by an evaluation of the constraints imposed by the accessibility or vulnerability of targets to such actions (Steinberg et al. 1998). Agent R can acquire opportunities actively (e.g., through purposeful activities such as gaining knowledge of the plans of Agent B, gaining an advantageous spatial position, performing deception actions, etc.) or passively (e.g., through the presence of environmental factors such as weather, cover, the presence of noncombatants, the terrain, etc.).

Similarly, threat analysis can be carried out through passive observation, but also proactively. For example, one can generate information through stimulative intelligence (Steinberg 2006), which is the systematic stimulation of red agents or their environment to elicit information (Steinberg 2007). Such stimulation can be physical (e.g., imparting energy to stimulate a kinetic, thermal, or reflective response), informational (e.g., providing false or misleading information), or psychological (e.g., stimulating perceptions, emotions, or intentions).

It must be added that the risk assessment of Agent B, when threatened by Agent R, is not only a function of the intent, capability, and opportunity of the latter to harm it, but also of the feasibility of Agent B's own options for defending itself. Thus, the risk would be low if these two balance out.

11.4.2 Threat Analysis in the Data Fusion Model

According to the data fusion model maintained by the Joint Directors of Laboratories' Data Fusion Group (JDL DFG), threat analysis comes under "impact assessment," which has to do with the estimation and prediction of effects on situations of planned or estimated/predicted actions by the participants (Steinberg et al. 1998). Per the revised JDL data fusion model, the level-3 data fusion process, originally called "Threat Assessment" (White 1988), has been broadened to that of impact assessment. Impact assessment, as formulated in the JDL DFG model, is the foundation of threat analysis (Roy et al. 2002).

Threat analysis in this framework involves assessing threat situations to determine whether adversarial events are either occurring or expected (Steinberg et al. 1998). Threat situations and threat events are inferred on the basis of the attributes and relationships of the agents involved. Estimates of physical, informational, and perceptual states of such agents are fused to infer both actual and potential relationships among agents. By evaluating and selecting hypotheses concerning agents' capability, intent, and opportunity to carry out an attack, the threat analysis system will provide indications, warnings, and characterizations of possible, imminent, or occurring attacks.

11.5 GOAL AND PLAN RECOGNITION

The process of threat analysis can be carried out by an Agent B trying to infer the goal of an Agent R from a sequence of observations. This can be viewed as an ongoing dynamic process of evidence gathering and hypotheses formulation, as illustrated in Figure 11.1.

Agent B observes the environment and the actions of Agent R (both agents' actions impact the environment) trying to infer its goals. It perceives evidences that confirm or contradict what he hypothesizes as being Agent R's goals. Hypotheses are put forth, strengthened, or discarded, based on Agent B's expectations regarding Agent R's current goals and the new evidence that is observed/sensed. To generate hypotheses about Agent R's goals (match expectations with observations), Agent B uses its model of the situation and its knowledge about the adversary. This model is fed by a priori knowledge (background knowledge on the potential behaviors and capabilities of Agent R, experience of past cases, high-level information, etc.). Agent B takes action on the basis of its hypotheses about the current situation, as generated by the use of the model and the observations. Solving the problem may involve not only reasoning about past behaviors, as indicated by the observations to date, but

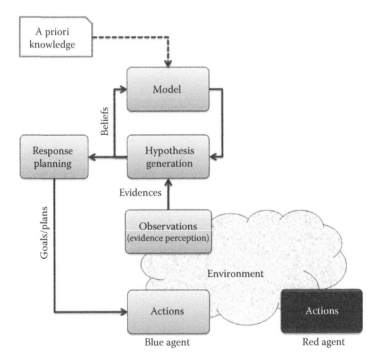

FIGURE 11.1   Threat analysis: the blue perspective.

also hypothesizing over likely future behaviors. Agent B's actions affect, in turn, the environment and the future actions of Agent R.

Goal recognition, on the basis of action observation, is a very complex problem. First, it must be noted that agents generally pursue consistent goals at different levels of abstraction. In some contexts, such as military operations, high-level goals are easier to figure out than low-level goals that must be recognized for a particular situation. For example, while the strategic goals of a nonfriendly country may be widely known—hence the identification of its assets as "hostile" prior to any intent assessment—the tactical goals (and the plans) or punctual objectives of that agent may be very difficult to discern in the field.

In a threat analysis context, Agent B, using its model of the situation and based on its observations, attempts to determine the relationship between the actions and goals of Agent R. Goal and/or plan recognition in this context, where Agent B attempts to discern a sequence of goal-directed actions, can be problematic in several regards. In effect, several misconceptions regarding the actual goal of the adversary Agent R are possible, which may be due either to Agent B's flawed perception of the situation or the reasoning it performs given its model of the situation (including its model of Agent R).

Various perception problems can arise. The action of interest may not be fully observed because of the imperfection that is inherent to the perception and identification of actions, whether by humans or nonhuman sensors. Agent B may fail to see some actions or may see arbitrary subsets of the actual actions (partial observability). It may not be able to distinguish actions of interest from clutter (activities of other agents or entities in the environment). It may also be reasoning on an action, a portion of which has not been observed yet.

Agent B can also make errors related to the use of the model (Figure 11.1):

1. *Missing goal*: A goal pursued by Agent R may be completely unknown to Agent B (B has no representation of that goal in its model). Another situation is when the goal is represented in Agent B's model but has been discarded because B assumes that such goal cannot be pursued. This occurs, for example, when B makes an assumption of rationality. We generally consider that other individuals follow the same line of reasoning as us and that agents behave based on decisions that are in accordance with their reason, i.e., their proper exercise of the mind. However, in the case of terrorism and asymmetric threats, one is often confronted with behavior that is based on decisions that can be qualified as irrational (Roy et al. 2002).

2. *Wrong inference on the structure of actions and goals*: The establishment of an action–goal relationship can be very complex. Consider the following cases: (1) an action of Agent R can contribute to several goals; (2) the goal of the action is rightly identified, but that action is only the initial phase of a higher-order action, and thus contributing to another goal; (3) a goal is dismissed because the action's conditions are not respected (duration, precondition, etc.); (4) Agent R is performing interleaved actions, i.e., a set of actions observed sequentially by Agent B are performed by Agent R in the

execution of different plans while pursuing different goals. For example, consider observing a person moving in a house and performing bits of different plans one after the other (e.g., clean up, write down two or three items on the grocery list, do some cooking, write down another item on the list, go back to cooking, etc.).

3. *Model manipulation:* Sometimes, agents whose plans and goals we attempt to identify may help us in our recognition process by making their goals as explicit as possible. However, in an adversarial context, deception is the rule. Agent R can attempt to dissemble, misdirect, or otherwise take actions to deliberately confuse Agent B. It does so by using its own model of Agent B's beliefs and reasoning process.

The cases discussed in item 2 come under the problem of plan recognition. In adversarial contexts, goal recognition generally implies some degree of plan recognition, as both parties achieve their goals through the accomplishment of a course of actions. If Agent B determines that the action of interest is not a single isolated action but is rather part of a plan, then it needs to organize the observed actions in a goal-oriented sequence. At the same time, if several opponents are involved, i.e., Agent R is member of a team, then, the role of each team member in the higher-level action must be determined.

Problems of perception can also be particularly problematic for plan recognition, which is an incremental inference process where the validity of a hypothetical plan can only be confirmed by the observation of significant elements or actions, or at least portions of them.

One of the major difficulties in goal and plan recognition, excluding those already mentioned, is that every situation is dynamic and constantly changing, and even more so in a battlespace. This has important consequences on Agent R's actions and on Agent B's interpretation of those actions and selection of defensive actions. Thus, Agent R may abandon its plan, change its initial plan (e.g., change resources, course of actions, etc.) to adapt it to the new circumstances (which may be the outcome of Agent B's actions), decide to act opportunistically, etc.

As previously discussed, Agent B has to assess the impact of Agent R's actions and goals on its own goals, plans, and on the environment (including neutral actors), whether Agent R's plan is recognized or not. More specifically, Agent B has to determine if Agent R's goal can be achieved given Agent R's capability and opportunity, and its own capability to defend itself.

11.6 THREAT ANALYSIS AS A PLAN RECOGNITION PROBLEM

Establishment of hostile intent may not be enough in the evaluation of a threat event, as the situation awareness needed for threat analysis requires that Agent B be able to organize the observed actions into a course of actions, and to some extent predict the evolution of the situation. This means that Agent B must engage in some kind of plan recognition. Note that recognizing the plan of a threat implies that, to a certain extent, its intent, capability, and opportunity have been recognized, but this implication is not true the other way around.

Unlike threat analysis, the plan recognition field is concerned with plans carried out by agents in general and not only non-friendly ones. Adversarial plan recognition can be brought close to threat analysis, although an adversary is not necessarily a threat. Also, while the plan recognition community is interested in the recognition of a plan as a process constituted of a sequence of goal-oriented actions (which can in certain cases be suspect or hostile), the threat analysis community is primarily concerned with the determination of what constitutes a threat and how to identify it.

11.6.1 Plan Recognition

The problem of plan recognition can be viewed as a case of abductive inference or of deductive explanation. This kind of explanation is concerned with the construction of theories or hypotheses to explain observable phenomena, thus requiring an abductive reasoning process. As Southwick (1991) observes, and this is the challenge of plan and/or goal recognition in general, "in order to arrive at a hypothesis, a person must first find some pre-existing model or schema, and try to interpret all data in terms of that model." As mentioned before, this abductive leap from a small amount of data to a working hypothesis is risky because of the incompleteness or uncertainty of the data and/or the use of a model that may be defective or wrong.

Plan recognition is used by everyone in everyday life to be able to manage a conversation, to avoid bumping into people in the corridors, or to guess what people around us are up to. It is used in cooperative, neutral, and adversarial settings. Two types of plan recognition can be distinguished: one type in which Agent W (neutral or cooperative) helps Agent B in its plan recognition (this is intended recognition), and another type where Agent R attempts to thwart recognition of its plan by Agent B (this is adversarial plan recognition). From the observer's viewpoint, the distinction is made between "keyhole" and "intended" plan recognition (Cohen et al. 1981). Keyhole means that the plan recognizer B is passively watching an Agent W execute its plans (W may not be aware of this observation) (e.g., story understanding). Intended means that the observed Agent W intends that the observing Agent B be able to infer its plan (e.g., tacit teamwork).

Plan recognition has long been established as one of the most fundamental and also challenging problems in human cognition. Through his psychological experiments, Schmidt provided evidence that humans do infer hypotheses about the plans and goals of other agents and use these hypotheses in subsequent reasoning (Schmidt 1976). Later, he positioned plan recognition as a central problem in the design of intelligent systems (Schmidt et al. 1978). Computational approaches to plan recognition have followed in various areas, such as story understanding (Bruce 1981) and natural language understanding (Allen 1983). The general problem can be described as the ability to infer, given fragmented description of the actions performed by one or more agents in a situation, a richer description relating the actions of the agents to their goals and future actions.

Automation of plan recognition means that the system contains a knowledge base, often called "plan libraries," of actions and recipes for accomplishing them (i.e., models of situations). These recipes include actions' preconditions, subgoals,

goals, and effects. To infer the agent's goal from the observed actions, the plan inference system constructs a sequence of goals and actions that connect the observed actions to one of the possible domain goals. This is accomplished by chaining from actions to goals achieved by the actions, from these goals to other actions for which the goals are preconditions or subgoals, from these actions to their goals, etc. (Carberry 2001). Knowledge engineering and computational efficiency remain the main challenges of plan recognition automation.

Plan recognition works differ both in how the problem is framed, and how the problem, once framed, is solved. Differences between different frameworks concern

1. *Plan representation:* Plans are commonly represented as straight line classical plans or Hierarchical Task Network (HTN) plans (see the example of Figure 11.2), which can capture complex, phased behaviors. The latter permit some partial ordering of actions, yet there is little extant work on temporal goals in plan recognition because there's little sense of how they would fit into plan libraries.

2. *The observer/observed relationship:* A distinction was made earlier between keyhole, intended, and adversarial plan recognition. The particularity of the latter is that it can involve deception. In an adversarial setting, behaviors may segment into deceptive/nondeceptive, which significantly increases the complexity of plan recognition.

3. *Observability:* Most plan recognition works assume full observation. This means that Agent B sees the full set of actions of Agent R or W. On the opposite, partial observability boils down to performing incremental plan recognition from some prefix of what Agent B expects to be a full plan.

There are ways of simplifying the plan recognition problem by performing only goal recognition (i.e., determine the "what" and not the "how" of what the agent is doing) or agent classification (do not figure out precise objectives, but only a category of goals or agents, such as hostile, friendly, etc.).

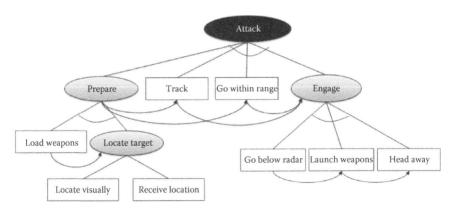

FIGURE 11.2 Example of HTN plan representation.

Activity modeling is another problem simplification approach, where instead of recognizing a particular plan, one recognizes an activity, i.e., a simple temporally extended behavior (Agent R is playing tennis is simpler to recognize than Agent R is preparing for an overhead smash).

11.6.2 PLAN RECOGNITION APPROACHES

Plan recognition approaches can be categorized into symbolic and probabilistic approaches. The latter treat uncertainty numerically, the former do not. Among probabilistic approaches, one can distinguish between temporal and nontemporal models.

11.6.2.1 Symbolic Approaches

One of the dilemmas of plan recognition is that of the Occam's razor or minimization, i.e., the principle according to which, of all plans explaining given observations, the minimalist one is the best explanation. Circumscription techniques which keep a minimal true set prevail here as they minimize the hypothesized plans. In *Generalized Plan Recognition*, Kautz and Allen (1986) define the problem as that of identifying a minimal set of top-level actions sufficient to explain the set of observed actions, representing it as plan graphs with top-level actions as root nodes expanded into unordered sets of child actions. Although efficient, this approach assumes that agents attempt one top-level goal at a time. Moreover, these techniques fail when likelihood matters. For example, in a medical diagnostic problem, HIV disease can be an explanation for virtually any symptom (best minimal explanation), but this hypothesis is very unlikely compared to a combination of likely hypotheses such as head cold and sinus infection.

Another symbolic approach in plan recognition is parsing, which is using a grammar (showing the decomposition of actions) and a parser (an algorithm which "reads" a given plan using the grammar). Based on Kautz and Allen's work, and taking advantage of the great amount of work in this area, Vilain (1991) investigates parsing as a way of exploring the computational complexity of plan recognition. The problem with this formalism is that the size of the grammar and the performance of the parser blow up in the presence of partially ordered grammars, where actions are represented as having a partial temporal order. Pynadath and Wellman (1995) first used probabilistic context-free grammars which suffer from the same problem. To overcome that, they proposed a probabilistic context-*sensitive* grammar. While handling state dependencies, this approach does not address the partial-ordering issues or the case of interleaved plans.

11.6.2.2 Nontemporal Probabilistic Approaches

Probability theory has imposed itself in plan recognition as it is the normative way of doing abductive reasoning. Probabilistic approaches include probabilistic decision trees, influence diagrams, and mainly Bayesian networks, which are directed graph models of probability distributions that explicitly represent conditional dependencies.

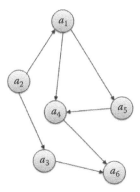

FIGURE 11.3 Bayesian network.

In a Bayesian network, nodes represent random variables and arcs between nodes represent causal dependencies captured by conditional probability distributions. When used for plan recognition, the nodes are propositions, the root nodes represent hypotheses about the plan of Agent R, and the probability assigned to a node represents the likelihood of a proposition given some observed evidence. A link from a variable a to a variable b could be interpreted as a causing b. This way, in the plan library (which specifies the potential plans an agent may execute), subgoals would be connected to goals, preconditions to actions, and actions to effects.

Figure 11.3 illustrates a simple Bayesian network for goal recognition. The variables shown in black circles represent the goals that an entity in the environment might have (transiting, reconnaissance, or attacking). The variables in gray circles are subgoals. For example, to attack a target, an entity must approach it, detect it, and engage it. The goal decomposition conveys some kind of a hierarchical plan that describes how to achieve a given task (goal) by decomposing it into subtasks (subgoals). The variables in white circles represent observable facts about the entity being observed. There is a conditional probability distribution for each variable given its parents. Using Bayesian inference, one can calculate the probability that the entity is committed to some goals given the observations that have been made.

Bayesian inference supports the preference for minimal explanations in the case of equally likely hypotheses (minimum cost proofs in symbolic logical approaches are equivalent to maximum a posteriori estimation), but also correctly handles explanations of the same complexity but with different likelihoods. Bayesian networks provide computational efficiency (avoid joint probability tables) and assessment efficiency (reduce the number of causal links that have to be modeled). However, like most diagrammatic schemes, Bayesian networks have only propositional expressive power (quantification and generalization are not possible). Another pitfall is that they do not explicitly model time, which is needed when it comes to reasoning about behaviors. However, it is possible to approximate the flow of time with causality.

Charniak and Goldman (1991) were among the first to use Bayesian inference for plan recognition. Their Bayesian network represented a hierarchical plan expressed as a decomposition of goals into subgoals and actions. It used a marker passing a form of spreading activation in a network of nodes and links to identify potential explanations for observed actions and to identify nodes for insertion into a Bayesian belief network.

Other works include Elsaesser and Stech (2007) where a Bayesian network is used to perform sensitivity analysis on the hypotheses generated by a planner. In Santos and Zhao (2007), a network represents the threat's beliefs on goals and high-level actions for both itself and its opponent, and an action network represents the relationship between the threat's goals and possible actions to realize them. Finally, Johansson and Falkman (2008) use a Bayesian network to calculate the probability of an asset being targeted by a threat.

Causality in Bayesian networks as exploited in these approaches is not enough to make inferences about complex behaviors. An explicit temporal model must be incorporated in order to make inferences on sequences of observations. Temporal dependencies between actions in plans must be modeled and related to goals. This is even truer for coordinated agents accomplishing arbitrary, temporally extended, complex goals.

11.6.2.3   Probabilistic Approaches with a Temporal Dimension

Temporal probabilistic models allow inferences over behaviors based on temporal sequences of observations. Different types of probabilistic queries can be made for threat analysis. A probabilistic explanation query would compute the posterior probability distribution of a given behavior (as a sequence of states) based on a sequence of observations. A probabilistic *filtering query* would compute the posterior probability distribution over the current goal or plan given the observations to date (e.g., the Kalman filter). This would require augmenting the state space with goals or plans that agents are pursuing. A probabilistic *prediction query* would compute the posterior probability over future goals given the observations to date.

Dynamic Bayesian networks (DBNs) are Bayesian networks where each "slice" represents a system state at a particular instant in time. Causal influences run from nodes in one time slice to the next (e.g., the state of Agent B at time $t+1$ is a probabilistic function of its state at time t and whether Agent R attacked it at time t). On the one hand, DBNs have the virtue of explicitly modeling the probabilistic dependencies among the variables. On the other hand, inference in DBNs has more computational complexity.

In the example of the Bayesian network in Figure 11.3, if an entity is observed with a heading toward a ship, then it follows that it has a high probability of approaching its target, regardless of its previous headings. In other words, the inference does not take into account the history or the past behavior of the entity. After all, the track could very well be in the process of turning and this particular heading may only be coincidental and temporary. It is possible to remedy the Bayesian network by including mega-variables capturing the history of events. However, that would be very tedious and error prone. A better approach consists in using a DBN that naturally captures the flow of time. Figure 11.4 shows a DBN obtained from the Bayesian network in Figure 11.3 by adding the temporal extension. With this DBN, the calculation of the

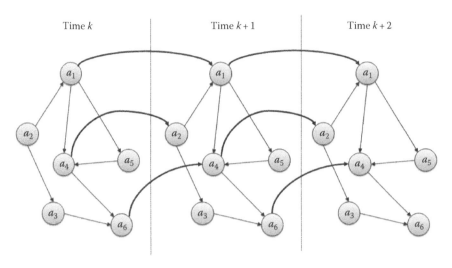

FIGURE 11.4 Dynamic Bayesian network.

probability of the subgoal *Approach* takes into account the previous heading, the previous distance, and the previous probability of the *Approach* subgoal. Using this approach, even if the probability for the subgoal *Approach* at time $k - 1$ was small, but the entity was/is heading toward the ship both at time $k - 1$ and k, the probability of *Approach* would nonetheless be higher at time k than at time $k - 1$.

An alternative to DBNs are hidden Markov models (HMMs). Actually, the latter are a particular form of DBNs in which the state of a process is described by a single discrete random variable. HMMs offer greater flexibility because one can specify the state transition and observation models using conditional probability tables. An HMM models the dynamics of only one variable and relates observations at time k only to the state of the variable at time $k + 1$. These approaches offer many of the efficiency advantages of parsing approaches, with the additional advantages of incorporating likelihood information and of supporting machine learning to automatically acquire plan models. However, because of the weak expressiveness of models (even weaker than that of grammars), state spaces can explode if complex plans are to be represented. Similarly, training can become very difficult.

Figure 11.5 illustrates an HMM for plan recognition. The hidden state is the current plan of an observed entity, where a plan is represented as a hierarchical

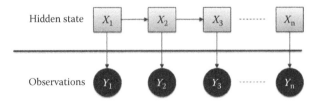

FIGURE 11.5 Hidden Markov model.

decomposition of goals (or tasks) into subgoals (or subtasks), which is more or less reminiscent of the causal structure underlying the Bayesian network in Figure 11.3. That is, a state is a graph in which the root node is a goal (task), the leaves are actions (primitive tasks), and the inner nodes are subgoals (subtasks). The observations are assumed to be caused by the actions of the observed entity while it executes the plan.

With HMMs, the dynamics related to the plan and plan execution can be modeled using plan libraries, as opposed to DBNs where they are explicitly modeled as variables. This keeps the inference mechanism related to the evolution of plans over time (conveyed by the plan libraries and their simulation) separated from the inference mechanism related to the generation and evaluation of competing hypotheses about the current plan (conveyed by the inferences within the HMMs).

HMMs have been commonly used in "activity recognition," specifically for recognizing behaviors of moving individuals for diverse purposes, such as eldercare (Liao et al. 2007), detection of terrorist activity (Avrahami-Zilberbrand and Kaminka 2007), and teamwork in sports and in the military (Sukthankar and Sycara 2006). An illustration of an HMM approach is the Probabilistic Hostile Agent Task Tracker (PHATT) introduced by Goldman et al. (1999) and later refined through successive improvements (Geib and Goldman 2003, 2005, Geib et al. 2008). A state of the HMM underlying PHATT is a set of concurrent plans the agent may be pursuing and the current points in the execution of these plans. From these current points, the next potential actions are derived (called pending sets) thereby constraining the model of observation (observations are mapped to effects of the pending actions to infer the probability distribution for the observed action). The states of the HMM are generated on the fly by simulating the execution of the plans in the current state based on the current observed action. By hypothesizing goals and plans for the agent, and then stepping forward through the observation trace, a possible sequence of pending sets is generated. When the end of the set of observations is reached, each observed action will have been assigned to a hypothesized plan that achieves one of the agent's hypothesized goals and a sequence of pending sets that is consistent with the observed actions. This collection of plan structures and pending sets is a single complete explanation for the observations.

A similar approach is taken by Avrahami-Zilberbrand and Kaminka (2005) who also maintain a set of hypotheses, but instead of using a model of plan execution and pending sets, they check the consistency of observed actions against previous hypotheses. Although solving some of the problems addressed by PHATT, the approach does not allow them to recognize those tasks that depend on pending sets, including negative evidence (actions not observed) (Geib and Goldman 2009). Kaminka et al.'s (2002) keyhole recognition for teams of agents considers the question of how to handle missing observations of state changes. However, it differs from PHATT significantly in using a different model of plan execution and by assuming that each agent is only pursuing a single plan at a time. On the other hand, it devotes a great deal of effort to using knowledge of the team and its social structures and conventions to infer the overall team behavior.

Finally, some "hybrid" works have used the theoretical framework of Bayesian Networks and HMMs but exploited research on parsing while mitigating the problems posed by partial orders. Such works include ELEXIR (Geib 2009) and YAPPR (Geib and Goldman 2009).

11.6.2.4 Mental State Modeling

Another way of approaching the problem of plan recognition is through adversary modeling. In domains where the tasks are radically unconstrained and it is too hard to build a database of possible plans, Agent B can simply invert the planning problem by putting itself in Agent R's shoes. For instance, Agent B can speculate on what Agent R may consider as critical to its mission. For Whitehair (1996), intent assessment is a model of the (adversarial) agent's internal utility/probability/cost assessment, by which the utility of particular states, the probability of attaining such states given various actions, and the cost of such actions are estimated.

In some areas, knowledge of Agent R's plans observed in the past may poorly predict its later plans. The military domain is one of these areas. The opposing forces' actions are nevertheless constrained by their doctrine and rules of engagement (at least in the case of conventional forces), their capabilities (weapons and resources), the environment in which they are operating, etc. All of these factors constrain what the opposing forces can do in practice.

In Glinton et al. (2005), field model prediction based on a priori knowledge is accomplished by the interpretation of opposing forces' disposition, movements, and actions within the context of their known doctrine and knowledge of the environment. Along the same reasoning line, TacAir-Soar, probably the most widely referenced expert system for tactical military operations (Jones et al. 1998), uses its knowledge of aircraft, weapons, and tactics to create a speculation space in which it pretends to be the opponent by simulating what it would do in the current situation. Such a methodology is generally known as *mental state modeling*, given that it literally consists in modeling the mental state of the opponent. TacAir-Soar is a symbolic rule-based system based on the Soar architecture for cognition (Laird et al. 1991). Its functionalities cover not just threat analysis, but also other command and control processes, including planning, and action execution (Jones 2010).

Game-theoretic methods must also be mentioned in this category, although they typically do not involve very complex iterative opponent modeling. Chen et al. (2007) discuss a mathematical framework for determining rational behavior for agents when they interact in multi-agent environments. The framework offers a potential for situation prediction that takes real uncertainties in enemy plans and deception possibilities into consideration. It can give an improved appreciation of the real uncertainty in the prediction of future development. However, prediction of the behavior of the other agents is based on an assumption of rationality.

11.6.3 ISSUES IN THREAT ANALYSIS

From a plan recognition perspective, threat analysis, or adversarial plan recognition, poses several challenges. Aside from the more general issues relative to the representation and interpretation of events and states, as discussed in Section 11.5, threat analysis can further complicate plan recognition frameworks because of the following:

1. *Model manipulation:* Generally, existing frameworks assume that the observed agent makes no use of deception.
2. *Plan revision:* Plan revision on the part of the observed agent is a serious challenge to plan recognition. Plan revision can be a consequence of a change in the operational environment or caused by a change in agents' inner motives. A particular case of plan revision is that of plan abandonment where an initial plan is abandoned and a new one developed. This can cause further multiplication of hypotheses. Also, it is difficult to determine when and on what basis previous active hypotheses must be terminated.
    a. *Multiple* interleaved plans: Where the observed agent attends to several tasks is another challenging area, as it can cause explosion of hypotheses generation.
    b. *State* representation: Systems typically observe actions rather than states of the world. Yet, states as much as actions are indicative of a threatening situation (e.g., a platform being in own forces' volume of interest).
    c. *Models* of *opponent actions:* It is very difficult to gather enough data to come up with concise and robust representations of the plans of the opponent.
    d. *Completeness of plan libraries:* No set of opponent plans will account for all possible scenarios.

Plan recognition, however, remains a very promising paradigm for the problem of threat analysis as it subsumes many of the elements necessary for the determination of the existence of threat and its evolution in time.

11.7 THREAT ANALYSIS IN MILITARY OPERATIONS

In this section, threat analysis is discussed from the perspective of the Command and Control (C2) process in an operational environment. This process can be decomposed into a set of generally recognized, accepted functions that must be executed within some reasonable delays to ensure mission success: picture compilation, threat analysis, engageability assessment, and combat power management (also referred to as weapons assignment).

The process of all actions and activities aimed at maintaining tracks on all surface, air, and subsurface entities within a certain volume of interest is referred to as *picture compilation*. It includes several subprocesses, the most important being *object localization* (or tracking), and *object recognition, and identification*. *Threat analysis* establishes the likelihood that certain entities within that volume of interest will cause harm to a defending force or its interests. The output of threat analysis, along with that of the engageability assessment process, which determines the defending force options against the threat, is used by the combat power management function to generate and optimize a response plan (Irandoust et al. 2010).

Threat analysis in an operational context such as the military setting is conducted based on a priori knowledge (e.g., intelligence, operational constraints and restraints, evaluation criteria, etc.), dynamically acquired and inferred information (e.g., kine-matics and identification of entities in a given volume of interest, as well as various indicators), and data received from complementary sources in relation to the mission objectives. Threat indicators are derived from the entity characteristics, tactical data, background geopolitical situation, geography, intelligence reports, and other data.

11.7.1 Task Complexity

Threat analysis is a highly demanding cognitive task for human analysts mainly because of the (typically) huge amount of data to be analyzed, the level of uncer-tainty characterizing these data, and the short time available for the task (Irandoust 2010).

The staff in charge of threat analysis must often process an important amount of data, of which only a small fraction is relevant to the current situation. The data come in multiple forms and from multiple sources. Analysts have to make difficult inferences from this large amount of noisy, uncertain, and incomplete data.

In a series of studies conducted by Liebhaber and his colleagues (Liebhaber and Feher 2002), it is shown that due to the multi-tasking, tempo, integration demands, and short-term memory requirements, threat analysis is cognitively challenging, even under normal conditions. It requires the mental integration and fusion of data from many sources. This integration/fusion requires a high level of expertise, includ-ing knowledge of the types of threats, the own force's mission, own and adversary doctrines, and assessment heuristics built from experience. The cognitive overload in a time-constrained environment puts the operators under a great amount of stress.

11.7.2 Contextual Factors

Threat analysis, like any other task, cannot be decoupled from the context in which it occurs. The context of operations greatly impacts the effective conduct of threat analysis through a set of fundamental factors that characterize any (tactical) military operation: the nature of the threat, the operational environment, uncertainty, and time. Moreover, these factors are inter-related and impact each other in many ways. For instance, the operational environment highly influences the nature of the threat, while both the former and the latter impact uncertainty and time.

11.7.2.1 Uncertainty

Uncertainty in the representation of the situation is mainly due to sensor limitations, the limited reliability of intelligence information, and the limited accuracy of inferences (by humans or systems) used to derive knowledge from this data. The individuals performing threat analysis have to deal with the unpredictability of (adversary) human behavior and the imperfection of the information sources on which they rely to observe the environment (including the adversary).

11.7.2.2 Time

Time is another key factor in threat analysis for three main reasons. Firstly, the information gathered and compiled during the picture compilation process, as well as the knowledge derived by the threat analysis process, remain valid for only a finite period of time. Secondly, time is a resource, both for own forces and the adversary, which is consumed as information is being gathered and processed. Thirdly, in an adversarial context, the high tempo of operations often limits the time available to understand the impact of the events on the situation at hand and to react to them. The high tempo imposes a requirement on responsiveness, i.e., critical agents (potentially red or harmful) must be assessed as early as possible so as to provide more reaction time to human decision makers. The responsiveness requirement involves reducing the decision process timeline while maintaining or increasing response quality.

Furthermore, time is also consumed by coordination requirements, including the requirement to liaise with a higher-echelon staff that may or may not possess the same appreciation of the situation, which is being driven by the dynamic actions of own force and the opposing force.

11.7.2.3 Nature of the Threat

Threats can be categorized along several dimensions such as predictability of the behavior, susceptibility to coercion, and symmetry. They can also be distinguished using the single/multiple dichotomy. The problem of multiple coordinated threats is addressed in Section 11.8.4.

11.7.2.3.1 Predictability of the Behavior

One possible classification is based on the predictability of the behavior. Deterministic threats are those which, once detected, can have their behavior determined without uncertainty on their intent, capability, future course of action, or trajectory (e.g., projectiles). Adaptive threats have the capability to adapt their behavior, making their evolution difficult to predict. In simple cases, this consists in the threat altering its trajectory, such as a cruise missile that adapts to the landscape features or follows waypoints. In more complex cases, threats can adopt various elaborate tactics. This is particularly obvious with manned or man-controlled threats, such as aircraft or seacraft, but is not exclusive to them. Unmanned vehicles (aerial, surface, or submarine) equipped with advanced technology can also exhibit sophisticated adaptive behaviors, involving a dynamic generation of goals and plans in reaction to changes in the environment. The capability of a threat to adapt its behavior is an important factor for the assessment of the threat opportunity, and even more so for the assessment of its intent. This factor increases the difficulty for these assessments.

11.7.2.3.2 Susceptibility to Coercion

Coercible threats, as opposed to unyielding threats, are threats which are equipped to potentially respond to deterrence. These are, in principle, manned or man-controlled threats which can respond to warnings, requests, and other deterrence actions, i.e., they have the capability to communicate, to reason, and to act. The capability of a threat to respond to deterrence is a factor in threat analysis as it provides options for assessing the intent of this threat through the observation of its reactions to its own force actions.

11.7.2.3.3   Symmetry

Symmetric versus asymmetric threats is another categorization that is very relevant to today's reality of conflicts. By asymmetric threats, we mean threats that adopt unconventional warfare strategies and tactics (e.g., conduct attacks using recognized civilian vehicles, such as boats, light aircraft, or cars). A wide disparity in military power between the parties leads the opponents to adopt strategies and tactics of unconventional warfare, the weaker combatants attempting to use strategies to offset deficiencies in quantity or quality (Stepanova 2008).

The potential presence of asymmetric threats imposes a nonuniform environment that cannot be pictured as a confrontation between friendly (blue) agents and enemy/undesirable (red) agents. Threat analysis, particularly intent and capability assessments, becomes even more challenging, as it is extremely difficult to anticipate the moves of an opponent who is no longer a crisp, well-defined entity and is determined to use unconventional means. Another challenge is about the sparse and ambiguous indicators of potential or actualized threat activity being buried in massive background data.

11.7.2.4   Operational Environment

A good example of the effects of changes in the environment on C2 operations and threat analysis in particular is the recent shift of emphasis toward congested environments such as urban and littoral areas. Contrary to the traditional maneuver space, urban and littoral areas are characterized by significant congestion due to the existence of nonmilitary activity. This activity complicates the process of picture compilation, and thereby necessitates increased efforts on the part of the analysts to generate and maintain a complete and clean operating picture.

In addition to the high number of background objects, modern warfare spaces impose nonuniform environments where blue and red, as well as neutral (white) agents are interspersed and overlapping, presenting a highly complex challenge with respect to discerning one type of agents from another.

The shift from open battlespaces to congested areas also increases the exposure of the forces to an adversary provided with the terrain advantage. Such environments are also very conducive to attacks from asymmetric threats. For example, modern navies face asymmetric threats such as suicide attacks from explosive-laden small boats, small and medium caliber weapons on small boats (individually or in swarms of many boats), low and slow flyers (civilian aircraft), and a wide range of underwater mines or improvised explosive devices (IEDs). While the ships are alongside, the threat may even be initiated by a dockside terrorist or a small boat. Increased traffic within the littoral environment can make discerning these threats from other traffic exceptionally complicated. These types of threats can also be more difficult to detect with sensors due to their reduced signatures.

11.8   THREAT ANALYSIS IN DISTRIBUTED ENVIRONMENTS

In distributed environments, entities, both red and blue agents, are physically dispersed over a wide geographic area. Own and friendly units operate conjointly to achieve mission objectives as a task force or task group. This configuration involves distributed teams on air, surface, and subsurface units cooperatively interacting to perform C2 activities. It also means that a global task must be decomposed into sub-components and communication channels and coordination mechanisms established so that these subcomponents can work together effectively, synergistically, and harmoniously. In such a configuration, information is shared and the threat analysis task is conducted collaboratively, in a distributed manner. The capability of the system as a whole becomes much greater than the sum of its subcomponents.

The geographical dispersal in a force operation offsets the vulnerability of individual units and improves the overall survivability of the force; however, distribution introduces additional C2 challenges. In the following, the problems of coordination inherent to distributed forces, as well as the cost and advantages of distribution are discussed. In a threat analysis context in particular, the complexity augments both from the blue and the red perspective with the multiplicity of agents.

11.8.1 CENTRALIZED AND DECENTRALIZED CONTROL

In force operations, C2 may be centralized or decentralized. This refers to the level of involvement, control, and responsibility exercised by the higher echelons and the subordinates during the conduct of operations. In decentralized C2, it is conceivable that the output of the C2 functions (picture compilation, threat analysis, engageability assessment, and combat power management) emerges from distributed cooperative interactions among the units rather than being directly consolidated by a central decision maker. This means that those units must develop shared situation awareness and coordinate their actions.

The concept of a decentralized C2 approach involves transferring the coordination function from the small hub of key decision makers to a larger group. In this situation, distributed units independently decide upon required actions based upon a shared tactical picture and common doctrine which sets the boundaries on approved behavior and in turn provides a coordination mechanism. A key to the decentralized approach is achieving *shared situation awareness* through the creation of a conflict-free force-level picture on all units and the development of a coherent understanding across the force by sharing information.

As part of the C2 process in a distributed environment, threat analysis can be carried out in a centralized or a decentralized manner. Centralized threat analysis implies that threats to the task force are identified, assessed, and prioritized by a central authority that uses the threat lists of individual units to derive a force-level threat evaluation. This includes the consolidation of the intent, capability, and opportunity assessments of each threat to the force, and the prioritization of all threats in terms of their relative threat ranking in order to generate a consistent force-level threat list.

Consistency in decentralized threat analysis is accomplished through a consolidation process of all the unit-level evaluations through a series of collaboration, information sharing, and communication mechanisms. Information is passed between units whereby the units converge to a conflict-free evaluation for the entire force. This is in stark contrast to the centralized approach whereby the force-level threat list is constructed by the central decision makers based on information from the individual units.

11.8.2 Advantages of Distribution

Distributed threat analysis inherently offers the following advantages of distributed systems:

- *Functional separation:* Distributed threat analysis spatially distributes entities that perform different tasks based on their capability and purpose. This function specialization simplifies the design of the system, as the latter is split into entities, each of which implementing part of the global functionality and communicating with the other entities.
- *Information superiority:* The main advantage of a distributed system is its ability to allow the sharing of information and resources. Information and knowledge provided by other sources and their fusion into a common picture enhances the quality of the assessment and supports informed decision making.
- *Enhanced real-time response:* Increased responsiveness is one of the major requirements of threat analysis. This can be achieved through distribution by deploying observers and processors close to the threat. In a networked environment, this has the potential of improving the flow of real-time information directly to decision makers, providing means of assessing rapidly changing situations and making informed decisions.
- *Robustness and resilience:* Distributed threat analysis has a partial-failure property since even if some blue agents fail, others can still achieve the task (at least partly). Such failure would only degrade, not disable, the whole evaluation outcome. If the blue multi-agent system has self-organization capabilities, it can also dynamically re-organize the way in which the individual agents are deployed. This feature makes the system highly tolerant to the failure and bias of individual agents.

11.8.3 Operational Challenges

The aforementioned advantages require that the components of the system performing threat analysis (including software and hardware agents) be able to exchange information clearly and in a timely manner. The lack of the following requirements can sometimes be an impediment to effective communication in a distributed context:

- *Interoperability:* This is the ability of two or more agents, systems or components to exchange information and to use the information that has been exchanged. Distributed threat analysis can encompass different autonomous, heterogeneous, distributed computational entities that must be able to communicate and cooperate among themselves despite differences in language, context, format, or content.
- *Connectivity:* Establishment of communications can be troublesome by itself. Provision of remote connectivity between the nodes in distributed

threat analysis is a major technical challenge which cannot be understated. Maintaining a communication channel is not guaranteed and, when it is, its quality can be degraded due to multiple environmental factors. Communications can also be hampered in an attempt by different units to use certain communication frequencies while remaining covert to minimize the detection, localization, and recognition by the opposing forces through the electromagnetic emissions (Athans 1987). Kopp (2009) listed security of transmission, robustness of transmission, transmission capacity, message, and signal routing and signal format and communications protocol compatibility as the main challenges of communication media in the military domain, although most of them apply also to nonmilitary domains.

- *Security:* Threat analysis represents a specific domain of interest that highly correlates with information system security. Although the use of multiple distributed sources of information can improve situational awareness, it can make the system more vulnerable to unauthorized access, use, disclosure, disruption, modification, or destruction.

Additional communication problems may arise in combined operations (Irandoust and Benaskeur [in press]), where the force units belong to different allied nations. Communication processes, technologies, codes, and procedures may be very different from one contingent to another. Moreover, the participating units may be reluctant to share sensitive information.

11.8.4 Analytical Challenges

Threat analysis poses several challenges relatively to the analysis of the situation by blue agents. These may be relative to the multiplicity of threats to analyze, or the change of perspective required by collaborative threat analysis (multiplicity of own force units). The following are some examples:

- *Multiplication of reference points:* When operating as a force, one should not only consider the own unit/platform as a potential target of the threat, but also the other units/platforms that are part of the force. Impact assessment must therefore be performed with regard to several reference points. This situation analysis issue entails a response planning problem, which is at the heart of the self-defense versus force-defense dilemma. It is quite conceivable that the highest priority threat from the unit's perspective does not equate to the highest priority threat for the force. As such, conflicts may arise with respect to applying defensive measures in response to the threat.
- *Recognition* of coordinated plans: In a distributed environment, a blue or defending force may have to deal with single or coordinated threats. Obviously, a group of threats acting in coordination is harder to comprehend in terms of its tactical capability. Moreover, it is not sufficient in this

case to determine the intent, capability, and possibly the plan of the adversary. One must also comprehend the spatial configuration of the different units (i.e., which unit is operating in which zone) and the temporal order of the actions carried out by each unit. One has to integrate the actions of different agents into a global plan.

- *Team recognition:* The identification of the structure (members, possibly subteams) and roles in the adversarial team is a difficult issue. A functional analysis of the different entities needs to be conducted to establish their respective roles in a coordinated action.
- *Spatial reasoning:* One must reason upon the operation of blue and red forces within a larger spatial environment.
- *Collaborative plan recognition:* In this configuration, different pieces of information are created and maintained by different agents. This information could be stored, routed through the network to be fused, analyzed, and used by other agents, which may or may not be aware of the existence of the agent(s) generating the information. Analysts must make sense out of this large amount of raw data that has been taken out of its context of observation.

11.8.5 COLLABORATION CHALLENGES

Remote collaboration is another challenging area for force-level threat analysis. In force operations, data and message sharing across several units is completed via networks. In turn, this information exchange is used to establish a common understanding of the task at hand. Yet, the inherent richness that accompanies face-to-face collaboration is not supported. As such, it is harder to effectively perform contentious discussions. A coalition context will further introduce miscommunications that can affect force-level threat analysis at different degrees (Irandoust and Benaskeur [in press]).

Furthermore, remote collaboration entails an additional coordination overhead. Within a dispersed force, there is a need for both inter-unit and intra-unit coordination. The task also becomes more complicated since there are numerous system interactions which may be dependent on the current disposition of the forces. Concurrency, whereby multiple units may be simultaneously performing similar and complementary activities, can result in conflicting conclusions. Moreover, delays in communication caused by limited bandwidth, interferences, and breakdowns can hamper force-level threat analysis.

Overall, the dependence on electronic communications, geographical distances, multiplication of parameters, and the impossibility of having direct face-to-face interactions on a regular basis are all obstacles to cohesion and effective collaboration in force threat analysis.

11.8.6 THREAT ANALYSIS AND NETWORK-CENTRIC OPERATIONS

Teamwork and collaborative decision making are critical elements of the military's vision of network–centric operations (Alberts et al. 1999). The main principle underlying the concept of a networked force is to allow individuals and/or groups the ability to leverage information both locally and globally to reach effective decisions quickly. Access to different perspectives and the widespread and timely collection and distribution of information around the battlefield will, it is anticipated, allow the more accurate and timely application of military force necessary to react to the ongoing situation. Advances in network technologies are augmenting the connectivity of military units, and automated sensors, and intelligence feeds provide an increased access to previously unavailable information. However, the electronic linkage of multiple units does not necessarily bring about automatic improvement in situation understanding, including threat analysis, collaboration, and the synchronization of defensive actions. While technology can offer C2 organizations a great information-processing capability, the need to consider and reconcile the variety and complexity of interpretations of information outputs generated by humans and computer systems remains. It is indeed incorrect to automatically assume that fusing information into a common operating picture will result in uniform interpretation of the information by the various users.

This is why great emphasis is put by the promoters of the network-centric approach on the social dimension of distributed operations. According to the conceptual framework of network-centric operations (Garstka and Alberts 2004), raw information must be transformed into actionable knowledge through collaborative sensemaking among the stakeholders. However, common understanding of a given situation requires that all participants use a common reference frame, i.e., use the same models, physical or mental, for interpreting the situation elements and "creating mutually intelligible representations" (Shum and Selvin 2000), which is the essence of collaborative sensemaking. Yet,

> there are not only gaps in the languages, frames of reference, and belief systems that people in the different communities of practice have, but gaps between their respective sensemaking efforts—their concepts in the representational situation are different. In many cases, different communities have mutually unintelligible sensemaking efforts, leading to mutually unintelligible representational effort (Shum and Selvin 2000).

Furthermore, it has been observed that a likely cause of failure for overall mission success is that the abilities of humans to access, filter, and understand information, to share it between groups, and to concur on their assessment of the situation are clearly limited, especially under stress and time-pressure (Scott et al. 2006).

Finally, Kolenda (2003) argues that shared situational awareness does not inevitably lead to "shared appreciation on how to act on the information" as different people, based on their experience, education, culture, and personalities will assess threat/risk and how to best "maximize the effectiveness of themselves and their organizations" differently. Simply providing people with access to the same information does not necessarily create a common understanding. The issue of how "common intent" can actually be promoted among network players, often from diverse backgrounds and cultures (both national and organizational) represents a major challenge for future operations.

11.9 DISCUSSION

In the preceding, the problem of threat analysis was addressed from different angles and at different levels of complexity. The use of primary concepts and defining features such as negativity, intentionality, potential, imminence, and relativity to a reference point allowed us to provide a framework in which the concept of threat is elucidated and distinguished from other goal conflict situations.

Threat analysis is a very challenging cognitive task that can involve different layers of reasoning when time allows it. Interference management, goal recognition, and plan recognition were extensively discussed, showing the complexity of the inferences which have to be made by an observing agent performing threat analysis. This provided a theoretical basis as the question of the automation of threat analysis was investigated.

By illustrating the problem in a military context, it was shown that threat analysis can be further complicated through contextual factors that characterize the warfare environment. These problems, described from a single unit perspective, remain valid at the force level, where new challenges are introduced. Collaborative threat analysis, while providing information superiority, was shown to impact situation analysis by multiplying the operational parameters and creating coordination overhead. Finally, distributed multi-threat scenarios were shown to significantly complicate the determination of intent, capability, opportunity, and the higher-level plan of adversary elements.

REFERENCES

Alberts, D.S., J.J. Garstka, and F.P. Stein. 1999. *Network Centric Warfare: Developing and Leveraging Information Superiority*. CCRP Publication Series, Department of Defense C4ISR Cooperative Research Program (CCRP), Washington, DC, www.dodccrp.org

Allen, J.F. 1983. Recognizing intentions from natural language utterances. In *Computational Models of Discourse*, M. Brady and R. Berwick eds. MIT Press, Cambridge, MA.

Athans, M. 1987. Command and control (C2) theory: A challenge to control science, *IEEE Transactions on Automatic Control*, AC-32(4), 286–293.

Avrahami-Zilberbrand, D. and G.A. Kaminka. 2005. Fast and complete symbolic plan recognition. *Proceedings of IJCAI 2005*, Edinburgh, U.K.

Avrahami-Zilberbrand, D. and G.A. Kaminka. 2007. Incorporating observer biases in keyhole plan recognition (efficiently!). *Proceedings of AAAI 2007*, Vancouver, British Columbia, Canada, pp. 944–949.

Benaskeur, A.R., A.M. Khamis, and H. Irandoust. 2010. Cooperation in distributed surveillance. *Autonomous and Intelligent Systems–First International Conference*, AIS 2010, Povoa de Varzim, Portugal, *Proceedings*, pp. 1–6, *IEEE*, 2010.

Bratman, M.E. 1987. *Intention, Plans, and Practical Reason*. Harvard University Press, Cambridge, MA.

Bruce, B. 1981. Plan and social action. In *Theoretical Issues in Reading Comprehension*, R. Spiro, B.C. Bruce, and W.F. Brewer (eds.). Lawrence Erlbaum, Hillsdale, NJ.

Carberry, S. 2001. Techniques for plan recognition. *User Modelling and User-Adapted Interaction*, 11(1–2), 31–48.

Castelfranchi, C. 1998. Modelling social action for AI agents. *Artificial Intelligence*, 103, 157–182.

Charniak, E. and R. Goldman. 1991. A probabilistic model of plan recognition. *Proceedings of AAAI'91*, Anaheim, CA.

Chen, G., D. Shen, C. Kwan, J. Cruz, M. Kruger, and E. Blasch. 2007. Game theoretic approach to threat prediction and situation awareness. *Journal of Advances in Information Fusion (JAIF)*, 2(1), 35–48.

Cohen, P.R., C.R. Perrault, and J.F. Allen. 1981. Beyond question answering. In *Strategies for Natural Language Processing*, W. Lehnert and M. Ringle (eds.), Bold, Beranek and Newman, Inc. Cambridge MA, pp. 245–274.

Elsaesser, C. and F.J. Stech. 2007. Detecting deception. In *Adversarial Reasoning: Computational Approaches to Reading the Opponent's Mind*, A. Kott and W.M. McEneaney (eds.). Chapman & Hall/CRC, Boca Raton, FL, pp. 111–124.

Garstka, J.J. and D.S. Alberts. 2004. Network centric operations conceptual framework—Version 2.0. Report prepared for the Office of the Secretary of Defense, Office of Force Transformation. Evidence Based Research, Vienna, VA.

Geib, C. 2009. Delaying commitment in plan recognition using combinatory categorial grammars. *Proceedings of IJCAI 2009*, Pasadena, CA, pp. 1702–1707.

Geib, C. and R. Goldman. 2003. Recognizing plan/goal abandonment. *Proceedings of IJCAI 2003*, Acapulco, Mexico, pp. 1515–1517.

Geib, C. and R. Goldman. 2005. Partial observability and probabilistic plan/goal recognition. *Proceedings of IJCAI 2005, Workshop on Modeling Others from Observations (MOO)*, Edinburgh, U.K.

Geib, C. and R. Goldman. 2009. A probabilistic plan recognition algorithm based on plan tree grammars. *Artificial Intelligence*, 173(11), 1101–1132.

Geib, C., J. Maraist, and R. Goldman. 2008. A new probabilistic plan recognition algorithm based on string rewriting. *Proceedings of ICAPS 2008*, Sydney, New South Wales, Australia, pp. 81–89.

Glinton, R., S. Owens, J. Giampapa, K. Sycara, M. Lewis, and C. Grindle. 2005. Intent inference using a potential field model of environmental influences. *Proceedings of Fusion 2005*, Philadelphia, PA.

Goldman, R., C. Geib, and C. Miller. 1999. A new model of plan recognition. *Proceedings of the Fifteenth Conference on Uncertainty in Artificial Intelligence*, Stockholm, Sweden.

Irandoust, H. and A. Benaskeur. (in press). *Political, Social and Command & Control Challenges in Coalitions—A Handbook*. Canadian Defence Academy Press, Kingston, Ontario, Canada.

Irandoust, H., A. Benaskeur, F. Kabanza, and P. Bellefeuille. 2010. A mixed-initiative advisory system for threat evaluation. *Proceedings of ICCRTS XV*, June 2010, Santa Monica, CA.

Johansson, R. and G. Falkman. 2008. A Bayesian network approach to threat evaluation with application to an air defense scenario. *Proceedings of Fusion 2008*, Cologne, Germany, pp. 1–7.

Jones, R. 2010. TacAir-Soar: Intelligent, autonomous agents for the tactical air control domain. www.soartech.com (accessed February 15, 2010).

Jones, M.R., O.M. Jones, J.E. Laird et al. 1998. Automated intelligent pilots for combat flight simulation. *AI Magazine*, 20, 27–41.

Kaminka, G., D. Pynadath, and M. Tambe. 2002. Monitoring teams by overhearing: A multi-agent plan-recognition approach. *Journal of Artificial Intelligence Research*, 17(1), 83–135.

Kautz, H.A. and J.F. Allen. 1986. Generalized plan recognition. *Proceedings of AAAI 1986*, Philadelphia, PA, pp. 32–37.

Kolenda, C.D. 2003. Transforming how we fight—A conceptual approach. *Naval War College Review*, LVI(2), 100–121.

Kopp, C. 2008. NCW101: An introduction to network centric warfare, AirPower Australia, Melbourne, Vic Australia.

Laird, J.E., A. Newell, and P.S. Rosenbloom. 1991. Soar: An architecture for general intelligence. *Artificial Intelligence*, 47, 289–325.

Liao, L., D. Patterson, D. Fox, and H. Kautz. 2007. Learning and inferring transportation routines. *Artificial Intelligence*, 171(5–6), 311–331.

Liebhaber, M. and B. Feher. 2002. Air threat assessment: Research, model, and display guidelines. *Proceedings of ICCRTS*, Quebec City, Quebec, Canada.

Little, E.G. and G.L. Rogova. 2006. An ontological analysis of threat and vulnerability. *Proceedings of Fusion 2006*, Florence, Italy, pp. 1–8.

Paradis, S., A. Benaskeur, M. Oxenham, and P. Cutler. 2005. Threat evaluation and weapons allocation in network-centric warfare. *Proceedings of Fusion 2005*, Philadelphia, PA.

Pynadath, D.V. and M.P. Wellman. 1995. Accounting for context in plan recognition, with application to traffic monitoring. *Proceedings of the Eleventh Conference on Uncertainty in Artificial Intelligence*, Montreal, Quebec, Canada, pp. 472–481.

Roy, J. 2009. A view on threat analysis concepts, Technical report, DRDC – Valcartier-SL-2009-384, July 6, Defence R&D Canada, Valcartier, Valcartier Quebec, Canada.

Roy, J., S. Paradis, and M. Allouche. 2002. Threat evaluation for impact assessment in situation analysis systems. *SPIE Proceedings: Vol. 4729, Signal Processing, Sensor Fusion, and Target Recognition XI*, Orlando, FL.

Santos, E. and Q. Zhao. 2007. Adversarial models for opponent intent inferencing. In *Adversarial Reasoning: Computational Approaches to Reading the Opponent's Mind*, A. Kott and W. McEneaney (eds.). Chapman & Hall/CRC, Boca Raton, FL, pp. 1–22.

Schmidt, C.F. 1976. Understanding human action: Recognizing the plans and motives of other persons. In *Cognition and Social Behavior*, J. Carroll and J. Payne (eds.). Erlbaum Press, Hillsdale, NJ.

Schmidt, C.F., N.S. Sridharan, and J.L. Goodson. 1978. The plan recognition problem: An intersection of psychology and artificial intelligence. *Artificial Intelligence*, 11, 45–83.

Scott, S.D., M.L. Cummings, D.A. Graeber, W.T. Nelson, and R.S. Bolia. 2006. Collaboration technology in military team operations: Lessons learned from the corporate domain. *Proceedings of CCRTS*, June 2006, San Diego, CA.

Shum, A.B. and A.M. Selvin. 2000. Structuring discourse for collective interpretation. In *Distributed Collective Practices: Conference on Collective Cognition and Memory Practices*, September 2000, Paris, France.

Southwick, R.W. 1991. Explaining reasoning: An overview of explanation in knowledge-based systems. *The Knowledge Engineering Review*, 6(1), 1–19.

Steinberg, A.N. 2005. An approach to threat assessment. *Proceedings of Fusion 2005*, Philadelphia, PA, 95–108.

Steinberg, A.N. 2006. Stimulative intelligence. *Proceedings of the MSS National Symposium on Sensor and Data Fusion*, McLean, VA.

Steinberg, A.N. 2007. Predictive modeling of interacting agents. *Proceedings of Fusion 2007*, Quebec City, Quebec, Canada.

Steinberg, A.N., C.L. Bowman, and F.E. White. 1998. Revision to the JDL data fusion model. *Joint NATO/IRIS Conference*, October 1998, Quebec City, Quebec, Canada.

Stepanova, E. 2008. Terrorism in asymmetrical conflict: Ideological and structural aspects. (Technical Report SPRI Research Reports 23) Stockholm International Peace Research Institute (SIPRI), Solna, Sweden.

Sukthankar, G. and K. Sycara. 2006. Robust recognition of physical team behaviors using spatio-temporal models. *Proceedings of the Fifth International Joint Conference on Autonomous Agents and Multiagent Systems*, Hakodate, Japan, pp. 638–645.

Vilain, M. 1991. Deduction as parsing. *Proceedings of AAAI*, Anaheim, CA, pp. 464–470.

White, F.E. 1988. A model for data fusion. *Proceedings of the First National Symposium on Sensor Fusion*, Vol. 2. GACIAC, IIT Research Institute, Chicago, IL, pp. 143–158.

Whitehair, R.C. February 1996. A framework for the analysis of sophisticated control. PhD dissertation, University of Massachusetts, Boston, MA, CMPSCI Technical Report 95.

12 Ontological Structures for Higher Levels of Distributed Fusion

Mieczyslaw M. Kokar, Brian E. Ulicny, and Jakub J. Moskal

CONTENTS

12.1 Introduction 327
12.2 Ontologies... 328
12.3 Querying the Net 330
12.4 Interoperability and Inference 334
12.5 Inferring Relevant Repositories... 336
12.6 Inferring Relevant Things... 337
12.7 Restricting Queries to Geographical Regions 338
12.8 Discussion: Inference of Suspicious Activities.................................. 341
12.9 Conclusion 343
References...344

12.1 INTRODUCTION

Today most, if not all, online information is either stored in various data stores (databases) or available in streaming form from either sensors or other sequential information providers. Both types of information sources can be termed *information producers*. On the other hand, this information is utilized by the various *information consumers* who require the information to achieve their informational goals or to support their decision processes. The producers and the consumers are all interconnected, resulting in what is called a *net-centric environment*, often referred to as Net-centric Enterprise Architecture in the business domain (cf. Network Centric Operations Industry Consortium n.d.) or as Net-centric operations in the military domain (cf. Cebrowski and Garstka 1998). Some of them play both roles at the same time. Information consumers have on-demand access to information producers. The ultimate goal is to have a user-defined operational picture on each consumer's screen.

A representation of the information producers and consumers nodes is shown in Figure 12.1. Here both information producers and consumers are viewed as services. Since our intent is to associate semantic descriptions with each such service, we call

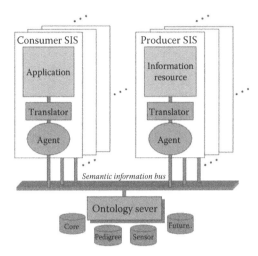

FIGURE 12.1   Information producers and consumers—Semantic Information Services (SIS).

those Semantic Information Services (SIS). All of the services are described using terms from an ontology. Here we use the term "ontology" as defined by Gruber (2009): "an ontology defines a set of representational primitives with which to model a domain of knowledge or discourse." Some representative ontologies are shown at the bottom of the figure.

In this chapter, we present some results of an effort to achieve the goals of net-centric operations, displaying necessary (and only necessary) information on the consumer's display when they need it and in a form that they can understand and act upon. In this approach, the consumer can query the network for information relevant to the consumer's current need. To explain how this goal can be achieved, we show a step-by-step process that starts with issuing a query to viewing a reply to the query. Each of the steps involves some ontological reasoning. To illustrate the approach, we provide fragments of ontologies that are necessary to derive the inference results.

This chapter is organized as follows. First, we provide a short explanation of ontologies. Then we show the particular activities of the whole process:

1. Annotation of information sources
2. Query formulation
3. Inferring the relevance of the particular information sources
4. Inferring the relevance of information based on the location of the objects being queried about
5. Inferring whether the particular objects are relevant to the queried situation

12.2   ONTOLOGIES

One of the basic principles of the approach to information integration described in this chapter is the representation of all the information in a common vocabulary. For the approach to be flexible, the vocabulary needs to be extensible and have formal,

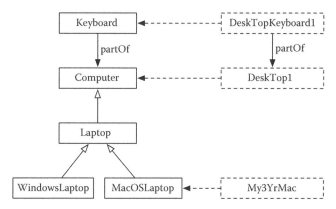

FIGURE 12.2   A simple ontology for the computer domain.

computer-processable semantics. This kind of vocabulary is known as *ontology*. As used in the knowledge representation domain, the term "ontology" stands for an explicit, formal, machine-readable semantic model that defines the classes, instances of the classes, interclass relations, and data properties relevant to a problem domain (Gruber 2009).

To introduce the basic concepts of ontology, we use a simple example in which we demonstrate how to represent those basic ideas for the computer domain. A very simple ontology for this domain is shown graphically in Figure 12.2. In this ontology, computers are represented as a class (a rectangle-labeled *Computer* in Figure 12.2). A specific computer, the one I bought 3 years ago (which we call here *My3YrMac*), is an instance of the class *Computer*. It is represented as dashed line rectangle. However, the computer I bought was a laptop. *Laptop* can be another class, which is subclass of *Computer*. My laptop thus is an *instance of* the class *Laptop*. This fact is shown by a dashed line connecting *My3YrMac* to *MacOSLaptop*. As we can see, *Laptop* is subclassified further so that *MacOSLaptop* and *WindowsLaptop* are subclasses of *Laptop*. Classes are interrelated. For instance, another class, *Keyboard*, is related by the *partOf* relation to *Computer*. The *partOf* relation may be used to capture the fact that a specific instance of *Keyboard* is part of a specific computer. In this example, *DeskTopKeyboard1* is part of *Desktop1*. Instances of particular classes can have various *data properties*. For example, every laptop has the data property of weight. Thus *weight* may be a data property of the class *Computer* (as well as of any class of physical objects). It can then be used to state the fact that the weight of my laptop is 2.8 lb.

The choice of the classes, relations and properties is obviously domain-dependent—if one wants to describe things in the domain of computers, one does not introduce classes like *Politician* or *Horse*. But the ontologist (the person who develops an ontology) has flexibility in terms of both the selection of terminology and the selection of concepts. For instance, the ontologist might want to classify laptops by the manufacturer, or by the price range. It all is a matter of what is more useful in the application that uses such an ontology.

Since the intent of having formal ontologies is to be able to process them by computers, ontologies need to be represented in a formal language. There are various approaches to representing formal ontologies including OWL (W3C 2009), conceptual graphs (Sowa 1992), topic maps (Pepper and Moore 2001), KIF (*Knowledge Interchange Format* 1992), and others. Our choice of language is OWL, a W3C standard. However, whenever OWL is not expressive enough to capture the meaning of a concept or a relationship, we will supplement it with rules and possibly some procedures (e.g., calls to a procedural language like Java).

Ontologies do not need to be developed from scratch. Recently, significant efforts went into the development of "data models" and "markup languages." Data models are represented either in a database schema description language, or in the Universal Modeling Language (UML). Markup languages are essentially vocabularies represented in XML. In some cases, data models are also represented in XML. For instance, the widely known data model JC3IEDM (the Joint Consultation, Command and Control Information Exchange Data Model 3.1) (2005), developed by the Multilateral Interoperability Programme (MIP), is a long-standing, NATO-supported model intended to foster international interoperability of command and control information systems through the development of a standard data model and exchange mechanism. The data model was first released in the mid-1990s as the Generic Hub (GH) Data Model. It captures information about 271 entities, 372 relationships between entities, 753 entity attributes, and over 10,000 value codes. Representations of this model in both ERwin and XML exist. In the past, we have developed an automatic translator for JC3IEDM (Matheus and Ulicny 2007), which takes an XML representation of JC3IEDM and converts it to OWL.

12.3   QUERYING THE NET

Now we will show how we make use of ontologies for the purpose of querying for information that is of interest to the user, finding out which of the information sources are relevant to the query, retrieving the information and integrating it into a homogeneous representation. We first start with the querying.

While it would be highly desirable to be able to express queries in natural language, because this would be the easiest way of formulating queries by humans, this option is both not quite achievable with the currently available technology and also not particularly advisable since it may cause all kinds of misinterpretation errors. In particular, it is very likely that the natural language expression presented by the user to the computer system may be either vague or imprecise. The computer's interpretation of such a query may be quite different than the user's intent. So from this point of view, providing a more structured language in which the user can formulate queries and interpret the results returned by the computer may be a better option.

Current querying technology provides a number of options for formulating queries. For instance, everybody uses search engines, like Google, to find information on the web. In doing so, the user typically provides a number of *keywords*, which are then used by the search engine to find documents that contain at least some of the provided keywords. Note, however, that the search engine returns a ranked list of tens of thousands of documents, but not the answers to a query. The users still need to open some of the documents found and figure out the answer to a query on their own.

Flexible querying, which is of interest here, can be viewed as a case of the activity known in the literature as "question answering." An overview and a comparison of a number of approaches and systems for question answering can be found in Ulicny et al. (2010). Since in this chapter we are interested in the methods that may be part of a number of different systems, rather than in the systems themselves, we provide just a short description of those approaches.

The vast majority of systems that store and retrieve data are based on representing the data in structured database formats, in which the structure of the tables, and the significance of each column, is specified in advance. Structured Query Language (SQL) commands and queries are then used to insert and retrieve data elements in tabular form. While it has become increasingly sophisticated over the years, SQL was initially envisioned as a natural language interface to databases. In web-enabled database applications, the SQL queries and commands are mostly hidden from the user and are dynamically constructed and executed when a user fills out and submits a form on a web page.

Wolfram Alpha represents a more sophisticated version of structured data query-ing. Wolfram Research is the producer of the major symbolic mathematical compu-tation engine Mathematica. The Wolfram Alpha engine sits on top of quantitative data and other reference works that have been "curated" from authoritative sources (Talbot 2009). When a user queries Wolfram Alpha, the engine attempts to interpret the query's intent so as to produce an output format that is the most likely to satisfy that query intention (sometimes providing both a geospatial overlay and timeline as output), without requiring the user to formulate the underlying Mathematica query him- or herself. While the curation process insures the trust relationship between the consumer and the producer, the recall of the retrieval process is still far from satisfactory.

Information producers may provide metadata for their data. For instance, a document might have metadata about the date when it was created, the author of the document, the location of the event described in the document, and such. This kind of metadata may be used in answering questions. MetaCarta's technology is an example of system that uses this approach. MetaCarta (2010) processes docu-ments in order to identify any expressions indicating locations (e.g., location name, postal code, telephone area codes), and marks up a representation of the document with geo-coordinates corresponding to those locations. The system can then be que-ried for documents that contain some combination of keywords and that have some geo-coordinates within a specified bounding box or radius. While this technology is pointing in the direction we are discussing in this chapter, it still does not go far enough to provide the flexibility of querying that is needed by today's users.

Logic-based systems, such as Powerset, recently acquired by Microsoft and incorporated into its Bing search engine, parse texts into a logical representation, using sophisticated natural language processing. After analyzing free text and con-verting it into a logic-based representation, questions can be formulated as queries over these logical clauses and returned as answers. While this is the direction we are interested in, we are not discussing this solution here for two reasons: first, because it still includes natural language processing, and second, because these are proprietary solutions to which we do not have access.

In the following, we show how a query is represented in the query language for the Semantic Web–SPARQL. As an example, we have selected the maritime domain, although most of the terminology and content is only loosely related to the terminology used in the U.S. Navy.

We are assuming that the user (an information consumer) would like to issue a query that in natural language could be expressed as follows:
Show me all watercraft located in Region1 that may be involved in a suspicious activity.

Such a query could be expressed in SPARQL as

```
Select ?vessel
where
   {?vessel rdf:type :Watercraft.
   ?vessel :locatedIn Region1.
   ?vessel :involvedIn ?Event.
   ?Event rdf:type :SuspiciousActivity
 }
```

To be interpretable by a SPARQL query engine, the terms in this query would need to be either SPARQL's keywords or be terms of a specific ontology. While this query will be the guide for the remaining discussion within this chapter, we will address various issues with this problem specification in small incremental steps. So, we begin by discussing the simpler query:

$$\text{``Show me all watercraft.''} \tag{1}$$

The main purpose of this exercise with such a simple query is to expose the issue of determining the relevance of particular repositories to a query. In other words, assuming that all the accessible repositories are described in terms of an ontology, the issue is to identify which of the repositories contain some information that is relevant to the query.

An example of a partial view of an ontology (we call it the Query Ontology) that would be needed to represent such a query is shown in Figure 12.3.

This ontology includes eight classes organized in a hierarchy. The top-level class *Object* represents anything that might be of interest to us. *Query* is a class of all possible queries. Whenever a query is generated, it is assumed that an instance of this class would be created. For this particular example, we are assuming that the earlier mentioned query is represented by an instance called "*CurrentQuery*" (shown in the next view of this ontology—Figure 12.4). *Repository* is a class whose instances are specific repositories available on the network. The *Vehicle* class has a subclass called *Watercraft*, which in turn has a subclass called *Boat*. The main reason for having this classification is to see whether the execution of query (1) distinguishes between repositories that contain information about boats, watercraft, and vehicles in general (e.g., automobiles).

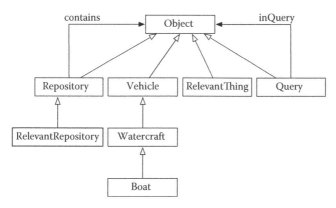

FIGURE 12.3   A Query Ontology.

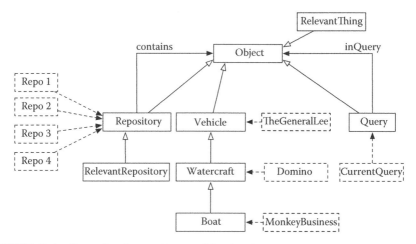

FIGURE 12.4   Query Ontology and some of the class instances.

The Query Ontology includes two relations (in OWL, relations are called *properties*). The property *inQuery* has *Query* as its domain and *Object* as its range. This implies that the relation *inQuery* will include pairs of instances—the first element being an instance of *Query* and the second an instance of *Object*. The second property is *contains*, with domain *Repository* and range *Object*. This property represents that a repository contains (a class of) objects.

Since in this chapter we use RDF and OWL for representing knowledge, we will also use the RDF notion of *triple* for representing instances of properties. A triple consists of three parts, called *predicate*, *subject*, and *object*. The predicate is the identifier of a property, the subject is the identifier of an element from the domain of the property, and the object is the identifier of element from the range of the property.

To complete the description of the Query Ontology, we need to describe the classes *RelevantThing* and *RelevantRepository*. So first of all, the intent of having the *RelevantThing* class is to capture those objects that may be relevant to a given

TABLE 12.1
Tabular Representation of the
Property *Contains*

contains	*Repo1*	*Watercraft*
contains	*Repo2*	*Domino*
contains	*Repo3*	*Boat*
contains	*Repo4*	*TheGeneralLee*

query. This class would be populated dynamically, based on a specific query. We will provide the mechanics of this dynamic construction after we discuss the instances of the ontology. Similarly, the class *RelevantRepository* is for capturing those repositories that are relevant to a given query. Again, this class is populated dynamically by the system since the designer of the ontology cannot be aware of every query the user might want to issue.

Figure 12.4 shows a view of the Query Ontology after some of the classes have been instantiated. The instances are represented as dashed line rectangles and connected to their classes by dashed line arrows annotated with an "*io*" (for instance of). As we can see from this figure, there are four repositories on the network: *Repo1*, *Repo2*, *Repo3*, and *Repo4*. Additionally, there are three instances of Vehicle. *TheGeneralLee* is an instance of the top class *Vehicle* (it's a car, but this is not inferable from this ontology). *Domino* is an instance of *Watercraft,* and *MonkeyBusiness* is an instance of *Boat.*

Although this is not visible in the graphical representation of the Query Ontology, this ontology also contains information about the properties *inQuery* and *contains.* Since query (1), represented in the ontology as *CurrentQuery*, mentions only *Watercraft*, the ontological description of the *inQuery* property also includes only one triple (predicate, subject, object):

$$\langle \text{inQuery CurrentQuery Watercraft} \rangle \tag{2}$$

We can also show the property information in tabular form. For example, Table 12.1 shows the property *contains.* Note that the repositories are described at two different levels of abstraction. *Repo1* and *Repo3* specify what classes of objects they contain (both *Watercraft* and *Boat* are classes). *Repo2* and *Repo3*, on the other hand, specify what instances they contain (*Domino* and *TheGeneralLee* are not classes, but rather instances of other classes—*Watercraft* and *Vehicle*, respectively).

12.4 INTEROPERABILITY AND INFERENCE

The two main advantages of using ontological representation are the interoperability and the ability of inferring facts that are only implicit in the representation. Interoperability is demonstrated when one network node sends information to another, and the other node "understands" the information. For instance, when a node sends information to a querying node that *Repo2* has information about *MonkeyBusiness*, which is a *Boat*, the querying node can take advantage of this information by inferring that *MonkeyBusiness* is a *Watercraft*, and thus is relevant to *CurrentQuery*. Thus interoperability is intimately related to inference. Without inference the information that *MonkeyBusiness* is a *Boat* might be useful to the querying node, but not to the specific query at hand. For instance, the query might actually ask explicitly about watercraft and boats. With the inference capability, however, the queries do not need to be that specific.

The term "inference" is often used in many different meanings. For instance, the running of an algorithm that computes the value of a function, given the values of the function's parameters, is often termed automatic inference. Note, however, that for such an inference to be possible, two conditions must be satisfied: (1) an algorithm for the function must be implemented and (2) the only query that this algorithm can execute is the query for the value of this function, given the values of the function's parameters. Obviously, one could implement a number of functions and then allow each of the algorithms to be queried by providing a function's name and the list of parameter values in a prespecified order. One can do this kind of thing in MATLAB®, or in any other function library.

Here, however, we are interested in the capability of formulating new queries expressible in a query language and deriving answers to such queries using *logical inference*. A full explanation of this kind of inference is beyond the scope of this chapter, so we only provide a simplified view of this kind of inference and then show examples of inferences throughout the rest of this chapter.

Logical inference is possible within a *formal system*, i.e., a system that includes a *formal language*, a *theory* (or axioms), and *inference rules*. Formal language is a language that has *formal syntax* and *formal semantics*. Formal syntax means rules for determining whether a given expression is in the language or not (sometimes referred to as *legal sentences* or *well-formed formulas*). Formal semantics refers to *interpretations*, which are mappings from the language to a mathematical domain (a set of individuals) and from sentences to *truth values*. Theories are then represented by *axioms*—sets of sentences in the language. Inference rules are rules that can be applied to the axioms of a theory to derive new sentences, which then become part of the theory. A formal system should be *sound*, i.e., given a consistent set of true sentences, it derives only true sentences, i.e., sentences that map to the value "true" by the interpretation function. Another desirable, but unachievable, feature of a formal system is *completeness*, i.e., the ability to infer all possible true sentences using the rules of inference.

An *inference engine* can then take a set of sentences in a formal language and apply the inference rules of the formal system to derive new sentences. The most important aspect of this process is that the inference engine is generic, i.e., it can be applied to any set of axioms expressed in the given language. Thus, referring back to the example of the calculation of the value of a function, the queries sent to the inference engine can be anything expressible in the formal language, rather than a predefined set. Thus the limit of inference is bound by the language, and not by a predefined set of functions and queries. While this discussion was abstract, we hope the reader will appreciate the value of logical inference from the rest of this chapter.

In this chapter, all of the examples are expressed in the Web Ontology Language (OWL) (W3C 2009a,b). OWL is a formal language with model theoretic semantics. A number of generic inference engines for this language exist. In our work, we use the

BaseVISor inference engine (Matheus et al. 2006). This engine is freely available for research purposes, as are some others. BaseVISor is implemented in Java. It supports the OWL 2 RL dialect of OWL. OWL 2 RL includes most of the constructs of OWL 2, but additionally, it also supports the expression of user-defined *policies* (collections of *rules*). The importance of rules stems from the fact that rules allow to express some more complicated relationships than just pure OWL can. BaseVISor is an inference engine applicable to OWL axioms and user-defined policies represented as rules. It is a forward-chaining rule engine since the rules are executed in the "forward" direction. That is, rules are applied for as long as there is new information that can be derived by rule applications. Since at the low level all the axioms are represented as triples, BaseVISor has been optimized for processing RDF- and OWL-expressed information.

12.5 INFERRING RELEVANT REPOSITORIES

As was shown in the previous discussion, each of the repositories is annotated (see Table 12.1) in terms of the *contains* property. However, it is not known which of the repositories are relevant to a given query. These facts must be inferred using an inference engine. The inference process needs to rely on the facts stored in the ontologies. Thus, first, the query needs to be represented in ontological terms, and then the inference can be carried out. In this section, we are assuming that the query representation has been done and that some initial inference on the query representation has been carried out. We assume that this inference process has populated the class RelevantThing (see Figure 12.3) with instances of the things that may be relevant to the query. The details of this inference will be presented in the next section.

In order to infer the relevant repositories, the ontology must contain some information about what it actually means for a repository to be relevant. In our example, we formulate the relevance as follows.

A repository is relevant if it contains information about relevant things.

To express this fact (axiom) in OWL, we need to use the concept of *OWL Restriction*. In short, an OWL Restriction is a class whose instances are defined based on the values of the properties for which this class is the domain. So in this example, the class in question is *RelevantRepository* (a subclass of *Repository*). The property we use here is the property of *contains*, whose range is *RelevantThing* (remember that we assume the instances of this class have already been inferred). So we define *RelevantRepository* as the class for which the property *contains* has some values in *RelevantThing*. A snippet of OWL code that captures the essence of this fact is as follows:

```
<owl:Class rdf:about = "#RelevantRepository">
  <owl:equivalentClass>
    <owl:Restriction>
      <owl:onProperty rdf:resource = "#contains"/>
      <owl:someValuesFrom rdf:resource = "#RelevantThing"/>
    </owl:Restriction>
  </owl:equivalentClass>
  <rdfs:subClassOf rdf:resource = "#Repository"/>
</owl:Class>
```

The result of running BaseVISor is
RelevantRepo found three results.

Variables (X)
    q:Repo1
    q:Repo2
    q:Repo3

This is obviously a correct result. BaseVISor was able to infer that *Repo1* was relevant simply because *Watercraft* was explicitly in the query. It was also able to infer that *Repo2* was relevant based on the fact that *Domino* is a *Watercraft*. Note that *Repo2* did not mention anything about *Watercraft* in its annotation (Table 12.1). *Repo3* was inferred to be relevant because it contained information about *Boat*, and because *Boat* is a subclass of *Watercraft*. *Repo4* was (correctly) not considered relevant to this query since *TheGeneralLee* is a *Vehicle* but not a *Watercraft*.

The earlier explanation of the BaseVISor inference is based on intuition and common sense. A more formal explanation will be provided after we discuss the inference of *RelevantThing*.

12.6   INFERRING RELEVANT THINGS

The definition of the class *RelevantThing* is a bit more involved; it includes OWL's notion of *inverse property*. An inverse property is defined in OWL as the inversion of domain and range of the property. So in this case, while the property *inQuery* has domain *Query* and range *Object*, the inverse has domain *Object* and range *Query*. The use of the inverse property is needed here because, similarly as in the case of *RelevantRepository*, we want to define the class *RelevantThing* as those objects for which the property—the inverse of *inQuery*—has the value of *CurrentQuery*.

Unfortunately, when we do just this, the reasoner only returns *Watercraft* as the relevant thing and *Repo1* as a relevant repository. Although this is a correct inference based on the restriction on the inverse property, this is not exactly what we want. We would also like to see that subclasses of *Watercraft* (in this case *Boat*) as well as all the instances of *Watercraft* (*Domino*) and of *Boat* (*MonkeyBusiness*) show up as relevant things. While some of this kind of inference could be achieved within OWL, here we use rules for this purpose. In particular, we add two rules—one for finding the instances of the classes that are already in *RelevantThing* and another for the subclasses of the classes that are in *RelevantThing*. These rules are expressed in English as shown in the following:

1. *If C is an instance of RelevantThing and C is of type Class and I is an instance of C, then I is an instance of RelevantThing.*
   The OWL code for the definition of *RelevantThing* is shown as follows:

```
<owl:Class rdf:about = "#RelevantThing">
  <owl:equivalentClass>
    <owl:Class>
      <owl:unionOf rdf:parseType = "Collection">
```

```
      <rdf:Description rdf:about = "#RelevantThing"/>
      <owl:Restriction>
      <owl:onProperty>
        <rdf:Description>
          <owl:inverseOf rdf:resource = "#inQuery"/>
        </rdf:Description>
      </owl:onProperty>
      <owl:hasValue rdf:resource = "#CurrentQuery"/>
      </owl:Restriction>
    </owl:unionOf>
  </owl:Class>
 </owl:equivalentClass>
 <rdfs:subClassOf rdf:resource = "&owl;Thing"/>
</owl:Class>
```

Note that we also used an additional OWL construct—*unionOf.* This construct is needed here so that the result of inference is complete. This construct puts together the already known instances of *RelevantThing* and those inferred.

   With this rule, the reasoner can infer that also *Domino* and *MonkeyBusiness* are relevant things and that *Repo2* is also a relevant repository. This inference still does not satisfy our expectation since it misses the fact that *Boat* is a relevant thing, too, and consequently *Repo3* is a relevant repository. To achieve this result, we need to add one more rule so that subclasses of relevant classes are included in *RelevantThing*. In English, this rule can be expressed as follows:

2. *If C is in RelevantThing and C is a class and S is a subclass of C then S is instance of RelevantThing.*
   The result of running BaseVISor on this ontology supplemented by the two rules is shown as follows:
   ThingsOfRelevantThings found five results.

Variables (X)
      owl:Nothing
      q:Watercraft
      q:Domino
      q:MonkeyBusiness
      q:Boat

This result captures our intent. One might wonder what *owl:Nothing* is doing in the answer, but the reader should not be too concerned with this result. *Owl:Nothing* is an instance of every class, just as the empty set is a subset of every set.

12.7   RESTRICTING QUERIES TO GEOGRAPHICAL REGIONS

To demonstrate the querying about objects located in particular geographical regions, we use an ontology shown in Figure 12.5. This ontology consists of two

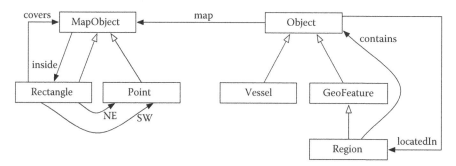

FIGURE 12.5 A geographical feature ontology.

parts—a sub-ontology for describing geographical features (on the right) and another sub-ontology for describing map objects. Each of these parts includes a top-level class—*Object* and *MapObject*, respectively.

GeoFeature is a subclass of *Object*; it has a subclass—*Region*. This part also includes the class *Vessel*, which is attached to this ontology just to indicate that the class of object we are interested in is Vessel, rather than, say, Person or School. The main property for this part is *contains*, as well as its inverse—*locatedIn*. These properties will be used to capture facts about which regions contain particular objects.

The sub-ontology on the left is essentially an ontology for describing maps, which are two-dimensional surfaces with some objects attached to particular locations. The top-level object, *MapObject*, has two subclasses—*Rectangle* and *Point*. This part also includes four properties. The properties *covers* and *inside* are inverses of each other. Both of them are *transitive properties*. This means that whenever two facts hold—⟨*inside A B*⟩ and ⟨*inside B C*⟩, this implies that the fact ⟨*inside A C*⟩ also holds. In other words, whenever the first two facts are known, an inference engine can infer the third fact.

Rectangle has two properties—*SW* and *NE*. As the reader may guess, these are the properties that allow one to uniquely locate a rectangle on a map by showing two points (instances of class *Point*) that represent two corners of the rectangle— the South-West corner and the North-East corner, respectively. The *SW* and *NE* properties are declared (in the OWL code, not shown in the figure) to be *functional*. This means that for each rectangle there may be declared at most one value of the *SW* and *NE* properties, which is equivalent to saying that each rectangle has unique corner points on the map.

The two parts of the ontology are linked by the *map* property. In a sense, this property relates two vocabularies—the vocabulary of geographical features and the vocabulary of geometry.

To demonstrate the use of the geographical feature ontology, we first show an example of an ontological representation of a geographical region, with some vessels located in the region, and a query that could be issued by a user.

Figure 12.6 shows an example of instance annotation in the geographical feature ontology.

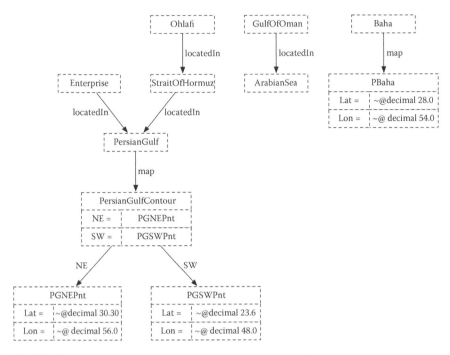

FIGURE 12.6 An example of an annotation of regions and objects.

The focus for this example is the region of Persian Gulf. It is mapped to a rectangle, called here *PersianGulfContour*, via the *map* property. This rectangle has two corners called here *PGSWPnt* and *PGNEPnt*. Each of the points has two data type properties called *Lat* and *Lon* (for latitude and longitude, respectively). The reader should not pay too much attention to the decimal values for this rectangle since the main issue here is to show how such an ontology can be used, rather than provide a precise geographical description of a particular region. This ontology also describes three other geographical features—the Strait of Hormuz (located in the Persian Gulf) and the Gulf of Oman, located in the Arabian Sea. Additionally, it includes three vessels—the *Enterprise*, located in the Persian Gulf, the *Ohlafi*, located in Strait of Hormuz, and the *Baha*, whose location is specified by its longitude and latitude.

All of this information has been given to the ontology explicitly. Running the reasoner on this ontology uncovers a few new facts, mainly because of the properties of the relations—transitive and inverse. And so the reasoner will infer that the *Ohlafi* is located in the Persian Gulf, the Arabian Sea contains the Gulf of Oman and that the Persian Gulf contains the Strait of Hormuz, as well as the *Ohlafi* and the *Enterprise*, and so on.

While all of these inferences may play a very significant role for interoperability, an even more interesting type of inference would need to go through combining the explicit information about the relationships among geographical features and objects with the reasoning about geometry. To achieve this, we developed a rule that relates the two sub-ontologies. In English, this rule can be expressed as follows:

3. *If an object O maps to a point OP on the map and OP's longitude and lati-*
   tude are within a rectangle R defined by its S-W and N-E corners, where R
   represents a region, then O is located in R.

   When we implement this rule in the rule language of BaseVISor,
   BaseVISor infers, in addition to all the facts inferred and shown earlier,
   that the *Baha* is located in the Persian Gulf. This inference was based
   on the information about the geo-coordinates of the *Baha*, as shown in
   Figures 12.5 and 12.6.

12.8   DISCUSSION: INFERENCE OF SUSPICIOUS ACTIVITIES

The information annotation and processing steps we have shown in this chapter could
be used for various purposes. For instance, one could use this kind of approach to
query for vessels in a specific region, where the vessels are involved in some activities that could be considered as "suspicious" by the user. Here we just briefly outline
the approach to achieve such a goal, rather than going into any details, since a full
description of the solution of this problem would take too much space.

The first step would be to develop an appropriate ontology that could be used
for annotating information, formulating definitions of what constitutes a suspicious
activity, querying a network of information sources in order to find relevant information, and finally running inference in order to determine who, where, and when
might be involved in such activities.

So the first question would be: what is a suspicious activity? Where does it stand
in relation to other concepts? While answering such a question might lead to a very
lengthy philosophical discussion, here we present our own proposal, without making any claims about the generality or correctness of our ontological interpretation.
Following our experiences with developing ontologies for various applications and
domains, we propose that suspicious activity, being a subclass of activity, can be
viewed as a "Situation." This view was promoted by Barwise and Perry (1983), and
then Devlin (1991). On the other hand, if one follows the philosophy of Davidson
(1967), one would call it an "Event," rather than a situation. We are calling this situation just because we already have developed an ontology, called Situation Theory
Ontology (or STO) (Kokar et al. 2009) which mainly incorporated the ideas from
Situation Theory by Barwise and Perry (1983). So our view of suspicious activity is
represented in the ontology shown in Figure 12.7.

According to this ontology, *SuspiciousActivity* is a subclass of *Situation*. *Situation*,
on the other hand, is a subclass of *Object* that is linked with other classes—*Goal*,
Individual, *Relation*, *Attribute*, and *Rule* via appropriate properties. So first of all, a
Situation, or a *SuspiciousActivity* in particular, is an entity, a "first class citizen" of
the ontology. It can have its own attributes and its own existence. *Goal* is what gives
focus to a situation. Sometimes this is called *relevance*, or *intent*. Otherwise, without a specific goal, anything could be related to a situation. *Goal* is expressed as a
relation. More specifically, for any situation, it is a single tuple that belongs to a relation, as expressed in set-theoretic terms. Perhaps the most important properties of
Situation are *relevantIndividual*, describing the objects that participate in a specific
situation, and *relevantRelation*, which describes the relations that must hold among

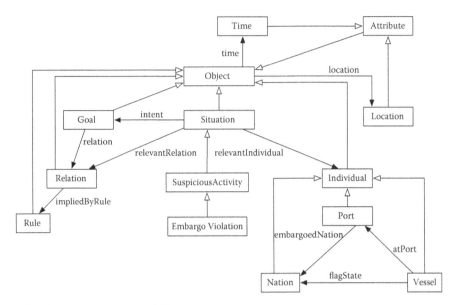

FIGURE 12.7   An extended STO.

the individuals, in order for the situation to take place. The existence of a specific relation is implied by a rule, either an axiom in the ontology or a rule specified in a rule language (e.g., the BaseVISor rule language). Situations, as any objects, can have their own attributes. Most typical attributes of situations include *Time* and *Location*.

If we are interested in a specific domain, we need to extend the STO with some domain-specific concepts. For example, if we are interested in the maritime domain, we show such an extension with the classes of *Vessel*, *Port*, and *Nation*, which are subclasses of the class *Individual*. To support descriptions of suspicious situations, we introduced the properties of *atPort* (to say that a given vessel is at a specific port), *flagState* (to say that a given vessel carries a flag of a given nation), and *embargoed-Nation* (to say that a given nation is on the embargo list).

To show an example of a situation of type *EmbargoViolation*, we show a case in which the information sources include data on a Cuban vessel being reported at Boston Harbor, while an embargo on Cuba is in effect. The dynamic construction of the situation of interest starts with the creation of an instance of *Situation* and an instance of *Goal*. An instance of Situation is generated by the system. Suppose it is represented as *BostonHarborAug19-1975*.

The goal is the same kind of entity as query in our previous examples. As can be seen from Figure 12.7, *Goal* is related to an instance of *Relation*. Suppose the query is: "*Is there a suspicious activity in Boston Harbor on August 19, 1975?*" Suppose this query is represented in OWL as *CurrentGoal*, which in turn is related to an instance of the relation (property), which is expressed here as a triple ⟨*rdf:type BostonHarborAug19-1975 SuspiciousActivity*⟩. Here *rdf:type* stands for the relation "instance of"; this is a term of the OWL language. Note, however, that even though this triple looks like a fact that is in the current ontology, it cannot be asserted into

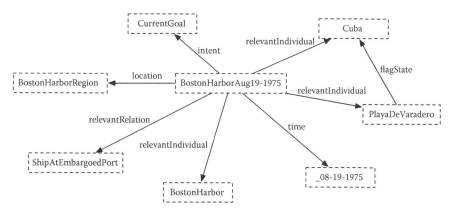

FIGURE 12.8   Representation of an Embargoed Port situation.

the ontology since it represents a query—a question rather than a fact. In the literature (Sowa 2005, Hayes n.d.), these kinds of statements are called "propositions," to
distinguish them from facts.

The next step in the processing of this query, as in the first example of this
chapter, is to derive the information that is relevant to this query. This may involve
various kinds of processing, which may also include logical inference. In particular,
the ontology of situations and suspicious activities may provide information on what
types of *Individual* may be relevant individuals and what types of *Relation* may be
relevant relations. For instance, our ontology in Figure 12.7 shows that the relevant
types of *Individual* are *Vessel*, *Port*, and *Nation*, while the relevant relations are
atPort, *flagState*, and *embargoedNation*.

The result of the processing of this situation is (partially) shown in Figure
12.8. In this figure, we can see the current situation (*BostonHarborAug19–1975*),
attributes of this situation (*location* and *time*), relevant individuals, and the relevant
relation *shipAtEmbargoedPort*, which is a sub-property of *relevantRelation*.
It is assumed that the class *EmbargoViolation* is defined by the relevant relation
shipAtEmbargoedPort. According to the ontology, this relation is implemented by a
rule, which then can be executed by BaseVISor.

12.9   CONCLUSION

The main objective of this chapter was to show the utility of ontologies in situation
awareness in particular and in information integration in general. The discussion
was focused on a network centric environment in which multiple information
producers and multiple information consumers exist. In this kind of scenario, it is
quite difficult for a user to know what information is available at particular network
nodes. A solution to this kind of a problem is to annotate all of the information
sources so that an automatic information collection system can decide whether a
particular information source has some potentially relevant information to a user's
query. Since the user queries are generated dynamically, and since the variety
of the types of queries may be infinite, it is not possible to list all of the possible

queries during the design of a net-centric system. Instead, a way of inferring the relevance of a specific information source to a given query is needed. We have shown both the inference of which sources are relevant and which things are relevant to a given query.

Another problem with situation awareness is to be able to narrow the queries to particular areas or geographical regions. In this chapter, we showed how this kind of knowledge can be represented using ontologies and how a generic inference engine can draw the appropriate inferences by means of OWL axioms and domain-specific rules.

Finally, to conclude this chapter, we provided a discussion of how much more complicated problems, like detecting suspicious activities in maritime domain scenarios, can be achieved within the same kind of framework.

Throughout the discussion, we stressed two aspects that are addressed by the use of ontologies: interoperability and inference. Interoperability means that particular nodes in a net-centric environment can understand what other nodes ask for or send. Inference means that facts that are only implicit in the information can be made explicit by the use of inference engines. Nodes can take advantage of their inference capability by inferring facts locally, rather than sending all the information over the communication links. Both the interoperability and the inference capability are limited only by the language used to represent ontologies and not by algorithms that need to be developed for any specific case.

REFERENCES

Barwise, J. and J. Perry. 1983. *Situations and Attitudes*. Cambridge, MA: MIT Press.

Cebrowski, A. K. and J. J. Garstka. 1998. Network-Centric Warfare: Its origin and future. *U.S. Naval Institute Proceedings*, 124(1):28–35.

Davidson, D. 1967. *The Logical Form of Action Sentences*. Pittsburgh, PA: University of Pittsburgh Press. Reprinted in *Essays on Actions and Events*. Oxford, U.K.: Clarendon Press (1980).

Devlin, K. 1991. *Logic and Information*. Cambridge, MA: Cambridge University Press.

Genesereth, M. R. and R. E. Fikes (Eds.). 1992. *Knowledge Interchange Format, Version 3.0 Reference Manual*. Computer Science Department, Stanford University, Stanford, CA, Technical Report Logic-92-1.

Gruber, T. 2009. Ontology. In *The Encyclopedia of Database Systems*, L. Liu and M. Tamer Özsu (Eds.). Berlin, Germany: Springer-Verlag.

Hayes, P. n.d. IKL Guide. Latest version available at: http://www.ihmc.us/users/phayes/IKL/ GUIDE/GUIDE.html (Accessed on June 28, 2012).

Kokar, M. M., C. J. Matheus, and K. Baclawski. 2009. Ontology-based situation awareness. *Information Fusion*, 10:83–98.

Matheus, C., K. Baclawski, and M. M. Kokar. 2006. BaseVISor: A triples-based inference engine outfitted to process RuleML and R-entailment rules. *Proceedings of the 2nd International Conference on Rules and Rule Languages for the Semantic Web*, Athens, GA.

Matheus, C. J. and B. Ulicny. 2007. On the automatic generation of an OWL ontology based on the Joint C3 Information Exchange Data Model. *12th ICCRTS*, Newport, RI, June.

MetaCarta. 2010. Available at: http://www.metacarta.com/products-platform-information-retrieval.htm (Retrieved January 2010).

MIP, Joint C3 Information Exchange Data Model (JC3IEDM Main), Greding, Germany, December 2005.

Network Centric Operations and Interoperability, Network Centric Operations Industry Consortium. n.d. Available at https://www.ncoic.org/home (Accessed on June 28, 2012).

Pepper, S. and G. Moore. 2001. XML Topic Maps (XTM) 1.0. Available at: http://www.topicmaps.org/xtm/1.0/

Sowa, J. F. 1992. Conceptual graphs summary. In *Conceptual Structures: Current Research and Practice*, P. Eklund, T. Nagle, J. Nagle, and L. Gerholz (Eds.), pp. 3–52. Chichester, U.K.: Ellis Horwood.

Sowa, J. F. 2005. Propositions. Available at: http://www.jfsowa.com/logic/proposit.htm

Talbot, D. 2009. Search Me: Inside the launch of Stephen Wolfram's new "computational knowledge engine." *Technology Review*, July/August. Available at: http://www,technologyreview.com/featured-story/414017/search-me/

Ulicny, B., C. J. Matheus, G. M. Powell, and M. M. Kokar. 2010. Current approaches to automated information evaluation and their applicability to priority intelligence requirement answering. *Proceedings of the 2010 International Conference on Information Fusion, Fusion'10*, ISIF, Edinburgh, UK.

W3C. 2009a. OWL 2 Web Ontology Language Direct Semantics. W3C Recommendation, October 27, 2009. Available at: http://www.w3.org/TR/2009/REC-owl2-direct-semantics-20091027/

W3C. 2009b. OWL 2 Web Ontology Language Document Overview. W3C Recommendation October 27, 2009. Available at: http://www.w3.org/TR/owl2-overview/

W3C. n.d. SPARQL Protocol and RDF Query Language Standards. Available at: http://www.w3.org/standards/techs/sparql#stds (Accessed on June 28, 2012).

13 Service-Oriented Architecture for Human-Centric Information Fusion

Jeff Rimland

CONTENTS

13.1 Introduction 347
13.2 Participatory Sensing and Sensor Webs 349
13.2.1 Participatory Sensing Campaigns...349
13.2.1.1 MobileSense..349
13.2.1.2 PEIR..350
13.2.1.3 "Voluntweeters" 350
13.2.1.4 DARPA Network Challenge 351
13.2.2 Sensor Webs...353
13.3 Service-Oriented Fusion Architecture..354
13.3.1 Service-Oriented Fusion Pyramid...354
13.3.1.1 Low-Level Operations 355
13.3.1.2 Composite Operations..356
13.3.2 High-Level Assessments...359
13.4 Hybrid Sensing/Hybrid Cognition over SOA 360
13.5 Conclusion 361
Acknowledgment 361
References...361

13.1 INTRODUCTION

Information is currently undergoing a paradigm shift that is radically changing how it is sensed, transmitted, processed, and utilized. One of the primary driving forces in this shift is the transformation of mobile device usage. The new mobile device user is an amazingly capable hybrid system of human senses, cognitive powers, and physical capabilities along with a suite of powerful physical sensors including high-definition (HD) video/still camera, global positioning satellite (GPS) positioning, and multi-axis accelerometers. Additionally, these devices are linked to the "hive mind" (Kelly 1996) of various social networks and the distributed power of ever-growing open-source information provided by nearly countless applications that have the potential to funnel specific geospatial and temporally appropriate bits of information to the user on a high-resolution display or through high-fidelity audio. The potential information gathering and processing power of massive social networks connecting these human/machine hybrids that we call "mobile device users" is unprecedented. However, advances in architecture and infrastructure are required to fully recognize this potential. The paradigm of service-oriented architecture (SOA) is evolving in exciting and largely unanticipated ways due to the convergence of these factors.

In O'Reilly and Battelle (2009), Tim O'Reilly discusses the potential for "collective intelligence" to exist in the World Wide Web, but states that the current web is somewhat like a newborn baby—having the basic facilities necessary to grow into an intelligent and conscious entity yet still "awash in sensations, few of which she understands." This analogy can also be applied to human-centric information fusion, which in many ways is a part of the second-generation "Web 2.0" that O'Reilly discusses.

The information fusion community has recently been shifting its emphasis toward network-centric and human-centric operations (see Castanedo et al. 2008, Fan et al. 2010, Keisler 2008, Kipp 2006). Distributed human-centric information fusion is proving indispensable in a broad variety of civilian and military applications. Team-based tasks that were formerly limited by geographic distance, siloed information, and inability to share mental models present great opportunities for a hybrid-sensing/ hybrid-cognition model. There has already been extensive research into "participatory sensing" campaigns that facilitate decentralized collaboration for disaster response, counterinsurgency efforts, and "citizen science." However, the potential for true *collective intelligence* remains largely unrealized. Addressing this need by combining the paradigms of participatory sensing and distributed sensor networks over an evolving human-centric SOA (Figure 13.1) is the focus of this chapter.

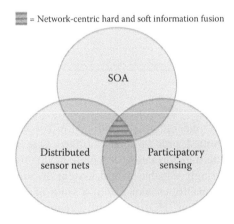

FIGURE 13.1 The intersection of distributed sensors, participatory sensing, and service-oriented architecture.

13.2   PARTICIPATORY SENSING AND SENSOR WEBS

Participatory sensing (Burke et al. 2006) enables the observation of both *people-centric* and *environment-centric* (Christin et al. 2011) information by leveraging the ubiquitous and increasingly powerful mobile devices already being carried by billions of people, the high-speed/high-reliability voice and data networks already in place to support these devices, and the uncanny human ability to capture, isolate, annotate, and transmit the data of interest.

Although this evolving technique of distributed "soft" sensing holds great promise, the field is still in its infancy and there are many challenges to planning and executing a successful participatory sensing campaign. There is a great deal of concern among participants and other citizens that the Orwellian mass surveillance mechanism of "Big Brother" (Orwell 1977) will effectively be realized by "Four Billion Little Brothers" (Shilton 2009) armed with smart phones. Without a mechanism to address this fear, the number of participants and the type of information shared will be limited. Additionally, new methods are needed for determining the veracity, objectivity, and "observational sensitivity" of the source (Schum and Morris 2007). Without sufficient quality control of the observers and the data provided, campaigns can be severely compromised through either intentional deception or simple lack of competence on the part of the observer. Incentivizing participation (Jiang 2010) is another challenge that can be especially difficult and critical to the success of campaigns that require longitudinal data. The following section will provide a sample of campaigns from a variety of disciplines.

13.2.1   PARTICIPATORY SENSING CAMPAIGNS

13.2.1.1   MobileSense

The MobileSense campaign (Lester et al. 2008) conducted by researchers at the University of Washington attempted to gather GPS, barometric pressure, and accelerometer data and use it to infer the type of activity that the subject is currently performing. Additionally, they used Geographical Information System (GIS) data layers in conjunction with the information that they gathered from each subject to determine specific locations where the subject had a tendency to dwell for various periods of time. This fusion of data allowed inferences to be made regarding where the subject lived, worked, shopped, and socialized.

They collected data from 53 test subjects over a period of 1 week using a device called the Mobile Sensing Platform (MSP) (MSP Research Challenge n.d.), which is a proprietary device that is worn using a belt clip. The subjects also manually recorded their activities every hour. For privacy, the users were allowed to switch the device on and off at will, resulting in a per-subject average of 53 h of data collected per week. Requiring the user to switch off the entire device for privacy is less than optimal in terms of user convenience and data gathering. In Christin et al. (2011), more advanced privacy schemes such as k-anonymity, identity blurring, and user-configurable granularity are introduced.

MobileSense is a useful early example of combining data gathered via participatory sensing with open-source GIS. Fusing a priori information with sensed data in this manner provides advantages that will be discussed in later sections.

13.2.1.2 PEIR

The Personal Environmental Impact Report (PEIR) (Mun 2009) is a long-running participatory sensing campaign led by the Center for Embedded Networked Sensing (CENS) at UCLA. PEIR uses a variety of GPS-enabled mobile devices to provide assessments of both how the individual impacts the environment, and how much exposure the individual has had to environmental threats and hazards. Although the system detects as variety of factors, the impact contributors can be summarized as carbon impact and sensitive site impact, and the environmental hazards can be summarized as smog exposure and (the somewhat controversial) fast food exposure.

The project has evolved somewhat since it entered "pilot production mode" in June 2008, but there are several aspects of this project that can serve as lessons for other participatory sensing campaigns. Rather than forcing the user to completely switch the sensor off when they are in a private location (as MobileSense does), PEIR instead allows the user to select locations that they would like to obscure from the public report. Additionally, when the user specifies that a location is private, the system uses an algorithm to create an alternate simulated path to avoid raising suspicion of unusual/illicit activity during the "blanked out" time period. While this synthetic generation of location data might not be appropriate for all participatory sensing projects, it adds to the user's convenience and therefore reduces the chances of them dropping out of the campaign.

Additionally, the system lets the user review their PEIR (and all contributing data) before uploading it to the server. After viewing the data, they are allowed to selectively delete any personal information that they would not like to share before uploading it to the CENS server. If they chose, the user may also share this information directly with their social network via a Facebook application.

Where some participatory sensing campaigns attempt to gather several modalities of data by equipping the users with advanced mobile sensor systems consisting of several integrated devices (Ishida et al. 2008), PEIR takes the approach of only gathering GPS data directly from the individual and using open-source information to obtain a myriad of other details related to that location. Rather than attaching a smog sensor to the mobile device and attempting to measure levels directly, they utilize the Emissions Factors Model (EMFAC) developed by the California Air Resources Board (CARB) to determine individual smog exposure based on their location at a given time. Although not always practical, this approach improves scalability by reducing dependency on nonstandard devices, reduces battery drain, and improves efficiency by offloading processing tasks to a remote server that is far more powerful than the mobile device.

13.2.1.3 "Voluntweeters"

While many crowdsourcing efforts are the result of a top-down, centralized campaign to achieve a specific purpose or collect specific data, the massive efforts of the "digital volunteers" who responded to the devastation that resulted from the 7.0 magnitude earthquake near the capitol city of Port-au-Prince Haiti on January 12, 2010, was largely self-organized and startlingly effective (Starbird 2011).

The importance of incentives is perhaps most relevant in this category of crowd-sourcing. The participants were taking part in these activities to save their own lives or the lives of others, or to determine whether their loved ones were safe.

Because of its low bandwidth and battery requirements, ease of use, ubiquity of compatible devices, and connection to a publicly searchable timeline, the micro-blogging service Twitter became the platform of choice. With the help of a few facili-tating organizations such as the CrisisCamp initiative (crisiscommons.org) and the ATLAS Institute (Starbird 2011), an augmented Twitter syntax called "Tweak the Tweet" (TtT) was designed to leverage the existing Twitter hashtag capabilities to further reduce some of the ambiguity inherent in 140 characters of free-form text.

If the Twitter users on the ground in Haiti were simply uploading messages to the public Twitter feed, that may have been somewhat helpful. However, the real utility came when social networks formed between people in need of assistance, people with information, people seeking information, and volunteers capable of translating the requests between both multiple languages and multiple networking methods.

The success of this grassroots effort shows the importance of incentive and the power of platforms that support and facilitate rapid self-organization of social networks.

13.2.1.4   DARPA Network Challenge

Due to the diversity of the goals, methods, and scopes of most participatory sensing campaigns, it is generally very difficult to obtain "apples-to-apples" comparisons for the test and evaluation (T&E) of these campaigns. The Defense Advanced Research Projects Agency (DARPA) Network Challenge (DARPA Report 2010) is one of the few instances where a direct quantitative comparison is possible between various campaigns that are competing to accomplish the same goal.

In December 2009, DARPA conducted an experiment that tested the combination of crowdsourcing, social network utilization, and viral marketing capability in a race to most quickly solve the distributed geo-location problem of finding ten 8 ft red bal-loons that were tethered at various locations throughout the United States. Locating the red balloons could be considered analogous to detecting the start of an epidemic, distributed terrorist attack, or massive-scale cyber attack—thus the importance of understanding the key elements of successfully undertaking this challenge is very high. Many lessons can be learned by the techniques that the top performing teams utilized to win the challenge.

The winning MIT team's strategy began with incentivizing participation. They used the $40,000 in potential prize money to construct a recursive incentive structure that rewarded not only those participants who actually spotted balloons, but also those who connected balloon spotters to the team (see Figure 13.2). This incentive structure rewards not only individuals who are good at performing the end task, but also those who are good at creating network connections that improve the overall odds of success at the given task. Since mobilization time was of the essence and there was insufficient time to develop an advanced machine-learning system to

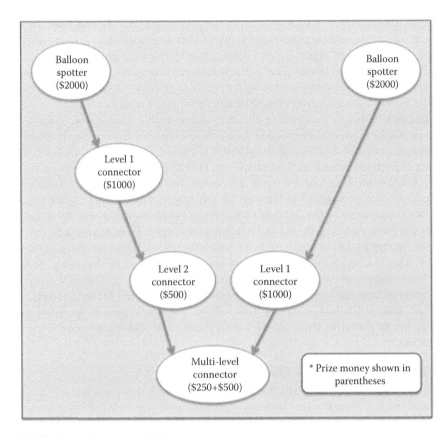

FIGURE 13.2   Example of MIT's recursive incentive structure.

determine which reports were valid and which were false, the winning MIT team relied on human reasoning of the data that was provided by the network (Tang et al. 2011). This exemplifies the effectiveness of modular hybrid-sensing, hybrid-cognition models (Rimland 2011) that enable an ad hoc combination of human and machine sensing as well as human and machine processing of the data. The hybrid approach used computerized Internet Protocol (IP) tracing to filter out obvious false reports (e.g., Pennsylvania IP addresses used to upload pictures of a balloon in Texas) and Web/GIS-assisted human analysis to verify that businesses, weather conditions, roadways, etc. that are shown in the reported pictures of the balloons are consistent with real-world features.

The tenth place iSchools team relied on open-source intelligence methods of searching cyberspace for potential leads and then confirming those leads by quickly activating the "hive mind" (Kelly 1996) of its extensive social network across Twitter, Facebook, and e-mail in an attempt to find direct confirmation of the observation.

There were many lessons about participatory sensing learned from the Network Challenge. The most successful teams were those that incentivized effectively, relied on their specific advantages (e.g., mass media coverage, wide geographical distribution, etc.), and—possibly most importantly—relied on the strengths of social networks, open web information and tools, and the power of an individual with a mobile device who is at the right place at the right time.

13.2.2 SENSOR WEBS

As sensors and networking capability improve in both performance and affordability, the interconnected network of sensors, or *sensor web*, is proving invaluable for tasks ranging from weather prediction (Hart and Martinez 2006) to tracking the behavior of endangered species (Hu et al. 2005). While the idea of using a distributed network of low-cost sensors instead of sending humans into conditions that may be hazardous or difficult/expensive to access is promising, there are several challenges.

Sensors record data in a wide variety of formats. Many of these are proprietary and designed to work with specific software tools. While this is not a problem in the case of a single organization accessing a small number of sensors, it makes it difficult or impossible to construct a network of distributed heterogeneous sensors or to share that sensor information between multiple organizations. Additionally, semantic standards describing exactly *what* is being observed by the sensors, as well as metadata related to accuracy, timeliness, and provenance, are required if the sensors are to be integrated in an extensible manner.

To address these issues, the Open Geospatial Consortium (OGC) has spearheaded improvements in Sensor Web Enablement (SWE), which is a series of standards and best practices for facilitating the connectivity and interoperability of heterogeneous sensors and devices that are connected to a network.

Among these innovations, Transducer Markup Language (TML) is of particular interest to the information fusion community. This XML-based language facilitates the storage and exchange of information between *sensor systems*—which may include actual sensors as well as transmitters, receivers, actuators, or even software processes. The data can be exchanged in either real-time streaming mode or archived form, and real-time data can be exchanged in various chunk sizes over multiple protocols (e.g., TCP or UDP) depending on network bandwidth and data integrity requirements. TML provides the capability to capture intrinsic (e.g., physical hardware) specifications as well as extrinsic (e.g., environmental) metadata that may be of interest to consumers of the data.

Sensor Planning Service (SPS) is another SWE tool that applies SOA principles to the identification, use, and management of sensor systems based on sensor availability and feasibility for successful completion of a specific task.

Other OGC standards for SWE include

1. Observations and Measurements (O&M)
2. Sensor Model Language (SensorML)
3. Sensor Observation Service (SOS)
4. Sensor Alert Service (SAS)
5. Web Notification Services (WNS)

The benefit of these openly available, consortium-designed standards is that they provide organizations with the ability to web-enable all manners of sensors, collections of sensors, sensor platforms, and even human observers via the web (Botts et al. 2007).

For the information fusion community, these advances in SWE translate into improved timeliness, coverage, metadata, and consistency. Perhaps most importantly, sensor net technology provides the potential for sensor tasking based on capability and not simply modality. For example, a weather forecaster might need to know if it is currently snowing in a given location. In some instances, a human observer will be the most cost-effective and accurate way of making this determination. In other instances (e.g., dangerous remote locations), a persistent physical sensor might be the best tool for the job. In either case, the consumer of the data requires a certain degree of precision, timeliness, and credibility, but is often not concerned with the modality of the sensor itself.

13.3 SERVICE-ORIENTED FUSION ARCHITECTURE

For several years, SOAs have been praised for providing loosely coupled and interoperable software services for accomplishing system requirements and goals. In the conventional SOA sense, these *services* are typically specific software functionalities made discoverable and accessible via a network. The service advertises its functionality, and then returns data and/or performs a task when called with the appropriate parameters. In a hybrid-sensing/hybrid-cognition framework, this paradigm is extended to not only allow software routines, but also sensors, sensor platforms (e.g., robots and Unpiloted Air Vehicles [UAVs]), and even humans with mobile devices to be queried and tasked in a manner analogous to software resources in conventional SOA.

This approach can be considered both a logical extension and a somewhat radical departure from existing methods. It is a logical extension in that it follows the SOA principle of encapsulating the inner workings of the service and selecting it based on its availability and capability. It is a radical departure in that it could result in a condition where human beings are receiving instructions from a software application. Although this may sound like cause for alarm, I will explain in further sections why this too is a natural progression of SOA.

The service-oriented computing (SOC) paradigm asserts that services are the fundamental and most basic elements with which applications and software systems may be created. Although these methodologies are conventionally applied to situations in which the service (or system of services) is provided purely by software, there is nothing about SOC or SOA that mandates this. In fact, when viewed in the context of creating systems for improving situational awareness or quality of knowledge via distributed hard and soft information fusion, these conventional machine-centric visions of SOC/SOA can be a limiting factor.

13.3.1 Service-Oriented Fusion Pyramid

When discussing the roles and components of conventional SOA, a pyramid diagram is often used (Georgakopoulos and Papazoglou 2008). The bottom third

Service oriented fusion pyramid

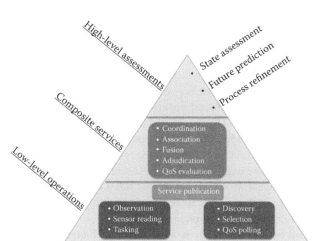

FIGURE 13.3 A perspective on SOA for human-centric information fusion.

of the pyramid represents the basic services provided by the architecture and the low-level "plumbing" tasks of the system including publication, discovery, selection, and binding. The middle or *composition* layer of the pyramid shows tasks related to the coordination, performance/integrity monitoring, and combination of various lower-level services to complete the task at hand. The top or *management* layer of the pyramid shows high-level concepts such as performance assurance and top-level status of the system. While this view is useful for visualizing the operational structure of conventional SOA, it does not offer insight into the utility of SOA for human-centric information fusion. The service-oriented fusion pyramid (SOFP) attempts to addresses this (see Figure 13.3).

In the SOFP, the levels of data fusion (as outlined by the Joint Directors of Laboratories [JDL] Fusion Model [Hall and Llinas 1997]) as well as relevant human factors are integrated with the corresponding levels of the SOA pyramid. The three levels of the SOFP are low-level operations, composite services, and high-level assessments.

13.3.1.1 Low-Level Operations

The bottom level is similar to existing SOA pyramids in that it contains service publication, discovery, and selection. These are fundamental SOA principles that make it possible for stateless, loosely coupled services to advertise their capabilities, locate other useful services, and perform basic communication. In conventional, first-generation SOA, services are described by the Web Services Description Language (WSDL), directory lookup is facilitated by the Universal Description, Discovery, and Integration (UDDI) framework, and the services eventually communicate via the XML-based Simple Object Access Protocol (SOAP) (Walsh 2002). Although it is beyond the scope of this chapter, it should be noted that the newer representational state transfer (REST) architecture eliminates much of this complexity. Fielding and

Taylor (2002) is recommended reading for details of REST. Polling data for Quality of Service (QoS) analysis is also performed at this level.

In addition to these SOA capabilities, the SOFP adds human observations, sensor data readings, and tasking at the bottom level of the pyramid. In the JDL data fusion model, data preprocessing and entity assessment correspond to levels 0 and 1, respectively. Tasking of sensors is typically considered a process refinement task that corresponds to JDL level 4. In one regard, tasking can be considered a higher-level operation because it relies on broad understanding of the system from a composite and multifaceted perspective (Bedworth and O'Brien 2000). However, in the context of the service-oriented model, physical sensors and even human observers can be considered as a service provider—although the ramifications of Humans as a Service (HaaS) require additional exploration (in a later section). From this perspective, tasking can often be a decentralized operation that relies on the localized needs of other services and entities in the system, as opposed to relying purely on high-level dictation from a centralized tasking mechanism.

Decentralized control has many benefits in a distributed heterogeneous system. In addition to increased robustness due to removing single points of failure and the performance advantage of parallel processing, decentralized control based on local inputs has been shown to have excellent potential for finding good solutions to problems that are otherwise considered intractable. For example, consider the Ant Colony Optimization (ACO) for solving the travelling salesman problem. In Dorigo et al. (2006), control in computational systems is modeled after control in ant colonies—which relies on both local stimulus (finding food) and localized messaging between ants (by modifying their environment with pheromones, which is known as *stigmergy*). Much as centralized control of an ant colony would be impossible and even undesirable, the same applies to the evolving paradigm of distributed crowdsourcing and information fusion. That is why tasking is presented as a low-level operation in this framework.

13.3.1.2 Composite Operations

Although the previous section extolled the virtues of decentralized control, most systems still require an element of logical hierarchy that can combine and compose basic services into more complex services, coordinate data flow between services, and ensure system integrity. Additionally, a primary task in designing a distributed information fusion system is ensuring that the data is available where it is needed and in a usable format. These are considered composite operations in the SOFP framework.

A meaningful task is seldom performed by a single service. For example, viewing a website and placing an order with an online retailer causes the elaborate orchestration of a multitude of services across multiple domains and corporate boundaries (Mallick et al. 2005). When everything works properly, the customer is presented with a coherent and appealing shopping experience in which product availability, detailed images, suggested complimentary purchases, customer reviews, shipping rates, and opportunities to save money (or spend more) through affiliate programs. Behind the scenes, however, this requires coordination between product databases, media services, retail supply, and delivery channels, and affiliate partners that are often geographically dispersed and have disparate representations of their data.

Another example that more closely reflects current concerns in the information fusion community is the Global Information Grid (GIG) model that the Department of Defense (DoD) has embraced for information superiority (Chang 2007). The next generation of battle applications relies heavily on network-centric and service-centric principles to provide a globally connected, highly secure, and extremely robust method for enabling operations that span multiple domains, varying security levels, a broad variety of HCI form factors (McNamee 2006). Accomplishing this feat requires an extensive orchestration and constant evaluation of system status.

Much like information exchange between various services relies on coordination and composition, information fusion relies on *association* of data from different sources before algorithms (such as Kalman filters, see Hall and McMullen [2004]) can be used to make predictions about future states. Techniques for data association include Nearest Neighbor (NN), Strongest Neighbor (SN), Probabilistic Data Association (PDA), and Multiple Hypothesis Testing (MHT) (Mitchell 2007). Additional details on data association and related aspects of data fusion can be found in Hall and Llinas (2001).

When making decisions based on fusion of information from a variety of sources, it often becomes necessary to perform adjudication over which sensors, observers, or fused combinations of these are most qualified to deliver accurate assessments. For example, in Tutwiler (2011), Flash LIDAR is fused with mid-wavelength infrared (MWIR) to deliver a product that provides both distance and thermal information about a scene or subject (see Figure 13.4). Under varying conditions and situations, each one of these modalities might prove more effective than the other for tasks such as identification, localization, and tracking. There are a variety of adjudication and voting methods (Parhami 2005) for physical sensors, yet there are very few that account for the introduction of humans as observers. This will be a ripe area for research over the coming years.

In complex systems that combine heterogeneous inputs from a broad variety of geographically distributed sources, providing adequate QoS is vital. It is informative to compare the approach to QoS taken by the SOA community with the T&E metrics that appear in the data fusion literature. Since SOA relies largely on aggregating component services (often from multiple providers) into a more complex composite service, the QoS metrics must also take into account the QoS of the services that it aggregates. This is typically done by looking at the following categories of metrics: (1) *provider-advertised metrics*, (2) *consumer-rated metrics*, and (3) *observable metrics* (Zeng et al. 2007).

Provider-advertised metrics are simply the claims or advertisements make by the service provider. Service cost, for example, is typically a provider-advertised metric. Consumer-rated metrics are based on the feedback and evaluations of past service consumers. This can be thought of as analogous to feedback left by buyers on online auction sites. Ratings for factors such as responsiveness or accuracy of information obtained can be averaged and supplied to future consumers. Finally, observable metrics can be obtained through direct measurement and the application of formulae that are typically specific to the domain in question.

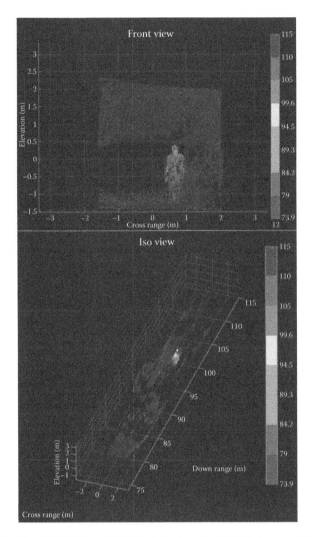

FIGURE 13.4 Fused LIDAR and MWIR data showing both distance and thermal information. (From Tutwiler, R., Hard sensor fusion for COIN inspired situation awareness, *Proceedings of the 14th International Conference on Information Fusion*, Chicago, IL, 2011.)

For years, the data fusion community has focused on measuring the quality of a fused information product (and the underlying fusion process) in terms of Measures of Performance (MOPs), Measures of Effectiveness (MOEs), and Measures of Force Effectiveness (MOFEs) (White 1999). MOPs include direct evaluation of performance factors such as

1. Detection probability
2. False alarm rate
3. Location estimate accuracy
4. Time from transmission to detection

MOEs evaluate fusion processes on their capability to contribute to the success of a task or mission. They typically include

1. Target nomination
2. Target leakage
3. Information timelines
4. Warning time

MOFEs look at the bigger picture of performance of a data fusion system as well as the larger force that it is a part of Hall and Llinas (2001). MOFEs are typically applied to military situations, but have other applications as well.

There is a good deal of literature and research related to QoS in SOA (Oriol 2009) and T&E of data fusion systems (Blasch 2004), but there is little research into the intersection of those areas—which occurs in the middle layer of the SOFP. Prior work in this area is especially sparse regarding human-centric factors.

13.3.2   High-Level Assessments

In the JDL data fusion model, levels 2 and 3 refer to situation assessment and threat or impact assessment. While the levels of the JDL model are not necessarily intended as sequential flowchart, fusion at these higher levels typically requires that information be represented at the *feature* or *entity attribute* level as opposed to raw data representations. Additionally, JDL level 4 is a meta-process in which the fusion process itself is evaluated. That is, in the JDL model, the level 4 process is a process that monitors the other ongoing fusion processes and seeks to optimize the processing results (e.g., by directing sensors, modulating algorithm parameters, etc.).

In SOA literature, the high-level *management* tasks include system evaluation through statistical analysis, delivering notification upon completion of high-level tasks, and telegraphing the results of high-level decision making. Since SOA supports *open service marketplaces* in which service providers can autonomously negotiate with each other to add value or help perform a task, *service-level agreements* (SLAs) are often provided to facilitate "fair trade" within these marketplaces. The management levels of SOA help to negotiate these agreements.

In this highest level of the pyramid, a significant change occurs from lower levels of SOA and data fusion. At lower levels, mathematical formulae, pattern matching, and various detection and tracking algorithms can refine signals and give estimates of attributes such as position, identity, and direction/velocity of motion. At higher levels, shared knowledge and understanding becomes necessary (Perlovsky 2007). Since software fusion systems lack the natural language capabilities that humans use to exchange understanding and knowledge between individuals, we rely on ontologies or other knowledge representations in an attempt to digitally describe and delineate properties that are often easily and intuitively understood by humans, yet poorly captured by machine representation. The resulting systems often work well for isolated "toy problems" or provide "one off" solutions, but may prove brittle in real-world applications.

13.4 HYBRID SENSING/HYBRID COGNITION OVER SOA

One approach to solving this problem is enlisting a "human-in-the-loop" to provide both information gathering and sense-making contributions to the system. While computers solve certain problems with speed and accuracy far exceeding that of any human, there are also tasks that are still poorly performed by the best sensors, software algorithms, and hardware. These tasks often require fusing multiple senses or applying innate reasoning or understanding.

In the simple act of stepping off a curb to cross a street, the human perceptual and cognitive systems are performing an amazing sequence of calculations that we are completely unaware of except on the relatively rare occasion that conscious intervention is needed on our part. The visual system is locating and tracking oncoming and passing traffic that we maintain record of just long enough for the visual inputs to be corroborated by our auditory system and finally for the sense of passing vibration and occasionally the rush of air against our skin. Our vestibular and proprioceptive systems maintain awareness of our balance and the location of each joint as it moves through its range of motion. In robotics, the inverse kinematics field (Tolani et al. 2000) is dedicated to calculating the proper angles and forces necessary to move robotic limbs into the correct position to perform a given task. Humans do this exquisitely and automatically.

Aside from advantages in perception, information fusion, and movement dynamics, humans have an amazing ability to make near-instantaneous assessments of current situation status or risk. Computer systems may be able to read license plates or even identify faces with superhuman speed and accuracy, but the best automated systems are still utterly incapable of judging individual intent. When one person is approached by another, perceptual elements and cues are integrated subliminally. We generally do not consciously notice the saccadic eye movements (Pelz et al. 2001) or subtle facial gestures, but we can quickly tell that a person in a crowd recognizes us, or that we have just made a statement that hurts a friend's feelings, or that someone is about to ask for a favor.

There are also tasks at which computers are at a clear advantage over humans. Performing complex numerical calculations, searching large volumes of text, rapidly matching patterns, and performing certain types of quantitative assessments (e.g., "the vehicle weighs 2819.56 pounds") are tasks that put computers and physical sensor systems at a clear advantage. However, the most significant task that computers are capable of is facilitating the connection of people in ways that were previously impossible.

Advances in several parallel technologies are now converging in a way that is poised to change how humans approach the most complex and difficult tasks that we undertake. This will happen through the following factors:

1. The mobile device user, capable of acting as a *sensor platform* to capture high-definition, high-fidelity digital information, is at the same time able to apply his or her innate human sensing and cognitive abilities to either annotate the captured digital information or share direct observations via speech or micro-blogging services such as Twitter. Additionally, the

mobile device user's capability as a sensor platform is enhanced through open-source information, geo-location, and group collaboration facilities available to them via the device.

2. Social networks allow members to readily identify and aggregate a "hive mind" that is ideally suited for the task at hand. Although this capability exists to some degree already, the concept of "friending" someone on a social network will evolve to include opportunistic sharing of data or cognitive ability for a specific task or type of task, as opposed to the current model of permissively sharing personal information with large numbers of friends or acquaintances.

3. Artificial intelligence and data fusion algorithms are improving—not only in their stand-alone capacities, but also through their increasing abilities to interact with a human-in-the-loop.

4. Service-oriented system methodologies not only connect computers, sensors, and mobile device users, but also facilitate abstraction and service description to allow sensing, information fusion, and cognition tasks to be performed by either computer, human, or a hybrid (e.g., mobile device user).

13.5 CONCLUSION

The last point mentioned earlier is the most important point of this chapter, so it is worth restating. The SOA, with its ability to provide access to data or processing algorithms via loosely coupled, rapidly reconfigurable, modular services, allows each of the earlier mentioned breakthroughs of mobile device technology, social networking, and advancing algorithms, to act as an effectiveness multiplier for each other. When the potential of this combination is fully realized, it will have broad implications for the sciences, prevention and recovery from natural and man-made disasters, medicine, and countless other aspects of human endeavor.

ACKNOWLEDGMENT

We gratefully acknowledge that this research activity has been supported in part by a Multidisciplinary University Research Initiative (MURI) grant (Number W911NF-09-1-0392) for "Unified Research on Network-based Hard/Soft Information Fusion," issued by the U.S. Army Research Office (ARO) under the program management of Dr. John Lavery.

REFERENCES

Bedworth, M. and J. O'Brien. 2000. The omnibus model: A new model of data fusion? *IEEE Aerospace and Electronic Systems Magazine*, 15(4):30–36.

Blasch, E. 2004. Fusion metrics for dynamic situation analysis. *SPIE Proceedings* 5429, Onlando, FL, pp. 1–11.

Botts, M., G. Percivall, C. Reed, and J. Davidson. 2007. OGC sensor web enablement: Overview and high level architecture (ogc 07-165). Open Geospatial Consortium White Paper, 28.

Burke, J. et al. 2006. Participatory sensing. *Workshop on World Sensor Web (WSW'06): Mobile Device Centric Sensor Networks and Applications*, Boulder, CO, pp. 1–5.

Castanedo, F., J. García, M. A. Patricio, and J. M. Molina. 2008. Analysis of distributed fusion alternatives in coordinated vision agents. *11th International Conference on Information Fusion*, Chicago, USA, pp. 1–6.

Chang, W. Y. 2007. *Network-Centric Service Oriented Enterprise* (illustrated ed. Dordrecht). Houten, the Netherlands: Springer.

Christin, D., A. Reinhardt, S. Kanhere, and M. Hollick. 2011. A survey on privacy in mobile participatory sensing applications. *Journal of Systems and Software,* doi:10.1016/j. jss. 2011.06. 073, 1–19.

DARPA. 2010. DARPA Network Challenge Project Report.

Dorigo, M., M. Birattari, and T. Stutzle. 2006. Ant colony optimization. *Computational Intelligence Magazine, IEEE 2006*, 1(4):28–39.

Fan, X., M. McNeese, B. Sun, T. Hanratty, L. Allender, and J. Yen. 2010. Human—Agent collaboration for time-stressed multicontext decision making. *Systems, Man and Cybernetics, Part A: Systems and Humans, IEEE Transactions*, 40(2):306–320.

Fielding, R. T. and R. N. Taylor. 2002. Principled design of the modern web architecture. *ACM Transactions on Internet Technology (TOIT) 2002*, 2(2):115–150.

Georgakopoulos, D. and M. P. Papazoglou. 2008. *Service-Oriented Computing*. Cambridge, MA: MIT Press.

Hall, D. L. and J. Llinas. 1997. An introduction to multisensor data fusion. *Proceedings of the IEEE 1997*, 85(1):6–23.

Hall, D. L. and J. Llinas. 2001. *Handbook of Multisensor Data Fusion*. Boca Raton, FL: CRC Press.

Hall, D. L. and S. A. H. McMullen. 2004. *Mathematical Techniques in Multisensor Data Fusion*. London, U.K.: Artech House Publishers.

Hart, J. K. and K. Martinez. 2006. Environmental sensor networks: A revolution in the earth system science? *Earth-Science Reviews 2006*, 78(3–4):177–191.

Hu, W. et al. 2005. The design and evaluation of a hybrid sensor network for cane-toad monitoring. *Information Processing in Sensor Networks, IPSN 2005, Fourth International Symposium*, Sydney, New South Wales, Australia.

Ishida, Y., S. Konomi, N. Thepvilojanapong, R. Suzuki, K. Sezaki, and Y. Tobe. 2008. An implicit and user-modifiable urban sensing environment. *Urbansense08*, 2008:36.

Jiang, M. 2010. Human-centered sensing for crisis response and management analysis campaigns. *Proceedings of the 7th International ISCRAM Conference*, Seattle, WA.

Keisler, R. J. 2008. Towards an agent-based, autonomous tactical system for C4ISR operations. *Proceedings of the Army Science Conference (26th)*, Orlando, FL, December.

Kelly, K. 1996. The electronic hive: Embrace it. *Computerization and Controversy: Value Conflicts and Social Choices*, 1996:75–78.

Kipp, J. 2006. The human terrain system: A CORDS for the 21st century. *Military Review*, September–October 2006:1–8.

Lester, J., P. Hurvitz, R. Chaudhri, C. Hartung, and G. Borriello. 2008. MobileSense-sensing modes of transportation in studies of the built environment. *Urbansense08*, 2008:46–50.

Mallick, S., A. Sharma, B. V. Kumar, and S. V. Subrahmanya. 2005. Web services in the retail industry. *Sadhana*, 30(2):159–177.

McNamee, D. 2006. Building multilevel secure web services-based components for the global information grid. *CROSSTALK The Journal of Defense Software Engineering*, May 2006:1–5.

Mitchell, H. B. 2007. *Multi-Sensor Data Fusion: An Introduction*. Berlin, Germany: Springer.

MSP Research Challenge. 2007. University Washington and Intel Research Seattle. http:// seattle.intel-research.net/MSP (accessed on August 2, 2011).

Mun, M. 2009. PEIR, the Personal Environmental Impact Report, as a platform for participatory sensing systems research. *Proceedings of the 7th International Conference on Mobile Systems, Applications, and Services*, Krakow, Poland, pp. 55–68.

O'Reilly, T. and J. Battelle. 2009. Web squared: Web 2.0 five years on. *Web 2.0 Summit 2009*, California, USA, pp. 1–13.

Oriol, M. 2009. Quality of Service (QoS) in SOA Systems. A systematic review. Master thesis, UPC, Departamento LSI [Biblioteca Rector Gabriel Ferraté de la Universitat].

Orwell, G. *1984: A Novel* (1977 edition). New York: Penguin.

Parhami, B. 2005. Voting: A paradigm for adjudication and data fusion in dependable systems. *Dependable Computing Systems: Paradigms, Performance Issues, & Applications*, 52:2, 87–114.

Pelz, J., M. Hayhoe, and R. Loeber. 2001. The coordination of eye, head, and hand movements in a natural task. *Experimental Brain Research*, 139(3):266–277.

Perlovsky, L. I. 2007. Cognitive high level information fusion. *Information Sciences*, 177(10):2099–2118.

Rimland, J. 2011. A multi-agent infrastructure for hard and soft information fusion. *SPIE Proceedings*, Orlando, FL.

Schum, D. A. and J. R. Morris. 2007. Assessing the competence and credibility of human sources of intelligence evidence: Contributions from law and probability. *Law, Probability and Risk*, 6(1–4):247.

Shilton, K. 2009. Four billion little brothers?: Privacy, mobile phones, and ubiquitous data collection. *Communications of the ACM*, 52(11):48–53.

Starbird, K. 2011. Voluntweeters: Self-organizing by digital volunteers in times of crisis. *Proceedings of the 2011 Annual Conference on Human Factors in Computing Systems*, Vancouver, BC, pp. 1071–1080.

Tang, J. C., M. Cebrian, N. A. Giacobe, H. W. Kim, T. Kim, and D. B. Wickert. 2011. Reflecting on the DARPA red balloon challenge. *Communications of the ACM*, 54(4):78–85.

Tolani, D., A. Goswami, and N. I. Badler. 2000. Real-time inverse kinematics techniques for anthropomorphic limbs. *Graphical Models*, 62(5):353–388.

Tutwiler, R. 2011. Hard sensor fusion for COIN inspired situation awareness. *Proceedings of the 14th International Conference on Information Fusion*, Chicago, IL, pp. 1–5

Walsh, A. E. 2002. UDDI, SOAP, and WSDL: The web services specification reference book. *Prentice Hall Professional Technical Reference*.

White, F. 1999. Managing data fusion systems in joint and coalition warfare; Signals, systems, and computers. Paper presented at the *Conference Record of the Thirty-Third Asilomar Conference on Signals, Systems, and Computers*, Pacific Grove, CA.

Zeng, L., H. Lei, and H. Chang. 2007. Monitoring the QoS for web services. *Service-Oriented Computing—ICSOC 2007*, 4749, 132–144.

14 Nonmyopic Sensor Management

Viswanath Avasarala and Tracy Mullen

CONTENTS

14.1 Introduction 365
14.2 Stochastic Dynamic Programming.. 366
14.3 Market-Oriented Programming.. 368
14.4 Performance Evaluation Using a Simulation Test Bed............................. 374
    14.4.1 Simulation Platform... 374
14.5 Conclusion 376
References.. 376

14.1 INTRODUCTION

Sensor management which may be defined as "a process, which seeks to manage or coordinate the use of sensing resources in a manner that improves the process of data fusion and ultimately that of perception, synergistically" (Manyika and Whyte 1994). The recent advances in sensor and communication technologies have led to a proliferation of sensor network applications (Balazinska et al. 2007, Hall and McMullen 2004). Traditional sensor management approaches apply network resources policies myopically so that allocation decisions are optimized only for the current time. Myopic sensor management is suboptimal since current network allocations have significant future consequences. To illustrate this example, consider the following scenario. Consider the following very simple scenario (*Scenario I*). A single sensor X is used to track targets of type A and type B. The utility of tracking targets of type A (U_A) is much greater than the utility of tracking targets of type B (U_B). The sensor X has energy reserves that are sufficient only for operating for T time schedules. Assume that the environment has only a target of type B in the first T time schedules, after which target A appears. If a greedy sensor management approach were used, the energy reserves of X would be exhausted by the time a high priority task becomes available. Nonmyopic behavior is suboptimal even when the only network resources that need to be considered are restricted to sensor schedules alone. For example, consider the scenario illustrated in Figure 14.1 (*Scenario II*). A sensor network spans a particular area with varying coverage. Two targets, A and B, exist in the environment. The sensor network's task is to reduce the uncertainty associated with estimating target positions to below a certain threshold. The tasks with tracking A and B have the same utility. Target A is expected to move along the

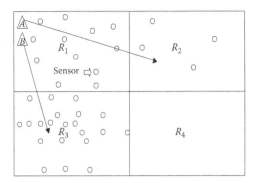

FIGURE 14.1 Sample scenario for nonmyopic sensor management.

path shown in Figure 14.1 from region $R1$ to $R2$. Target B is expected to move from region $R1$ to $R3$. Since $R3$ has much higher sensor coverage then $R2$, an optimal sensor should consider allocating more tracking resources to A than B, when both A and B are in $R1$. However, myopic sensor managers (SM) consider optimization only for the current schedule and therefore do not prioritize between A and B in $R1$.

As Figure 14.1 illustrates using a simple scenario, an optimal sensor should be nonmyopic. That is, it should have a far-sighted strategy where the sensor allocations are optimized over a time horizon. However, the resource allocation problem is exponentially complex in the length of time horizon. Furthermore, modeling uncertainty associated with future possibilities is also a challenging problem. As a result, traditionally sensor management problems have used myopic approaches where the optimality for the current is considered (refer to Avasarala [2006] for a good review of greedy techniques). However, recent breakthroughs in stochastic optimization techniques combined with increases in computational speed have made nonmyopic sensor management techniques real-time feasible.

In this chapter, we review two of the most promising techniques for nonmyopic sensor management, stochastic dynamic programming, and market-oriented programming. This chapter is organized as follows. Section 14.2 introduces the stochastic dynamic programming approach and elaborates on a few crucial aspects of this approach. Section 14.3 introduces market-oriented programming. Our own work in this domain is presented as an example implementation in Section 14.4. Finally, Section 14.5 provides some concluding remarks.

14.2 STOCHASTIC DYNAMIC PROGRAMMING

Stochastic dynamic programming (Bertsekas and Tsitsiklis 1996) is based on the following principle (Gelly et al. 2006): "Take the decision at time step t such that the sum 'cost at time step t due to your decision' plus 'expected cost from time steps $t + 1$ to T from the state resulting from your decision' is minimal. Bellman's optimality principle states that this strategy is optimal. Unfortunately, it can only be applied if the expected cost from time steps $t + 1$ to T can be guessed, depending on the current state of the system and the decision."

These methods solve the equation of the form:

$$V_t = g_t\left(\Lambda_t\right) + \alpha\, E\left(\frac{V_{t+1}\left(S_{t+1}\right)}{S_t}, \Lambda_t\right) \tag{14.1}$$

where

$g_t(\Lambda_t)$ is the utility of taking action Λ_t at time t
S_t is the "*state*" of the system at time t
$E(V_{t+1}(S_{t+1})/S_t)$ is the expected future value of current actions
α is the time discount factor

In recent years, stochastic dynamic programming approaches have been applied to sensor management problems. In 1995, Castañon considered a multigrid, single sensor detection problem. Under certain assumptions about the target distributions and probability distribution of sensor measurements, Castañon solved the problem to optimality. The optimal allocation policy was to search either of the two most likely target locations during each round of scheduling. Castañon (1997) further demonstrated swing a simulation study that this optimal policy outperforms a greedy information-theoretic approach. However, except for trivial cases, solving stochastic dynamic programming to optimality is not possible because of their complexity. As a result, more generally, researchers have used approximation techniques to solve stochastic dynamic programming problems. Researchers have used various approximation methods for solving the sensor management problem that is concerned with tracking targets. Washburn et al. (2002) formulate a single-sensor, multitarget scheduling problem as a stochastic scheduling problem and use the Gittin's index rule to develop approximate solutions. Williams et al. (2005) consider a single-target, multisensor allocation problem with communication constraints and use adaptive Lagrangian relaxation to solve the constrained dynamic programming problem. Schneider et al. (2006) have used approximate dynamic programming to allocate gimbaled radars for detecting and tracking tracks over a multihorizon time period. As explained earlier, these more complex stochastic dynamic programming problems have been solved used using greedy approximations.

As a sample illustration, consider Schneider et al.'s (2006) rollout approximation-based approach. In this problem, a base sensor allocation policy is adopted for myopic allocation. The authors have used certain heuristics as the base policy but more generally any greedy single-period scheduling problem like the information-theoretic sensor management (Kastella 1996, Walsh et al. 1998). Assuming a base policy, the stochastic dynamic programming problem is solved as follows: find the optimal action at time t_k by assuming that all allocations for times $t > t_k$ are calculated using the base policy. This method can be applied straightforwardly when the only resources available are sensor schedules. However, for scenarios that involve other network resources like energy (as explained in Scenario I), the rollout algorithm-based results can be inadequate. This is because, by definition, a greedy approach for nonmyopic optimization implies that a sensor should be used for tasks in the current round schedule if its energy reserves permit it. Therefore, using a greedy base policy for future schedules leads to very conservative energy utilization allocations for the current time. An alternative approach to avoid this pitfall when heterogeneous network resources have to be modeled is to use a weighted sum of performance on multiple objectives like track accuracy, energy reserves, and communication bandwidth. This approach is similar to the approach presented in Williams et al. (2005).

So far, the research approaches for sensor management on stochastic dynamic programming have considered on a single type of target. For example, Castañon has considered the detection problem in 1995 (Castañon 1995), classification problem in McIntyre and Hintz (1996), and the tracking problem has been considered in Washburn et al. (2002), Williams et al. (2005), and Schneider and Chong (2006). Research studies that analyze the performance of a more generic multisensor, multitask sensor network environment are not currently available in the public domain.

14.3 MARKET-ORIENTED PROGRAMMING

Market-oriented programming techniques use market algorithms for resource allocation in distributed environments (Wellman et al. 2001). The genesis of market-oriented programming was in Artificial Intelligence (AI) community's work in developing economic mechanisms for distributed problem solving and can be first traced to the contract net protocol. Davis and Smith (1983) and Sandholm (1993) extended the contract net protocol by integrating the concepts of cost and price. For a detailed review of market-oriented programming, refer to Wellman et al. (2001) and Wellman (1991). We have developed market architecture for sensor management (MASM) based on market-oriented programming techniques for multitask, multiconsumer, multisensor sensor management. MASM assumes that all sensors belong to a single platform where a centralized SM is responsible for allocating resources to task. For smart dust environments, Mainland et al. (2005) have proposed similar price-based mechanisms that do not require a central resource allocation agent.

The design of MASM is shown in Figure 14.2. The Mission Manager (MM) is responsible for allocating task responsibilities and budgets to the various agents in the market. The actual details of the sensor network are invisible to the consumer agents. Instead, consumers are allowed to bid on high-level tasks, like "*Track Target X to an accuracy Y*." The SM stipulates the type of tasks that the consumers can bid on, so that the consumer bids are restricted to tasks SM knows how to accomplish. Consumer bids are of the type $<t, p>$ where t is the task description that includes the task type and final task quality desired by the consumer, and p is the price that the consumer is willing to pay. For example, the task description for a bid to search a particular grid, x, so that the uncertainty of target presence (as measured by entropy) <0.001 is as follows:

(type: search
entity: grid no x
quality: (entropy < 0.001))

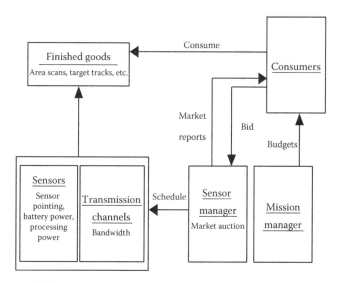

FIGURE 14.2   MASM architecture.

Since MASM models network resources as commodities that need to be sold in the market, it has to set prices for these resources during each round of scheduling. In the current version of MASM, we use a pricing protocol similar to the tatonement process for determining resource prices. *Tatonement* is an iterative procedure for finding equilibrium prices based on the search parameter (e.g., price or quantity) (Samuelson 1947, Walras 1954, Waldspurger et al. 1992). The price adjustment process starts with an auctioneer communicating an arbitrary price set to the users. The users compute their demand for the first good at the given prices and communicate it to the auctioneer. Depending on whether the aggregate demand for the first good is positive or negative, the auctioneer either increases or decreases its price. This process continues until a price at which aggregate demand for the first good equals zero is reached. This process is then repeated for the second good and so on. At the end of the first cycle, only the last good is guaranteed to have a zero demand, but assuming gross substitutability (i.e., when the price of good j goes up, there is a positive increase in the demand for every other good by each user) the price set arrived at after each cycle is closer to equilibrium than the previous one. More refined algorithms using partial derivatives of the demand functions have been developed to search for equilibrium in parallel (Shoven and Whalley 1992, Ygge 1998). Though the gross substitutability assumption is often violated (as in sensor networks), the tatonement process has been found to give satisfactory results (Cheng et al. 2005).

To use the tatonement process in MASM, we model the supply and demand functions for a resource at a particular price. MASM estimates these functions using the current resource usage rate. Prices for individual resources are initialized to zero during the sensor network initialization. After each round of scheduling, the prices $\left(\vartheta_S^{t+1}\right)$ for the resource S for the next round of scheduling are calculated based on the current usage rate of the resource $\left(r_S^t\right)$ and the available usage rate of a_S^t:

$$\vartheta_S^{t+1} = \max(0, \vartheta_S^t + \tau * (r_S^t - a_S^t))$$

$$\vartheta_S^0 = 0 \qquad (14.2)$$

where τ is the constant which determines the rate at which prices are updated.

The definitions of r_S^t and a_S^t are dependent on the resource being modeled. For example, for sensors, we have used the available battery power. Let sensor A be endowed with initial battery power b_i and assume that sensor A needs to be available for a total operating time of T. At time t, if the available battery power is b_t, then

$$r_A^t = \frac{(b_i - b_t)}{t} \quad \text{if } t > 0, \quad 0 \text{ otherwise}$$

$$\qquad (14.3)$$

$$a_A^t = \frac{(b_t)}{(T - t)} \quad \text{if } t < T, \quad 0 \text{ otherwise}$$

Once the resource prices are formulated, SM uses a combinatorial auction mechanism to find the optimal allocation given the consumer bids. However, the consumer bids are on high-level tasks and the sensor resources are raw sensor schedules or communication bandwidth, etc. Thus, some method for bundling goods produced by various sellers is essential to create commodities that consumers are interested in and can bid for. The SM uses a special-purpose protocol, continuous combinatorial auction (CCA), to disintegrate the high-level consumer bids to bids for sensor resources. CCA has been designed to decrease computational and communication complexity of the market. The details of CCA are available in Avasarala (2006). We briefly describe the salient features of CCA. CCA uses discrete time slots to schedule resources. For each time slot, a combinatorial auction is held to determine the optimal allocation of resources. As explained earlier, there is a dichotomy in what the consumers bid on (high-level tasks) and what resources the sensor network has. To address this dichotomy, the CCA uses a bid translation function to create low-level resource bids from high-level task bids. For each time slot t, the auctioneer constructs bids on each resource set, S, that can be allotted to task X. To construct these bids, the auctioneer needs to calculate the price associated with the bid on resource set S, for task X, based on the consumer bid price P. For this purpose, a novel mechanism of price calculation has been devised. For the task X, the auctioneer computes the bid price for a resource set S as the percentage of the consumer task completed by the resource set multiplied by the actual consumer bid price P. Computation of the percentage of task completed by a resource set S is in terms of readings of a canonical sensor, A, as follows. Let task X require, on average, n_a consecutive schedules of the standard sensor A to be completed. Instead, if a resource bundle S is allocated to task X during the current round of scheduling, the expected number of standard sensor readings required changes to n'_a. The percentage of task completed by resource set S is equal to the percentage savings in the required number of sensor A readings:

$$f_{S,X} = \frac{(n_a - n'_a)}{n_a} \qquad (14.4)$$

Then, the bid price for resource set S at time t for task X with bid price P is

$$p_{S,X} = f_{S,X} * P - \vartheta^t_s \qquad (14.5)$$

where ϑ^t_s is the price of the bundle during the tth round of scheduling. It is calculated as the sum of the prices of the individual resources comprising S. Once the resource bids and their corresponding bid prices are formulated, MASM uses a standard combinatorial auction winner determination algorithm (Andersson et al. 2000) to find the optimal resource allocation.

The number of resource combinations that can be allocated to a task is exponential in the number of resources. For large sensor networks, we have formulated a special genetic algorithm that solves the bid translation problem in polynomial time (Avasarala 2006). The representation schema used for the algorithms resembles the SGA algorithm that we developed for determining combinatorial auction winners (Avasarala et al. 2006). First, let the number of bids be k and the number of sensors be n. The representation schema used by the genetic algorithm is a string of length n, where each string element is a real-valued number between 1 and k. The jth string member represents the bid to which sensor j is allocated. We obtain this string's overall utility as the sum of utilities obtained from individual bids, which are calculated as the sum of individual bid prices calculated as shown in Equation 14.5. The genetic algorithm is anytime and has polynomial complexity.

Clearly, the efficiency of the market algorithm depends to a great extent on the bidding strategies of the consumer. Formulating an optimal bidding strategy for MASM consumers is not straightforward because of the following reasons. In MASM, the SM accepts bids only on a set of predefined tasks. The consumer agent is responsible for deconstructing its high-level tasks or goals into a sequence of SM acceptable subtasks on which it can bid. Furthermore, it is responsible for calculating appropriate bid prices for these subtasks in order to bid in the market. The resource requestors have utility for tasks that they are trying to accomplish but not necessarily for the tasks the SM accepts bids for. Therefore, the consumer agents have to formulate a bidding strategy to formulate the optimal prices for their auction bids. For example, assume that a consumer has utility u_t for destroying a target T. The high-level task of destroying T might consist of the following two subtasks, for which SM accepts bids: (a) search for target T and (b) track target T so that the uncertainty about its position is reduced. The consumer then has to divide its overall utility, u_t into utilities for the two subtasks, so that it can formulate bid prices. The optimal weights of the individual subtasks are dependent on the system conditions. For example, it might be difficult to search for targets in some environments whereas in others tracking tracks to high accuracy might be the bottleneck. Utilities for the subtasks should take into consideration the competition for resources. For example, sensors that can be used for velocity estimation, like Doppler sensors, might be abundant in the network, making the tracking task a less competitive one. Moreover, MASM uses a combinatorial auction–based mechanism for resource allocation and therefore predicting the optimal bidding strategies is a difficult problem (Cramton et al. 2005). Furthermore, future market conditions depend on a number of unpredictable variables including future competing consumer task load, their bidding strategies. As an initial study, we implemented a consumer bidding strategy that uses some simplified market assumptions (Avasarala et al. 2009).

We frame this problem as follows. Let φ_i ($i = 1$ to m) be the set of tasks that the consumer has a utility for. Let u_i be the consumer utility of accomplishing the ith consumer task. Let ϕ_j {$j = 1$ to n} be the set of tasks that the sensor network accepts bids for and can accomplish. We assume that each consumer task φ_i consists of a collection of sensor network tasks ϕ_j accomplished in a certain sequence. The sequence of sensor tasks for the ith consumer task is denoted by χ_i. We also assume that there is one to many mapping between consumer tasks and sensor network tasks. That is, each sensor network task can be a subtask for one and only one consumer task. For example, in the MASM simulation scenario, consumer has a utility for destroying targets ($\varphi_1 =$ "destroy *targets*"). To accomplish this task, the consumer has to use the sensor network resources to first search for and detect targets {$\phi_1 =$ "*search for targets*"}. Then, the consumer has to track targets, {$\phi_2 =$ "*track targets*"}. In this case,

$$\chi_1 = \{\text{"search for targets", "track targets"}\}.$$

Clearly, establishing the optimal bid prices involves solving a stochastic, multiperiod optimization. However, we make the following assumptions to make consumer bidding optimization for a single-period, deterministic optimization problem.

1. Consumers model the market as a fixed-price market. In a fixed-price market, the different commodities have fixed prices and the only choice consumers have is whether to buy them or not (Cliff and Bruten 1997). For fixed-price markets, consumers can construct the price–quality of service (QoS) mapping as a deterministic relationship.
2. Consumers model the price–QoS to be independent of time. That is, consumers assume that the current price–QoS mapping will persist throughout the sensor network operation.
3. Consumers optimize bidding prices under the assumption that they use a constant price for each sensor network task throughout the sensor network operation.

Since the consumers assume that the market is a fixed-price market, they can model the market behavior with a series of monotonically increasing functions, γ_{ϕ_j} {$j = 1$ to n} where $\gamma_{\phi_j}(P_{\phi_j})$ represents the fraction of sensor task φ_j that will be completed in any round of scheduling if the consumer pays a price P_{ϕ_j}. The number of schedules for completing the sensor network task φ_j using P_{ϕ_j} as the bid price is

$$\tau_{\varphi_j} = \left[\frac{1}{\gamma_{\varphi_j}(P_{\varphi_j})} \right] \tag{14.6}$$

where $[x] = \min\{n \in Z | n \geq x\}$, Z being the set of integers.

The consumer earns a utility u_i for completing a consumer task φ_i. If χ_i is the set of sensor tasks that comprise φ_i, then the estimated number of bidding cycles to complete consumer task φ_i can be calculated as

$$\Gamma_{\varphi_i} = \sum_{k=1}^{|\chi_i|} \tau_{\chi_i^k} \tag{14.7}$$

where χ_i^k is the kth element of χ_i.

Let p_{χ_i} be the vector of bidding prices $P_{\chi_i^k}$ ($k = 1$ to $|\chi_i|$) for the sensor network tasks that comprise φ_i. To calculate the optimal set of bidding price for the ith consumer task, consumers have to solve the optimization problem:

$$\max_{P_{\chi_i}} \left(\frac{u_i - \sum_{k=1}^{|\chi_k|} P_{\chi_i^k} \tau_{\chi_i^k}}{\Gamma_{\varphi_i}} \right) \tag{14.8}$$

In the earlier equation, we have assumed that consumers are profit-maximizing agents. Instead, in a cooperative environment where consumers are unselfish and intend to maximum the number of tasks completed subject to budget constraints, the optimization problem faced by the consumers is of the form:

$$\min_{P_{\chi_i}} (\Gamma_{\varphi_i})$$

subject to the condition that

$$\left(u_i \geq \sum_{k=1}^{|\chi_k|} P_{\chi_i^k} \tau_{\chi_i^k} \right) \tag{14.9}$$

For the earlier formulation, we assumed that the consumer can simultaneously pursue multiple consumer tasks and one sensor task pertaining to each consumer task is active at any time. For example, in the MASM simulation scenario involving target destruction explained previously, this translates to the assumption that consumers can either search for targets or track a particular target at any given time, but cannot do both simultaneously. These assumptions were guided by the design of consumer agents in the MASM simulation environment (see the next section). For alternate scenarios, Equations 14.8 and 14.9 have to be reformulated accordingly. Since both Equations 14.8 and 14.9 are deterministic optimization problems, straightforward heuristic-based approaches like genetic algorithms can be used to solve them. However, the results of this optimization are dependent heavily on the price–QoS mapping Y_{ϕ_i}, which is constructed based on current market conditions.

Directly using the values obtained by maximizing Equation 14.6 leads to extensive reliance on current market conditions. For example, a particular consumer agent that is tracking a highly important target might bid aggressively for tracking resources. As a result, the estimate of price–QoS mappings for the track subtask might show a temporary shift. If all the consumer agents adapt rapidly to the new auction outcomes, they cause a permanent price increase in the market. To avoid speculative behavior, we implemented Widrow–Hoff learning to buffer consumer responses, similar to the approach used by Cliff and Bruten's Zero Intelligence Plus (ZIP) traders (Cliff and Bruten 1997).

ZIP traders were originally designed to extend the zero intelligence agent–based simulations of Gode and Sunder (1997). The Widrow–Hoff learning rule with momentum (Widrow and Hoff 1960) is used by the consumers to adapt their bid prices, based on their bid prices in the previous auction round and the calculated optimal prices. Assume that the bidding price used by the consumer for task φ_i in the previous round of scheduling is $p_{\varphi i}^{t-1}$. Let the optimal bidding price calculated according to Equation 14.6 or 14.8 be p_{φ_i}. The current bidding price for task budget, $p_{\varphi i}^t$, is calculated as follows:

$$e = (p_{\varphi i} - p_{\varphi i}^{t-1})$$

$$\tau_{updated} = \kappa * \lambda \tau_{current} + (1 - \lambda\kappa) * e \qquad (14.10)$$

$$p_{\varphi i}^t = p_{\varphi i}^{t-1} + \lambda * \tau_{updated}$$

The learning rate parameter, $\lambda\kappa$ determines the rate at which the budget is changed. During each iteration, the search budget is updated using the current momentum, $\tau_{current}$, which is a weighted sum of the momentum during the previous iteration and the current error. Current error is defined as the difference between the current search budget and optimal search budget. The momentum rate parameter, $\kappa\tau$ determines the weight given to the past changes in the calculation of momentum.

14.4 PERFORMANCE EVALUATION USING A SIMULATION TEST BED

14.4.1 SIMULATION PLATFORM

We developed a multisensor, multiconsumer, multitarget platform to serve as a test bed for MASM. The design of the sensor network and the communication channel are derived from McIntyre and Hintz (1997). The complete details of the simulation environment are available in Avasarala (2006). The simulation environment represents a two-dimensional area where targets are uniformly distributed. These targets move around the search area with constant velocity. The search area has several different kinds of sensors, including sensors that provide range and bearing, bearings only sensors, electronic support measure (ESM) sensors. The simulation has a set of software agents that search for and destroy targets. The agents are not provided any sensing resources and they depend on the sensor network for obtaining information about the environment. They bid for sensor resources and update their status based on information provided by the SM. They use the sensor network's resource to search for potential targets and if the probability of target existence within their range exceeds a certain threshold, initialize target tracks. Once a target track is initialized, the agents attack the target if their confidence in the target position is greater than a certain threshold. This is again accomplished by buying sensing resources from the sensor network. Agents are assumed to have a utility u_t (=1.0) for destroying a target. To divide the overall utility into utilities for search and track tasks, agents initially use equal priorities. During the simulation run, agents update the search to track budget ratio using the learning previously described. The consumer agents converged to a budget of 0.95 for tracking task and 0.05 for search task (Avasarala 2006). In this simulation environment, we found that MASM outperforms information-theoretic sensor management. For a detailed review of these results, refer to Avasarala (2006). Here, we elaborate on the nonmyopic nature of MASM.

We found that MASM uses resource prices to implicitly reserve resources for future use by high priority tasks, even if no high priority tasks are currently in progress. For example, consider a situation where the first user is tracking a target and the rest of the users are in search mode. Both MASM and ITSM give highest priority to the track task. The first user has a high budget for a track bid and bids accordingly. However, during the tracking task, the prices associated with the sensing resources increases since the rate of their battery power usage during tracking is high. After the tracking task is completed and when only detection tasks are in progress, prices of the sensor schedules would have increased. Consequently, sensors will be used at a slower rate during the detection phase, effectively reserving sensors for future higher-priority tasks. However, ITSM has no method of prioritizing between two tasks, except when both the tasks are currently in progress. Figure 14.3 shows the number of sensors used during different rounds of scheduling using MASM, where the number of sensors used when tracking tasks are in progress is higher than the number of sensors used when only detection tasks are in progress. When only detection tasks are present, a significant percent of sensors are resting, thereby preserving their battery power for future use.

This explains how MASM addresses the situation presented in Scenario I.

Although the current simulation is not equipped to generate situations similar to Scenario II, it is easy to how MASM would generate nongreedy behavior for this situation, even while using the simple consumer learning behavior explained earlier. If MASM is used for resource allocation in Scenario II, it will learn to price the sensors in region $R2$ higher than sensors in region $R3$ after a certain period of operation, since they are scarcely available. Consequently, the price–QoS relationship for tracking A constructed using historic data will yield a lower QoS for the same price as compared to the price–QoS relationship for tracking B. Therefore, bids on task B will be priced lower than bids on task A since bidding agents will learn that task B can be completed at a overall lower cost. Consequently, in $R1$, the tracking tasks involving targets that are expected

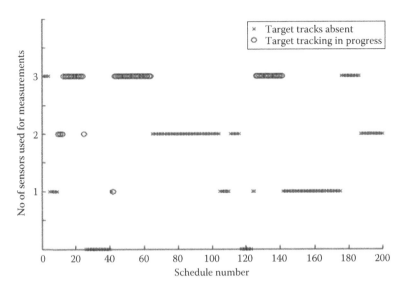

FIGURE 14.3 A comparison of the number of sensors used for measurements, based on whether target tracks are currently in progress or not.

to move to $R2$ get preferential resource allocation than tracking tasks that involve targets expected to move to $R3$.

14.5 CONCLUSION

In this chapter, we presented two promising techniques for implementing nonmyopic sensor management. Approximate dynamic programming approaches offer a rigorous framework for this problem but require elaborate problem-specific formulation. Market-oriented programming approaches offer a comprehensive framework for multisensor, multiuser, multitask nonmyopic sensor management. However, the implementation involves using a few heuristics that might have to be optimized specifically for the concerned domain.

REFERENCES

Andersson, A., M. Tenhunen, and F. Ygge. 2000. Integer programming for combinatorial auction winner determination. Presented at *Fourth International Conference on Multi-agent Systems (ICMAS)*, Boston, MA.

Avasarala, V. 2006. Multi-agent systems for data-rich, information-poor environments. PhD College of Information Sciences and Technology, The Pennsylvania State University, State College, PA.

Avasarala, V., T. Mullen, D. Hall, and S. Tumu. 2009. An experimental study on agent learning for market-based sensor management. *IEEE Symposium on Computational Intelligence in Multicriteria Decision-Making (MCDM 2009)*, Nashville, TN, pp. 30–37.

Avasarala, V., H. Polavarapu, and T. Mullen. 2006. An approximate algorithm for resource allocation using combinatorial auctions. *Proceedings of IAT 2006*, Hong Kong, China, pp. 571–578.

Balazinska, M., A. Deshpande, M. Franklin et al. 2007. Data management in the Worldwide Sensor Web. *IEEE Pervasive Computing*, 6(2), 30–40.

Bertsekas, D. P. and J. N. Tsitsiklis. 1996. *Neuro-Dynamic Programming*. Belmont, MA: Athena Scientific.

Castañon, D. A. 1995. Optimal search strategies for dynamic hypothesis testing. *IEEE Transactions on Systems, Man, and Cybernetics*, 25, 1130–1138.

Castañon, D. A. 1997. Approximate dynamic programming for sensor management. *Proceedings 36th Conference on Decision and Control, IEEE*, December, San Diego, CA, Vol. 2, pp. 1202–1207.

Cheng, E., K. Leung, K. Lochner et al. 2005. Walverine: A Walrasian trading agent. *Decision Support Systems*, 39, 169–184.

Cliff, D. and J. Bruten. 1997. Minimal-intelligence agents for bargaining behaviors in market-based environments. Technical Report, HPL-97-91, HP Labs, Bristol, U.K.

Cramton, P., Y. Shoham, and R. Steinberg, eds. 2005. Introduction to combinatorial auctions. *Combinatorial Auctions*. Cambridge, MA: MIT Press.

Davis, R. and R. G. Smith. 1983. Negotiation as a metaphor for distributed problem solving. *Artificial Intelligence*, 20, 63–109.

Gelly, S., J. Mary, and O. Teytaud. 2006. Learning for stochastic dynamic programming. *11th European Symposium on Artificial Neural Networks, ESANN 2006*, April, Bruges, Belgium, pp. 191–196.

Gode, D. K. and S. Sunder. 1997. Allocative efficiency of markets with zero-intelligence traders: Market as a partial substitute for individual rationality. *Journal of Political Economy*, 101, 119–137.

Hall, D. L. and S. A. H. McMullen. 2004. *Mathematical Techniques in Multisensor Data Fusion*, 2nd edn. Norwood, MA: Artech House Publishers.

Kastella, K. 1996. Discrimination gain for sensor management in multi-target detection and tracking. *Proceedings IEEE-SMC and IMACS Multi-Conference CESA'96*, July, Lille, France, Vol. 1, pp. 167–172.

Mainland, G., D. C. Parkes, and M. Welsh. 2005. Decentralized, adaptive resource allocation for sensor networks. Presented at *2nd USENIX/ACM Symposium on Networked Systems Design & Implementation (NSDI '05)*, Boston, MA.

Manyika, J. M. and H. F. D. Whyte. 1994. *Data Fusion and Sensor Management: A Decentralized Information Theoretic Approach*. New York: Ellis Horwood.

McIntyre, G. A. and K. J. Hintz. 1996. An information theoretic approach to sensor scheduling. *Proceedings of SPIE*, 2755, 304–312.

McIntyre, G. A. and K. J. Hintz. 1997. Sensor management simulation and comparative study. *Proceedings of SPIE*, 68, 250–260.

Samuelson, P. A. 1947. *Foundations of Economic Analysis*. Cambridge, MA: Harvard University Press.

Sandholm, T. 1993. An implementation of the contract net protocol based on marginal cost calculations. *Proceedings of the National Conference on Artificial Intelligence*, Washington, DC, pp. 256–262, Association for Advancement of Artificial Intelligence (AAAI), Palo Alto, CA.

Schneider, M. K. and C.-Y. Chong. 2006. A rollout algorithm to coordinate multiple sensor resources to track and discriminate targets. *Proceedings of the SPIE Conference on Signal Processing, Sensor Fusion and Target Recognition*, April, Orlando, FL, Vol. 6235, 62350E1–62350E10.

Shoven, J. B. and J. Whalley. 1992. *Applying General Equilibrium*. New York: Cambridge University Press.

Waldspurger, C. A., T. Hogg, B. A. Huberman, J. O. Kephart, and S. Stornetta. 1992. Spawn: A distributed computational economy. *IEEE Transactions on Software Engineering*, 18, 103–117.

Walras, L. 1954. *Elements of Pure Economics.* Homewood, IL: Irwin.

Walsh, W. E., M. P. Wellmen, P. R. Wurman, and J. K. MacKie-Mason. 1998. Some economics of market-based distributed scheduling. *Proceedings of the 18th International conference on Distributed Computing Systems*, Amsterdam, pp. 612–621.

Washburn, R., M. Schneider, and J. Fox. 2002. Stochastic dynamic programming based approaches to sensor resource management. *Proceedings of the 5th International Conference on Information Fusion*, Annapolis, MD, pp. 608–615.

Wellman, M. P. 1991. Review of Huberman. *Artificial Intelligence*, 52, 205–218.

Wellman, W., W. Walsh, P. Wurman, and J. MacKie-Mason. 2001. Auction protocols for decentralized scheduling. *Games and Economic Behavior*, 35, 271–303.

Widrow, B. and M. E. Hoff. 1960. Adaptive switching circuits. *IRE WESCON Convention Record*, 4, 96–104.

Williams, J. L., J. W. Fisher III, and A. S. Willsky. 2005. An approximate dynamic programming approach to a communication constrained sensor management problem. *Proceedings of the 8th International Conference Information Fusion*, July, Philadelphia, PA, pp. 1–8.

Ygge, F. 1998. Market-oriented programming and its application to power load management. PhD, Department of Computer Science, Lund University, Lund, Sweden.

15 Test and Evaluation of Distributed Data and Information Fusion Systems and Processes

James Llinas, Christopher Bowman,
and Kedar Sambhoos

CONTENTS

15.1 Brief Remarks on Abstract Concepts Related to Test and Evaluation.......... 379
15.2 Understanding Distributed Fusion System Concepts
     and Implications for Test and Evaluation 382
     15.2.1 Implications for Test and Evaluation 384
     15.2.2 Measures and Metrics in the Network Value Chain......................... 386
     15.2.3 Fusion Estimates and Truth States... 388
     15.2.4 Notion of a Performance Evaluation Tree 389
     15.2.5 Complexities in Error Audit Trails 393
     15.2.6 Formal Experimental Design and Statistical Analyses................... 393
15.3 Summarizing Impacts to and Strategies for Distributed
     Fusion System T&E 398
15.4 Remarks from a DDIFS Use Case... 399
15.5 Summary and Conclusions 404
References.. 407

15.1 BRIEF REMARKS ON ABSTRACT CONCEPTS RELATED TO TEST AND EVALUATION

Since both the first and second editions of this handbook for multisensor data fusion have addressed various of the fundamental and abstract concepts related to test and evaluation (T&E henceforth), we will only briefly remark and summarize some of these still-important notions (see Hall and Llinas [2001], chapter 20 and Liggins et al. [2009], chapter 25 for more extended remarks).

A still-valid and important assertion is that generalized methods and procedures as well as formalized statistical/mathematical analysis techniques for improved T&E of data and information fusion systems and processes are still very understudied. Partially as a result of this deficiency, transition of seemingly capable data fusion systems has been problematical. Among factors needing additional study to smooth the path toward reliable operational use are new, user-oriented techniques, for example, involving new ideas on metrics such as trustworthiness metrics as well as those related to dealing with the inherent uncertainties involved with fusion processes so that a clear understanding of these effects can be realized by typical users. As the field of multisensor data fusion moves its focus to the higher levels of abstraction and inference (Levels 2, 3, and 4 of the Revised JDL Model, see Figure 15.1 [Steinberg et al. 1999, Bowman and Steinberg 2001, Bowman 2004, Llinas et al. 2004, Steinberg and Bowman 2004]), efforts have been made toward defining new types of metrics for these fusion levels (Blasch 2003; Blasch et al. 2004, 2010; Haith and Bowman 2010) with extensions for dual resource management levels, but much more needs to be done in regard to efficient T&E techniques for high-level state estimates, since these are much more complex and of higher dimensionality than Level 1 type states.

Following Hall and Llinas (2001) and Liggins et al. (2009), we still emphasize the remarks having to do with test philosophy and context. A *philosophy* is that line of thinking that establishes or emphasizes a particular point of view for the tests and/or evaluations that follow. Philosophies primarily establish points of view or perspectives for T&E that are consistent with, and can be traced to, the goals and objectives: they establish the *purpose* of investing in the T&E process. T&E philosophies, while generally stated in nonfinancial terms, do in fact establish economic philosophies for the commitment of funds and resources to the T&E process. The simplest example of

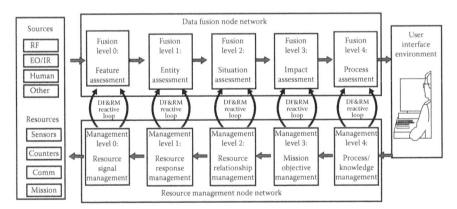

FIGURE 15.1   Dual functional levels of the data fusion and resource management (DF&RM) dual node network (DNN) technical architecture (Data from Bowman, C.L, The dual node network (DNN) DF&RM architecture, *AIAA Intelligent Systems Conference*, Chicago, IL, 2004; Bowman, C.L. and A.N. Steinberg, A systems engineering approach for implementing data fusion systems, In *Handbook of Multisensor Data Fusion*, D. Hall and J. Llinas, J. (Eds.), Chapter 16. Boca Raton, FL: CRC Press, 2001; Steinberg, A.N., et al., *Proceedings of SPIE Conference Sensor Fusion: Architectures, Algorithms, and Applications III*, 3719, 430, 1999; Steinberg, A. and C.L. Bowman, Rethinking the JDL data fusion levels, National symposium on sensor and data fusion (NSSDF). Johns Hopkins Applied Physics Lab (JHAPL), Laurel, MD, 2004; Llinas, J. et al., Revisiting the JDL data fusion model II. *International Conference on Information Fusion*, Stockholm, Sweden, 2004.)

this notion is reflected in the so-called black box or white box viewpoints for T&E, from which either external (I/O behaviors) or internal (procedure execution behaviors) are examined. Another point of view revolves about the research or development goals established for the program. The philosophy establishes the high-level statement of the context for testing and is closely intertwined with the program goals and objectives. Assessments of delivered value for defense or other critical systems must be judged in light of system or program goals and objectives. In the design and development of such systems, many translations of the stated goals and objectives occur as a result of the systems engineering process, which both analyzes (decomposes) the goals into functional and performance requirements and synthesizes (reassembles) system components intended to perform in accordance with these requirements. Throughout this process, however, the program goals and objectives must be kept in view because they establish the context in which value will be judged.

Context, therefore, reflects what the program (i.e., the DF&RM process or a function within it) is trying to achieve—(e.g., what the research or developmental goals [the purposes of building the system at hand] or the learning intelligent DF&RM system are). Such goals are typically reflected in the program name, such as a "Proof of Concept" program or "Production Prototype" program. Many recent programs involve "demonstrations" or "experiments" of some type or other, with these words reflecting in part the nature of such program goals or objectives.

Once having espoused one or another of the philosophies, there exists a perspective from which to select various criteria, which will collectively provide a basis for evaluation. There is, in the most general case, a functional relationship as

$$\text{Criterion} = \text{fct} \, [(\text{Measure}_i = \text{fct} \, (\text{Metric}_i, \text{Metric}_j \ldots),$$

$$\text{Measure}_j = \text{fct} \, (\text{Metric}, \text{Metric}_i \ldots), \text{etc.}]$$

that defines how each criterion is dependent on certain measures that are in turn derived hierarchical functions of higher-level metrics (e.g., from probability of mission success on down), which are the quantities that are (importantly) observable in an experiment. Each metric, measure, and criterion also has a scale that must be considered. Moreover, the scales are often incongruent so that some type of normalized *figure of merit* approach may be necessary in order to integrate metrics on disparate scales and construct a unified, quantitative parameter for making judgments.

Another important element of the T&E framework is the *approach* element of the T&E process. In this sense, approach means a set of activities, which are both procedural and analytical, that generates the "measure" results of interest (via analytical operations on the observed metrics) as well as provides the mechanics by which decisions are made based on those measures and in relation to the criteria. The approach consists of two components as described in the following:

- A *procedure*, which is a metric-gathering paradigm; it is an experimental procedure.
- An *experimental design*, which defines (1) the test cases, (2) the standards for evaluation, and (3) the analytical framework for assessing the results.

Aspects of experimental design include the formal methods of classical, statistical, experimental design. Few if any fusion T&E research efforts in the literature have applied this type of formal strategy, presumably as a result of cost limitations or other unstated factors. Nevertheless, there are the serious questions of sample size and confidence intervals for estimates, among others, to deal with in the formulation of any T&E program, since simple comparisons of mean values, etc. under unstructured test conditions may not have very much statistical significance in comparison to the formal requirements of a rigorous experimental design. Any fusion-based T&E program, because all fusion systems and processes are inherently dealing with random variables and stochastic behaviors, should at least recognize the risks associated with such simplified analyses.

The T&E process contains a Level 4 fusion performance evaluation (PE) process as per the DNN technical architecture; see Haith and Bowman (2010), Bowman (2008), Gelfand et al. (2009), and Bowman et al. (2009). The PE process architecture typically involves a network of interlaced PE fusion and T&E Process Management (PM) DF&RM nodes. Each PE node performs data preparation, data association, and state estimation where the data are DF&RM outputs (e.g., track estimates) and estimates of truth and the output state estimates are the Measures of Performance (MOPS) such as described earlier.

Much more is said in Hall and Llinas (2001) and Liggins et al. (2009) on the general considerations of T&E for data and information fusion systems, and the interested reader is directed to those sources for additional commentary and insight, along with many references.

15.2 UNDERSTANDING DISTRIBUTED FUSION SYSTEM CONCEPTS AND IMPLICATIONS FOR TEST AND EVALUATION

Testing and evaluating anything requires that the item to be tested, the "test article," be clearly defined. It is also important to understand the role or purpose of the test article in the context of its use or setting in a larger system framework. To stimulate this discussion, let us characterize what a Distributed Data or Information Fusion System (DDIFS) is; it is appreciated that other chapters in this book may have other characterizations of a DDIFS, but we feel it is important to review these, even if redundant, in relation to developing thoughts about testing and evaluating such systems and functions. So, our local characterization describes a DDIFS as follows:

1. It is first of all "distributed," meaning that its components (which immediately implies that it comprises a number of components) are spread apart somehow; very often this is a geographical separation, or for defense/security applications, a platform separation where DDIFS components are hosted on an aircraft or ship or satellite, etc. (that could, in turn, be geographically separated)—thus we can also have a kind of local distribution embedded in a larger distributed-system context.

2. The components are interconnected (informationally) according to the design of a specified communication/datalinking network, and share information and/or processing results according to some supported protocol.

   a. Note that this makes the components interdependent in some way.

3. The components also may have local resources of various description or type to include sensors, processors, and manageable resources; representative component functionalities can include sensor nodes, processing nodes, fusion nodes, communication nodes, etc.—not every component in a DDIFS is necessarily a fusion node in the sense of producing state estimates, as some may perform functions that contribute to the formation of estimates.

4. In this framework, the components can only fuse two things: local ("organic") information from resources that they "own" (i.e., for which they have control and design authority), and information that comes to them "somehow" (i.e., according to the inter-component information-sharing strategy (ISS) or protocol) from other components in the networked system (we interchange the terms distributed and networked).

5. Metadata must also be shared across components along with the shared information in order that sender-components can appropriately inform receiver-components of certain information necessary to subsequent processing of the sent "message" or data-parcel by receiver-components. (It can also be the case that receiver-components can request information from various other components, and such requests may have metadata and the requesting-component may ask that certain metadata be contained in the reply.)

   a. Another reason for metadata is due to the generally large size of most distributed systems that prevents any given component from knowing much about "distant" components and their (dynamic) status.

6. The topology of the DDIFS is very important since it affects a number of overall system properties to include connectivity, failure vulnerability, etc. Table 15.1, drawn largely from Durrant-Whyte (2000) and Utete (1994), shows a subjective characterization of some DDIFS properties as a function of topological type.

7. It can be expected that in larger, complex systems any given fusion components or nodes may have to have two fusion processes operating, one to process local, organic data as described earlier—since these data are best understood by the local node, allowing optimal fusion processes to be developed—and one to process received network information, about which only the metadata are known, restricting the realization of optimal methods for this "external" data. Such separation may also be required because of distinct differences in the nature of the fusion operations, requiring different algorithmic techniques.

We will restrict our discussion to DDIFSs that are coherently designed toward some bounded set of overarching purposes and capabilities, as distinct from loosely coupled sets of components that may operate opportunistically or some ad hoc manner, in the fashion of a federated system. Such a restriction makes development of a T&E scheme or plan more controllable, but nevertheless requires a clear partitioning

TABLE 15.1

Subjectively Judged Properties of a DDIFS as a Function of Topology-Class

DDIFS Topology	Inherent Redundancy/ Failure Protection	Scalability	Ability to Manage Redundant Information (Double Counting)	Practicality
Fully connected	Good	Very poor	Good	Poor
Tree	Poor; branch failures can lead to tree splitting	Poor	Limited to most recent transaction	Reasonable
Decentralized	Moderate	Good	Possible but requires careful design	Good
Dynamically managed	Good	Moderate	Moderate	Complex

Source: Adapted from Durrant-Whyte, H.F., A beginner's guide to decentralized data fusion, Technical report, Australian Centre for Field Robotics, University of Sydney, Sydney, New South Wales, Australia, 2000; Utete, S., Network management in decentralized sensing systems, PhD thesis, The University of Oxford, Oxford, U.K., 1994.

of requirements and capability specifications to parts of the DDIFS. Note that this framework is not unlike the system design approach characterized by Bowman for centralized and distributed fusion systems (Bowman 1994).

There is another feature of DDIFSs that needs to be mentioned, although this attribute is applicable to all information fusion systems and many intelligence and surveillance systems that may not even employ IF methods. That attribute is that the fundamental nature of any IF or DDIFS is that the data and even often the knowledge employed in system design and in system operation has a stochastic quality.* This immediately raises the question about how to define and develop a T&E approach, in the sense of questioning how to account for and measure statistically assured confidence in the test results. More is said about this in the later sections of the chapter.

15.2.1 IMPLICATIONS FOR TEST AND EVALUATION

There are various implications that the features of a DDIFS impute onto the nature of and methods for T&E. For example, there are two broad types of testing used in the development of defense systems: developmental test & evaluation (DT&E) and operational test & evaluation (OT&E). DT&E is oriented to a bounded system as a test article, the "system under test or SUT," and verifies that the system's design is satisfactory and that all technical specifications and contract requirements have been met. It is kind of a check-list process of examining whether defined SUT requirements

---

* Sensibly every sensor has embedded thermal noise and other factors that attach randomness to the measured/observed data obtained by the sensor; much system design knowledge is imperfect and draws from world models that have inherent stochastic features, and some such knowledge is drawn from knowledge elicitation from humans that of course involve imperfect and random effects.

have been met, one-by-one, as determined by T&E processes that address these requirements either singly or in combination. As noted, it is typically a process that is checking that the delivered system satisfies contractual requirements and so is closely related to the acquisition process. DT&E is usually managed by the governmental client but can be conducted by the government, by the contractor, or by a combined test team with representatives from both government and industry. Most early DT&E in a program will likely be done at the contractor's facilities under controlled, laboratory conditions. OT&E follows DT&E and validates that the SUT can satisfactorily execute its mission in a realistic operational environment including typical operators and representative threats. The difference between DT&E and OT&E is that DT&E verifies that the system is built correctly in accordance with the specification and contract, and OT&E validates that the system can successfully accomplish its mission in a realistic operational environment. Another way to think of these differences is that DT&E is concerned chiefly with attainment of engineering design goals, whereas OT&E focuses on the system's operational effectiveness, suitability, and survivability.

For DDIFSs, it can be seen that these differences can become cloudy and problematical, due to the underlying nature of various interdependencies between nodes or platforms in such a system. To define a SUT in a DDIFS, one must cut the connectivity to the network at some points so that a standalone, bounded system can be tested as an integrated deliverable within a contract framework. We will later in this chapter discuss our work in supporting the U.S. Major Test Range at Edwards Air Force Base, California, in their preparations for testing new tactical aircraft that have embedded datalinking and data fusion capabilities. These platforms are designed to share sensors and data, as well as locally computed parameters and target tracks, for example. The fundamental mission sortie envisions multiple aircraft flying cooperatively together in the execution of a mission. However, Edwards has historically been a DT&E facility, testing aircraft against single-platform requirements. With the evolution of fusion-capable aircraft and purposefully cooperative mission plans, the nature of what comprises a SUT and how to do DT&E gets muddy. It may be that some new type of T&E activity that bridges between DT&E and OT&E will need to be defined and developed. Such issues also raise the question of the costs of such boundary activities, for example, the very high cost of flying multi-aircraft "SUTs," or the corresponding technical challenge of developing real-time capable surrogate aircraft simulation capabilities as virtual wingmen as one alternative strategy for a cost-effective approach. So it can be seen that there are some subtle but nontrivial issues to deal with when deciding on a scheme for DDIFS DT&E and OT&E.

We are discussing here automated DDIFSs, where the core technical and functional capabilities are enabled in software, so another core issue in thinking about DDIFS T&E is the domain of software testing. By and large, software testing is the process of executing a program or system with the intent of finding errors. Software is not unlike other physical or functional processes where inputs are received and outputs are produced, but where software differs is in the manner in which software processes fail. Most physical systems fail in a fixed and bounded set of ways. By contrast, software, ironically because of a wide variety of interdependencies (analogous to DDIFSs in the large), can fail in many bizarre ways. Detecting all of the different failure modes for software is generally infeasible because the complexity of software is generally intractable. Unlike most physical systems, most of the defects in software are design errors, and once the software is shipped, the design defects—or bugs—will be buried in and remain latent until activation.

The transition to network-centric capabilities has introduced new T&E challenges. Network functional capabilities can reside in both nodes and links, and various common system capabilities can reside in, for example, service-oriented architecture (SOA) infrastructures. The T&E of capabilities in this type of framework involving specialized and common functionalities requires new thinking and a new strategy; this is another SUT-defining challenge. In the same way that using live/real nodes or platforms in testing adds great expense as was discussed previously, evaluating the performance of the software network itself is probably not going to be accomplished without extensive use of modeling and simulation because the expense of adding live nodes in a laboratory increases dramatically with the number of nodes added to the test apparatus. A T&E strategy that mitigates risk in the development of a network infrastructure that will support network-centric warfare requires a balance of theoretical analysis and laboratory testing.

15.2.2 MEASURES AND METRICS IN THE NETWORK VALUE CHAIN

Chapter 3 addressed the topic of the network-centric value chain. The value chain has three major quality dimensions: data/information quality, quality of shareability/reachability, and quality of interactions. All of these dimensions will occur to varying degrees in any net-centric operation (NCO), and the degrees to which they occur form the basis for the wide range of metrics suggested in Garstka and Alberts (2004).

This viewpoint is shown in Figure 15.2, from Garstka and Alberts (2004). One way then to develop a measures and metrics framework for a DDIFS is to simply shift the labeling from the NCO application to DDIFS, as there are more or less one-to-one equivalencies in the applicability of these notions as a basis for T&E and a basis of measurement. One distinction would be that fusion processes do not inherently yield a Sensemaking capability but they can be key to realizing such capability. The fusion–Sensemaking interdependency is expressed and actualized via well-designed human–computer interfaces.

In the same way that in Section 15.1 we defined criteria, their dependency on measures, and the dependency of measures on metrics (the ultimate parameters measured in a T&E experiment or trial), Garstka and Alberts (2004) define top-level concepts (the three we have discussed), the attributes upon which they depend, and the measures and metrics used to quantify them. These dependencies are shown in Figure 15.3, from Garstka and Alberts (2004). Four categories of attributes are defined (excerpted literally from Garstka and Alberts [2004]):

Objective Attributes measure quality in reference to criteria that are independent of the situation. For example, the *currency* of a given data element indicates the age of the information available and can be expressed in units like minutes, hours, days, etc.

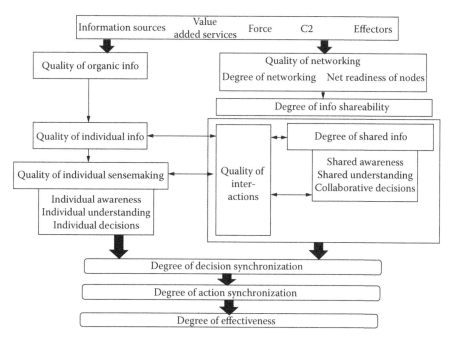

FIGURE 15.2 The NCO framework with quality and degree measures. (Adapted from Garstka, J. and Alberts, D., Network Centric Operations Conceptual Framework Version 2.0, U.S. Office of Force Transformation and Office of the Assistant Secretary of Defense for Networks and Information Integration, Vienna, VA, 2004.)

Fitness-for-Use Attributes measure quality in reference to criteria that are determined by the situation. For example, the *timeliness* of a given data element indicates the extent to which the information is received in a time that is appropriate for its intended use. What is appropriate is context dependent. In some contexts a currency of two hours is adequate, whereas in other contexts a currency of two minutes is what is needed. Fitness-for-use attributes allows one to capture information that is context dependent.

Agility Attributes measure the aspects of agility across the six dimensions. These attributes inherently are comparative, i.e., agility implies an ability to change over time and, as such, the values of the metrics for these attributes have to be compared to some baseline values.

Concept Specific Attributes measure unique aspects of some concepts. For instance, synchronicity is an attribute of the Quality of Interactions concept that measures the extent to which C2 processes are effective across time (synchronous vs. asynchronous) and space (collocated vs. distributed). This attribute is appropriate in determining the extent to which elements in a C2 organization can interact simultaneously in time and space but is not necessarily relevant to other concepts.

In the same way that we ported the NCO evaluation concepts to the DDIFS application, these attribute categories can also be ported to DDIFS applicability.

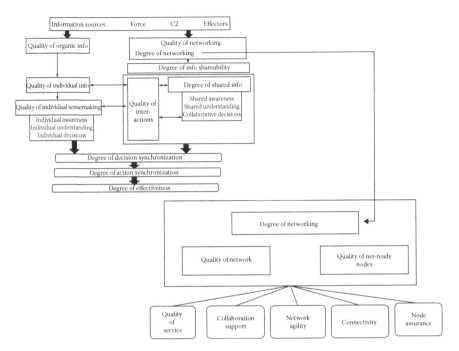

FIGURE 15.3 Concepts-to-quality/degree dimensions-to-attributes and measures/metrics. (Adapted from Garstka, J. and Alberts, D., Network Centric Operations Conceptual Framework Version 2.0, U.S. Office of Force Transformation and Office of the Assistant Secretary of Defense for Networks and Information Integration, Vienna, VA, 2004.)

15.2.3 FUSION ESTIMATES AND TRUTH STATES

Information fusion can generally be thought of as an association and estimation process, yielding estimates ranging from attributes of an entity to an estimate of a complex, dynamic, multi-entity situational picture. The entities can be physical objects, events, relationships, and courses of action (COAs). The entity estimates are based upon the association of the data together over space, time, type, etc. as an entity. A rational basis for evaluating the performance of such an estimation process is to compare the estimates to the underlying truth states of either the entity attributes or situations, or whatever estimation product is sought from the fusion process. When the fusion-based estimation process involves multiple entities of various types (physical objects, events, behaviors, informational entities, etc.), there can be a combinatoric complexity in determining which fused estimate should be compared with which truth entity; this is an issue that is known in the fusion community and typically called the "track-to-truth" problem, as the question arose in the application to evaluating multitarget tracking systems. The problem gives rise to essentially a separate fusion type problem, in which an adjunct data association function is required to reconcile which estimate-to-true associations are correct in order to support subsequent computation of estimation errors. These PE functions are part of the Level 4 Process Assessment Fusion which lies outside the L0–3 Fusion SUT.

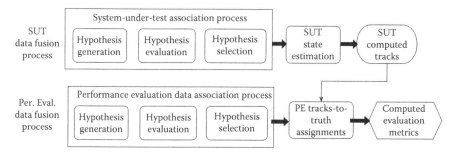

FIGURE 15.4 Multiple data fusion/association processes in fusion process performance evaluation.

The idea is shown in Figure 15.4, for a multitarget tracking application. On the top we have the SUT fusion process (where we have removed the data preparation common referencing processes for both SUT and PE for simplicity), involving the traditional three-step data association process supporting the production of the SUT state estimates and computed multitarget tracks. Below we have the PE data association process that determines, using an association score and an assignment algorithm, which SUT tracks should be compared to which truth tracks. The assertion of these associations is a core functionality that supports, in turn, the computation of evaluation metrics. It can be seen that, the specific details and nature of the evaluation metrics are clearly interdependent on the methodological details of the PE data association process. Such considerations become yet more complex in the distributed fusion (DDIFS) application since there are different state estimates being produced at various nodes and shared and further fused across the intermodal network.

The DNN technical architecture specified PE Level 4 fusion node is shown in Figure 15.5.

15.2.4 NOTION OF A PERFORMANCE EVALUATION TREE

These PE nodes are processed in networks (e.g., trees) that are interlaced with the SUT DF&RM nodes such as shown in Figure 15.6 for RF, electronic support

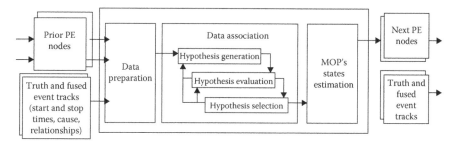

FIGURE 15.5 Exploded view of PE node processes.

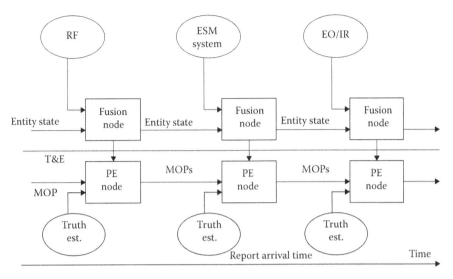

FIGURE 15.6 PE process in context of data fusion and resource management architecture.

measures (ESM), and EO/IR SUT fusion nodes. These PE nodes can also be inter-laced with PM nodes that manage how these nodes are applied over time, mission, space, etc.

The DNN technical architecture helps to break the PE and PM processes into more manageable design steps as follows for a PE process architecture.

Step 1 is PE role optimization. This step defines the role for PE as a blackbox to include all its inputs and outputs and measures of success. An example of the role for PE in a T&E system is shown in Figure 15.7. In this example, there is a SUT with multiple subsystems, with the fusion subsystem being only one of them. The role for fusion here is to support certain SUT Effector systems such as weapon systems, and to support a user, say a pilot. The evaluation focus is on the fusion system in this particular context or role only. This is also the step where baseline accuracy and

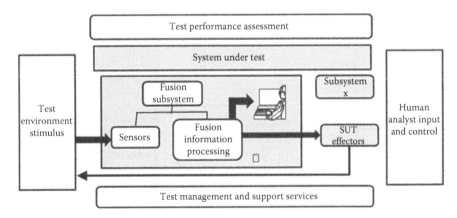

FIGURE 15.7 Sample role for PE in a T&E system.

timing MOPS based on PE requirements are established. By providing an objective evaluation, the MOPS help

- Determine whether the fusion algorithms meet engineering and operational requirements
- Compare alternative algorithms and approaches
- Optimize configuration parameters for a given algorithm

Step 2 is PE fusion network optimization. This step determines how to divide and conquer the PE problem over space, mission, time, etc.

Step 3 is the PE fusion node optimization. While the details of the PE algorithms are tailored to each data analysis step, the overall process follows the canonical fusion node steps, namely data preparation, data association, and state estimation as described earlier. Examples of these processes are given in the remaining sections in this chapter.

Building upon the previous remarks, it can be seen that in a DDIFS there are estimates being produced at different nodes and these estimates are also evolving in time. In turn, additional fusion operations are occurring at certain receiving nodes that combine the estimates sent to them from various sending nodes. Thus, there is also a temporal dimension to the T&E functions (true for most fusion processes whether distributed or not, so long as the problem space is dynamic), and there can be a need to compute both the evolving real-time performance as well as to compute cumulative performance. Hence, in the same way that a typical fusion process can be viewed as a kind of tree or in general a network (see Bowman [1994]), one can also envision a PE Tree, as introduced earlier. The PE Tree will have various computational modules, nodes that accumulate evaluation-related computations, and a network of such nodes that gather the computations in a coordinated way according to the PE Tree design, framed to satisfy the overall role of the PE/T&E process for the evaluation process. As one example, the PE nodes could be the places where, for example, the evaluation calculations for a given platform in a multiplatform DDIFS system are gathered. The rationale for arranging or batching the PE nodes can be drawn from the same considerations given to batching fusion processes in multi-sensor type systems; this idea is shown in Figure 15.8, where one could think of batching the PE nodal processes according to individual sources or sensor—these could also be thought of as nodes of a given type in a DDIFS—or according to a PE sampling-time—or according to important events from an evaluative point of view. Conceptually, any of the nodes in a PE Tree can be performing the SUT-to-Truth calculations shown in Figure 15.4.

A simple, time-batched PE Tree is shown in Figure 15.9 (Rawat 2003) for a notional simple three-node DDIFS performing target tracking, being tested in a simulation environment. At each time slice, the network simulation data are sent to the fusion/tracking nodes according to whatever data-to-node and internodal communication protocol exists (these details not shown), and each node computes its track estimates accordingly. As mentioned earlier, the PE process for each node would have a track-to-truth association process that determines the local best associations for PE at the given time, according to whatever MOPS are being used. It is typical

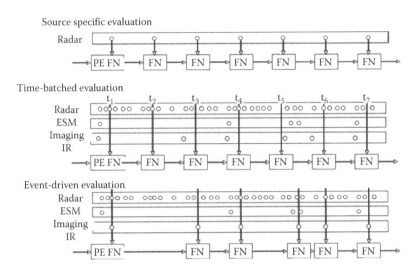

FIGURE 15.8 Alternative strategies for computing evaluative metrics in a DDIFS PE process.

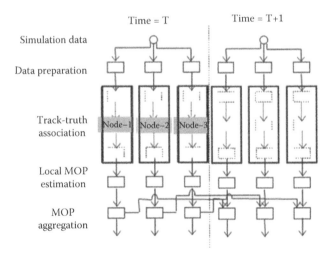

FIGURE 15.9   Notional time-based PE Tree.

that cumulative performance would want to be computed, and separate PE functions perform these separate calculations as shown. This is a simple case but it can be appreciated that PE Tree (network) design can be relatively complex for more complex network topologies; and when the network information flow protocols (the ISSs discussed previously) are more complex, along with further complexities such as separate local and network fusion operations being done at any network node, a fair (accurate, unbiased) yet affordable PE is needed. Engineering guidelines for achieving the knee-of-the-curve in PE "fairness versus complexity" have been developed based upon the DNN technical architecture. Namely,

- The PE solution space is organized as a network of PE functional nodes.
- Each PE node performs fusion and truth data preparation, data association, and MOP state estimation.

In addition, the PE process "fairness" can be improved by the Level 4 PM function as described in Haith and Bowman (2010). For example, the distributed fusion test article and the PE process functional parameters can be optimized for each test scenario to insure a "fair" comparison of alternative distributed fusion systems under test. This optimization can be based upon the top level Measure of Success (e.g., probability of meeting mission requirements) or on the selected MOPS. In this latter case, a "Pareto optimal front" (Haith and Bowman 2010) of parameter values can be derived (i.e., a boundary in parameter space where all other parameters values will yield a lesser performance in at least one MOPS).

15.2.5 COMPLEXITIES IN ERROR AUDIT TRAILS

T&E is performed in part to understand the causes of errors and it is typical that an error audit trail would be developed to understand where improvements can be or need to be made, i.e., to discern the error-producing operation and how to repair it. In the same way that pedigree metadata tags are needed for certain DDIFS fusion functions, it may be necessary to incorporate pedigree tagging to track certain network processing operations for the purpose of error tracking. At the design level, there is both a complexity and tension in developing an optimized DDIFS design between the two major functions of a DDIFS: the nodal fusion operations and the network ISS. Similarly, tracking causal errors is also problematical since, for example, the fusion processes at receiving nodes can only operate on data sent to them, so asserting a cause of a fusion deficiency can be difficult; that is, determining if there was a lack of appropriate data sent to a node or a defect in the nodal association/estimation processes can be difficult and if nothing else adds to T&E analysis complexity. This is not very different than the error audit trail complexities in other fusion systems that have any type of adaptive operation such as dynamic sensor management.

15.2.6 FORMAL EXPERIMENTAL DESIGN AND STATISTICAL ANALYSES

There is usually little argument that any fusion process produces estimates and that those estimates have a stochastic character. This is because, in the strictest sense, the inputs to the fusion processes are the statistically noisy sensor or other data having stochastic properties. These features have yet other implications for the T&E methodology, namely that the stochastic nature of the process needs to be recognized and dealt with in any T&E approach. At least when conducting any simulation-based T&E, this implies that (1) the experiments should be designed through the employment of the methods of statistical experimental design (a.k.a. design of experiments or "DOE") and (2) in conjunction with this that Monte Carlo based replications of any given test condition should be done. Further, given the execution of such planned experiments, the analysis processes would employ methods that can frame the statistical quality of the analysis results, such as methods from analysis of variance or ANOVA, as well as other formal statistical techniques.

It is recognized by the way that such rigor comes at a price, even when using simulations, and especially when doing field tests and the like. It is likely that there has been limited application of these formal methods because of the cost implications. However, DDIFSs are used in life-critical and other important applications, and it would seem that the cost of rigorous testing is a price that should be paid to assure that the best understanding of system performance is being achieved. It is only through the use of such methods that assertions about the computed metrics can be made with statistical confidence. These remarks are not only applicable to DDIFSs but also to any fusion system.

At any given phase in both the SUT fusion process design and the PE Tree process design, there is the consideration of the independent variables or Factors in that layer's design, and the Effects of each of those Factors, or perhaps even the composite Effect of certain Factor combinations that might be of interest to evaluate.* It is convenient to think of Factors as independent variables related in part to the "problem space" (e.g., in tracking problems these can be related to the nature and behaviors of the targets, or also the tracking environment, meaning both weather (which affects the nature of sensor observations used for tracking) or clutter, such as the nature and extent of "confuser" objects, etc. (we mean this in the wide sense). Factors or independent variables can also be related to the "solution space," meaning the Factors that affect the performance of particular fusion algorithms (e.g., the nature and number of models in an interacting multiple model tracker), or for the case of a DDIFS, the choice of topological structure. Thirdly, and peculiar to the nature of the overall PE process being suggested here, there are Factors involved in the PE approach itself, such as the choice of technique for track-to-truth assignment, or the Factors upon which a specific PE Tree might be partitioned, etc. Thus, in this overall approach, there are *three classes of Factors* around which the PE process revolves: Problem-space Factors, Solution-space Factors, and PE process Factors. Said otherwise, Factors are those parameters whose influence on performance is sought; in nonfusion applications the PE Factors would not normally be present, but note that now we have a new class of Factors of interest. The influence of any Factor on a performance/effectiveness measure is labeled as the "Effect" in the statistical literature, and is in essence defined explicitly by a given measure or metric. The notion of an Effect can be thought of as the change in response (i.e., in an MOP) resulting from a change in the level of that Factor. For example, we might inquire as to the Effect of a change in SUT nodal tracker type from Kalman to Alpha-Beta on a given MOP, or the difference in an MOP resulting from different inter-target spacing. At any given level of a Factor, we conduct a number of Monte Carlo *replications*, so we really

---

* We capitalize the words "Factor" and "Effect" purposely here as we are soon to introduce the language of statistically designed experiments and the associated analysis processes.

examine whether there is a statistically significant difference on a *mean* MOP value resulting from these changes in factor levels, and whether the results really reflect the significance of a statistical hypothesis test. It can happen of course that *combinations* of Factors can cause Effects; this is called an "interaction" in the statistical literature. Interactions among Factors can occur in a combinatoric sense; if there are three Factors say, then there are three 2-way interactions and one 3-way interaction (ab, ac, bc, and abc, assuming order is unimportant, as is usually done).

What is sought in determining a PE approach is a statistically sound yet cost-effective way to gather the metrics and/or measures. The statistical DOE is a formal and highly quantitative way to develop a test plan that gathers the metrics in a provably cost-effective manner. That is, a DOE-based test or experimental plan extracts the maximum statistically significant information from the minimum number of test runs. DOE is a quite-mature area of study in the field of statistics, and its specific use to perform the PE function in the overall PE Tree methodology can yield the best rigorous framework for T&E.

We believe there are two major reasons for formal experimental designs and formal methods of data analysis: statistical validation of a nominated DDIFS fusion solution for some important real-world application or statistical validation of some knowledge gained about fusion processes in a range of applications for the advancement of science in an in-depth sense (i.e., "laws" as validated, explainable empirical generalizations). This latter rationale can in fact be important for empirically learning design laws for DDIFSs, and we argue in fact that the only way to develop design guidelines for DDIFSs is empirically, due to the combinatorial complexities in choosing design variables.

Designed experiments reflect a notion of a phased learning process, in which a succession of hypotheses are confirmed or denied and knowledge is gained sequentially. The need for a phased process is typically driven by the "curse of dimensionality" and the qualification problem, i.e., that there are too many Factors whose Effects need to be understood or isolated, so that a divide-and-conquer type approach must be employed to achieve in-depth understanding. The details of a phased approach, i.e., the staging of hypotheses of inquiry, are a case-dependent choice, and are of course influenced by the stage-by-stage outcomes.

The dominant analysis methodology for statistically designed experiments is analysis of variance or ANOVA. ANOVA is an analysis technique that determines whether the *mean values of an MOP* or MOE, for the several "treatments" or set of experimental conditions (as depicted in the Factor-level combinations of the set of independent variables both in the problem-space and the solution [or fusion-process]-space) are equal or not, by examining the estimated population variances across these conditions, often using Fisher's F-statistic (the "test statistic" can change in various cases). The treatments can be the result of changing problem-domain independent variables or design-domain (fusion process) variables or PE design variables, and/or the associated levels of each variable, or, as noted previously, the Factors that influence the nature of the PE approach. The F-statistic is based on the estimates of the population variance as drawn from the sample variance of the data. ANOVA basically compares two estimates of this variance, one estimate drawn from the variance exhibited *within (all) treatment* *conditions.* That is, for any given treatment, say a given tracker design for a given problem condition, the variance of, say, position errors across the "n" Monte Carlo replications for this treatment condition is a "within treatment" variance, and only exists because of the collective errors in this tracker estimation process. As a result, this variance is called the variance due to error in the statistical DOE literature. When these within-treatment variances are properly pooled across all treatments of the experiment,* they form a pooled estimate of the (supposedly) common variance within each of the treatments. The other estimate is drawn from the variance exhibited *between (all) treatment conditions*—if we were concerned with position error, for example, this would be the variance of the mean position errors of two different trackers from the global mean position error. However, these two estimates of variance are only equal if in fact there is no difference in position error variance for each tracker. The ANOVA process and the F-statistic are the means by which a hypothesis test that in effect tests the equality of these variance estimates is performed.

When employing DOE test-planning methods, one issue that can arise is the complexity involved in designing efficient test plans if there are many independent variables (Factors) whose Effects on the DDIFS process under test or SUT want to be known. Using traditional DOE experimental designs, the number of runs that have to be made will grow exponentially when the number of Factors is large, and the number of "levels" (specific value settings of the Factors) is large; these go as the number of levels raised to the number of Factors, or L^F. This exponential growth is associated with the type of experimental design being employed, called a "factorial" design, which not only allows the so-called main effects to be discerned from the experiments but also what are called "interaction" effects, where knowledge is gained about the Effects on the metrics of interest due to interacting Effects among the Factors. A representative "2^k" factorial DOE design of test runs for a case involving studying the Effects of target maneuverability, tracker type, target spacing, track-truth association technique, and error in truth tracks is shown in Table 15.2; recall these combinations of test run conditions represent the most cost-effective strategy to gain the information desired.

If the desire to learn about the interaction Effects is relaxed, using a type of experimental design called a "fractional factorial" design, the severity of the exponential growth is lessened but can still be an issue to deal with. One notion of a phased but still DOE-based approach is shown in Figure 15.10, where the fractional designs are used initially as a screening step to determine those Factors which are most influential on the metrics, and then the factorial designs to better understand the main and interaction Effects of the key variables and, if necessary, what are called "response surface" methods to understand the broad Effects of the Factors across the levels of interest for the application.

Various alternative strategies may be possible, since there are also many types of DOE techniques, each designed for environments involving varying numbers of Factors and where prior information/knowledge may suggest the level

---

* Meaning, in a two-tracker comparative experiment, the within-treatment variances for both trackers across the varied problem conditions.

TABLE 15.2

Table Showing DOE Experimental Design for Two Levels of Each Factor ("2^k Factorial Design")

Run Order	Maneuverability	Filter	Target Spacing	Track/Truth Association	% Error in Truth
1	Non-Man	Kalman	Low	Switch	0
2	Man	Kalman	Low	Switch	0
3	Non-Man	Alpha-Beta	Low	Switch	0
4	Man	Alpha-Beta	Low	Switch	0
5	Non-Man	Kalman	High	Switch	0
6	Man	Kalman	High	Switch	0
7	Non-Man	Alpha-Beta	High	Switch	0
8	Man	Alpha-Beta	High	Switch	0
9	Non-Man	Kalman	Low	Window	0
10	Man	Kalman	Low	Window	0
11	Non-Man	Alpha-Beta	Low	Window	0
12	Man	Alpha-Beta	Low	Window	0
13	Non-Man	Kalman	High	Window	0
14	Man	Kalman	High	Window	0
15	Non-Man	Alpha-Beta	High	Window	0
16	Man	Alpha-Beta	High	Window	0
17	Non-Man	Kalman	Low	Switch	10
18	Man	Kalman	Low	Switch	10
19	Non-Man	Alpha-Beta	Low	Switch	10
20	Man	Alpha-Beta	Low	Switch	10
21	Non-Man	Kalman	High	Switch	10
22	Man	Kalman	High	Switch	10
23	Non-Man	Alpha-Beta	High	Switch	10
24	Man	Alpha-Beta	High	Switch	10
25	Non-Man	Kalman	Low	Window	10
26	Man	Kalman	Low	Window	10
27	Non-Man	Alpha-Beta	Low	Window	10
28	Man	Alpha-Beta	Low	Window	10
29	Non-Man	Kalman	High	Window	10
30	Man	Kalman	High	Window	10
31	Non-Man	Alpha-Beta	High	Window	10
32	Man	Alpha-Beta	High	Window	10

of concern for expected interaction Effects. In Kleijnen (2005), a plot of DOE technique (many of these have person's names associated with them) for different magnitudes of the number of Factors and the expected degree of Factor interactions (or analysis complexity, shown here as response surface complexity) is presented, shown here as Figure 15.11. So, in addition to a phased/layered approach as in Figure 15.10, a direct approach using these special DOE designs can be an alternative approach.

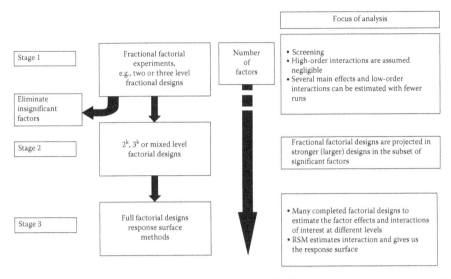

FIGURE 15.10 Notional layered experimental design/DOE strategy for large numbers of Factors and levels.

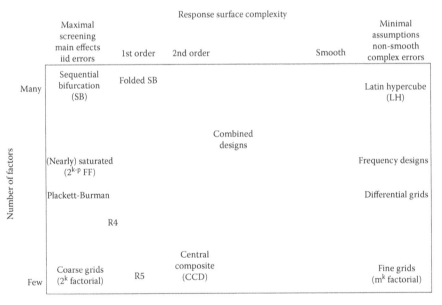

FIGURE 15.11 Suggested DOE strategies according to numbers of factors and interaction complexity. (Adapted from Kleijnen, J.P.C. et al., *INFORMS J. Comput.*, 17(3), 263, 2005.)

15.3 SUMMARIZING IMPACTS TO AND STRATEGIES FOR DISTRIBUTED FUSION SYSTEM T&E

Much more could be said about these various high-level thoughts regarding how to approach the topic of T&E for distributed fusion systems; there is a large body of literature that can be accessed to further explore the ideas offered here as well as yet more issues on this topic. It can be seen that there are some very basic issues that need to be addressed; just defining the test article or the SUT may not be so easy. In the practical world where a team of contractors may have come together to build a DDIFS, establishing responsibilities for various parts of a DDIFS during the T&E phase, and understanding causal effects and audit trails of errors to determine responsibility (and imputed costs) for corrective actions can be problematical and can create complexities in writing equitable contracts. The Network Centric Operations Conceptual Framework of Garstka and Alberts (2004) forms one reasonable basis from which to develop top-level ideas on DDIFS T&E, but as always the devil is in the details. Many of the subtleties such as the fused estimate-to-truth association issue and the various statistical aspects discussed here are often not adequately addressed in much fusion literature. Any given R&D or development program of course only has a given number and amount of resources, and the role for and value of the T&E phase of the program has to be weighed in terms of overall cost-effectiveness, but the ramifications of poor/inadequate T&E lead to poor transition and receptivity of any fusion prototype. The worse outcome of course is that poor/inadequate T&E results in some type of disastrous outcomes, possibly involving loss of life.

15.4   REMARKS FROM A DDIFS USE CASE

This section describes the ideas and a number of details of the project the authors were involved with for the U.S. Edwards Air Force Base (EAFB) that formed a basis for T&E of advanced tactical fighter aircraft that had integrated Information Fusion capabilities and were linked to concepts of employment that set them in a networked/distributed mission context. EAFB is nominally a DT&E test facility, but staff there have agreed that there is an issue in DDIFS applications as to the atypical nature of DT&E and the tendency toward what is more like an OT&E test environment, as we have previously remarked. EAFB is a large test range in the California desert where prototype tactical aircraft are tested in near-operational conditions. To explore some of the T&E issues and ideas, a simple use case involving two friendly aircraft in a test scenario was defined; each platform has three on-board sensors: Radar, ESM, and IRST (infrared search and track). The focus was on target tracking and threat estimation or fusion Levels 1 and 2 type capability in a two-node network where the aircraft exchanged tracking estimates, as shown in Figure 15.12; "CTP" in the figure means common tactical (track) picture.

The problem scenario was suggested by staff at EAFB and comprised a two versus six offensive sweep problem as shown in Figure 15.13. During the scenario, there are simulated missile launches and various flight dynamics emulating a plausible scenario of this type. The PE Tree for this problem was defined to have seven PE nodes, performing the following evaluative operations: (1) three individual sensor nodes, two ownship nodes (friendlies only), one distributed fusion track-to-truth PE node, and (2) one internetted platforms track-to-track PE node. The PE nodes described in Section 15.2.4 perform three necessary fusion functions:

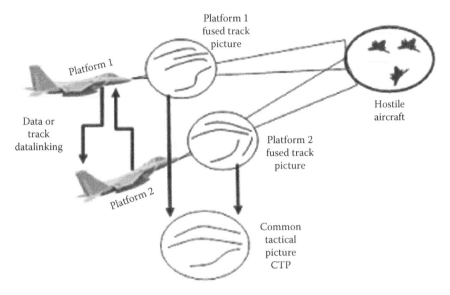

FIGURE 15.12   Use case two-aircraft configuration.

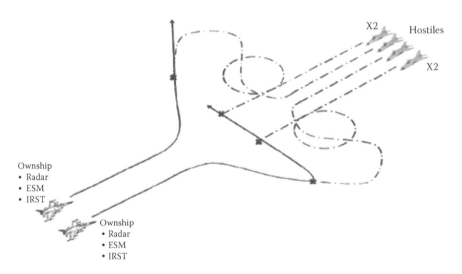

FIGURE 15.13   Two versus six offensive sweep scenario.

(1) data preparation, (2) data association, and (3) estimation of the metrics or MOPS. During data preparation the PE node puts tracks and truth information in [x, y] coordinates and common time. Data association performs deterministic track-to-truth association and track-to-track association.

In this case study there are two platforms which have their own view of the truth picture called "common" and "unique" pictures. The common tracks are seen by both the platforms while unique tracks are uniquely seen by platforms 1 and 2. From the point of view of supporting the tactical mission, one critical issue of course if whether there is a consistent "track picture" across the two aircraft. It can be seen in Figure 15.12 that it is typical that there are differences in the local target track pictures on each platform which need to be reconciled for mission application. In this study then, one focus of analysis was the fused track picture consistency as a function of certain factors, looking at both track-to-truth and track-to-track consistency metrics. Each of the platforms exchange their track files and data fusion is done upon receipt of this information at each platform. We explain below how this information is exchanged where we assume that there are no bandwidth limitations in communication. The baseline distributed fusion output is the consistent tactical picture (CTP). The sensor track file "consistency" is computed at each time point as the average over time of the percentage of matching CTP tracks in the track files of each platform. In addition to these, the following metrics are computed:

1. Track-to-track consistency
2. Track-to-truth consistency
3. Percentage of tracks from first (or second) platform that are not associated with truth tracks (PFT) (this is just a track-to-truth MOP)
4. Percentage of tracks from first platform that is not associated with tracks from second platform (PFT1)
5. Percentage of tracks from second platform that is not associated with tracks from first platform (PFT2)
6. The average number of standard deviations of error in the associated tracks at each time point
7. The average location error standard deviation of associated tracks at each time point
8. Percentage of correct classification for both platforms
9. Range to correct ID for both platforms

In addition to the above consistency PE metrics, the corresponding performance metric of each of the platform track files relative to truth is computed.

In relation to Figure 15.12, we defined three "tiers" of processing:

- Tier 0: here, each friendly platform generates fusion-based but sensor-specific tracks; that is, each of the radar, ESM, and IRST sensor data streams are locally associated and used to generate tracks.
- Tier 1: here, the above sensor-specific tracks are associated (track-to-track association) and fused, but this fused picture is still local to the "ownship," or unique tracks as seen by the particular friendly aircraft.
- Tier 2: here, each of the ownship Tier 1 track files are fused at each Tier 1 track file update time (again, track-to-track association and fusion).

Within this framework, we defined a simple but executable statistical experimental design or DOE (partially driven by scope limitations) that was a 2^k type full factorial design with three main Factors with two levels each, as shown in the following:

DOE Factors and Levels:

		Levels	
PE factors	Association algorithm gating factor	Vogel approximation 3	Hungarian algorithm 5
SUT design factors	Gating factor	5	15

1. PE Factors:
   a. Two alternative track-to-truth association schemes
   b. Two alternative association gate sizes
2. SUT design factors:
   a. Two alternative association gate sizes

The factorial experiment is analyzed using ANOVA from the MINITAB statistical analysis package. The factors and interactions that are significant for various MOPS are denoted by "S" in Table 15.3. In Tier 0, we have three sensors on two platforms and they do not fuse any data within or across platform. Hence we have to only analyze track-to-truth associations for each of the MOPS. The summary of the results is shown in Table 15.3. For each MOP we have the normal probability plot and Pareto chart which recapitulates the significant factors. Then for the significant factors we plot the main effects plot which tells us how the change in factor affects the MOP. For the significant interactions we plot the interaction plot which shows the effect of change in factor level combination on MOPS. After taking a look at the summary Table 15.3, we can say that SUT design gating factor is comparatively more significant than PE gating factor and PE association algorithm. SUT design gating factor appears to be a significant factor in nearly all the Tier 0 DOE runs. So at Tier 0 we must be sensitive toward selection of SUT design gating factor.

In Tier 1, we have three sensors on two platforms and they fuse data within platform (not across platform). So we have to analyze track-to-truth and track-to-track associations for each of the MOPS. The summary of the results is also shown in Table 15.3. After taking a look at the summary Table 15.4, we can say that all the three factors are very significant. All the three factors appear to be significant in nearly all the Tier 1 DOE runs. The interaction between SUT design gating factor and PE gating factor is mostly significant for all the MOPS.

In Tier 2, we have three sensors on two platforms and they fuse data within and across platforms. So we have to analyze track-to-truth and track-to-track associations for each of the MOPS. The summary of the results is shown in Table 15.5. After taking a look at the summary Table 15.5, we can say that none of the three factors are significant. In this case only some of the two- and three-way interactions are significant, which suggests that fusing data across platforms reduces the discrepancies in the input data.

In addition to these DOE runs, we ran another set of full factorial runs to see the effect of communication tiers on the various MOPS. We added another factor, (D)

TABLE 15.3
Tier 0 Analysis of Variance (ANOVA) Results

		A	B	C	AB	AC	BC	ABC
Track 1 to truth radar	Consistency	S						
	PFT	S	S				S	
	Mean location error	S			S	S		
	Average standard deviation location error	S	S			S		
	Average standard deviation	S		S	S			
Track 2 to truth radar	Consistency	S		S				
	PFT	S	S	S			S	
	Mean location error	S						
	Average standard deviation location error	S		S		S		
	Average standard deviation	S		S				
Track 1 to truth ESM	Consistency							S
	PFT	S						
	Mean location error	S						
	Average standard deviation location error	S			S			
	Average standard deviation	S			S	S		
Track 2 to truth ESM	Consistency	S	S	S	S			
	PFT							
	Mean location error	S		S				S
	Average standard deviation location error	S						S
	Average standard deviation	S						S
Track 1 to truth IRST	Consistency	S						
	PFT	S						S
	Mean location error	S			S		S	
	Average standard deviation location error	S				S		
	Average standard deviation	S			S	S	S	
Track 2 to truth IRST	Consistency	S	S					
	PFT	S						
	Mean location error	S	S					S
	Average standard deviation location error	S					S	
	Average standard deviation	S	S	S	S	S		S

S = statistically significant.

TABLE 15.4
Tier 1 Analysis of Variance (ANOVA) Results

		A	B	C	AB	AC	BC	ABC
Track-to-track	Consistency	S	S	S	S			
	PFT$_1$	S	S	S	S			
	PFT$_2$		S	S	S			
	Mean location error	S	S	S	S		S	
	Average standard deviation location error	S	S	S			S	
	Average standard deviation	S	S	S	S		S	
Track 1 to truth	Consistency	S	S	S	S			
	PFT	S	S	S	S	S		
	Mean location error	S			S			
	Average standard deviation location error			S	S			
	Average standard deviation	S	S	S	S	S	S	S
Tract 2 to truth	Consistency	S	S	S	S	S		S
	PFT	S	S		S	S		S
	Mean location error	S	S	S				
	Average standard deviation location error	S	S					
	Average standard deviation	S	S	S	S	S	S	S

S = statistically significant.

Tier, with two levels: Tier 1 and Tier 2. Table 15.6 shows the significant factors and their interactions for the various MOPS. Table 15.6 shows that factor D is significant for all the MOPS, which confirms the intuitive result that fusing data across platforms reduces the input data inconsistency.

15.5   SUMMARY AND CONCLUSIONS

Moving information fusion processes and algorithms into the context of a distributed or networked architecture has many potential operational benefits but can add considerable complexity to the framing of a T&E activity. This chapter has offered some discussion on these complicating factors, to include

- The fundamental question of defining what is being tested, i.e., the test article or system under test
- The fuzzification of the boundary between DT&E and OT&E
- The question of functional boundaries between application functions or services and the generic-service infrastructure, for example, in an SOA

TABLE 15.5

Tier 2 Analysis of Variance (ANOVA) Results

		A	B	C	AB	AC	BC	ABC
Track-to-track	Consistency							S
	PFT_1							
	PFT_2							
	Mean location error					S		
	Average standard deviation location error							
	Average standard deviation					S		
Track 1 to truth	Consistency							
	PFT							
	Mean location error							
	Average standard deviation location error							
	Average standard deviation							
Track 2 to truth	Consistency						S	
	PFT							S
	Mean location error						S	
	Average standard deviation location error							
	Average standard deviation						S	

S = statistically significant.

- The question of degree of investment in supporting test facilities and simulation environments
- The need to think about architecting a PE Tree structure to support analysis of the various and many types of functions, processes, and metrics involved in DDIFSs
- The challenge of employing statistically rigorous experimental designs and post-test data analysis techniques to improve the statistical sophistication, but more importantly the effective and efficiency insights into how a DDIFS is functioning

and some other considerations. The user community and R&D community need to come to grips with the challenges of DDIFS T&E, and to carefully examine how to allocate funding and resources to find a best cost-effective path through these challenges in achieving a "fair" PE system. In support of this, we offer the DNN technical architecture providing problem-to-solution space guidance for developing distributed DF&RM PE systems as a Level 4 fusion process that includes PE functional components, interfaces, and engineering methodology, see also Bowman (2004), Bowman and Steinberg (2001), Steinberg et al. (1999), Haith and Bowman (2010), and Bowman et al. (2009).

TABLE 15.6
Inter-Tier (Tiers 1 and 2) Analysis of Variance (ANOVA) Results

		A	B	C	D	AB	AC	AD	BC	BD	CD	ABC	ABD	ACD	BCD	ABCD
Track-to-track	Consistency									S	S					
	PFT$_1$	S	S							S	S					
	PFT$_2$									S	S	S				
	Mean location error				S			S		S		S				
	Average standard deviation location error	S	S		S			S	S	S	S					
	Average standard deviation		S		S			S			S					
Track 1 to truth	Consistency	S	S		S			S			S	S		S		S
	PFT$_1$	S			S			S				S				S
	Mean location error				S											
	Average standard deviation location error				S											
	Average standard deviation	S		S	S			S			S	S	S	S		S
Track 2 to truth	Consistency	S			S			S	S			S		S	S	S
	PFT		S		S	S						S	S		S	S
	Mean location error				S											
	Average standard deviation location error	S	S		S								S			S
	Average standard deviation	S		S	S			S	S		S	S		S	S	S

S = statistically significant.

REFERENCES

Blasch, E.P. 2003. Performance metrics for fusion evaluation. *Proceedings of the MSS National Symposium on Sensor and Data Fusion (NSSDF)*, Cairns, Queensland, Australia.

Blasch, E.P., M. Pribilski, B. Daughtery, B. Roscoe, and J. Gunsett. 2004. Fusion metrics for dynamic situation analysis. *Proceedings of the SPIE*, 5429:428–438.

Blasch, E.P., P. Valin, and E. Bossé. 2010. Measures of effectiveness for high-level fusion. *International Conference on Info Fusion—Fusion10*, Edinburgh, U.K.

Bowman, C. 1994. The data fusion tree paradigm and its dual. *Proceedings of 7th National Symposium on Sensor Fusion*, Invited paper, Sandia Labs, NM, March.

Bowman, C.L. 2004. The dual node network (DNN) DF&RM architecture. *AIAA Intelligent Systems Conference*, Chicago, IL.

Bowman, C.L. 2008. Space situation awareness and response testbed (*AIAA Intelligent Systems Conference*, Chicago, IL) performance assessment (PA) framework, Technical Report for AFRL/RV, January.

Bowman, C.L. and A.N. Steinberg. 2001. A systems engineering approach for implementing data fusion systems. In *Handbook of Multisensor Data Fusion*, D. Hall and J. Llinas, J. (Eds.), Chapter 16. Boca Raton, FL: CRC Press.

Bowman, C., P. Zetocha, and S. Harvey. 2009. The role for context assessment and concurrency adjudication for adaptive automated space situation awareness. *AIAA Conference Intelligence Systems*, Seattle, WA.

Durrant-Whyte, H.F. 2000. A beginner's guide to decentralized data fusion. Technical report, Australian Centre for Field Robotics, University of Sydney, Sydney, New South Wales, Australia.

Garstka, J. and D. Alberts. 2004. Network Centric Operations Conceptual Framework Version 2.0, U.S. Office of Force Transformation and Office of the Assistant Secretary of Defense for Networks and Information Integration, Vienna, VA.

Gelfand, A., C. Smith, M. Colony, and C. Bowman. 2009. Performance evaluation of distributed estimation systems with uncertain communications. *International Conference on Information Fusion*, Seattle, WA.

Haith, G. and C. Bowman. 2010. Data-driven performance assessment and process management for space situational awareness. *Journal of Aerospace Computing, Information, and Communication (JACIC)* and *2010 AIAA Infotech@Aerospace Conference*, Atlanta, GA.

Hall, D.L. and J. Llinas, Eds. 2001. *Handbook of Multisensor Data Fusion*. Boca Raton, FL: CRC Press.

Kleijnen, J.P.C. et al. 2005. State-of-the-art review: A user's guide to the brave new world of designing simulation experiments. *INFORMS Journal on Computing*, 17(3):263–289.

Liggins, M.E., D.L. Hall, and J. Llinas, Eds. 2009. *Handbook of Multisensor Data Fusion: Theory and Practice*, 2nd edn. Boca Raton, FL: CRC Press.

Llinas, J. et al. 2004. Revisiting the JDL data fusion model II. *International Conference on Information Fusion*, Stockholm, Sweden.

Rawat, S., J. Llinas, and C. Bowman. 2003. Design of a performance evaluation methodology for data fusion-based multiple target tracking systems. *Presented at the SPIE Aerosense Conference*, Orlando, FL.

Steinberg, A. and C.L. Bowman. 2004. Rethinking the JDL data fusion levels, National symposium on sensor and data fusion (NSSDF). Johns Hopkins Applied Physics Lab (JHAPL), Laurel, MD.

Steinberg, A.N., C.L. Bowman, and F.E. White. 1999. Revisions to the JDL data fusion model. *Proceedings of SPIE Conference Sensor Fusion: Architectures, Algorithms, and Applications III*, 3719:430–441.

Utete, S. 1994. Network management in decentralized sensing systems. PhD thesis, The University of Oxford, Oxford, U.K.

16 Human Engineering Factors in Distributed and Net-Centric Fusion Systems

Ann Bisantz, Michael Jenkins, and Jonathan Pfautz

CONTENTS

16.1 Introduction 409
16.2 Characterizing the Domain to Drive Fusion System Design,
Development, and Evaluation 412
16.3 Identifying Fusion System Capabilities to Mitigate Domain Complexities 418
16.4 Identification of Touch Points within a Hard-Soft Fusion Process for
Intelligence Analysis 421
    16.4.1 Touch Point 1 424
    16.4.2 Touch Point 2 424
    16.4.3 Touch Point 3 425
    16.4.4 Touch Point 4 426
    16.4.5 Touch Point 5 427
    16.4.6 Touch Point 6 428
16.5 Conclusion 430
Acknowledgment 431
References...431

16.1 INTRODUCTION

Successful deployment and use of distributed fusion systems require careful integration of human and automated reasoning processes. While fusion systems exploit sophisticated and powerful algorithms that can rapidly and efficiently transform masses of data into situational estimates, human operators remain a crucial component of the overall fusion process. Not only are humans information consumers, but they are also participants throughout the fusion process (Blasch and Plano 2002), acting individually or collectively as sensors or, more critically, as in-the-loop guides for specific computations. Because of the complex interaction between human and automation inherent in fusion systems, attention to human-system integration issues during design, development, and evaluation is critical in the ultimate success of these systems.

For example, as information is delivered by the fusion system, human operators must make a judgment about its pertinence, as well as about the means by which it was generated (e.g., what data sources were used, which algorithms with what assumptions and error characteristics) and how that may affect the quality of the information (e.g., whether a remote sensor is reliable, whether a model for the situation is valid, whether a correlation tool has sufficient data to reach desired levels of accuracy and precision). These judgments determine how the human as information consumer will use, interpret, and appropriately value or trust the fused information, and are based in part on the information that qualifies the outputs of the fusion system, or meta-information (e.g., uncertainty, source quality or pedigree, timeliness, pertinence) (Bisantz et al. 2009, Pfautz 2007). These judgments may be more challenging in distributed or net-centric situations where the operator is removed (spatially, jurisdictionally, and/or hierarchically) from the sources of data or nodes performing the data processing.

In well-designed fusion systems, operators can work in concert with the fusion process, using their knowledge of the situational context and environmental constraints to:

- Provide qualitative data as a sensor—for example, judge the location or count of visible entities
- Guide the selection of input data—for example, choose sources, filter sources, filter certain types of data, or prioritize data (e.g., as a function of source)
- Provide corrections to intermediate results—for example, manually provide detections, "hand-stitch" tracks to correlate objects across frames, or select most-apt situation models based on mission context
- Adjust or "tune" fusion system parameters—for example, manipulate detection thresholds, define training sets for learning models, or specify processing stages or algorithms

Operators can perform some or all of these tasks to improve the fusion system's performance, particularly in mixed-initiative, dynamic fusion systems, and "hard/soft" fusion systems, which combine data from human or "soft" sources with those from physical sensors (Hall et al. 2008). Figure 16.1 illustrates the range of actions operators can take with respect to fusion system performance. Operators can also interact with the fusion system to pose questions and test their own hypotheses about the implications of the fused information. These tasks, like those of an operator simply consuming the fused information, are also directly affected by the meta-information inherent in the fusion process (e.g., Algorithm B performs in real-time, Algorithm A can only perform in near-real-time, but has a higher rate of detection; therefore, given a critical need for timely information, Algorithm B is the better choice, despite lower detection rates, because it performs in real-time).

The need to make these judgments raises important human factors challenges regarding appropriate methods for coordinating mixed human-automated control over fusion processes, communicating system processes and states to the human operator, and visualizing key information and meta-information. These challenges are exacerbated in net-centric environments, in which humans distributed in time and space provide input, interact with, and make decisions based on fusion processes.

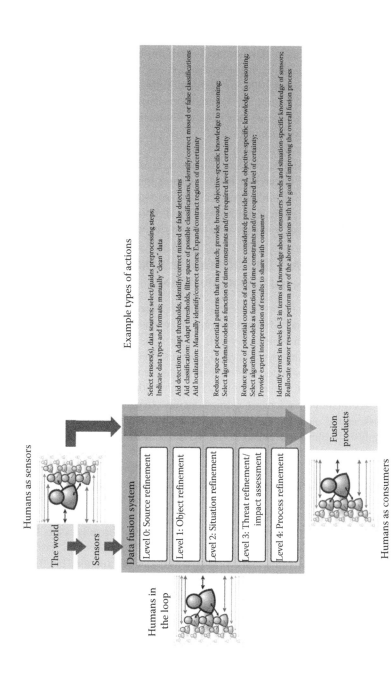

FIGURE 16.1   Characterizing potential points of human–fusion system interaction.

To further explore the role of human operators within a complex fusion process as well as to illuminate various challenges and potential human factors solutions, this chapter uses the design of a fusion system that aids a hypothetical intelligence analysis process as a case study. This case study demonstrates that by combining knowledge of both human and system strengths and limitations with an understanding of the work domain (i.e., through the application of cognitive systems engineering methods [Bisantz and Burns 2008, Bisantz and Roth 2008, Crandall et al. 2006, Hollnagel and Woods 2005, Vicente 1999]), engineers can design fusion systems that successfully integrate human operators at critical interaction points ("touch points") to enhance overall system performance.

This chapter begins with a discussion of military intelligence analysis, its characteristics as a work domain, and the challenges it poses for human analysts. Subsequent sections highlight potential features of fusion systems that can mitigate these challenges for different stages of the intelligence analysis process. The chapter ends with design characteristics and rationale for a set of touch points for human control and interaction with a distributed, hard-soft information fusion system.

16.2 CHARACTERIZING THE DOMAIN TO DRIVE FUSION SYSTEM DESIGN, DEVELOPMENT, AND EVALUATION

One of the first requirements for identifying and supporting beneficial interactions between a fusion system and a human operator is to understand the relevant characteristics of the domain in which they will perform, as well as the requirements and constraints that rise from environmental, computational, operational, and socio-organizational factors. This understanding can be derived via formal systems engineering practices (e.g., requirements analysis [Laplante 2009]), but these approaches often fail to adequately characterize human factors. Cognitive systems engineering and its associated methodologies for characterizing work domains (e.g., cognitive work analysis [Bisantz and Burns 2008, Bisantz and Roth 2008, Vicente 1999]; human-centered system engineering [Crandall et al. 2006, Hollnagel and Woods 2005]) represent an approach that focuses on both human and system performance and their interrelationships. In our case study, we demonstrate how this approach reveals specific features of human-system interaction that should influence design, development, and evaluation of fusion systems.

The intelligence analysis domain: "Intelligence analysis" broadly refers to reasoning over available data with the goal of making a coherent whole of the past, present, and/or future states of some real-world environment or situation. Intelligence analysts review and process (i.e., filter, validate, correlate, and summarize) as much data as feasible under constraints such as externally imposed deadlines, availability of data, and availability of data collection resources. This typically massive amount of data crosses spatio-temporal scales (from a street corner to vast regions, from microseconds to decades) (Mangio and Wilkinson 2008). It is interpreted by the analysts who use their experience to compare their interpretations with existing hypotheses (Tam 2008–2009). The analyst revises and generates hypotheses defining the current situation (which continues to evolve during the analysis) until a threshold of confidence is reached, or, more often, a deadline is reached. At this point, the analyst creates useful products for decision-makers, often in distributed locations. This entire process may be completed in only seconds (e.g., to answer requests such as "Where did the target go?") or months (e.g., for questions such as "What are the historic and economic drivers behind the regional instability?").

Prior literature provides a variety of characterizations of intelligence analysis, including some from the human factors community (Cook and Smallman 2008, Elm et al. 2004, Grossman et al. 2007, Heuer 1999, Patterson et al. 2001a, Pfautz et al. 2005b, Pfautz et al. 2006, Pirolli and Card 2005, Powell et al. 2006, Powell et al. 2008, Trent 2007, Zelik et al. 2007). An extensive review of this literature is out of the scope of this chapter; however, there are several characteristics worth highlighting. In terms of the tasks that analysts commonly perform, Hughes and Schum (2003) point out that regardless of the specific form or goal of an analysis, the intelligence analysis process always involves three parts: (1) hypotheses generation, (2) evidence gathering and evaluation, and (3) generation and evaluation of arguments linking evidence and hypotheses. Incomplete and uncertain information combined with the need for timely solutions mean that success in this domain is best characterized by convergence (Grossman et al. 2007). Elm et al. (2005) Potter et al. (2006) define convergence as "a stable balance of applying broadening and narrowing … to focus on a reduction of the problem towards an answer." Therefore, a successful analysis effort will be one that applies multiple cycles of broadening and narrowing based on available resources (e.g., time, information, cognitive capacity, etc.) to avoid premature closure and arrive at a set of final hypotheses that best explain the substantive problem.

Kent's early model describing seven intelligence analysis stages (1965) includes activities which can be associated with more recent models, as shown in Figure 16.2.

These models help provide a general definition of the steps of intelligence analysis and support identification of the characteristics of the domain that make successful performance challenging. By leveraging human factors engineering techniques, several research efforts identified factors that contribute to the complexities that commonly arise during the analysis process. The majority of these factors relate to the unique characteristics of the data that feed the intelligence analysis process and the dynamic structure of the problems that the intelligence analysis process attempts to solve. Table 16.1 presents a nonexhaustive list of factors that create challenges during problem analysis and references that illustrate the methods used to uncover and characterize each factor. In subsequent sections, we demonstrate how this analysis can contribute to the design of supportive fusion systems.

Finally, in addition to understanding the characteristics of the domain that make performance challenging, cognitive engineering analyses typically characterize expert performance within the domain. Here, that focus is on identifying the knowledge and strategies required for successful intelligence analyses. The iterative broadening and narrowing across data collection, synthesis, and hypothesis evaluation analysis phases as described in Elm et al. (2004, 2005) is one example strategy. Multiple cognitive task analyses of the intelligence analysis process support the fact that analysts employ simplifying strategies as a result of high cognitive demands (particularly in terms of large amounts of data with varying characteristics, the need to simultaneously consider multiple hypotheses, and time

Elm et al.'s Support Function Model	Hughes and Schum's Three-Stage Model	Kent's Seven-Stage Model	Characteristics and Challenges
	Hypothesis generation	Stage 1: Appearance of substantive problem	• Analysts receive/construct substantive problem through a direct request for information, through the emergence of something unusual in their analysis, or for general situation development based on mission/objectives • Source of the problem often affects the degree of complexity and difficulty of subsequent analysis
	Evidence gathering and evaluation	Stage 2: Analysis of the substantive problem	• Focus on framing and understanding the substantive problem • Starting point for evidence gathering and evaluation (Pfautz et al. 2006) • Goal is to reduce ambiguity and focus solution to be applicable to consumer(s) (Hutchins et al. 2004) • Problem breadth often influences overall ambiguity since problems that are both too broad or too narrow create additional challenges
Hypothesis exploration / Down collect		Stage 3: Data collection	• Focus on gathering as much potentially relevant data from as many sources as possible, within resource constraints • Problems of information overload; premature narrowing of data collection (to relieve overload) (Elm et al. 2005); structure of data collection and repositories can make access difficult (Flynn et al 2010) • Challenges may arise from the need to direct collection managers or tasks across physical, language, domain expertise, or organizational boundaries (Grossman et al. 2007), particularly if collection assets are needed to service operational as well as intelligence goals

	Stage	Characteristics
	Stage 4: Data evaluation	• In-depth review of collected information (both content and pedigree) to determine impact on developing hypotheses and previously reviewed information • Reasoning over uncertainty and assessing source and information pedigree are common challenges
Generation and evaluation of arguments linking evidence and hypotheses	Stage 5: The moment of hypothesis	• A stage or moment where the analyst forms the inklings of an explanation for the evaluated data • Numerous explanations should be considered in parallel, but due to the limitations of human cognitive capabilities, the ideal comparison set is rarely considered
Evidence gathering and evaluation	Stage 6: More collecting and more testing of hypotheses	• Similar to Stage 4, with a greater focus on supporting/refuting developing hypotheses • Often dependent on the availability of time and/or new/updated information • Multiple cognitive biases appear that pose challenges to analysts (e.g., confirmation bias)
Conflict and collaboration Down collect	Stage 7: Presentation	• Creation of an artifact to communicate the established hypothesis (or ideally, competing hypotheses [Greitzer 2005]) to the intelligence analysis consumer(s) • Artifact should establish a "new and better approximation of the truth" (Hutchins et al. 2004) regarding the situation of interest • Analysts challenged with the need to provide "actionable" intelligence given the context of use for the information requested by the consumer

FIGURE 16.2 Comparison of models of intelligence analysis and implied characteristics and challenges of intelligence analysis tasks. (From Elm, W. et al., Designing support for intelligence analysis, Presented at the *48th Annual Meeting of Human Factors and Ergonomics Society*, New Orleans, LA, 2004; Hughes, F.J. and Schum, A., Preparing for the future of intelligence analysis: Discovery-proof-choice, Joint Military Intelligence College, Unpublished Manuscript, 2003; Kent, S., *Special Problems of Method in Intelligence Work. Strategic Intelligence for American World Policy*, Archon Books, Hamden, CT, 1965.)

TABLE 16.1
Nonexhaustive List of Factors Contributing to the Challenges of Problem Analysis

Factor	Description	References
Ambiguous requests for information (RFI)	Analysts may receive RFIs from a third party they have no means of contacting. In these cases, if information needs are poorly expressed or communicated, or if the operational context surrounding the RFI is unclear, then the analyst is faced with the problem of first defining the problem that he or she is supposed to be solving. Information, such as why the question is being asked, what are the boundaries for investigation, and what the consumer is really trying to accomplish, are all critical pieces of information for successful analysis that are often missing	IW JOC (2007) Elm et al. (2005) Heuer (1999) Hewett et al. (2005) Hutchins et al. (2004) Johnston (2005), Kent (1965) Pfautz et al. (2006) Roth et al. (2010)
Complex data characteristics	The complexity of the data that analysts must interpret is increased by numerous factors that are all commonly present during analysis. Factors, such as multiple data formats (e.g., images, written/verbal reports, technical data, etc.), operational context (i.e., multiple lines of operation require multiple skill sets for interpretation), massive volumes of data, data set heterogeneity, data set complexity (i.e., number of relationships within the data set), and data class (e.g., HUMINT, SIGINT) all increase the overall complexity of the data. As these factors become more prominent, the difficulty of the analysis increases and analysts become more susceptible to problems such as data overload	MSC 8/06 (2006), 2FM 3-24 12/06 (2006) IW JOC (2007) Drucker (2006) Elm et al. (2004) Greitzer (2005), Heuer (1999), Hutchins et al. (2006) Larson et al. (2008) Patterson et al. (2001a,b), Pfautz et al. (2006), Roth et al. (2010), Taylor (2005)
Distributed work structure	Analysts may not be in the environment or participants in the situation they are investigating. Analysts may or may not be co-located with their "customers." Socio-organizational structure may impose several levels (and geospatial distance) between the analyst and the source of information, information requestor, and other analysts or experts, creating communication barriers that impede analysis. Analysts may be required to work on requests that overlap with current or past projects of analysts in other organizations. The lack of inter-organizational information sharing in the intelligence community (even within the same organization) creates information access challenges, and results in a failure to consider multiple perspectives	Hewett et al. (2005) Johnston (2005) Kent (1965) Pfautz et al. (2006)

TABLE 16.1 (continued)
Nonexhaustive List of Factors Contributing to the Challenges of Problem Analysis

Factor	Description	References
Multiple perspectives for consideration	Analysts face the difficult challenge of simultaneously integrating considerations from multiple perspectives when performing their analysis, including the perspective of the information sources; the situation/culture/area being assessed; and the objectives, goals, and situational context of the consumer or user of the information being developed	Johnston (2005) Taylor (2005)
Information may be deceptive, unreliable, or otherwise qualified	Issues of deception are common in the intelligence analysis domain and add complexity to each piece of information that analysts must interpret. Information may come from human sources which are biased or unreliable due to characteristics of the situation or the human observer. Analysts must understand the impact these different levels of meta-information have on information utility, as well as maintain and communicate qualifiers throughout the analysis process, a task which can be particularly challenging in distributed systems	Bisantz et al. (2009), Hardin (2001), Heuer (1999), Jenkins et al. (2011), Pfautz et al. (2005a), Pfautz et al. (2007), Pfautz et al. (2005b), Zelik et al. (2007)
Dynamic nature of domain	As time passes during the course of an analysis, analysts are faced with the challenge of constantly re-evaluating their hypotheses and supporting/refuting evidence to ensure they are still relevant and valid. This is a requirement resulting from the dynamic nature of the real-world. For example, the location of a person of interest is likely to be highly dynamic and must be continuously monitored and updated by the analyst. Additional considerations must also be given to meta-data (e.g., the source of data), which can evolve over time as real-world events play out	Pfautz et al. (2006), Roth et al (2010)
Unbounded problem space	During the broadening process, when analysts are attempting to uncover novel explanations for a set of information, they are faced with the challenge of considering a theoretically infinite problem space. The "open world" of possible explanations and events creates a wide range of unpredictable hypotheses for consideration. This unbounded range of possibilities is likely to result in a high degree of mental workload if analysts attempt to consider every hypothesis they can think of. Further, even in the case when they consider a large set of possible hypotheses, asking "Am I missing something?" is a very difficult question to answer, especially after an extensive attempt to avoid missing any relevant information	Hutchins et al. (2006), Pfautz et al. (2006)

pressure) (Endsley 1996, Jian et al. 1997), and that these strategies may introduce biases into the analysis process (see Heuer 1999, Hutchins et al. 2004, 2006, Johnston 2005, Kahneman et al. 1982, Mangio and Wilkinson 2008, Patterson et al. 2001a,c, Pirolli 2006, Pirolli and Card 2005, Roth et al. 2010, Tam 2008–2009, Trent et al. 2007). Examples of decision biases which may impact intelligence analysis include:

- *Selectivity bias:* Information is selectively recalled as a function of how salient the information is to the individual analyst.
- *Availability bias:* Frequency of an event is predicted based on how easily it is recalled from long-term memory.
- *Absence of evidence bias:* Failure to recognize and incorporate missing data into judgments of abstract problems (e.g., "What am I missing?").
- *Confirmation bias:* Interpreting information to confirm existing beliefs or hypotheses.
- *Overconfidence bias:* Sureness that one's own hypotheses are correct when most of the time they are wrong.
- *Oversensitivity to consistency bias:* Placing too much reliance on small samples and failure to discern that multiple reports come from the same source information.
- *Discredited evidence bias:* Persistence of beliefs or hypotheses even after evidence fully discrediting those hypotheses is received.

Systems designed to enhance analysis performance must provide support to reduce or overcome the impacts of these biases.

16.3 IDENTIFYING FUSION SYSTEM CAPABILITIES TO MITIGATE DOMAIN COMPLEXITIES

Information fusion systems can be designed to support information analysis processes. Considering domain characteristics and complexities allows identification of fusion system capabilities which support human performance in this challenging environment. After identifying the challenges that intelligence analysts are likely to face over the course of their analyses, the domain complexities that create these challenges can be characterized and high-level fusion system capabilities that address the complexities can be determined. Fusion system capabilities and their mapping to domain complexities are provided in Table 16.2.

In many situations, fusion systems are designed to integrate into existing workflows, as is the case with our fusion system to support a hypothetical intelligence analysis process. In these cases, it is often helpful to expand the capabilities map to show where in the existing workflow the end-user will access each capability. This map is useful for two reasons. First, it allows fusion engineers and system designers to understand where in their existing workflow end-user will access the system, and what goals they will have in mind at that time. This is important because it allows system designers to better understand the end-user's information seeking, system control or monitoring, and/or other interaction-related tasks.

TABLE 16.2

Intelligence Analysis Complexities and the High-Level Fusion System Capabilities Selected to Mitigate Them

Intelligence Analysis Complexity	Fusion System Capabilities	Justification
Ambiguous RFIs	Support multiple searches (stored or ad hoc) for situations of interest	Allow analysts to explore multiple possibilities simultaneously to cover multiple interpretations or contexts. Facilitate systematic interaction between analyst and requestor to enable clarification and tie to specific fusion system products and processes
Data characteristics	Support manual or automated data association	Challenges relating to volume, heterogeneity, complexity, uncertainty, etc. of data can all be mitigated by providing external tracking of meta-data and automating data association to alleviate the analyst's burden
	Support manual/ automated situation assessment	Given the volume of data that analysts receive on a daily basis and the often high degree of complexity, having the fusion system perform automated situation assessments that can be highlighted for the analyst based on some pre-defined criteria will allow the analyst to handle a significantly larger volume of data during the analysis process
Distributed Structure	Store access credentials to multiple data sources	Analysts need to access different databases and information sources across departments and with different access requirements that need to be remembered. Having the fusion system store these credentials and automatically pull in data from these sources as needed will help mitigate this issue
	Automate language translation	Feasibility may be an issue, but translating incoming natural language messages and reports into the analyst's native language will open up the analysis to a range of sources that may otherwise be ignored due to the difficulty of overcoming the language barrier
Lack of data	Set up custom search alerts to notify when data or situation is available/appears	Lack of key data or data sets may require analysts to repeatedly return to a source or data set to check for updates or appearance of the missing data. Allowing the analysts to set expectancies in terms of missing data and be automatically notified by the fusion system on its appearance/update will save time/effort during an analysis
	Maintain and fusing pedigree data during data association processes	Information meta-data or pedigree is critical for analysts to evaluate the quality of the data and weight it appropriately to support or refute their hypotheses. Having the fusion system maintain this data in an accessible format before, during, and after processing will help analysts to quickly evaluate different pieces of data

(continued)

TABLE 16.2 (continued)
Intelligence Analysis Complexities and the High-Level Fusion System Capabilities Selected to Mitigate Them

Intelligence Analysis Complexity	Fusion System Capabilities	Justification
Potential for deception	Support what-if scenarios (e.g., what if this piece of data were false/true)	Allowing analysts to quickly change the characteristics of a piece of data (e.g., set uncertainty to 100% or 0%) and see how it affects networks of information and situations of interest will allow analysts to see how important different pieces of data are so they understand the degree to which their hypotheses rely upon them
		Characterize and compute meta-information as part of the fusion process
		Augment or enhance the algorithms in the fusion process to expose factors that allow for reporting of confidence, certainty. Similarly expose and maintain the qualities of the data feeding the fusion process. Provide an overall characterization of the state of the fusion process, its components, etc.
Real-world dynamics	Update knowledgebase in near real-time with most recently received data	Analysts need access to the most recent data to perform the most accurate analysis possible. System should update its database with new information as quickly as possible
	Create temporal boundaries on situation assessment and data association	The theoretical dataspace for an analysis is infinite and impossible for a single analyst to fully consider. Allowing the analyst to set temporal boundaries on what data should be considered helps to limit the data under consideration and ensures that the data being considered falls within a specific time-frame, an important feature for historical analyses
	Maintain temporal reference meta-data	All real-world events play out on the same timeline so analysts can drill-down to data or organize sets of data to view the order in which they occurred
Unbounded problem space	Support custom boundary ranges	For seemingly unbounded problems where theoretical data space is near infinite allowing the analyst to set custom boundary ranges (e.g., temporal, regional, cultural, etc.) will significantly reduce the volume of data for consideration. It also allows the analyst to provide top-down feedback to the system on where relevant data are likely to be found based on prior experience

The second reason is that the capabilities map allows system designers to see if they are designing capabilities that do not readily integrate into existing work-flows. If, during the capabilities-to-workflow mapping process, it is found that a capability does not have a clear location within the existing workflow, then the capability needs to be augmented or the workflow must change to incorporate the new capability.

Figure 16.3 provides an example of the capabilities map created for our fusion system to support intelligence analysis with the capabilities from Table 16.2 mapped to different common stages of the intelligence analysis process. The stages of analysis are based on Kent's overview of the intelligence analysis process (described earlier) (1965). Cells shaded with grid backgrounds in Figure 16.3 represent stages of analysis where the respective intelligence analysis complexities are most likely to present challenges to analysts. This complexity-to-workflow analysis mapping was uncovered during the characterization of the domain and its complexities during the analysis of the work domain. Ideally, all highlighted cells will have associated capabilities that mitigate the challenges that arise due to the respective complexity at that stage of the intelligence analysis process. However, due to technological or other limitations, some of these capabilities may not be feasible. For example, Figure 16.3 shows that the challenges arising from the distributed structure of the domain during the analysis of the substantive problem (Stage 2) are not being addressed by any planned capabilities. Finally, capabilities which do not directly map to the intended purpose of the system (i.e., those that appear in cells with white backgrounds in Figure 16.3) can also be identified and reviewed regarding their respective utility (for example, to provide redundancy across stages).

16.4   IDENTIFICATION OF TOUCH POINTS WITHIN A HARD-SOFT FUSION PROCESS FOR INTELLIGENCE ANALYSIS

The domain and capabilities analysis is combined with information regarding the structure of a proposed hard-soft information system being developed to support intelligence analysis (Jenkins et al. 2011). Human-system interaction touch points represent stages during processing where the human operator will interact with the fusion system. Touch points can be included to accomplish a variety of goals; however, the overall suite of touch points should provide the information, interfaces, and control needed for the fusion system to support the human operator in the fusion loop. Subsequently, the selection and design of respective touch points are critical to the overall success of the fusion system because they will serve as the windows to the fusion system for the end-user. If the human operators do not have access to the information they need, there is a strong possibility they will develop an inappropriate understanding of the system's true capabilities, which can lead to under- or overreli-ance on the fusion system (Lee 2008, Parasuraman and Riley 1997).

This example presents the location and early definition of six touch points identi-fied for inclusion in the hard-soft fusion system. For this example, touch point loca-tions and definitions were based on an early-stage fusion architecture to illustrate how touch points can be identified early in the overall system development process. Early identification of touch point locations is important as touch point locations

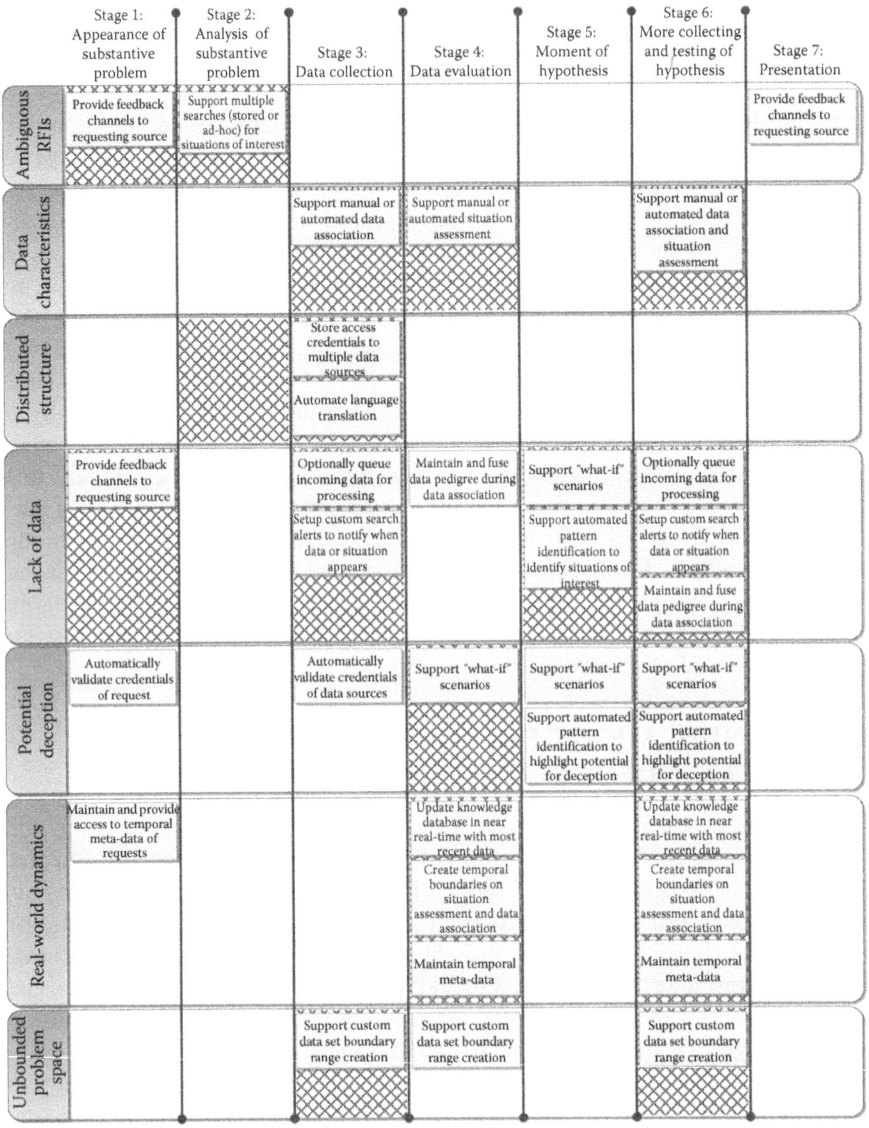

	Stage 1: Appearance of substantive problem	Stage 2: Analysis of substantive problem	Stage 3: Data collection	Stage 4: Data evaluation	Stage 5: Moment of hypothesis	Stage 6: More collecting and testing of hypothesis	Stage 7: Presentation
Ambiguous RFIs	Provide feedback channels to requesting source	Support multiple searches (stored or ad-hoc) for situations of interest					Provide feedback channels to requesting source
Data characteristics			Support manual or automated data association	Support manual or automated situation assessment		Support manual or automated data association and situation assessment	
Distributed structure		Store access credentials to multiple data sources; Automate language translation					
Lack of data	Provide feedback channels to requesting source		Optionally queue incoming data for processing; Setup custom search alerts to notify when data or situation appears	Maintain and fuse data pedigree during data association	Support "what-if" scenarios; Support automated pattern identification to identify situations of interest	Optionally queue incoming data for processing; Setup custom search alerts to notify when data or situation appears; Maintain and fuse data pedigree during data association	
Potential deception	Automatically validate credentials of request		Automatically validate credentials of data sources	Support "what-if" scenarios	Support "what-if" scenarios; Support automated pattern identification to highlight potential for deception	Support "what-if" scenarios; Support automated pattern identification to highlight potential for deception	
Real-world dynamics	Maintain and provide access to temporal meta-data of requests			Update knowledge database in near real-time with most recent data; Create temporal boundaries on situation assessment and data association; Maintain temporal meta-data		Update knowledge database in near real-time with most recent data; Create temporal boundaries on situation assessment and data association; Maintain temporal meta-data	
Unbounded problem space			Support custom data set boundary range creation	Support custom data set boundary range creation		Support custom data set boundary range creation	

FIGURE 16.3 Extended capabilities map showing intended capabilities, the associated intelligence analysis complexity they are intended to mitigate, and the stage of intelligence analysis where they are expected to be accessed. Stages of intelligence analysis are based on Kent's model (1965). Highlighted cells indicate stages in the workflow where a complexity is likely to appear.

and requirements influence the design of the fusion system. For example, a touch point requiring the display of specific information to the human operator influences the format and type of information the fusion system must maintain. This approach is consistent with the type of formative cognitive systems modeling and analysis described by Vicente (1999), in which modeling the work domain complexities can lead to new requirements for data sensing and processing capabilities necessary for successful performance.

The fusion system architecture used for this example is limited by user interface mediums (e.g., keyboard/mouse/monitor), system input/output formats, and availability of source and data meta-information not yet being defined in terms of both format and availability. Subsequently, the touch points were defined by first identifying their location in the fusion processing stream given the capabilities they were intended to support. They were then refined by focusing on feature level details needed to support those capabilities. Each touch point can then be mapped to the stage(s) of analysis where it will be leveraged and the domain complexities that it is expected to mitigate. These details help designers understand the justification of respective touch points and link the capabilities map to the touch points list. Figure 16.4 provides the outline of the fusion architecture that was used to select the location of touch points (for additional details on this fusion architecture, see [Jenkins et al. 2011]) along with their location in the fusion stream. Figure 16.4 shows each of the touch points, the capabilities they support, the domain complexities they mitigate, and the interaction features required to support the human operator's anticipated goals during the touch point interaction. In addition to these details, each fusion touch point provides the expected stage or stages of analysis (based on Kent's seven-stage process mentioned previously [Kent 1965]) where the

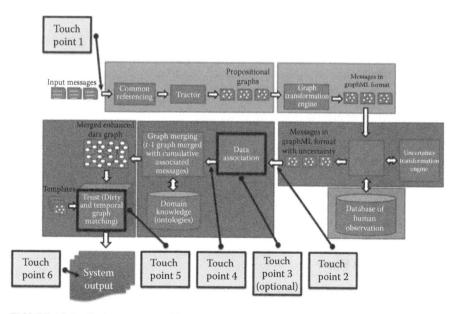

FIGURE 16.4 Fusion system architecture to support the intelligence analysis process.

analyst is expected to access the touch point. Understanding when the analyst will leverage the touch points within their current (or future) workflow allows system designers to anticipate the intelligence analyst's goals and design more effective supporting capabilities.

16.4.1 TOUCH POINT 1

Description: While analysts are researching the problem(s) they are responding to, they must begin by setting boundaries on the data for consideration. This touch point allows the analyst to select data sets for inclusion in the final fused data set that they will access throughout the course of their analysis. Analysts can revisit this touch point to change the boundaries of data being considered during analysis if new information becomes available, which needs to be integrated with the already processed data or if their understanding of the substantive problem changes in a way that affects what data is likely to be relevant, creating a need for previously excluded data sets to be integrated into the fused data.

Location: Before incoming data set processing, but after incoming data sets have been received and are accessible to the fusion system for processing.

Expected intelligence analysis stage(s) when accessed:

- Stage 2—Analysis of substantive problem
- Stage 3—Collection of data
- Stage 6—Additional data collection and hypothesis testing

Capabilities to be supported:

- Multiple searches (stored or ad hoc) for situations of interest
- Queue incoming data for processing
- Custom boundary ranges

Required features:

- Create custom data sets
- Prioritize multiple data sets for processing
- Scan/review data sets pre-processing

Complexities addressed:

- Ambiguous requests for information
- Lack of data
- Unbounded problem space

16.4.2 TOUCH POINT 2

Description: After the fusion system has performed initial operations on incoming data to prepare it for data association, intelligence analysts can review the data and potentially make changes to data association settings before processing occurs. Analysts will want to ensure that respective data sets (including the primary fusion database) are appropriate for identifying associations, given their contextual understanding of the substantive problem and the goals of the analysis. Potential settings that analysts may want to alter include values such as the threshold level for associating two pieces of data or for merging together co-referenced entities. Settings can include thresholds specific to entity type, general thresholds, or other levels of granularity to ensure that associations are identified based on the analyst's preferences. The analyst can also manually set the uncertainty/probability/reliability value associated with a piece of data or with a data source, which will override system-generated values. This feature will support what-if scenarios where analysts may wish to see what happens if a piece of data is true/false, and let analysts leverage their past experiences with the reliability of respective data sources or classes/types of information.

Location: After initial processing and uncertainty alignment of incoming data sets, but before data associations and co-references are identified and resolved.

Expected intelligence analysis stage(s) when accessed:

- Stage 3—Collection of data
- Stage 6—Additional data collection and hypothesis testing

Capabilities supported:

- Manual/automated data association
- What-if scenarios (e.g., what if this piece of data were false/true)
- Custom boundary ranges

Required features:

- Review of data sets being considered for data association
- Filtering of data sets to determine custom boundaries to utilize for data association
- Selection of additional data sets (previously processed) to be included in data association
- Custom threshold levels set to determine when manual approval is needed for a system-proposed merge to be carried out
- Custom uncertainty values to be assigned to individual data elements (e.g., what if scenarios)

Complexities addressed:

- Data characteristics
- Potential for deception
- Unbounded problem space

16.4.3   TOUCH POINT 3

Description: Depending on analysts' preferences and requirements, they can approve potential data merges during the data association process. This needs to be done during data association processing because subsequent processing decisions may be affected by the decision to merge or not merge two pieces of data. Given the large volume of data that can be processed over the course of an analysis, the high number of associations likely to be uncovered within the volume of data, and the rapid updating of data elements as new and/or updated information becomes available, analysts need the ability to monitor fusion processes to maintain their contextual understanding of what the volume of data represents. A touch point during fusion processing can serve as a window to the underlying fusion processes to allow the analyst to understand how the system is manipulating the data before it presents its final output.

Location: Available throughout data association processing, either on demand or as needed by the system based on the system's authority to autonomously carry out data association.

Expected intelligence analysis stage(s) when accessed:

- Stage 3—Collection of data

Capabilities supported:

- Manual/automated data association
- What-if scenarios (e.g., what if this piece of data were false/true)
- Maintain and fuse pedigree data during data association processes

Required features:

- Review of system proposed merges based on predetermined threshold level
- Approval/rejection of system-proposed merges

Complexities addressed:

- Data characteristics
- Potential for deception

16.4.4   TOUCH POINT 4

Description: The system will maintain a large-scale database that represents a network of associated entities and their respective attributes. This database will be a primary source of relevant information for analysts. Analysts can rapidly search through the database and set boundaries to focus on sections of the database for consideration. However, analysts could prefer to perform fusion operations on two or more data sets and analyze the results before they are merged and associated with the larger database of previously fused information. This touch point was incorporated to support this anticipated analyst requirement. It serves as a pre-integration window to analyze incoming data before it is associated and merged with previously processed, older information. This touch point is also especially critical for distributed fusion environments, where multiple analysts and fusion processes may be simultaneously working on similar problems. It will also allow analysts to highlight/flag/annotate different pieces of data or relationships before integration into the larger fusion system database to allow analysts to understand where pieces of data they—or other analysts—viewed as relevant ended up in the larger network of information.

Location: After data association processing, but before the integration of merged and associated data set into the overall fusion entity-association database

Expected intelligence analysis stage(s) when accessed:

* Stages 3 and 4—Collection and evaluation of data

Capabilities supported:

* Automated language translation
* Manual or automated data association
* Maintain and fuse pedigree data during data association processes
* Maintain temporal reference meta-data
* Support automated pattern identification to highlight potential situations of interest
* Custom search alerts that notify when data or situation is available/appears

Required features:

* Browsing/reviewing of incoming data sets' entity-association network
* Selection of entities within the network to highlight or annotate prior to fusing with the overall fusion database
* Review of executed data merges
* Search input for the incoming data sets' network for situations of interest
* Expand entity or association meta-data, source data, and data association log

Complexities addressed:

* Distributed structure of the intelligence analysis domain
* Potential for deception
* Real-world dynamics

16.4.5   TOUCH POINT 5

Description: The fusion system can automatically pull in available data sets to be associated and integrated into the active (current bounded volume for analysis) or static (overall fusion database) volume of data maintained by the fusion system. When this occurs, analysts can review updates to the database so they can rapidly become aware of what is going on in the scenario represented by the incoming data (i.e., focus their attention on what is changing). To support the analysis of multiple hypotheses, the analyst can create data milestones that, if they occur, would help to confirm or reject one of the analyst's current hypotheses. In this situation, the analyst can view the incoming data that signaled the milestone "alert" within its original context before it is merged into the larger fusion data set. This view would include the data's meta-data and associations, which may be removed during the merge.

Location: After any update to the fusion system information database

Expected intelligence analysis stage(s) when accessed:

- Stages 3 and 4—Collection and evaluation of data

Capabilities supported:

- Store access credentials to multiple data sources
- Set up custom search alerts to notify when data or situation is available/appears
- Maintain and fuse pedigree data during data association processes
- What-if scenarios (e.g., what if this piece of data were false/true)
- Custom boundary ranges
- Update knowledge database in near real-time with most recently received data
- Maintain temporal reference meta-data

Required features:

- Browsing/reviewing of the fusion database
- Selection of database boundaries with respect to search/browsing capabilities
- Review of highlighted/annotated entities and/or associations
- Manual editing/addition of entities/associations/attributes
- Search the database using custom criteria
- Creation of entity/association placeholders that indicate expected hypotheses not yet incorporated/observed
- Drill-down to entity or association meta-data, source data, update log, weighting, edit precedence, and data association log

Complexities addressed:

- Distributed structure of intelligence analysis domain
- Potential for deception
- Real-world dynamics
- Unbounded problem space

16.4.6 TOUCH POINT 6

Description: Touch Point 6 allows interaction with the fusion system's primary output. Depending on the output capabilities of the system, this touch point can help analysts generate their presentation artifact(s) by providing network association diagrams, key pieces of data supporting/refuting their hypotheses, meta-information on data and/or source, or other components in a consumable format. Because this touch point supports a number of analyst goals, the list of required features generated even at this high-level of abstraction is already lengthy; however, maintaining an understanding of the end-user's goals as the touch point and associated features are defined will provide the foundation to ensure the final suite of features provide a positive and beneficial user experience.

Location: Outside of the fusion processing architecture, this touch point serves as a global touch point which provides access to the output of the fusion system.

Expected intelligence analysis stage(s) when accessed:

- Stage 3—Collection of data
- Stage 4—Evaluation of data
- Stage 6—Additional data collection and hypothesis testing
- Stage 7—Presentation

Capabilities supported:

- Multiple searches (stored or ad hoc) for situations of interest
- Manual data association
- Manual situation assessment
- Store access credentials for multiple data sources
- Automate language translation
- Set up custom search alerts to notify when data or situation is available/ appears
- Maintain pedigree data
- What-if scenarios (e.g., what if this piece of data were false/true)
- Automated pattern identification to highlight potential situations of interest
- Create temporal boundaries on situation assessment and data association
- Maintain temporal reference meta-data
- Support custom boundary ranges

Required features:

- Filtering of data sets to determine custom boundaries
- Custom uncertainty values to be assigned (e.g., what-if scenarios)
- Review/approval of system proposed data merges
- Browse/review individual or group data sets' entity-association database
- Select entities within the entity-association database to highlight or annotate
- Review executed data merges
- Search input to the entity-association database for situations of interest
- Select entity-association database boundaries/filters for search or other capabilities
- Review highlighted/annotated entities
- Manual editing/addition of entities and/or attributes
- Creation of entity/association placeholders that indicate expected hypotheses not yet incorporate/observed
- Drill-down to entity or association meta-data, source data, update log, weighting, edit precedence, and data association log
- Revision to data set boundaries

Complexities addressed:

- Data characteristics
- Distributed structure

- Lack of data
- Potential for deception
- Real-world dynamics
- Unbounded problem space
- Integration of additional data sets
- Review of data set and fusion entity-association database updates since pre-
  vious milestone

16.5   CONCLUSION

The case study presented in this chapter describes the high-level challenges and
system capabilities relevant to the support of human operators in fusion systems.
These examples were presented within the context of developing a hard-soft fusion
system to aid in intelligence analysis; however, the techniques used to analyze the
domain and identify the challenges and posit design requirements are generalizable.
Similarly, the process by which touch points for humans (and human organizations)
allow interaction with data fusion systems may also be generalized from the example
presented here. Our process involved three phases.

First, an extensive literature review was used to identify common stages of intelligence analysis and factors that commonly add complexity to the task(s) facing the
analyst. This review formed the foundation of understanding the role of humans in
a distributed data fusion system, and could be expanded to include a review of work
products and system or process documentation, ethnographic observations, study of
artifacts in the work domain, structure interviews, and other knowledge elicitation
activities (as formalized in Cognitive System Engineering practices), as well as a
domain-specific review of research literature.

Second, fusion system capabilities were identified that could support analysts in
situations where factors that commonly contribute to complexity are present. In this
phase, our generalized process rapidly focuses a potentially broad analysis on areas
of greatest need, and identifies where more design (and development and evaluation)
efforts are required. Finally, by overlaying this capabilities mapping to the common
stages of analysis, recommend touch points were identified that allow analysts to
best leverage the fusion system capabilities throughout their analysis. While further
effort is still required to refine the capabilities mapping and to define the specifics
of each interaction touch point, focusing on these critical interactions facilitates
the appropriate calibration of analyst perceptions to match the fusion system's
true capabilities, which will lead to appropriate reliance and, in turn, successful
integration into analyst workflows. This critical final step of the process is focused
on where and how human interaction with the data fusion system will result in
overall improvement in unified human-system performance. Explicit definition of
the touch points (and where they are required) allows for design and development
investment to address specific human-related challenges, whether the human is a
sensor, in the loop, or a consumer of fusion products. This same process applies to
socio-organizational challenges inherent in distributed human/automated systems
(see [Pfautz and Pfautz 2008] for a treatment of the different approaches to analysis
of socio-organizational challenges)—as touch points can be considered not only

between the human and the fusion system but also among humans (e.g., a human operating in the loop with a data fusion system may report on confidence to the decision-maker consuming fusion products). Clearly, the role of humans in data fusion cannot be trivialized nor simplified—the deep complexity in human-system interaction should and must guide the design and development of fusion systems.

ACKNOWLEDGMENT

Authors Jenkins and Bisantz were supported by Army Research Office MURI grant W911NF-09-1-0392 for "Unified Research on Network-based Hard/Soft Information Fusion" for the conduct of this work.

REFERENCES

Bisantz, A. M. and C. M. Burns, eds. 2008. *Applications of Cognitive Work Analysis*. Boca Raton, FL: CRC Press.

Bisantz, A.M., R. Stone, J. Pfautz, A. Fouse, M. Farry, E. M. Roth, E.M. et al. 2009. Visual representations of meta-information. *Journal of Cognitive Engineering and Decision Making*, 3(1):67–91.

Bisantz, A. M. and E. M. Roth. 2008. Analysis of cognitive work. In: *Reviews of Human Factors and Ergonomics*, Boehm-Davis, D.A., ed. Santa Monica, CA: Human Factors and Ergonomics Society, pp. 1–43.

Blasch, E. and S. Plano. 2002. JDL Level 5 fusion model: User refinement issues and applications in group tracking. *Proceedings of the SPIE 2002*, 4729(Aerosense):270–279.

Cook, M.B. and H. S. Smallman. 2008. Human factors of the confirmation bias in intelligence analysis: Decision support from graphical evidence landscapes. *Human Factors: The Journal of the Human Factors and Ergonomics Society*, 50:745–754.

Crandall, B., G. A. Klein, and R. R. Hoffman. 2006. *Working Minds: A Practitioner's Guide to Cognitive Task Analysis*. Cambridge, MA: The MIT Press.

Drucker, S.M. 2006. Coping with information overload in the new interface era. In the *Proceedings of the Computer-Human Interaction (CHI) 2006 Workshop "What is the Next Generation of Human Computer Interaction?"* Montreal, Quebec, Canada.

Elm, W. et al. 2004. Designing support for intelligence analysis. Presented at the *48th Annual Meeting of Human Factors and Ergonomics Society*. New Orleans, LA.

Elm, W. et al. 2005. Finding decision support requirements for effective intelligence analysis tools. *Human Factors and Ergonomics Society Annual Meeting Proceedings*, 49:297–301.

Endsley, M.R. 1996. Automation and situational awareness. In *Automation and Human Performance: Theory and Applications*, Parasuraman, T.V. and Mouloua, M., eds. Mahwah, NJ: Lawrence Erlbaum Associates.

Flynn, M. T., M. Pottinger, and P. D. Batchelor. 2010. Fixing intel: A blueprint for making intelligence relevant in Afghanistan. Working Papers, Washington, DC: Center for New American Security, accessed at: http://www.cnas.org/node/3924. (Accessed on June 2011).

FM 3-24 12/06. 2006. Counterinsurgency, U.S.D.o.t. Army, Editor Headquarters, Department of the ARMY, Washington DC, Field Manual (FM) number (No.) 3-24. http://www.fas.org/irp/doddir/army/fm3–24.pdf (accessed June 2011).

Greitzer, F.L. 2005. Methodology, Metrics and measures for testing and evaluation of intelligence analysis tools. Technical Report PNWD-3550. Richland, WA: Battelle-Pacific Northwest Division.

Grossman, J. B., D. D. Woods, and E. S. Patterson. 2007. Supporting the cognitive work of information analysis and synthesis: A study of the military intelligence domain. *Proceedings of the Human Factors and Ergonomics Society Annual Meeting*, 51(4):348–352.

Hall, D.L., M. McNeese, and J. Llinas. 2008. A framework for dynamic hard/soft fusion. In *11th International Conference on Information Fusion*. Cologne, Germany: Proc. ICIF.

Hardin, R. 2001. Conceptions and explanations of trust. In *Trust in Society*, Cook, K.S., ed. New York: Russell Sage Foundation, pp. 3–39.

Heuer, R. J. J. 1999. *The Psychology of Intelligence Analysis*. Washington, DC: Center for the Study of Intelligence.

Hewett, T. et al. 2005. *An Analysis of Tools That Support the Cognitive Processes of Intelligence Analysis*. Philadelphia, PA: Drexel University.

Hollnagel, E. and D. D. Woods. 2005. *Joint Cognitive Systems: Foundations of Cognitive Systems Engineering*. Boca Raton, FL: Taylor & Francis.

Hughes, F. J. and A. Schum. 2003. Discovery-Proof-Choice, The art and Science of the process of Intelligence analysis preparing for the future of intelligence analysis: Discovery-proof-choice. Joint Military Intelligence College.

Hutchins, S. G., P. L. Pirolli, and S. K. Card. 2004. A new perspective on use of the critical decision method with intelligence analysts. *Ninth International Command and Control Research and Technology Symposium*. San Diego, CA: Space and Naval Warfare Systems Center.

Hutchins, S. G., P. L. Pirolli, and S. K. Card. 2006. What makes intelligence analysis difficult? A cognitive task analysis of intelligence analysts. In *Expertise out of Context: Proceedings of the Sixth International Conference on Naturalistic Decision Making*, Hoffman, R.R., ed. New York: Lawrence Erlbaum Associates, pp. 281–316.

IW JOC 9/07. 2007. Irregular Warfare (IW) Joint Operating Concept (JOC), Department of Defense, Editor U.S. Department of Defense, Irregular Warfare (IW) Joint Operating Concept (JOC), Version 1.0, September 11, 2007. http://www.fas.org/irp/doddir/dod/iw-joc.pdf (accessed on June 2011).

Jenkins, M. P., G. Gross, A. M. Bisantz, and R. Nagi. 2011. Towards context-aware hard/soft information fusion: Incorporating situationally qualified human observations into a fusion process for intelligence analysis. *Cognitive Methods in Situation Awareness and Decision Support (CogSIMA), 2011 IEEE First International Multi-Disciplinary Conference on Situation Awareness and Decision Support*. Miami Beach, FL, pp. 74–81.

Jian, J.-Y. et al. 1997. *Foundations for an Empirically Determined Scale of Trust in Automated Systems*. Buffalo, NY: State University of New York at Buffalo, Center of Multisource Information Fusion, p. 48.

Johnston, R. 2005. *Analytic Culture in the US Intelligence Community: An Ethnographic Study*. Washington, DC: U.S. Government Printing Office.

Kahneman, D., P. Slovic, and A. Tversky. 1982. *Judgment under Uncertainty: Heuristics and Biases*. New York: Cambridge University Press.

Kent, S. 1965. *Special Problems of Method in Intelligence Work. Strategic Intelligence for American World Policy*. Hamden, CT: Archon Books.

Laplante, P. 2009. *Requirements Engineering for Software and Systems*. Redmond, WA: CRC Press.

Larson, E. V. et al. 2008. *Assessing Irregular Warfare: A Framework for Intelligence Analysis*. Santa Monica, CA: RAND Corporation.

Lee, J. D. 2008. Review of a pivotal human factors article: "Humans and automation: Use, misuse, disuse, abuse." *Human Factors: The Journal of the Human Factors and Ergonomics Society*, 50(3):404–410.

Mangio, C. A. and B. J. Wilkinson. 2008. Intelligence analysis: Once again. Interim report. Mason, OH: SHIM Enterprise Inc.

MSC 8/06, Multi-Service Concept for Irregular Warfare,include U.S. Marine Corps Combat Development Command, Quantico, VA and U.S. Special Operations Command Center for Knowledge and Future, MacDill AFB, FL, August 2006. https://www.mccdc.usmc. mil/CIW/ER/Multi%20Service/Multi-Service%20Concept%20for%20Irregular%20 Warfare%20-%20DistributionC.pdf (accessed on June 2011).

Parasuraman, R. and V. Riley. 1997. Humans and automation: Use, misuse, disuse, abuse. *Human Factors: The Journal of the Human Factors and Ergonomics Society*, 39(2):230–253.

Patterson, E., E. M. Roth, and W. Woods. 2001a. Predicting vulnerabilities in computer-supported inferential analysis under data overload. *Cognition, Technology & Work*, 3:224–237.

Patterson, E. S. et al. 2001b. Using cognitive task analysis (CTA) to seed design concepts for intelligence analysts under data overload. *Human Factors and Ergonomics Society Annual Meeting Proceedings*, 45:439–443.

Patterson, E. S. et al. 2001c. *Aiding the Intelligence Analyst in Situations of Data Overload: From Problem Definition to Design Concept Exploration*. Columbus, OH: Ohio State University.

Pfautz, J., A. Fouse, K. Shuster, A. Bisantz, and E. M. Roth. 2005a. Meta-information visualization in geographic information display systems. *Digest of Technical Papers—SID International Symposium*, 2nd edn. Boston, MA: Society for Information Display.

Pfautz, J. and S. Pfautz. 2008. Methods for the analysis of social and organizational aspects of the work domain. In *Cognitive Work Analysis: Current Applications and Theoretical Challenges*, Bisantz, A. and Burns, C., eds. Boca Raton, FL: CRC Press.

Pfautz, J., E. M. Roth, A. M. Bisantz, A. Fouse, S. Madden, and T. Fichtl. 2005b. The impact of meta-information on decision-making in intelligence operations. *Proceedings of the Human Factors and Ergonomics Society 49th Annual Meeting*. Santa Monica, CA.

Pfautz, J. et al. 2006. Cognitive complexities impacting army intelligence analysis. *Proceedings of the Human Factors and Ergonomics Society Annual Meeting*, 50(3):452–456.

Pfautz, J., A. Fouse, M. Farry, A. Bisantz, and E. Roth. 2007. Representing meta-information to support C2 decision making. In *Proceedings of the International Command and Control Research and Technology Symposium (ICCRTS '07)*. Newport, RI.

Pirolli, P. L. 2006. *Assisting People to Become Independent Learners in the Analysis of Intelligence*. Palo Alto, CA: Palo Alto Research Center, Inc., p. 100.

Pirolli, P. L. and S. K. Card. 2005. The sensemaking process and leverage points for analyst technology as identified through cognitive task analysis. In *International Conference on Intelligence Analysis*. McLean, VA.

Potter, S. S., W. C. Elm, and J. W. Gualthieri. 2006. Making sense of sensemaking: Requirements of a congintive analysis to support C2 Decision support system design. MANTECH SMA, pittsburgh, PA, Cognitive systems engineering center.

Powell, G. M. 2006. Understanding the role of context in the interpretation of complex battlespace intelligence. In *9th International Conference on Information Fusion*. Florence, Italy.

Powell, G. M. et al. 2008. An analysis of situation development in the context of contemporary warfare. In *Proceedings of 13th International Command and Control Research and Technology Symposium*. Bellevue, WA.

Roth, E. M., J. Pfautz, S. M. Mahoney, G. M. Powell, E. C. Carlson, S. L. Guarino, T. C. Fichtl, and S. S. Potter. 2010. Framing and contextualizing information requests: Problem formulation as part of the intelligence analysis process. *Journal of Cognitive Engineering and Decision Making*, 4(3):210–239.

Tam, C.K. (2008–2009). Behavioral and psychosocial considerations in Intelligence Analysis: A Preliminary review of literature on critical thinking skills. Interim report, May 2008–October 2008, 2009, Report ID: AFRL-RH-AZ-TR-2009-0009, Air Force Research Lab Mesa, AR. Human Effectiveness Directorate. p. 17. http://www.dtic.mil/cgi-bin/ GetTRDoc?AD=ADA502215 (accessed June 2011).

Taylor, S. M. 2005. The several worlds of the intelligence analyst. In *International Conference on Intelligence Analysis*. McLean, VA: MITRE Corporation.

Trent, S. A., E. S. Patterson, and D. D. Woods. 2007. Challenges for cognition in intelligence analysis. *Journal of Cognitive Engineering and Decision Making*, 1:75–97.

Vicente, K. J. 1999. *Cognitive Work Analysis*. Mahwah, NJ: Erlbaum.

Zelik, D. J., E. S. Patterson, and D. D. Woods. 2007. Judging sufficiency: How professional intelligence analysts assess analytical rigor. *Human Factors and Ergonomics Society Annual Meeting Proceedings*, 51:318–322.

17 Distributed Data and Information Fusion in Visual Sensor Networks

Federico Castanedo, Juan Gomez-Romero,
Miguel A. Patricio, Jesus Garcia, and
Jose M. Molina

CONTENTS

17.1 Introduction 436
17.2 Visual Sensor Networks... 436
    17.2.1 Requirements and Issues 437
        17.2.1.1 Communication.. 437
        17.2.1.2 Camera Calibration.. 437
        17.2.1.3 Object Detection 438
        17.2.1.4 Object Tracking 438
        17.2.1.5 Classification... 439
        17.2.1.6 Process Enhancement 439
    17.2.2 Related Research 439
    17.2.3 Context-Based Approaches to High-Level Information Fusion 441
17.3 Multi-Agent Systems in Visual Sensor Networks 443
    17.3.1 Belief–Desire–Intention Paradigm 445
    17.3.2 Communication and Coordination 445
17.4 Multi-Agent Approach to Manage Data in VSN 446
    17.4.1 Sensor Agents: Object Tracking 448
    17.4.2 Fusion Agents: Low- and High-Level Data Fusion, Context
        Exploitation, Feedback 449
17.5 Application Example: Indoor Surveillance 452
    17.5.1 Framework Configuration: Camera Calibration and Context
        Definition 453
    17.5.2 Low-Level Information Fusion... 455
    17.5.3 Contextual Enhancement to Tracking 456
    17.5.4 Scene Interpretation.. 459
17.6 Summary and Future Directions 459
References.. 460

17.1 INTRODUCTION

Computer vision, and in particular multi-camera environments, has been widely researched over the recent years, thus leading to several proposals of multi-camera or visual sensor networks (VSNs) architectures (Valera and Velastin 2005). The aims of these systems are very different; to name some of them, there are examples in surveillance applications (Regazzoni et al. 2001), sport domains (Chen and De Vlesschouwer 2010), or ambient intelligence applications for elderly care (Zhang et al. 2010). Despite the specific goal of each system, all of them have to cope with a distributed architecture of visual sensors to acquire and process information from the environment. The obtained information must then be fused in order to generate a meaningful global picture of the environment. Since a distributed VSN can be applied to different domains/scenarios, a specific ontology provides meaning and sense of the information that the system uses for interpretation purposes.

This chapter explores the use of the multi-agent paradigm and ontology-based knowledge representation formalisms to perform distributed data and information fusion (DIF) in VSNs. The multi-agent paradigm, which has been widely applied in distributed systems, provides a theoretical and practical framework to allow communication and cooperation among the components of the system. For instance, in Lesser et al. (2003) several multi-agent protocols are presented to solve the task allocation problem in distributed sensor networks, but without visual capabilities.

Classical distributed visual systems work well for monitoring and surveillance tasks, but they can be improved using a multi-agent paradigm and ontology-based mechanisms. The underlying idea is to provide autonomous elements of the system with standard communication capabilities compliant to a content ontology in the process to achieve high-level data fusion.

The remainder of this chapter is organized as follows. The next section describes the main requirements and issues that should be taken into account when building VSNs. Section 17.3 introduces the application of multi-agent systems in visual sensor domains. Section 17.4 provides a description of a specific architecture to fuse data in a VSN. An example using this architecture is shown in Section 17.5. Finally, Section 17.6 presents some open research problems and prospective directions for future work.

17.2 VISUAL SENSOR NETWORKS

Modern VSNs involve the deployment of a number of cameras in a wide area and the management of these geographically distributed monitoring points. Third-generation video systems apply techniques that resemble the human intelligent process of surveillance, which activates certain cognitive abilities, to satisfy the challenges posed to modern security applications (Regazzoni et al. 2001). The most characteristic aspect of third-generation video systems is the use of physically distributed cameras able to locally run image-processing algorithms. Due to the huge amount of data, the natural processing architecture for a VSN is distributed (hierarchical or decentralized) with processors dedicated to each visual data stream in a first level, before the information is communicated through the network. The combination of multiple viewpoints brings potential improvements to the reliability and accuracy of the results, although the existence of multiple cameras inevitably increases the complexity of the system. Although it is conceivable to achieve real-time performance with centralized processing, sending raw video streams to centralized servers is not practical, especially if the communication costs between nodes are accounted. Hence, local processing is necessary. Moreover, distribution increases system robustness and fault tolerance, since the same information may be captured and replicated at different points of the network.

17.2.1 REQUIREMENTS AND ISSUES

Two main requirements usually arise in distributed visual systems. First, it is necessary to implement suitable procedures to fuse local data (captured by single cameras) in order to obtain an integrated view of the situation while reducing bandwidth consumption. Second, coherence and scalability of the global system must be guaranteed with independence of the specific sensors and their configuration. This objective is difficult to accomplish when new heterogeneous cameras are incorporated to build a large and scattered network. Consequently, local data acquired by distributed video cameras must be combined to obtain a global understanding of the current scenario. Therefore, distributed systems for VSNs require techniques, algorithms, and procedures to solve the following issues.

17.2.1.1 Communication

Information acquired from each camera should be shared with others cameras and processing nodes, usually over a wired or wireless Internet Protocol (IP) network. The first decision in a multi-camera system is the physical installation of cameras. The number and placement of individual cameras have a great impact on system cost and capabilities. Since the main objectives are precise tracking of interesting objects, maximizing reliability and continuity of tracks, thus target-to-target or background-to-target occlusions must be minimized by using multiple cameras monitoring the same area from different viewpoints.

17.2.1.2 Camera Calibration

Information in the VSN must be expressed in a common reference frame. Camera calibration, or common referencing, is the process of transforming from the local coordinates of each camera to a global coordinate space. Calibration and synchronization can be done during an offline phase prior to system operation. This process is necessary to have a correspondence between the objects captured by different cameras. The resulting translation may include a reconstruction step to obtain a 3D representation of the 2D image. The most employed methods for camera calibration are those proposed by Tsai (1987), Heikkila (2000), and Zhang (2000). When the cameras have significant overlapping fields of views, the homograph between two corresponding image ground planes from two cameras can be computed by using target footprint trajectories and optimization techniques (Lee et al. 2000, Black and Ellis 2001). Typically, images of a calibration target (an object whose location and geometry are known) are first acquired. Then, correspondences between 3D points on the target and their image pixels are obtained. This involves estimating the intrinsic and extrinsic parameters of the camera by minimizing the projection error of the 3D points on the calibration object. The Tsai camera calibration technique was popular in the past, but it requires a nonplanar calibration object with known 3D coordinates. Zhang (2000) proposed a more flexible planar calibration grid method in which either the planar grid or the camera can be freely moved. For multi-camera surveillance applications with little or no overlap areas between cameras, research has focused on automatically learning camera topology.

Some authors have proposed online calibration techniques. For instance, Javed et al. (2008) exploited the redundancy in paths that humans and cars tend to follow (e.g., roads, walkways, and corridors) by using motion trends and appearance of objects to establish correspondence. In Ellis et al. (2003) and Makris et al. (2004), authors used learned entry and exit zones to build the camera topology by exploiting temporal correlation of objects transiting between adjacent camera fields of view. In Pollefeys et al. (2009), a method is proposed to simultaneously compute the epipolar geometry and synchronization of cameras after considering the epipolar constraints that need to be satisfied by every camera pair.

17.2.1.3 Object Detection

Interesting objects must be identified in the sequence of images provided by the camera. There are various approaches to the detection of moving objects. *Temporal differencing* is based on calculating the pixel-by-pixel difference of various consecutive frames (Lipton et al. 1998). *Background subtraction* is based on subtracting the current snapshot pixel values with a predefined background image (Piccardi 2004). *Statistical methods* are a variation of basic background subtraction method. They are based on the difference of additional statistical measures (Wang et al. 2003). *Optical flow*, in turn, is based on the computation of the flow vectors of moving objects over time (Barron et al. 1994).

17.2.1.4 Object Tracking

Detected objects should be tracked over time by matching the detections between consecutive frames. Object tracking, which involves state estimation and data association, has been traditionally tackled by applying statistical prediction and inference methods. Some tracking methods in general DIF are distributed multiple hypothesis tracking (MHT) (Chong et al. 1990), distributed joint probabilistic data association (JPDA) (Chang et al. 1986), covariance intersection (CI)/covariance union (CU) (Julier and Uhlmann 2001), and distributed Kalman filter (Olfati-Saber 2007).

In the case of video data association, it is necessary that objects are robustly tracked in time, even though the image processing algorithms may fail to segment them as single foreground regions (blobs) in some intervals. Problems with object segmentation often occur (Genovesio and Olivo-Marin 2004) when (1) the object is occluded by another region, a fixed object in the scene, or other moving object; (2) the object image is split into fragments during image segmentation; (3) the images from different objects are merged because of their close or overlapping projections on the camera plane.

Classical data association techniques have been adopted and extended by computer vision researchers. The JPDA filter has been applied to 3D vision reconstruction (Chang and Aggarwal 1991, Kan and Krogmeier 1996). Cox (Cox and Hingorani 1996) proposed the first adaptation of Reid's MHT (Reid 1979) to visual data association problems. In more recent approaches (Khan et al. 2005, Cai et al. 2006, Liu et al. 2008), a Markov Chain Monte Carlo strategy is applied to explore the data association space in order to estimate the maximum a posteriori joint distribution of multiple targets. Other recent approaches (Fleuret et al. 2008) are based on a discretized occupancy maps in the real world onto which the objects are projected. As we shall explain in the following, the estimation process is very sensitive to particular conditions of the scenario. Thus statistical methods may be insufficient in VSNs, which requires the incorporation of additional information and knowledge in the process.

17.2.1.5   Classification

Object and activity recognition aim to determine the type of an object (e.g., car, human, aircraft) or the type of an activity (e.g., approaching, walking, manoeuvring). Depending on the specific application, classification can involve object type classification (car, human, aircraft, etc.) or activity classification based on the object movements. Recognition can be viewed as a probabilistic reasoning problem, in which case it is tackled through probabilistic models (Markov models, Bayesian networks, etc.) (Hongeng et al. 2004). It can also be modeled as a classification problem, in which case pattern recognition techniques (neural networks, self-organizing maps, etc.) (Hu et al. 2004) are employed.

17.2.1.6   Process Enhancement

Process enhancement, also known as active fusion, focuses on the implementation of suitable mechanisms that use the more comprehensive interpretation of the current situation obtained after fusing data to improve the performance of the previous tasks. Generally speaking, process enhancement improves a fusion procedure by using feedback generated at a more abstract level. For instance, the behavior of a tracking algorithm can be changed once a general interpretation of the scene has been inferred. When the system recognizes that an object is moving out of the camera range through a door, the tracking procedure will be informed to be ready to delete this track in the near future.

17.2.2   RELATED RESEARCH

A wide range of alternative architectures and algorithms for distributed camera systems have been proposed in the last decade. Cai and Aggarwal (1999) proposed a multi-camera framework for people tracking in outdoor environments. Mittal and Davis (2003) developed a multi-camera system for people tracking and action analysis.

Video surveillance and monitoring (VSAM), developed by Collins et al. (2001), is a system that addresses the problem of tracking multiple objects in a multi-camera scenario. VSAM presents the global picture of the environment to a human operator through a unified graphical user interface.

Snidaro et al. (2003, 2004) described a system for outdoor video surveillance in which data are acquired from different types of sensors (optical, IR, radar). In the first level, data are fused to perform the tracking of objects in each zone of the monitored environment. Next, this information is sent to higher levels to obtain the global trajectories of the objects. They employed an aspect ratio metric obtained for each detected object over all the sensors. The fused result is obtained by weighting each sensor's aspect ratio measurement. Analogously, Besada et al. (2005) proposed a distributed solution for airport surface traffic control based on a video network.

Matsuyama and Ukita (2002) developed a real-time multi-camera vision system in which the cameras are moved automatically with three degrees of freedom (pan, tilt, and zoom) according to the situation.

Typical examples of commercial surveillance systems are DETEC (DETEC Online) and Gotcha (GOTCHA Online). For outdoor applications, a representative example is the DETER system (Pavlidis et al. 2001). DETER reports unusual movement patterns of pedestrians and vehicles in outdoor environments such as car parks. In these conditions, the systems typically require a wide spatial distribution that implies camera management and data communication. Nwagboso (1998) proposes combining existing surveillance traffic systems based on networks of smart cameras. The term "smart camera" is normally used to refer to a camera that has processing capabilities (either in the same casing or nearby) and can autonomously perform event detection and event video storage.

In general, third-generation surveillance systems provide highly automated information, as well as alarms and emergencies management. This is the aim of CROMATICA (CROMATICA Online), a system for crowd monitoring and its successor, PRISMATICA (Velastin et al. 2005), a pro-active integrated system for security management. PRISMATICA, which is one of the most sophisticated surveillance systems of the recent years, is a wide area multi-sensory, multimodal distributed system. It receives inputs from closed-circuit television (CCTV), local wireless camera networks, smart cards, and audio sensors. Intelligent devices in the network process sensor inputs and send/receive messages to/from a central server module. Another important project is ADVISOR (Siebel and Maybank 2004), which aims to assist human operators by automatically selecting, recording, and annotating images containing events of interest. Although both systems are classified as distributed architectures, they have a significant difference: PRISMATICA employs a centralized approach which controls and supervises the whole system, whereas ADVISOR can be considered a semi-distributed architecture. In Yuan et al. (2003), an intelligent video-based visual surveillance system (IVSS) is presented. This system aims to enhance security by detecting certain types of intrusion in dynamic scenes. The system involves object detection and recognition (pedestrians and vehicles) and tracking, with an architecture similar to ADVISOR (Siebel and Maybank 2004).

Scalability has been specifically addressed by including new security devices or analysis modules after the initial deployment of the surveillance system. Within this context, service-oriented computing has been used to design a framework to deploy video surveillance applications (Enficiaud et al. 2006). The authors used this framework to detect and count people in monitoring environments.

One disadvantage of most classical systems is that they rely on expensive computational costs. This high processing load may be impossible to accomplish in real-time video applications, since image processing introduces a bottleneck due to the foreground/background subtraction algorithms. A second problem is that the employed algorithms usually rely on very strong statistical assumptions (such as Gaussian linear dynamic models of targets and noise), which unfortunately do not hold in several application domains. In video processing, statistical techniques have encountered practical limitations mainly due to the difficulty of obtaining analytical models of the source errors.

Researchers have proposed solutions to overcome the problems that usually arise when dealing with visual information. There is a growing interest in the design of open and flexible DIF software architectures and techniques that improve the classical approaches. One of the main challenges for achieving enough reliability in the information inferred from a visual network is the use of appropriate context representation and management formalisms in the fusion process. Also, the coherence in the network requires communication and coordination mechanisms to share information and carry out the necessary adjustments in the information derived.

Besides, distributed visual data fusion must address problems that are common to any distributed data fusion application. First of all, when dealing with images as an input source, it is very difficult to have a predefined model of sensor error and a priori detection probabilities (visual information may have problems with illumination changes, occlusions, etc.) Other problems with distributed solutions are the need of clock synchronization between sources, the presence of out of sequence measurement and data incest problems.

For these reasons, in this chapter we explore the use of multi-agent architectures in distributed fusion with specific reasoning procedures at the low-level (contextual) and high-level to obtain an appropriate interpretation of the environment. The use of ontologies is also considered to represent the exchanged information and formalize the exploitation of contextual information.

17.2.3   CONTEXT-BASED APPROACHES TO HIGH-LEVEL INFORMATION FUSION

Broadly speaking, high-level information fusion (HLIF) refers to those inferences developed by IF systems which correspond to a higher level of abstraction. Cognitive approaches to HLIF propose building a symbolic model of the world, expressed in a logic-based language, to abstractly represent the scene objects, events, and behaviors, as well as the relations among them (Vernon 2008). Such a model can be regarded as the mental representation of the scene gained by cognitive software agents. It may include both perceptions and more complex contextual information. Cognitive approaches are robust and extensible, but they require the development of suitable interpretation and reasoning procedures.

The use of symbolic models to acquire, represent, and exploit knowledge in IF, and particularly in visual IF, has increased in the last decade. Lambert (2003) highlights three requirements that are crucial to the implementation of model-based IF systems: (1) to discern what knowledge should be represented, (2) to determine which representation formalisms are appropriate, (3) to elucidate how acquired and contextual inputs are transformed from numerical measures to symbolic descriptions, which is known as the grounding problem (Pinz et al. 2008).

Regarding selection of knowledge to be represented, there is a consensus about the importance of context knowledge in visual IF. Recently, researchers in IF have recognized the advantages of cognitive situation models, and have pointed out the importance of formal context knowledge to achieve scene understanding. Specifically, the last revision of the Joint Directors of Laboratories (JDL) specification highlights the importance of context knowledge (Steinberg and Bowman 2009), especially when visual inputs are to be interpreted (Steinberg and Rogova 2008). Henricksen (2003) defines context as *the set of circumstances surrounding* a task *that are potentially of relevance to its completion.* Kandefer and Shapiro (2008) extend this definition and state that context is *the structured set of variable, external constraints to some (natural or artificial) cognitive process that influences the behavior of that process in the agent(s) under consideration.*

To be consistent with this definition, we can consider that context in visual applications includes any external piece of knowledge used to complement the quantitative data about the scene computed by straightforward image-analysis algorithms. Context information (CI) is therefore an "external constraint" (because it is not directly acquired by the primary system sensors) that "influences the behavior" of the fusion process (since it is used to guide and support visual IF). Adapting the characterization by Bremond and Thonnat (1996), four sources of CI must be taken into account in visual DIF: (1) the scene environment: structures, static objects, illumination, and other behavioral characteristics, etc.; (2) the parameters of the sensor: camera, image, and location features; (3) historic information: past detected events; (4) soft information provided by humans.

Several representation formalisms have been proposed to be used in IF problems. Nevertheless, logic-based languages have received modest interest, in spite of their notable representation and reasoning advantages. Moreover, in this case most approximations have used ad hoc first-order logic representation formalisms (Brdiczka et al. 2006), which have certain drawbacks: they are hardly extensible and reusable, and reasoning with unrestricted first-order logic models is semi-decidable. Recently, there is a special interest in ontologies (Nowak 2003), since they overcome these problems. Current approaches are using ontologies to combine contextual and perceptual information, but there is still a lack of proposals that describe in detail how context knowledge can be characterized and integrated in general fusion applications.

At the low-level IF (i.e., JDL levels 0 and 1), one of the most important contributions is the Core Ontology for Multimedia (COMM). COMM is an ontology to encode MPEG-7 data at image level (i.e., JDL L0) (Arndt et al. 2007). It is represented with the Ontology Web Language (OWL), the standard proposed by the World Wide Web Consortium (W3C) (Hitzler et al. 2009). COMM does not represent high-level entities of the scene, such as people or events. Instead, it identifies the components of a MPEG-7 video sequence in order to link them to semantic web resources. Similarly, the Media Annotations Working Group of the W3C is working in an OWL-based language for adding metadata to web images and videos (Lee et al. 2009).

Other proposals are targeted at modeling video content at object level (i.e., JDL L1). For example, a framework for video event representation and annotation is described in François et al. (2005). This framework includes two languages, namely the Video Event Representation Language (VERL) and the Video Event Markup Language (VEML). VERL defines the concepts to describe processes, such as entities, events, time, and composition operations; and VEML establishes an XML-based vocabulary to markup video sequences, such as scenes, samples, streams, etc. VEML 2.0 has been partially expressed in OWL. Other authors have discussed and improved this approach to support the representation of uncertain knowledge (Westermann and Jain 2007). Halfway between data and object level is the research work by Kokar and Wang (2002), who present a symbolic representation for the data managed by a tracking algorithm. In this approach, the data managed by a tracking algorithm are represented symbolically to solve the grounding problem and to support further reasoning procedures. The low-level ontologies presented in Section 17.4.2 are based in this notion. In addition, higher-level knowledge inferred by abductive reasoning is also considered in our proposal.

High-level IF issues (i.e., JDL L2 and L3) are being dealt with ontologies as well. Little and Rogova (2009) study the development of ontologies for situation recognition, and propose a methodology to create domain-specific ontologies for information fusion based on the upper-level ontology Basic Formal Ontology (BFO), and its sub-ontologies SNAP and SPAN, used for endurant (*snapshot*) entities and perdurant (*spanning*) processes, respectively. In Neumann and Möller (2008), the authors present an ad hoc proposal for scene interpretation based on Description Logics and supported by the reasoning features of the Renamed Abox and Concept Expression Reasoner (RACER) (Häarslev and Möller 2001). The authors also distinguish between lower-level representations and higher-level interpretations to avoid the grounding problem. The representation of high-level semantics of situations with a computable formalism is also faced in Kokar et al. (2009), where an ontology encoding Barwise's situation semantics is developed. The approach in Aguilar-Ponce et al. (2007) defines a multi-agent architecture for object and scene recognition in VSNs. In addition, the later authors propose the use of an ontology to communicate information between task-oriented agents, in a similar way as the proposal described in Section 17.4.1. A practical approach to surveillance is shown by Snidaro et al. (2007), who developed an OWL ontology enhanced with rules to represent and reason with objects and actors.

All these works focus on contextual scene recognition, but it is also interesting to apply this knowledge to refine image-processing algorithms (which corresponds to JDL L4), as described in Section 17.1. An approach to this topic is presented in Gómez-Romero et al. (2011). In this paper, the authors describe an ontology-based framework to support scene recognition and fusion process enhancement, and discuss contributions and drawbacks from an architectural and knowledge management point of view.

17.3 MULTI-AGENT SYSTEMS IN VISUAL SENSOR NETWORKS

Multi-agent systems have been proposed as a solution for distributed surveillance, since they naturally support coordination of multiple tasks aimed at the analysis of object behaviors in dynamic and complex situations. Multi-agent systems are arguably well suited for the development of distributed systems in dynamic environments as VSNs. Agents have been applied in several approaches to identify faces and adapt the segmentation process in monitoring context, as discussed in Lee (2003).

Solving tracking tasks is one of the most studied problems by approaches that use agents to monitor objects. It is possible for agents to communicate and cooperate to monitor multiple objects simultaneously. A representative example of this approach was proposed by Remagnino et al. (2004), where they design the camera agent to calibrate the camera, track objects, and learning their behavior. The authors proposed a multi-agent architecture for visual monitoring where the agents are dynamically created when a new object is detected in order to cast the concept of agent to the detected objects. Similar proposals were later discussed in Garcia et al. (2005), which focuses on the communication messages exchanged between agents. The work in Castanedo et al. (2010) is also based on the application of multi-agent systems in a VSN. Recently, Albusac et al. (2010) also proposed a multi-agent architecture to incorporate expert domain knowledge into automatic monitoring and to provide a scalable and flexible solution tested in an urban traffic scenario.

As a matter of fact, the notion of agent suits very well to the concept of intelligent camera, since each software agent acquires and processes the visual images. On the one hand, nodes in the VSN are *autonomous*, in the sense that they have processing capabilities to acquire and process information in its field of view. On the other hand, the *social abilities* of agents provide the necessary means to share the visual information across the network and cooperate in the overall objective of the VSN. In order to avoid errors due to local knowledge of the world, nodes (developed as agents) establish social relations to build a global fused result depicting a more accurate and abstract view of the scenario.

In addition, agent-based *standard communication protocols* are the support to achieve interoperation with other systems at a high abstraction level. Last but not least, the existence of several multi-agent *frameworks*, which hide particular communication details, provides an easy way for developing distributed systems due to the loosely coupled architecture of multiple agents.

Ontologies can be used in such architecture to define the content language of agents' messages. The use of a common communication ontology facilitates agent interoperability, since the messages are expressed in the same well-defined language. This allows systems to be flexible, extensible, and independent of the implementation technologies. Moreover, sharing and reusing features of ontologies make them especially suitable for DIF in VSN. As mentioned before, VSN applications are highly context-dependent, but ontologies can be reused or extended to suit specific domain requirements. The agent communication ontology defines a set of concepts to describe the tracking information interchanged by the agents of the VSN. It behaves as an agreed vocabulary that allows tracking data to be represented in an abstract, common, and understandable way. Agents manage a local instantiation of the ontology, where individual ontologies corresponding to runtime scenario data are created. As we explain in the next section, ontologies are used in the architecture not only as a message content language but also to represent fused data and contextual knowledge.

17.3.1 Belief–Desire–Intention Paradigm

Multi-agent systems (Weiss 1999) can be divided into three different types: reactive, deliberative, and hybrid. The belief–desire–intention (BDI) paradigm is considered a hybrid architecture, since it divides the execution time of the system between deliberation and execution. The main difference with respect to the purely reactive architectures is that hybrid architectures spend more time reasoning to choose the next plan for execution. On the contrary, purely deliberative architectures follow a pure logic representation that requires an agent to manipulate symbols, and the percentage of time spent on the execution of the actions is less than the hybrid ones.

BDI paradigm has an explicit representation of the agent's notion following Bratman's theory of practical reasoning (Bratman 1987). The knowledge of an agent at any given time is based on the state of the BDI data structures. The belief data structure stores facts in a belief base acquired from the environment. Desire represents the final affairs that an agent wants to achieve. Finally, Intention describes specific plans that an agent has committed to execute in order to achieve its desires. Therefore, intentions should be consistent with the agent's desires. The BDI reasoning cycle must choose those plans for execution that match with the agent's desires, given the current belief. In this sense, the BDI architecture follows a similar reasoning process as the rule-based planning systems. However, multi-agent architectures also implement the social and communication capabilities required in any distributed system.

One of the advantages of using a multi-agent architecture is the separation between the management of the execution control and the reasoning mechanism, and plan execution is clearly separated inside the architecture. Therefore, there is no need to have an external management process.

17.3.2 Communication and Coordination

Agent communication in the VSN is the cornerstone to more complex DIF procedures. Communication mechanisms and protocols employed by the agents are usually based on the speech act theory (Searle 1970). To the speech act theory, spoken sentences in natural language are actions that produce changes in the receiver. Thus, in agent-based models, utterances are actions that result in changes in the internal state of the agents involved in the conversation. The messages sent by the agents are labeled using specific intention identifiers (e.g., *query* or *inform*). Exchanged information may range from essential data to complete acquired sequences, and from raw data to processed information. Besides, communication protocols can be based on pull messages (*ask* for information) or push messages (*provide* information).

The current standards for communication in multi-agent systems are defined in the Foundation of Intelligent Physical Agents (FIPA) specifications. Regarding message-passing, FIPA defines Agent Communication Language (ACL), a transport language that defines the format of the messages' envelope, a set of communicative acts, and a set of interaction protocols. ACL allows specifying the vocabulary to be used to encode agent contents. Traditionally, message semantics have been expressed in the FIPA Semantic Language (SL), a first-order logic derived language. The main drawback of SL is that it is undecidable in its general form; i.e., it is not guaranteed that all the inferences are computable in a finite time. Therefore, there is a growing interest in using formal ontologies as content languages (Hendler 2001, Schiemann and Schreiber 2006, Erdur and Seylan 2008), since they have appropriate computational properties and several supporting tools.

Ontologies can be accordingly defined to describe visual information exchanged by the agents of the VSN. In the simplest case, a suitable ontology can be created to represent tracking information. Such ontology would define a vocabulary including a set of concepts, relations, and axioms to describe tracking data. Agents manage a local instantiation of the ontology, where individual ontologies corresponding to the runtime data provided by the low-level tracking procedure are represented. Thus, the agents use the same vocabulary to interchange beliefs, which internally can be represented by using the ontology or not. Decoupling internal and external belief representations and the use of formal and standard languages facilitate the incorporation of heterogeneous elements to the VSN. In the most complex case, this ontology can include more abstract terms to represent objects, situations, or threats, and be the support of more sophisticated high-level fusion procedures, as described in the next section.

Besides communication, multi-agents also support the implementation of coordination schemes along communication protocols, in order to promote cooperation between agents and achieve better solutions. One of the most employed protocols for agent coordination is the contract-net (Smith 1980), which is mainly focused on task allocation problems. In a VSN, coordination mechanisms can be used to form smart camera coalitions, i.e., groups of sensors able to carry out complex processing tasks and collaborate with their neighbors. Another typical example of the application of agent cooperation in VSNs is camera handover (Patricio et al. 2007).

17.4 MULTI-AGENT APPROACH TO MANAGE DATA IN VSN

In the multi-agent approach for DDF in VSN, we can distinguish two main types of agents: sensor agents and fusion agents. Since the sources are completely distributed, but the fusion process is carried out by a centralized process level, a hierarchical and partially distributed architecture is proposed as is shown in Figure 17.1.

The figure shows two sensor agents and one fusion agent. However, it is possible to deploy several agents of each specific type. The only constraint is that a set of sensor agents are managed by a fusion agent following a hierarchical scheme. That is, the whole system has to include fewer fusion agents than sensor agents.

Sensor agents obtain the tracking information from the sensed environment through the acquired images and communicate the detected tracks to the fusion agent. So the external perception of each sensor agent is based on the processed images. The local perception of each sensor agent's environment is stored in the belief base as agent's beliefs. The obtained images are processed following the previous steps: object detection, data association, and state estimation. On the other hand, the fusion agent receives the track information from sensor agents and fuses it to obtain a global view of the scenario. The more comprehensive knowledge of the current situation obtained after fusing data can be used to provide sensor

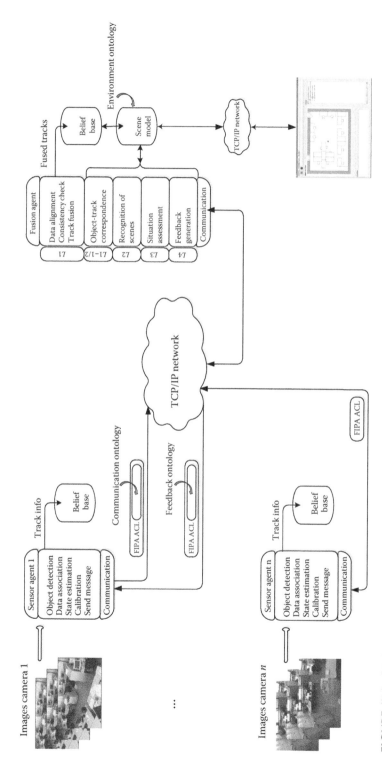

FIGURE 17.1   High-level hierarchical and partially distributed architecture.

agents with additional information, allowing them to correct their local knowledge. Communication between each sensor agent and the corresponding fusion agent is carried out by using the defined ontology as the content language in the FIPA ACL messages. Each agent (both sensor and fusion) is uniquely identified through its agent ID, which is composed of the IP address of the computer plus the agent platform and agent name. Next, the overall process is described in more detail.

17.4.1   Sensor Agents: Object Tracking

VSN data processing is performed by agents at two logical levels: (1) the tracking layer and (2) the BDI layer. First, each camera is associated with a tracking process. It sequentially executes various image-processing algorithms to detect and track all the targets within the local field of view. The tracking layer is arranged in a pipelined structure of several modules, as shown in Figure 17.2, which corresponds to the successive stages of the tracking process (Besada et al. 2005): (1) detection of moving objects, (2) blob-to-track multi-assignment, (3) track initialization/deletion, and (4) trajectory analysis.

The BDI layer uses an ontological model to encode these perceptions acquired by the agent. At this level, the purpose of the ontology is to serve as a symbolic representation of the numerical estimates from tracking. Therefore, the ontology is used for belief representation. This ontology, representing track information, can be also used for agent communication, as described in Section 17.3.2. Agent beliefs are represented as instances of the ontology, whereas desires and intentions are defined as plans following the JADEX format (Pokahr et al. 2005). We identify the following beliefs, desires, and intentions of camera-agents in a VSN:

Beliefs: Agent beliefs include information about the outside world, like objects that are being tracked (storing the location, size, trajectory, etc.), and geographic

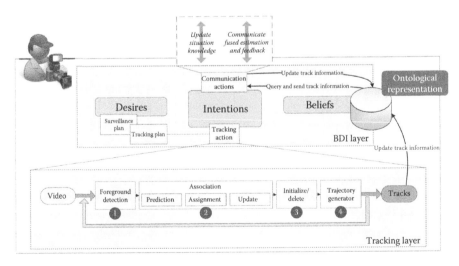

FIGURE 17.2   Sensor agent.

information about the camera itself, such as location, neighbor cameras, etc. The belief base of the agent is updated with the new perceived information. It may also be convenient to constrain the stored beliefs in a temporal window, in order to avoid the overhead of keeping all past knowledge. Therefore, the ontology will include convenient classes to describe tracks and track properties changing in time.

Desires: Since the final goal of agents is tracking the moving objects correctly, they have two main desires: permanent surveillance and temporary tracking. The surveillance plan is continuously executed. Sensor agents continuously capture images from the camera until an intruder is detected or announced by a warning from another agent. In this case, the tracking plan is triggered. The tracking plan runs inside a tracking process (implemented at the tracking layer), using the images from the camera until it is no longer possible. The tracking plan includes suitable actions to update beliefs of the agent, that is, to provide the track estimates to the BDI layer.

Intentions: Agents perform two types of actions: internal and external. Internal actions are related to video processing and tracking, and involve the issue of commands to the tracking subsystem or the camera. External actions correspond to communication acts with other agents. Agents send and receive messages carrying beliefs, which are represented with the ontology. Communication between sensor agents and fusion agents is performed by interchanging FIPA-compliant messages. The use of standard FIPA messages with content represented with the defined ontology promotes interoperability in the platform, as well as the incorporation of new heterogeneous agents. Two main types of interaction dialogs or conversations can happen between agents in the framework.

Update situation knowledge dialog: This interaction dialog sends to the fusion agent information about moving objects in the sensor agent field of view. The messages from the sensor agents include their local perceptions expressed as tracks and track properties represented in the communication ontology.

Communicate-fused estimation dialog: This interaction dialog sends to the sensor agent information and feedback about the global situation after data fusion is performed, according to the updates provided by the sensor agents.

17.4.2   FUSION AGENTS: LOW- AND HIGH-LEVEL DATA FUSION, CONTEXT EXPLOITATION, FEEDBACK

The fusion agent processes the *update situation knowledge* messages which are sent by sensor agents and initiates the fusion process. The fusion agent first extracts suitable data from this formal representation and starts a low-level fusion process based on existing DIF algorithms. From this formal representation of the low-level fused tracks, a high-level fusion process is developed. High-level information fusion in the fusion agent has two objectives: (1) to obtain a high-level interpretation of the scene from the perceptions of the distributed sensors—i.e., to perform L1 to L3 fusion; and (2) to determine how the fusion processes might be changed to improve their performance—i.e., to perform L4 fusion.

Essentially, HLIF in the fusion agent is a model-building procedure, which results in the construction of an ontological instantiation that abstractly represents the fused scene. We envision a knowledge model structured in five layers, from tracking data to impacts and threats:

Tracking data (L1). Output of the basic fusion algorithm represented symbolically. Examples include frames, tracks, and track properties (color, position,
velocity, etc.)

Scene objects (L1 – L1/2). Objects resulting from making a correspondence
between existing tracks and possible scene objects. For example, a track
can be inferred to correspond to a person (possibly by applying CI). Scene
objects include static elements which may be defined a priori and dynamic
objects, which may be defined a posteriori. Examples include person, door,
column, window, etc.

Activities (L2). Description of relations between objects that persist in time.
Examples include grouping, approaching, picking/leaving an object, etc.

Impacts and threats (L3). Cost or threat value assigned to activities.

Feedback and process improvement (L4). Abstract representation of the sug-
gestions provided to the tracking procedure.

An ontology of an upper abstraction level is based upon an ontology of a lower abstraction level. For example, the ontology for scene objects defines a property to associate instances of scene objects (e.g., people) to the actual track instances stored as agent's beliefs. Thus, information at this level is described in terms of objects instead of tracks, but the association between them is purposely represented. In the same way, a more abstract ontology is defined to represent scene situations. These situations would be inferred from the relevant objects represented in the lower-level scene objects ontology, which in turn is related to the track information ontology. Therefore, the communication ontology is the lowest level ontology and allows for making a correspondence between cognitive and perceived entities.

The fusion process in the fusion agent is depicted in Figure 17.3. This figure represents the information processing flow: first from bottom to top, to interpret the scene; and second, from top to bottom, to generate feedback.

Scene interpretation is a paradigmatic case of abductive reasoning, in contrast to the Description Logics classical deductive reasoning. Abductive reasoning takes a set of facts as input and finds a suitable hypothesis that explains them (sometimes with an associated degree of confidence or probability). This is the case of scene interpretation: the objective is to figure out what is happening in the scene from the observations and the contextual facts. In terms of the fusion agent architecture, scene interpretation is an abductive transformation from instances of a lower-level ontology (representing perceived or contextual entities) to instances of a higher-level ontology. Abductive reasoning is not directly supported by ontologies (Elsenbroich et al. 2006), since monotonicity of ontology languages forbids adding new knowl-edge to the models while reasoning. Nevertheless, it can be simulated by using customized procedures or preferably by defining transformation rules in a suitable query language. The RACER inference engine, presented in Section 17.2.3, allows abductive reasoning, and therefore it may be a good choice to implement the reason-ing procedures within the ontologies.

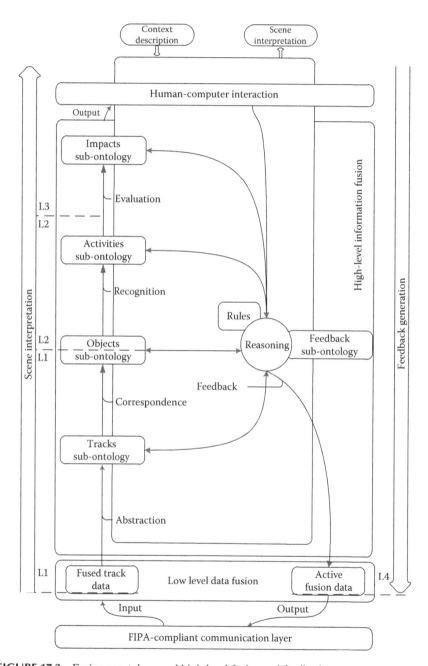

FIGURE 17.3   Fusion agent: low- and high-level fusion and feedback.

In the proposed architecture, abductive rules formally represent contextual, heuristic, and common sense knowledge to accomplish HLIF and low-level tracking refinement. Accordingly, we have two types of rules: bottom-up rules and top-down rules. On one hand, *bottom-up* rules are used in scene interpretation and obtaining instances of an upper-level ontology from instances of a lower-level ontology. For instance, some rules can be defined to identify objects from track measurements, i.e., to obtain instances of the scene objects ontology from instances of the tracking data ontology. An example rule may be: "create a person instance when an unidentified track larger than a predefined size is detected inside a region of the image." On the other hand, *top-down* rules create suggested action instances from the current interpretation of the scene, the historical data, and the predictions. These actions are used to adapt hypothesis at a lower-level to interpretations of a higher-level, which means the creation of instances of a lower-level ontology from instances of an upper-level ontology.

Eventually, top-down rules may create instances of the feedback ontology, which can be asynchronously returned to the sensor agent to update its knowledge. As a result of reasoning with the scene interpretation, active fusion information can be asynchronously returned to the sensor agent by starting a *communicate-fused estimation* dialog. These active fusion messages are also transmitted with the FIPA protocol and encoded with the communication ontology presented in Section 17.4.1.

17.5   APPLICATION EXAMPLE: INDOOR SURVEILLANCE

In this section, we will show how the framework presented in Section 17.4 is implemented in a specific application domain. Let us suppose an indoor surveillance system inside the university facilities aimed at tracking people and detecting interesting situations. We will focus on the computer laboratory, where three cameras are installed to cover the room area (see Figure 17.4). In this example, we have three sensor agents and one fusion agent. For the sake of simplicity, we will not consider additional cameras located at the nearby corridor. However, they can easily be incorporated to the framework and provide support for information handover when an individual enters the computer laboratory.

Before starting the processing, the framework must be configured. More precisely, the fusion agent must be informed of the positions of the cameras and provided with contextual information to be used in the fusion procedure. Once the framework has been configured, sensor agents start the execution of the *continuous surveillance* plan; i.e., agents process frames until the tracker detects a moving person in the room. Tracking data are encoded in the communication ontology and sent to the fusion agent by starting an *update situation knowledge* dialog. The fusion agent processes the tracking data obtained by the three cameras and combines them by applying a classical low-level fusion algorithm. This procedure results in updating the track data ontology, which triggers higher-level and contextual fusion procedures. Scene interpretation may lead to feedback generation to the sensor agents, which is returned back by starting a *communicate-fused estimation* dialog.

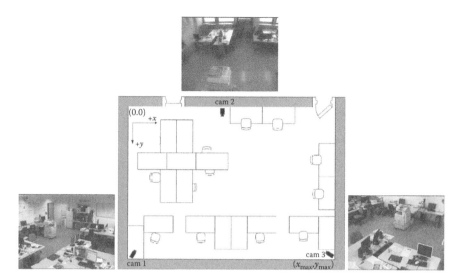

FIGURE 17.4   Computer laboratory scenario and cameras.

In the remainder of this section, we describe in more detail how these procedures are performed in the framework. This is not a comprehensive explanation of the implementation of such a system. Instead, we will make several assumptions to simplify the explanation of the system features in order to provide a general overview of the benefits of the approach and the open problems that remain to be solved in the future.

17.5.1   FRAMEWORK CONFIGURATION: CAMERA CALIBRATION AND CONTEXT DEFINITION

Camera calibration is achieved by applying the Tsai technique (1987). We manually mark some distinct points on the ground plane situated inside the overlapping area of the cameras. The homography matrix is calculated from the position of the distinct points in global and local coordinates. Linear optimization techniques are used to numerically calculate the values of the matrix. The homography matrix is used by the agents to transform from camera coordinates (as seen by sensor agents) to global coordinates (as seen by the fusion agent). Dynamic calibration techniques can be also applied, but for the sake of simplicity we will assume pre-calibration of the cameras (Figure 17.5).

After defining the common reference space, we use the ontological model to represent CI applicable to the scenario. Positions of the contextual entities are defined in global coordinates. To do this, we develop a specific ontology for surveillance based on the generic model presented in Section 17.4.2 to represent interesting entities of the surveillance domain, namely, the SURV ontology. This ontology defines the extensional knowledge of the application (i.e., concepts and relations). The intensional

FIGURE 17.5 An example of point correspondence in the three different views employed for the offline camera calibration phase.

knowledge (i.e., instances) will be created as a result of the fusion procedure. The SURV ontology in this example imports the sub-ontologies of the generic model and specializes them, for instance, with additional

- Concepts:
    Objects: *Door, Person, Table, CopyMachine, MeetingArea*
    Scenes: *Approach, Meeting*
- Relations:
    inMeeting
- Axioms:
    Person ⊑ DynamicObject (a person is a dynamic object)
    CopyMachine ⊑ OccludingObject (a copy machine is an occluding object)
    Table ⊑ OccludingObject (a table is an occluding object)

The SURV ontology is used to annotate the scenario. Annotating the scenario means to create instances of the ontology describing static objects. Therefore, we initially insert instances in the ontology to indicate the position of the door, the tables, the copy machine, and the meeting area. Figure 17.6 depicts the correspondence between ontology instances and scenario information. We also show the OWL code corresponding to the definition of *copymachine1* as an instance of *CopyMachine* at position (695, 360) in global coordinates. Unfortunately, annotation must be performed manually. Further tools to support scenario annotation should be developed and learning procedures could be considered. These are interesting directions for future work.

After initialization, the SURV ontology is loaded into the reasoning engine (e.g., RACER). Contextual rules (abductive and deductive) must also be created in this step. Some simple example rules, expressed in plain text, are presented in the following. These rules are represented in a suitable rule language such as the previously mentioned nRQL.

- Object association:
    [Rule 1] If a track is bigger than (50×50) pixels, then it corresponds to a person
- Activity recognition:
    [Rule 2] If there are more than one person inside the meeting area for a while, a meeting is being held

FIGURE 17.6 Scenario annotation.

- Process enhancement and feedback:
    [Rule 3] If a person is close to an occluding object, sensor agents must be warned about a possible future occlusion
    [Rule 4] If a meeting is being held, do not care about the tracks associated to the people in the meeting to avoid confusion

17.5.2 LOW-LEVEL INFORMATION FUSION

Figure 17.7 depicts a scenario in which we have an individual moving around the following a predefined path (the ground truth is known a priori). The picture show the frames captured by the cameras at time $t=200\,\mathrm{s}$ and the result of the background subtraction

Camera 1   Camera 2   Camera 3

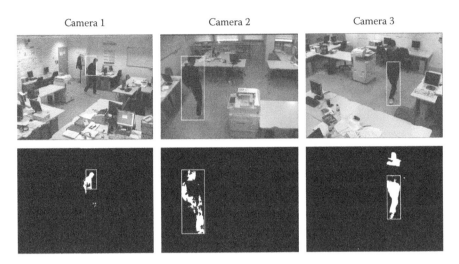

FIGURE 17.7   Local tracking results obtained by sensor agents ($t = 200$ s).

procedure. The frames also include the bounding box calculated by each sensor agent as a result of the tracking procedure based on the local data in its field of view.

It can be seen that the results obtained by sensor agent 1 are not very accurate at this frame. Regarding sensor agent 1, while the x position of the center of the track is correctly calculated, the y position is moved up (in local coordinates). Regarding camera 3, both x and y positions of the track are misplaced, but this has no effect on the projection, since the individual's feet are correctly detected and positioned on the floor. The projection of the track position to the ground plane clearly shows this malfunctioning (Figure 17.8). The graphs depict the (x, y) positions in global coordinates estimated at each frame of the sequence with respect to the ground truth. Positions corresponding to the frames at $t = 200$ s are highlighted with a square.

Tracking information obtained by sensor agents is sent to the fusion agent, which performs a low-level fusion procedure to combine the tracks and correct sensor errors. We have used the algorithm presented in Castanedo et al. (2007). As explained, tracking information is encoded with the communication ontology and wrapped in FIPA-compliant messages. In this case, the results of the Fusion Agent outperform the local estimates, as depicted in Figure 17.9 where fused (x, y) positions on global coordinates at each frame are shown.

Fused tracking information is inserted into the HLIF knowledge model as instances of the tracking sub-ontology. This update may trigger further reasoning processes in the contextual layer, as described in Section 17.5.4. In addition, after detecting a deviation between local and fused estimates, the fusion agent may initiate an active fusion process and send appropriate feedback to sensor agents.

17.5.3   CONTEXTUAL ENHANCEMENT TO TRACKING

In the previous example, estimation errors were the consequence of the limited information available. Thus, fusion significantly increased the accuracy of the system.

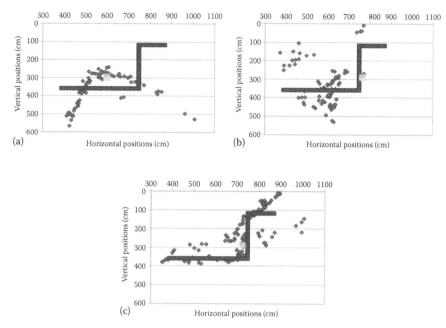

FIGURE 17.8 Local tracking results obtained by sensor agents compared to the ground-truth positions. (a) Sensor agent 1 (camera 1) (b) Sensor agent 2 (camera 2) (c) Sensor agent 3 (camera 3).

Nevertheless, in other cases classical data fusion procedures are insufficient to solve local tracking errors due to the inherent limitations of statistical tracking methods to adapt to complex situations.

For example, in Figure 17.10 we show the frames captured by the cameras at time $t = 180$ s and the (x, y) positions estimated at this frame in global coordinates.

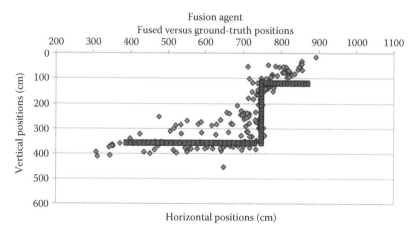

FIGURE 17.9 Fused tracking results obtained by fusion agent (from $t = 0$ to 200 s).

Camera 1   Camera 2   Camera 3

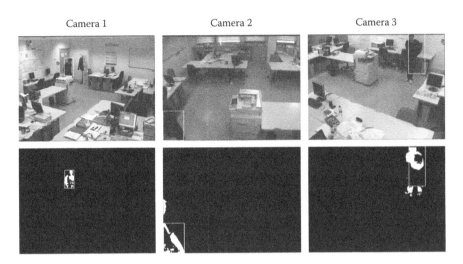

FIGURE 17.10 Local tracking results obtained by sensor agents ($t = 180$ s).

It can be seen that there is a significant error in the estimates of the three sensor agents. In this case, besides the previous difficulties (the individual is outside the field of view of cameras 2 and 3), there is an additional issue: a partial occlusion in camera 1. Partial occlusions result in track discontinuity, since hidden parts of the moving entities are not considered by the tracker and, therefore, track positions are misplaced.

Representation and reasoning with context knowledge in the fusion agent are applied to handle these situations. Scenario annotation is used to identify potential occlusive objects, contextual rules are fired when the conflictive situation is about to happen, and feedback is provided to the sensor agents to handle errors appropriately.

As a matter of example, let us suppose that the individual is being correctly detected by the tracker before $t = 180$ s. Fused information corresponding to this track would be consequently inserted into the HLIF knowledge model as instances of the track information sub-ontology. Rule 1 is triggered, and the track is identified as a person object by creating a proper instance in the object sub-ontology. In the next few frames, as the individual approaches the copy machine, the corresponding track information is updated, and eventually rule 3 is triggered. Consequently, an *expected occlusion* situation is created as an instance of the feedback sub-ontology. Subsequently, low-level fusion procedures and sensor agents may be notified about the situation by initiating a proper *communicate fusion estimate* dialog. If necessary, fused track information, encoded in the communication ontology, is sent back to the sensor agents by using FIPA-compliant messages. Low-level fusion procedures and sensor agents are responsible for handling the information properly. For instance, an appropriate action will be to incorporate track information to correct the Kalman filter matrix in order to avoid misplacing of the track position when the occlusion happens.

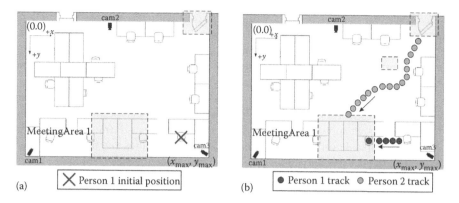

FIGURE 17.11   An example of detecting a meeting situation. (a) Person 1 is working without generating tracking updates and (b) activity in the meeting area results in a new detected situation.

17.5.4 SCENE INTERPRETATION

Let us suppose a situation in which we have an individual working on a desk of the computer laboratory (see Figure 17.11a). Tracking updates for these individuals are not sent to the fusion agent, because slight movements are not considered by the sensor agents. Next, one of the individuals (*person1*) stands up and moves into the meeting area. During this trajectory, sensor agents send information to the fusion agent, which updates the scene model. Some abductive and deductive reasoning procedures may be triggered as a result of ontology instance assertions, as explained before. Similarly, a second individual (*person2*) enters the room and moves to the meeting area. At this point, the current situation reflected in the ontological model is the one depicted in Figure 17.11b: we have two individuals labeled as persons who have entered the meeting area.

Consequently, rule 2 is triggered. A new *Meeting* instance is created in the activities sub-ontology, with *person1* and *person2* associated through the *inMeeting* property. This new *Meeting* instance fires rule 4. The aim of the rule is to prevent the agents from missing tracks corresponding to people who are close and probably overlapping. This feedback can be sent back to the sensor agents, which can handle this recommendation by stopping tracking in this area and storing track identifier and additional interesting track properties (e.g., predominant color), in order to identify tracks coming out of the meeting area.

17.6 SUMMARY AND FUTURE DIRECTIONS

More research works and implementations of general frameworks for visual DIF are needed to foster the creation of competitive solutions while cutting development costs in critical application areas. The first step toward domain-independent frameworks is to develop operational prototypes and to test them with existing data sets. The architecture proposed in Section 17.4 presents the overall picture of the system, but real implementations will have to deal with several specific problems that are identified in the description. In Llinas (2010), the author envisions a possible approach to a general IF multi-layer framework with a front-end that manages hard and soft sensor inputs; an initial layer for detection, semantic labeling, and flow control, based on an intelligent repository of pluggable algorithms; a fusion layer, composed of several interrelated fusion nodes that process information at different JDL levels and incorporate CI to the process; and a presentation layer to convey the results through appropriate visualization interfaces. Such IF frameworks should provide an adaptable infrastructure where specific procedures can be easily reused and/or integrated, especially those based on artificial intelligence techniques, which are likely to play a key role in the next-generation fusion applications. We strongly believe that the multi-agent paradigm and ontologies as representation formalisms can be the theoretical support of such frameworks.

As for the specific design of the presented architecture, it is important to notice that we have proposed a hierarchical schema for DIF. We have limited data alignment at tracking level, but it should be possible to combine estimations performed by fusion agents at different levels in such a way that the system will be able to obtain a combined view of the scenario from the detected objects or the recognized situations, instead of only the track data. This will require further investigations both at data and process level, since it involves the formation of local coalitions of coordinated agents. Reputation mechanisms should also be taken into account to measure the confidence in the data provided by different sources, in order to achieve conflict resolution.

Another interesting research area is the incorporation of uncertain and vague information representation formalisms and reasoning procedures into the framework for visual HLIF. Classical ontologies do not provide support for this kind of knowledge, which is inherent to vision applications, and extensively, to IF applications. There are three main sources of uncertainty and imprecision in HLIF applications. Firstly, we have errors due to the imprecise nature of sensor data. They can be statistically modeled, but are affected by physical conditions. Secondly, there is uncertainty resulting from scene interpretation procedures; for example, when there is more than one object in the scene or the situation cannot be clearly discerned. Finally, there is uncertainty resulting from fusion procedures; for instance, data combination may be trusted to a certain degree. In addition, it may be interesting to add imprecise knowledge management features to the reasoning model in order to deal with vague spatiotemporal relations such as close, far, before, after, etc.

REFERENCES

Aguilar-Ponce, R., A. Kumar, J. L. Tecpanecatl-Xihuitl, and M. Bayoumi. 2007. A network of sensor-based framework for automated visual surveillance. *Journal of Network and Computer Applications*, 30(3):1244–1271.

Albusac, J., D. Vallejo, J. J. Castro-Schez, P. Remagnino, C. Gonzalez, and L. Jimenez. 2010. Monitoring complex environments using a knowledge-driven approach based on intelligent agents. *IEEE Intelligent Systems, Special Issue on Intelligent Monitoring of Complex Environments*, 25(3):24–31.

Arndt, R., R. Troncy, S. Staab, L. Hardman, and M. Vacura. 2007. COMM: Designing well-founded multimedia ontology for the web. *Proceedings of the 6th International Semantic Web Conference (ISWC 2007)*, pp. 30–43, Busan, South Korea.

Barron, J., D. Fleet, and S. Beauchemin. 1994. Performance of optical flow techniques. *International Journal of Computer Vision*, 12(1):42–77.

Besada, J., J. Garcia, J. Portillo, J. M. Molina, A. Varona, and G. Gonzalez. 2005. Airport surface surveillance based on video images. *IEEE Transactions on Aerospace and Electronic Systems*, 41(3):1075–1082.

Black, J. and T. Ellis. 2001. Multi camera image tracking. *Proceedings of the 2nd IEEE International Workshop on Performance Evaluation of Tracking and Surveillance*. Kauai, HI.

Bratman, M. E. 1987. *Intentions, Plans and Practical Reason*. Cambridge, MA: Harvard University Press.

Bremond, F. and M. Thonnat. 1996. A context representation for surveillance systems. *Proceedings of the Workshop on Conceptual Descriptions from Images at the 4th European Conference on Computer Vision (ECCV'96)*, Cambridge, U.K.

Brdiczka, O., P. C. Yuen, S. Zaidenberg, P. Reignier, and J. L. Crowley. 2006. Automatic acquisition of context models and its application to video surveillance. *Proceedings of the 18th International Conference on Pattern Recognition*, pp. 1175–1178, Hong Kong, China.

Cai, Q. and J. K. Aggarwal. 1999. Tracking human motion in structured environments using a distributed camera system. *IEEE Transactions on Pattern Analysis and Machine Intelligence*, 21(11):1241–1247.

Cai, Y., N. de Freitas, and J. Little. 2006. Robust visual tracking for multiple targets. *European Conference on Computer Vision*, pp. 107–118, Graz, Austria.

Castanedo, F., J. García, M. A. Patricio, and J. M. Molina. 2010. Data fusion to improve trajectory tracking in a cooperative surveillance multi-agent architecture. *Information Fusion*, 11(3):243–255.

Castanedo, F., M. A. Patricio, J. García, and J. M. Molina. 2007. Robust data fusion in a visual sensor multiagent architecture. *The 10th International Conference on Information Fusion (FUSION 2007)*, Quebec, Canada.

Chang, Y. L. and J. K. Aggarwal. 1991. 3D structure reconstruction from an ego motion sequence using statistical estimation and detection theory. *IEEE Workshop on Visual Motion*, pp. 268–273, Princeton, NJ.

Chang, K., C. Y. Chong, and Y. Bar-Shalom. 1986. Joint probabilistic data association in distributed sensor networks. *IEEE Transactions on Automatic Control*, 31(10):889–897.

Chen, F. and C. De Vlesschouwer. 2010. Personalized production of basketball videos from multi-sensored data under limited display. *Computer Vision and Image Understanding*, 114(6):667–680.

Chong, C. Y., S. Mori, and K. C. Chang. 1990. Distributed multitarget multisensor tracking. In *Multitarget-Multisensor Tracking: Advanced Applications*, Y. Bar-Shalom, Ed., Vol. 1, pp. 247–295. Norwood, MA: Artech House.

Collins, R. T., A. J. Lipton, H. Fujiyoshi, and T. Kanade. 2001. Algorithms for cooperative multisensor surveillance. *Proceedings of the IEEE*, 89:1456–1477.

Cox, J. and S. L. Hingorani. 1996. An efficient implementation of Reid's multiple hypothesis tracking algorithm and its evaluation for the purpose of visual tracking. *IEEE Transaction on Pattern Analysis and Machine Intelligence*, 18(2):138–150.

CROMATICA, Crowd Monitoring with Telematic Imaging and Communication Assistance. http://dilnxsrv.king.ac.uk/cromatica/ (accessed September 22, 2012).

DETEC, Detec Video Surveillance Software. http://www.detec.no/ (accessed September 22, 2012).

Ellis, T. J., D. Makris, and J. Black. 2003. Learning a multi-camera topology. *Proceedings of the Joint IEEE International Workshop VS-PETS*, Nice, France.

Elsenbroich, C., O. Kutz, and U. Sattler. 2006. A case for abductive reasoning over ontologies. *Proceedings of the OWL Workshop: Experiences and Directions (OWLED'06)*. Athens, GA.

Enficiaud, R., B. Lienard, and N. Allezard. 2006. Clovis—A generic framework for general purpose visual surveillance applications. *IEEE Workshop on Visual Surveillance*, pp. 177–184, Graz, Austria.

Erdur, R. C. and I. Seylan. 2008. The design of a semantic web compatible content language for agent communication. *Expert Systems*, 25(3):268–294.

Fleuret, F., J. Berclaz, R. Lengagne, and P. Fua. 2008. Multicamera people tracking with a probabilistic occupancy map. *IEEE Transactions on Pattern Analysis and Machine Intelligence*, 30(2):267–282.

François, A., R. Nevatia, J. Hobbs, R. Bolles, and J. Smith. 2005. VERL: An ontology framework for representing and annotating video events. *IEEE Multimedia*, 12(4):76–86.

Garcia, J., J. Carbo, and J. M. Molina. 2005. Agent-based coordination of cameras. *International Journal of Computer Science and Applications*, 2(1):33–37.

Genovesio, A. and J. C. Olivo-Marin. 2004. Split and merge data association filter for dense multi-target tracking. *17th International Conference on Pattern Recognition*, Vol. 4, pp. 677–680. Cambridge, U.K.

Gómez-Romero, J., M. A. Patricio, J. García, and J. M. Molina. 2011. Ontology-based context representation and reasoning for object tracking and scene interpretation in video. *Expert Systems with Applications*, 38(6):7494–7510.

GOTCHA, Video Surveillance Software. http://www.gotchanow.com/ (accessed September 22, 2012).

Häarslev, V. and R. Möller. 2001. Description of the RACER system and its applications. *Proceedings of the International Workshop on Description Logics (DL2001)*, Stanford University, Stanford, CA.

Heikkila, J. 2000. Geometric camera calibration using circular control points. *IEEE Transactions on Pattern Analysis and Machine Intelligence*, 22(10):1066–1077.

Hendler, J. 2001. Agents and the semantic web. *IEEE Intelligent Systems*, 16(2):30–37.

Henricksen, K. 2003. A framework for context-aware pervasive computing applications. PhD thesis, University of Queensland, St. Lucia, Queensland, Australia.

Hitzler, P., M. Krötzsch, B. Parsia, P. F. Patel-Schneider, and S. Rudolph. 2009. OWL 2 web ontology language primer. In http://www.w3.org/TR/owl2-primer/ (Online, accessed on October 2011).

Hongeng, S., R. Nevatia, and F. Bremond. 2004. Video-based event recognition: Activity representation and probabilistic recognition methods. *Computer Vision and Image Understanding*, 96(2):129–162.

Hu, W., D. Xie, T. Tan, and S. Maybank. 2004. Learning activity patterns using fuzzy self-organizing neural network. *IEEE Transactions on Systems, Man and Cybernetics*, 34(3):1618–1626.

Javed, O., K. Shafique, and Z. Rasheed. 2008. Modeling inter-camera space-time and appearance relationships for tracking across non-overlapping views. *Computer Vision and Image Understanding*, 109(2):146–162.

Julier, S. and J. Uhlmann. 2001. General decentralized data fusion with covariance intersection. In *Handbook of Multisensor Data Fusion*, 2nd edn., M. E. Liggins, D. Hall, and J. Llinas, eds., pp. 319–344. Boca Raton, FL: CRC Press.

Kan, W. and J. Krogmeier. 1996. A generalization of the pda target tracking algorithm using hypothesis clustering. *Signals, Systems and Computers*, 2:878–882.

Kandefer, M. and S. C. Shapiro. 2008. A categorization of contextual constraints. *Biologically Inspired Cognitive Architectures: Papers from the AAAI Fall Symposium*, pp. 88–93. Menlo Park, CA: AAAI Press.

Khan, Z., T. Balch, and F. Dellaert. 2005. Multitarget tracking with split and merged measurements. *Proceedings of the IEEE Conference on Vision and Pattern Recognition*, 1:605–661.

Kokar, M., C. Matheus, and K. Baclawski. 2009. Ontology-based situation awareness. *Information Fusion*, 10(1):83–98.

Kokar, M. and J. Wang. 2002. Using ontologies for recognition: An example. *Proceedings of the 5th International Conference on Information Fusion*, Vol. 2, pp. 1324–1330, Annapolis, MD.

Lambert, D. 2003. Grand challenges of information fusion. *Proceedings of the 6th International Conference of Information Fusion*, Vol. 1, pp. 213–220, Cairns, Australia.

Lee, R. S. T. 2003. iJADE surveillant—An intelligent multiresolution composite neuro-oscillatory agent-based surveillance system. *Pattern Recognition*, 36(6):1425–1444.

Lee, W., T. Bürger, and F. Sasaki. 2009. Use cases and requirements for ontology and API for media object 1.0. Retrieved from W3C Working Draft: http://www.w3.org/TR/media-annot-reqs/

Lee, L., R. Romano, and G. Stein. 2000. Monitoring activities from multiple video streams: Establishing a common coordinate frame. *IEEE Transactions on Pattern Analysis and Machine Intelligence*, 22(8):758–767.

Lesser, V., C. Ortiz, and M. Tambe. 2003. *Distributed Sensor Networks: A Multiagent Perspective*. Berlin, Germany: Springer.

Lipton, A., H. Fujiyoshi, and R. Patil. 1998. Moving target classification and tracking from real-time video. *Proceedings of the 4th IEEE Workshop Applications of Computer Vision (WACV 98)*, pp. 129–136, Princeton, NJ.

Little, E. G. and G. L. Rogova. 2009. Designing ontologies for higher level fusion. *Information Fusion*, 10(1):70–82.

Liu, J., X. Tong, W. Li, T. Wang, Y. Zhang, and H. Wang. 2008. Automatic player detection, labeling and tracking in broadcast soccer video. *Pattern Recognition Letters*, 30(2):103–113.

Llinas, J. 2010. A survey and analysis of frameworks and framework issues for information fusion applications. *Proceedings of the 5th International Conference on Hybrid Artificial Intelligence Systems conference (HAIS'10)*, LNCS 6076, San Sebastián, Spain.

Makris, D., T. J. Ellis, and J. Black. 2004. Bridging the gaps between cameras. *Proceedings of the Computer Vision and Pattern Recognition*, Washington, DC.

Matsuyama, T. and N. Ukita. 2002. Real-time multi-target tracking by a cooperative distributed vision system. *Proceedings of the IEEE*, 90(7):1136–1150.

Mittal, A. and L. Davis. 2003. M2 Tracker: A multi-view approach to segmenting and tracking people in a cluttered scene. *International Journal of Computer Vision*, 51(3):189–203.

Neumann, B. and R. Möller. 2008. On scene interpretation with Description Logics. *Image and Vision Computing*, 26(1):82–101.

Nowak, C. 2003. On ontologies for high-level information fusion. *Proceedings of the 6th International Conference on Information Fusion (FUSION 2003)*, pp. 657–664, Cairns, Australia.

Nwagboso, C. 1998. User focused surveillance systems integration for intelligent transport systems. In *Advanced Video-Based Surveillance Systems*, C. S. Regazzoni, G. Fabri, and G. Vernazza, Eds., Chapter 1.1, pp. 8–12. Boston, MA: Kluwer Academic.

Olfati-Saber, R. 2007. Distributed Kalman filtering for sensor networks. *Proceedings of the 46th Conference in Decision and Control*, pp. 5492–5498, New Orleans, LA.

Patricio, M. A., J. Carbó, O. Pérez, J. García, and J. M. Molina. 2007. Multi-agent framework in visual sensor networks. *EURASIP Journal on Advances in Signal Processing*, 2001:1–21.

Pavlidis, I., V. Morellas, P. Tsiamyrtzis, and S. Harp. 2001. Urban surveillance systems: From the laboratory to the commercial world. *Proceedings of the IEEE*, 89(10):1478–1497.

Piccardi, M. 2004. Background subtraction techniques: A review. *Proceedings of the IEEE International Conference on Systems, Man and Cybernetics*, Vol. 4, pp. 3099–3104, The Hague, the Netherlands.

Pinz, A., H. Bischof, W. Kropatsch, et al. 2008. Representations for cognitive vision, ELCVIA. *Electronic Letters on Computer Vision and Image Analysis*, 7(2):35–61.

Pokahr, A., L. Braubach, and W. Lamersdorf. 2005. Jadex: A BDI reasoning engine. In *Multi-Agent Programming*, J. Dix, R. Bordini, M. Dastani, and A. Seghrouchni, eds. Dordrecht, the Netherlands: Kluwer.

Pollefeys, M., S. N. Sinha, L. Guan, and J. S. Franco. 2009. Multi-view calibration, synchronization, and dynamic scene reconstruction. In *Multi-Camera Networks: Principles and Applications*, H. K. Aghajan and A. Cavallaro, eds. Amsterdam, the Netherlands: Elsevier.

Regazzoni, C., V. Ramesh, and G. Foresti. 2001. Scanning the issue/technology: Special issue on video communications, processing and understanding for third generation surveillance systems. *Proceedings of the IEEE*, 89(10):1355–1366.

Reid, D. B. 1979. An algorithm for tracking multiple targets. *IEEE Transactions on Automatic Control*, 24(6):843–854.

Remagnino, P., A. Shihab, and G. Jones. 2004. Distributed Intelligence for multi-camera visual surveillance. *Pattern Recognition*, 37(4):675–689.

Schiemann, B. and U. Schreiber. 2006. OWL-DL as a FIPA-ACL content language. *Proceedings of the Workshop on Formal Ontology for Communicating Agents*, Malaga, Spain.

Searle, J. R. 1970. *Speech Acts: An Essay in the Philosophy of Language*. Cambridge, U.K.: Cambridge University Press.

Siebel, N. T. and S. J. Maybank. 2004. The advisor visual surveillance system. *Proceedings of European Conference on Computer Vision. Workshop Applications of Computer Vision*, Prague, Czech Republic.

Smith, R. G. 1980. The contract net protocol: High level communication and control in a distributed problem solver. *IEEE Transactions on Computers*, 29(12):1104–1113.

Snidaro, L., M. Belluz, and G. L. Foresti. 2007. Domain knowledge for surveillance applications. *Proceedings of the 10th International Conference on Information Fusion (FUSION 2007)*, Quebec, Canada.

Snidaro, L, G. Foresti, R. Niu, and P. Varshney. 2004. Sensor fusion for video surveillance. *Proceedings of the 7th International Conference on Information Fusion (FUSION 2004)*, Stockholm, Sweden.

Snidaro, L., R. Niu, P. Varshney, and G. Foresti. 2003. Automatic camera selection and fusion for outdoor surveillance under changing weather conditions. *IEEE Conference on Advanced Video and Signal Based Surveillance*, Miami, FL.

Steinberg, A. N. and C. L. Bowman. 2009. Revisions to the JDL data fusion model. In *Handbook of Multisensor Data Fusion*, pp. 45–67. Boca Raton, FL: CRC Press.

Steinberg, A. N. and G. Rogova. 2008. Situation and context in data fusion and natural language understanding. *Proceedings of the 11th International Conference on Information Fusion (FUSION 2008)*, Cologne, Germany.

Tsai, R. 1987. A versatile camera calibration technique for high accuracy 3d machine vision metrology using off-the-shelf TV cameras and lenses. *IEEE Journal of Robotics and Automaton*, 3(4):323–344.

Valera, M. and S. A. Velastin. 2005. Intelligent distributed surveillance systems: A review. *IEE Proceedings—Vision, Image, and Signal Processing*, 152(2):192.

Velastin, S. A., B. A. Boghossian, B. P. L. Lo, J. Sun, and M. A. Vicencio-Silva. 2005. Prismatica: Toward ambient intelligence in public transport environments. *IEEE Transactions on Systems, Man and Cybernetics, Part A*, 35(1):164–182.

Vernon, D. 2008. Cognitive vision: The case for embodied perception. *Image and Vision Computing*, 26(1):127–140.

Wang, L., W. Hu, and T. Tan. 2003. Recent developments in human motion analysis. *Pattern Recognition*, 36(3):585–601.

Weiss, G. 1999. *Multi-Agent Systems. A Modern Approach to Distributed Artificial Intelligence.* Cambridge, MA: The MIT Press.

Westermann, U. and R. Jain. 2007. Toward a common event model for multimedia applications. *IEEE Multimedia*, 14(1):19–29.

Yuan, X., Z. Sun, Y. Varol, and G. Bebis. 2003. A distributed visual surveillance system. *Proceedings of the IEEE Conference on Advanced Video and Signal Based Surveillance (AVSS 2003)*, Miami, FL.

Zhang, Z. 2000. A flexible new technique for camera calibration. *IEEE Transactions on Pattern Analysis and Machine Intelligence*, 22(11):1330–1334.

Zhang, S., S. Mcclean, B. Scotney, X. Hong, C. Nugent, and M. Mulvenna. 2010. An intervention mechanism for assistive living in smart homes. *Journal of Ambient Intelligence and Smart Environments*, 2(3):233–252.

Index

A

Ad hoc observers, information, 1–2
AI, *see* Artificial intelligence (AI)
Analysis of variance (ANOVA)
    factorial experiment, 402
    tier 0 analysis, 402, 403
    tier 1 analysis, 402, 404
    tier 2 analysis, 402, 405
ANOVA, *see* Analysis of variance (ANOVA)
Architectures, distributed estimation
    appropriate architectures selection, 104–105
    fusion graph, *see* Fusion architecture graph
    information communicated and common
        prior knowledge, 104
    information graph, 100–104
    measurements and processors, 97–98
    sensors, processors and users, 97
Artificial intelligence (AI), 287–288

B

Bar-Shalom-Campo and Speyer fusion rules
    cross-covariance, 130
    defined, 129–130
    fused estimation error, 130
    likelihood function, 130
    naïve fusion rule, 130
    repeated track fusion, *see* Repeated
        track fusion
    simple convex combination rule, 130
Bayes filter
    Markov transition density, 203
    multisensor measurements, 204
    normalization factor, 203
Bayesian distributed fusion algorithm, 259–260
    equation, 107
    Gaussian random vectors, *see* Gaussian
        random vectors
    goal and measurements, 105
    node 1 and 2, 106
    private and common information, 106
    probability distribution, discrete variable,
        105–106
Bayesian Formulation, 69–70
Bayesian maximum-likelihood fusion
    (BML) rule
    defined, 132–133
    likelihood function, 133
    weight matrices, 133

BDI, *see* Belief–desire–intention (BDI)
Belief–desire–intention (BDI)
    communication and coordination, 445–446
    description, 445
    practical reasoning, 445
Belief network (BN) model
    distributed parts, 274, 275
    junction tree, 271
    situation assessment, 273, 274
Bounded covariance inflation
    covariance ellipses, 20, 21
    description, 20
    inflated covariance matrix, 22
    joint covariance matrices, 20
    Kalman filter update equation, 21
    linear transform, 21
    upper and lower bounds, 22
Burst communications, 180

C

Camera calibration
    contextual rules, 454–455
    definition, copymachine1, 454
    description, 453
    homography matrix, 453
    point correspondence, offline camera,
        453, 454
    SURV ontology, 453–454
Cardinalized PHD (CPHD) filter fusion
    double-counting, 236–237
    known double-counting, 227
    measurement-update equations, 219–220
    T^2F, independent sources, 224–226
    T^2F, mathematical derivations, 234–235
    time-update equations, 218–219
    XM fusion, 228–229, 237–238
CCA, *see* Continuous combinatorial auction (CCA)
Central processing, 126
Channel filter
    asynchronous operation, 169, 170
    cache *vs.* filter algorithms, 167–168
    common and contributed information, 168
    common information *vs.* contributed
        information, ideal pair, 169
    description, 167
    lost transmissions, 169, 171
    miscommunication, 169
    operations, 167–168

CI, *see* Covariance intersection (CI)
Classifier fusion
    algorithm-combining approach, 255
    MAP, 255
    methods, taxonomy, 254
    product rule, 256
    taxonomic categorization, 253
Closest point of approach (CPA), 301
Common tactical picture (CTP), 272, 276
Communication and decentralized data
        fusion (DDF)
    channel filter approaches, *see* Channel filter
    description, 164–165
    dynamic systems, *see* Dynamic systems, DDF
    global agreement, nodes, 170
    log-likelihood/information form, 169
    tree network topology, channel cache,
        165–167
Communications algorithm, k-tree
    data-tag set elimination, 191
    description, 190
    separator, 190
Complementary sensor
    local measurement error covariance
        matrices, 141
    noise intensity level and initial state
        accuracy, 142–143
    normalized initial position standard
        deviation, 142, 143
    normalized process noise intensity, 141, 142
    position estimation performance, 142
Consistent tactical picture (CTP), 401
Contextual enhancement, tracking
    accuracy, system, 456
    communicate fusion estimate dialog, 458
    expected occlusion situation, 458
    sensor agents, 457, 458
Continuous combinatorial auction (CCA), 370
Correlated decisions
    Chair-Varshney rule, 83
    copula theory, 84
    description, 81
    LRTs, 82
    marginal distributions, 83
    M-ary hypothesis, 83
    ratio-based tests, 82
    suboptimal binary quantizers, 83
Counting rule, 74–76
Coupling scalars, 22–23
Covariance intersection (CI)
    Chen-Arambel-Mehra fusion rule, 135
    Chernoff fusion rule, 134
    cross-covariance matrix, 133–134
    error hyper-ellipsoids, 208
    fused target-localizations, 209
    Shannon fusion rule, 134–135
    Speyer fusion rule, 134

CPA, *see* Closest point of approach (CPA)
CTP, *see* Common tactical picture (CTP);
        Consistent tactical picture (CTP)

D

DARPA, *see* Defense Advanced Research
        Projects Agency (DARPA)
Data fusion and resource management
        (DF&RM), 380
DBNs, *see* Dynamic Bayesian networks (DBNs)
DDF, *see* Decentralized data fusion (DDF);
        Distributed data fusion (DDF)
Decentralized data fusion (DDF)
    characterization, constraints, 162
    and communication, *see* Communication and
        decentralized data fusion (DDF)
    communications latencies and failures, 193
    description, 162
    dynamic systems, *see* Dynamic systems, DDF
    information, 163–164
    k-tree topologies, redundant and dynamic
        networks, *see* K-tree topologies
    marginalization, information form, 193–194
    trajectory information form equivalence,
        194–196
Defense Advanced Research Projects
        Agency (DARPA)
    social network utilization, 351
    web information and tools, 353
Delayed and asequent observations
    description, 179
    destructive prediction estimation, 179
    N timesteps, 179
    prediction step, 178
    small delay, 180
    trajectory information matrix, 179
Design of experiments (DOE)
    factors and interaction complexity, 397, 398
    statistical, 395
    test-planning methods, 396
    types, 396–397
Deterministic dynamics, object tracking, 117–118
Developmental test & evaluation (DT&E)
    bounded system, 384
    engineering design goals, 385
    types, testing, 384
DF&RM, *see* Data fusion and resource
        management (DF&RM)
Distributed data and information fusion systems
        (DDIFS), *see* Test and evaluation
        (T&E), DDIFS
Distributed data fusion (DDF)
    applications, 1
    APPs and HCI, 2
    cognitive and information processes, 6
    content, 18

description, 1, 18
design, 19
Endsley's situation awareness model, 8, 9
frameworks, 6–8
HAC design concerns, 42–43
HAC opportunities, 43–44
implications, 12–13
information, 9, 11, 12
information recycling, *see* Information
    recycling
IT, 9–11
JDL, *see* Joint Directors of
    Laboratories (JDL)
mathematical sense meaning and redundant
    observations, 2
military applications and intelligence, 8
multiple observers/sensors, 2–3
net-centric generation, 4
OODA, 8
probabilistic model, computational
    trust, 35
sensor coordination, *see* Sensor coordination
stakeholders, *see* Stakeholders
state estimation, 2
system concept, 18, 19
traditional information, 2
Distributed detection, wireless sensor networks
binary distributed detection, 66
decision-making structures, 65
decision theory/hypothesis testing, 65–66
FDR-based decision, 89
ideal communication channels, *see* Ideal
    communication channels
nonideal communication channels, *see*
    Nonideal communication channels
NP and FDR, 66–67
ROI, 90
WSNs, 66–67, 90
Distributed estimation
architectures, *see* Architectures, distributed
    estimation
Bayesian distributed fusion algorithm, *see*
    Bayesian distributed fusion algorithm
description, 96–97
fusion architecture and best
    performance, 96
Gaussian distributions/error covariances,
    see Error covariances, distributed
    estimation
MAP, BLUE and cross-covariance
    fusion, 121
multiple sensors, 96
object classification, *see* Object classification
object tracking, 117–119
optimal Bayesian distributed fusion, *see*
    Optimal Bayesian distributed fusion
optimal fusion algorithm, 121

suboptimal Bayesian distributed fusion
    algorithms, *see* Suboptimal Bayesian
    distributed fusion
Distributed fusion environments
centralized, hierarchical, peer-to-peer and
    grid-based, 277
network, distributed fusion nodes, 276
process observations, 274
Distributed high-level fusion
algorithm, situation assessment, 277–282
BN model, situation assessment, 273, 274
CTP, 271–272
decentralized processing environment, 273
distributed fusion environments, 274–277
distributed Kalman filter, 282–285
GIG, 273
NCW, 285–286
role, intelligent agents, 286–290
SA, 271
Distributed Kalman filter
fusion nodes, 285
JSTARS, 283
target tracking with and without
    feedback, 284
UAV, 283
Distributed processing, 126
Distributed situation assessment, algorithm
junction tree, 279
junction tree construction and
    inference, 280–282
pairwise communication-link
    information, 278
sensor network, 278
spanning tree, 278
DNN, *see* Dual node network (DNN)
DOE, *see* Design of experiments (DOE)
Double-counting
CPHD/PHD filter distributed fusion, 227
multitarget distributed fusion, 223
T2F, single-target distributed fusion,
    206–207
DT&E, *see* Developmental test & evaluation
    (DT&E)
Dual node network (DNN), 380
Dynamic Bayesian networks (DBNs), 311, 312
Dynamic systems, DDF
application, trajectory state approach, 177
burst communications, 180
common process noise problem, 177–178
delayed and asequent observations,
    178–180
delayed states, 171
description, 170, 177
filtering, stored filter, 181–182
filtering, trajectory state system, 181
observation/communication
    interruptions, 171

operation, channel caches and trajectory
    states, 183
solution, trajectory states, 180–181
state dynamics, *see* State dynamics
trajectory state formulation, 183

E

Electronic support measure (ESM), 374–375
EPP, *see* Expected posterior probability (EPP)
Error covariances, distributed estimation
    cross-covariance fusion, 116–117
    posteriori fusion/best least unbiased estimate,
        115–116
ESM, *see* Electronic support measure (ESM)
Expected posterior probability (EPP)
    classification *vs.* number of communications,
        267, 268
    classification *vs.* time, 264, 265
    object class separation *vs.* correct
        classification, 264, 266
    transition probability *vs.* classification,
        266, 267
Exponential mixture (XM) fusion
    CPHD filter distributed, 228–229
    GM-PHD tracks, T^2F, 232–234
    multitarget distributed, 223–224
    particle-PHD tracks, T^2F, 234
    PHD filter distributed, 229–230
    single-target distributed fusion, 209–212

F

False discovery rate (FDR)
    algorithm to control, 78–79
    defined, 78
    description, 76
    distributed detection system, 79–81
    MCPs, 77
    nonidentical decision thresholds, 76
    SNRs, 76
    in statistics, 77–78
FDR, *see* False discovery rate (FDR)
Finite-set statistics
    CPHD Filter, 218–220
    multitarget calculus, 214–216
    multitarget recursive Bayes filter, 212–214
    PHD filters
        constant-gain Kalman filters, 216
        Markov transition density, 217
        SMC, 218
    sensor-bias estimation, 221
    SLAM, 220
Formal experimental design and statistical
        analyses
    DOE, 396, 397
    dominant analysis methodology, 395

factors and interaction complexity, DOE
    strategies, 397, 398
Monte Carlo replications, 394–395
notional layered experimental design,
    396, 398
statistical experimental design, 393–394
stochastic properties, 393
topological structure, 394
types, DOE techniques, 396–397
Fusion
    "a priori" and "a posteriori", 59
    characterizations, sense-making, 60
    computer-based information fusion, 57
    nonmonotonic logic, 59
Fusion agents
    description logics, 450
    HLIF, 449–450
    HLIF and low-level tracking
        refinement, 452
    knowledge model structure, 450
    low-and high-level fusion and feedback,
        450, 451
    top-down rules, 452
    update situation knowledge messages, 449
Fusion algorithms
    EPP, *see* Expected posterior
        probability (EPP)
    RMS, *see* Root mean square (RMS)
Fusion architecture graph
    multiply connected fusion, 99–100
    singly connected fusion, 98–99
Fusion rules
    Chair-Varshney, 72
    composite hypotheses, 72
    description, 71
    LRT, 72
    SR noise, 73
    UMP and GLRT, 73
Fusions system
    identification, domain complexities, *see*
        Intelligence analysis
    system design, development and evaluation,
        see Intelligence analysis

G

Gaussian random vectors
    description, 107
    error covariance and filter equations,
        108–109
    fusion equation, 107
    fusion node observation equation, 108
    information matrix, 107–108
    information matrix fusion equations, 109
Generic Hub (GH) data model, 330
GIG, *see* Global information grid (GIG)
Global information grid (GIG), 273, 286

H

HACs, *see* Human-agent collectives (HACs)
Hard-soft fusion process
    human-system interaction, 421
    system architecture, 423
    touch point 1, 424
    touch point 2, 424–425
    touch point 3, 425–426
    touch point 4, 426–427
    touch point 5, 427–428
    touch point 6, 428–430
HCI, *see* Human-computer interaction (HCI)
Heterogeneous contracts
    inflated independent beta distributions, 37–38
    Kalman filter trust model, 38–39
Hidden Markov Models (HMMs), 312
Hierarchical architecture
    fusion with feedback, 110
    fusion without feedback, 109
Hierarchical task network (HTN), 308
High-level information fusion (HLIF); *see also*
            Fusion agents
    CI and DIF, 442
    context knowledge, visual IF, 442
    description, 441
    IF systems, 441
    MPEG-7, 442
    OWL ontology, 443
    RACER, 443
    SNAP and SPAN, 443
    VERL and VEML, 443
HMMs, *see* Hidden Markov Models (HMMs)
HTN, *see* Hierarchical task network (HTN)
Human-agent collectives (HACs)
    design concerns, 42–43
    individual and collective goals, 42
    opportunities, 43–44
Human-computer interaction (HCI), 2
Human engineering factors
    automated processes, 409
    characterization, human-fusion system
            interaction, 410, 411
    cognitive system engineering, 430
    design and development, 431
    human-system integration, 409
    identification, fusions system, 418–421
    military intelligence analysis, 412
    system design, development and evaluation,
            412–418
    touch points, hard-soft process, 421–430
Hybrid sensing/hybrid cognition, SOA
    artificial intelligence and data fusion
            algorithm, 361
    mobile device user, 360–361
    service-oriented system methodologies, 361
    social networks, 361

I

Ideal communication channels
    asymptotic regime, 73–74
    Bayesian Formulation, *see* Bayesian
            Formulation
    correlated decisions, *see* Correlated decisions
    counting rule, 74–76
    decision rule partitions, 68
    design, fusion rules, 71–73
    false discovery rate-based sensor decision
            rules, Sensor decision rules
    hypothesis testing problem, 67
    joint density, 69
    K sensors, 67–68
    Neyman–Pearson Formulation, *see* Neyman–
            Pearson Formulation
    nonidentical decision rules, 69
    parallel configuration, 68
Indoor surveillance, VSNs
    communicate-fused estimation dialog, 452
    computer laboratory, 452, 453
    contextual enhancement, tracking, *see*
            Contextual enhancement, tracking
    continuous surveillance, 452
    framework configuration, *see* Camera calibration
    low-level information fusion, *see* Low-level
            information fusion
    scene interpretation, 459
    update situation knowledge dialog, 452
Inflated independent beta distributions, 37–38
Informational transactions, 127
Information graph
    description, 100–101
    distributed architectures, 103–104
    multiply connection, hierarchical fusion,
            101–103
    singly connected graph, singly connected
            fusion architectures, 101
Information recycling
    bounded covariance inflation, 20–22
    coupling scalars, 22–23
    decentralized tracking, 23, 24
Information-sharing strategy (ISS), 383
Information technology (IT)
    description, 9
    "invisible" computers, 9
    technology trends impacts, data fusion, 9–11
    traditional sensing/computing networks, 1
Intelligence analysis
    comparison, models, 413–415
    complexities and fusion system capabilities,
            418–420
    decision biases, 418
    hard-soft fusion process, *see* Hard-soft fusion
            process
    information processes, 418

military, 412
nonexhaustive factors, 413, 416–417
stages and extended capabilities map, 421, 422
system/process documentation, 430
Intelligent agents
agent-based application, 286
agent properties and data fusion, 287, 288
AI, 287
decentralized data fusion system, 288
description, 287–288
graphical Bayesian belief networks, 289
knowledge-based, 290
MADSNs, 290
military hierarchical organizations, 291
NCW, 287
real-time distributed tracking, 289
ISS, *see* Information-sharing strategy (ISS)
IT, *see* Information technology (IT)

J

JDL, *see* Joint Directors of Laboratories (JDL)
Joint Directors of Laboratories (JDL)
ASAS, 4
defined high-level processes in, 5–6
description, 4
subprocesses and functions, 5
target's kinematics, 6
top level model, 4
Joint multitarget (JoM) estimator, 214
Joint surveillance target attack radar system
(JSTARS), 283
JoM estimator, *see* Joint multitarget (JoM)
estimator
JSTARS, *see* Joint surveillance target attack
radar system (JSTARS)

K

Kalman filter trust model, 38–39
K-tree topologies
allowable links, 186
allowable topology, 184–186
communications algorithm, *see*
Communications algorithm, k-tree
data-tagged decentralized algorithm, 192
data-tagging sets, 187–188
DDF on, 187
description, 184
fully connected topology, 186–187
link and node failure robustness, 191–193
local neighborhood property, 189–190
scalability and correctness, 184
separator property, 188–189
spanning-tree algorithms, 184
treewidth, graph, 186

L

Link and node failure robustness, 191–193
Low-level information fusion
fused tracking, fusion agent, 456, 457
ground-truth positions, 456, 457
HLIF knowledge, 456
local tracking, sensor agents, 455, 456
tracking information, 456

M

MADSNs, *see* Mobile agent-based distributed
sensor networks (MADSNs)
MAP, *see* Maximum posterior probability
(MAP)
Marginalization, information form, 193–194
Market architecture for sensor management
(MASM)
architecture, 368, 369
market-oriented programming
techniques, 368
models network resource, 369
optimal bidding strategy, 371
optimal resource allocation, 370
target destruction, 373
tatonement process, 369
Widrow–Hoff learning rule, 374
Market-oriented programming
CCA, 370
current error, 374
deterministic optimization, 372
genetic algorithms, 373–374
market algorithms, resource allocation,
368
MASM, 368
multiperiod optimization, 372
optimal resource allocation, 371
sensor networks, 371
target destruction, 373
Tatonement, 369
Markov transition density, 217
MAS, *see* Multi-agent systems (MAS)
MASM, *see* Market architecture for sensor
management (MASM)
Maximum posterior probability (MAP), 255
Max-sum algorithm
art approximate algorithms, 27
decentralized coordination algorithm, 27
defined messages, 26
description, 25
factor graph, 25
function to variable, 26
sensor network, 25
variable node, 26
variable to function, 26

Measurement-to-track fusion (MTF), 200
Measures of Effectiveness (MOEs), 358
Measures of Performance (MOPs), 382
Military operations, threat analysis
    operational environment, 318
    predictability of the behavior, 317
    susceptibility to coercion, 317
    symmetry, 318
    task complexity, 316
    time, 316–317
    uncertainty, 316
Minimum-variance (MV) fusion rule, 132
Mobile agent-based distributed sensor networks
        (MADSNs), 290
MOEs, *see* Measures of Effectiveness (MOEs)
MOPs, *see* Measures of Performance (MOPs)
MTF, *see* Measurement-to-track fusion (MTF)
Multi-agent systems (MAS)
    computational models, trust, 34
    data fusion and decision-making node, 18
    stakeholders, 18
    VSNs
        autonomous and social abilities, 444
        BDI, *see* Belief–desire–intention (BDI)
        description, 443–444
        detected objects, 444
        FIPA ACL messages, 448
        fusion agents, *see* Fusion agents
        high-level hierarchical and partially
                distributed architecture, 446, 447
        ontologies, 444
        sensor agents, *see* Sensor agents
        standard communication protocols, 444
Multiple-target tracking problems, 125
Multiple trajectory states
    banded matrix, 175
    information matrix and vector, 174–175
    nonadditive form, 176
    timestep, 176
    total information, trajectory system, 175
    trajectory state system propagation, 176
Multitarget calculus
    integral-differential, 214
    multitarget probability distribution, 216
    Poisson process, 215
Multitarget distributed fusion
    known double-counting, T²F, 223
    T²F, independent sources, 222–223
    XM fusion, 223–224
Multitarget recursive Bayes filter
    cardinality distribution, 213
    defined, 213
    JoM, 214
Multitarget T²F
    independent sources, 222–223
    known double-counting, 223

N

Network-centric concepts
    description, 47
    operational advantages, 48
    role, fusion, 47, 52–53
    self-organization and self-synchronization,
        60–61
    sense-making, 52–53
    "share-ability", 48
    value chain, 48–49
    value of information, decision-making,
        51–52
Network-centric warfare (NCW)
    cognitive domain, 286
    conceptual vision, 285, 286
    GIG, 286
    information domain, 285
    physical domain, 285
Network value chain, measures and metrics
    degree dimensions-to-attributes and
            measures/metrics, 386, 388
    NCO, 386
    quality and degree measures, 386, 387
Neyman–Pearson Formulation, 70–71
Nondeterministic dynamics, object tracking
    augmented state vector and
            approximation, 119
    cross-covariance, single time, 119
    description, 118–119
Nonideal communication channels
    Chair-Varshney fusion rule, 86
    distributed detection and no channel state
            information, 88–89
    distributed detection and partial channel state
            information, 87–88
    ECG fusion rule, 86–87
    MRC fusion rule, 86
    optimal likelihood ratio-based fusion rule,
            85–86
    parallel fusion model, 84, 85
    signal model and fusion center, sensor, 85
    WSNs, LPI/LPD, 84
Nonmyopic sensor management
    defined, 365
    market-oriented programming, 368–374
    network resources policies, 365
    PE, simulation test bed, 374–376
    stochastic dynamic programming, 366–368
Nontemporal probabilistic approaches, threat
        analysis
    Bayesian network, 310
    decision trees, 309
    sensitivity analysis, 311
Normalized standard error (NSE), 41
NSE, *see* Normalized standard error (NSE)

O

Object classification
    algorithms, 120–121
    architectures, 119–120
    Bayesian, 256–268
    classifier fusion, 253–256
    declaration fusion process, 248
    discriminative approaches, 247, 248
    explicit double-counting, 251
    generative approaches, 247, 248
    imaging techniques, 259
    implicit double-counting, 251–252
    information-sharing strategies, 250
    legacy systems, 252
    mathematical/formal integrity, 250
    mixed uncertainty representations, 253
    NCTR methods, 247
    notional multisensor object classification
        process options, 248, 249
    support vector machines, 247
    types, data, 250
Object tracking
    deterministic dynamics, 117–118
    nondeterministic dynamics, 118–119
Observe-orient-decide-act (OODA), 7, 8
OGC, *see* Open Geospatial Consortium (OGC)
One-time track fusion
    Bar-Shalom-Campo and Speyer fusion,
        129–130
    BML rule, *see* Bayesian maximum-likelihood
        fusion (BML) rule
    calculation, cross-covariance matrix,
        133–134
    characterization, Ornstein-Uhlenbeck
        model, 138
    CI methods, *see* Covariance intersection (CI)
    complementary sensor case, *see*
        Complementary sensor
    constant-velocity model/small-white-noise
        model, 137
    description, 127–128
    estimation error covariance, 129
    fusion rules, 129
    Gaussian approximation, 128
    joint probability density function, 128
    linear combination, 129
    linear Gaussian estimation, 128
    local data processor, 128
    local estimation error, 128
    maneuvers approaches, 139
    MV fusion rule, *see* Minimum-variance
        (MV) fusion rule
    optimality, 135–137
    Ornstein-Uhlenbeck model, 137
    supplementary sensor case, *see*
        Supplementary sensor

    tracklet fusion rule, 130–132
    zero-mean Gaussian random vector, 127
Ontological structures, distributed fusion
    annotation, regions and objects, 339, 340
    computer domain, 329
    embargoed Port situation, 343
    geographical feature ontology, 338, 339
    geographical regions, 338–341
    GH, 330
    inferring relevant repositories, 336–337
    information annotation and processing, 341
    information integration, 328
    information producers and consumers, 327, 328
    interoperability and inference, 334–336
    net querying, 330–334
    ontological reasoning, 328
    OWL, 330
    RelevantThing, 337–338
    SIS, 327, 328
    STO, 341, 342
    types, information sources, 327
    UML, 330
Ontology web language (OWL), 442–443
OODA, *see* Observe-orient-decide-act (OODA)
Open Geospatial Consortium (OGC), 353
Optimal Bayesian distributed fusion
    arbitrary distributed fusion architecture,
        110–111
    hierarchical architecture, 109–110
Optimal Bayesian object classification
    centralized algorithm, 258
    communication, local agent, 261
    comparison, fusion algorithms, 264–268
    DBN, 257
    distributed fusion algorithm, 259–260
    extrapolation, high-level agent, 261–262
    fusion, high-level agent, 262
    performance evaluation approach, 263–264
    probability distance measures, 263
    sensor measurements, 262
    simulation scenario and data generation, 263
Optimality track fusion
    covariance matrix, 136
    description, 135
    extrapolation step, 136
    Koch-Govaers fusion rule, 136
    linear-Gaussian systems and off-line
        information, 137
    local variance matrices, 136
    MAP, 135
OWL, *see* Ontology web language (OWL)

P

Participatory sensing and sensor webs
    DARPA, 351–353
    data networks, 349

information fusion community, 354
mobile sense, 349–350
OGC, 353
PEIR, 350
SPS, 353
TML, 353
voluntweeters, 350–351
PDA, *see* Probabilistic Data Association (PDA)
PE, *see* Performance evaluation (PE)
PEIR, *see* Personal Environmental Impact
    Report (PEIR)
Performance evaluation (PE)
    comparison, sensors, 375, 376
    data fusion and resource management
        architecture, 389–390
    ESM, 374–375
    evaluative metrics, 391, 392
    MASM, 375
    notional time-based, 391, 392
    Pareto optimal front, 393
    T&E system, 390
Personal Environmental Impact Report (PEIR),
    350–354
Person-by-Person optimization (PBPO)
    approach, 70
PHATT, *see* Probabilistic Hostile Agent Task
    Tracker (PHATT)
Plan recognition
    automation, 307
    dynamic process, hypotheses formulation, 304
    evolution, time, 315
    goal recognition, 305
    HTN plan representation, 308
    mental state modeling, 314
    model manipulation, 315
    nontemporal probabilistic approaches, 309–311
    perception, problems, 306
    plan revision, 315
    probabilistic approaches, temporal
        dimension, 311–314
    symbolic approaches, 309
Probabilistic Data Association (PDA), 357
Probabilistic Hostile Agent Task Tracker
    (PHATT), 313
Probability hypothesis density (PHD) filters
    Chernoff information, 239–240
    independent sources, T^2F, 226
    known double-counting, 227
    T^2F fusion, mathematical derivations, 236
    XM fusion, 229–230
Process management, 382, 393

Q

Querying technology
    information producers, 331
    logic-based systems, 331

MetaCarta's technology, 331
natural language expression, 330
query ontology, 332, 333
SQL, 331

R

RACER, *see* Renamed Abox and Concept
    Expression Reasoner (RACER)
Renamed Abox and Concept Expression
    Reasoner (RACER)
    inference engine, 450
    scene interpretation, 443
    SURV ontology, 454
Repeated track fusion
    architectures, distributed tracking systems, 143
    categorization, fusion rules, 144
    description, 143
    estimation error covariance matrices and
        sensors, 144
    with feedback
        Bar-Shalom-Campo, Speyer and CI rules,
            149–150
        description, 148
        information graphs, 148, 149
        MV fusion rule, 150
        normalized process noise intensity, 149
        optimal distributed fusion algorithm, 150
        tracklet rule, 150
    without feedback
        Bar-Shalom-Campo, Speyer and CI fusion
            rules, 145
        decorrelation form, 148
        and decorrelation method, 146
        deterioration, distributed tracking, 148
        informational transactions, 145
        information graphs, processing
            architectures, 144
        MV fusion rules, 145–146
        normalized process noise intensity, 147
        Ornstein-Uhlenbeck model, 147
RMS, *see* Root mean square (RMS)
Root mean square (RMS)
    classification probability error *vs.* object class
        separation, 264, 266
    defined, 264
    probability error *vs.* average number of
        communications, 267, 268
    time *vs.* classification probability error,
        264, 265
    transition probability *vs.* classification
        probability error, 266, 267

S

Semantic information services (SIS), 327–328
Sense-making

"community of interest", 57
    definitions, 53
    dynamics, 56, 57
    forms of ignorance, 54, 55
    frame-building process, 53
    problem characteristics, 54, 56
    process characterization, 57–59
    utility-type function, 54
Sensor agents, 448–449
Sensor coordination
    description, 24
    max-sum algorithm, 25–27
    target tracking, 27–28
Sensor Planning Service (SPS), 353
Sequential Monte Carlo (SMC), 218
Service-oriented architecture (SOA)
    distributed human-centric information
        fusion, 348
    distributed sensors, participatory sensing, 348
    GPS, 347
    high-level assessments, 354–359
    hybrid sensing/hybrid cognition, 360–361
    information fusion community, 348
    mobile device usage, 347
    participatory sensing and sensor webs,
        349–354
    pyramid, *see* Service-oriented fusion pyramid
Service-oriented fusion pyramid
    composite operations, 356–359
    human-centric information fusion, 355
    low-level operations, 355–356
Simultaneous localization and mapping
        (SLAM), 220
Single-target distributed fusion
    Bayes filter, 202–204
    covariance intersection, 207–209
    exponential mixture fusion, 209–212
    independent sources, T^2F, 204–205
    T^2F, known double-counting, 206–207
SIS, *see* Semantic information services (SIS)
Situation assessment (SA)
    BN model, 272
    NCW, 271
Situation theory ontology (STO), 341, 342
SLAM, *see* Simultaneous localization and
        mapping (SLAM)
SMC, *see* Sequential Monte Carlo (SMC)
SOA, *see* Service-oriented architecture (SOA)
SPS, *see* Sensor Planning Service (SPS)
SQL, *see* Structured query language (SQL)
Stakeholders
    communication network, 30
    computational mechanism, 29
    description, 28, 29
    governmental and nongovernmental
        organizations, 29
    MAS, 31

    mechanism, 33–34
    multisensor network target tracking, 30
    ROI, sensor, 34
    sensor network system and communication
        allocation, 34, 35
    sensor-net-work topology, 31
    valuation function, 31–32
State dynamics
    dynamic transformation, 172
    equivalence, conventional approach, 173–174
    linear discrete time state dynamic model, 172
    multiple trajectory states, *see* Multiple
        trajectory states
    trajectory information approach, 172–173
    trajectory state approach, 171
STO, *see* Situation theory ontology (STO)
Stochastic dynamic programming
    adaptive Lagrangian relaxation, 367
    approximation techniques t, 367
    Bellman's optimality principle, 366
    research approach, 368
    sensor measurements, 367
Stored filter, 181–182
Structured query language (SQL), 331
Suboptimal Bayesian distributed fusion
    Bhattacharyya fusion, 114–115
    channel filter fusion, 112–113
    Chernoff fusion, 113–114
    description, 111–112
    naïve fusion, 112
Supplementary sensor
    Bar-Shalom-Campo rule, 141
    BML rule, 140
    conditional probability density, 140
    description, 139
    inter-sensor cross-covariance matrix, 140
    normalized initial position standard
        deviation, 140, 141
    normalized process noise intensity, 139, 140
    process noise intensity and stationary velocity
        covariance, 141

T

Temporal probabilistic approaches, threat
        analysis
    DBN, 311, 312
    HMMs, 312
    PHATT, 313
    types, 311
Test and evaluation (T&E), DDIFS
    ANOVA, 402–405
    complexities, error audit trails, 393
    CTP, 401
    DF&RM, 380
    DNN, 380
    DT&E, 384

experimental design, 381
formal experimental design and statistical
    analyses, 393–398
fusion estimates and truth states, 388–389
information fusion processes and
    algorithms, 404
inter-tier(tiers1and2), ANOVA, 402, 406
ISS, 383
MOPs, 382
network value chain, 386–388
notion, PE tree, 389–393
PE process, 382
"production prototype" program, 381
SOA, 386
software testing, 385
statistical/mathematical analysis
    techniques, 379
strategies, T&E, 398–399
subjectively judged properties, 383, 384
two-aircraft configuration, 399, 400
two *vs.* six offensive sweep scenario, 399, 400
Theoretical foundation, distributed fusion
CPHD/PHD filter distributed fusion, 224–230
exact T²F formulas, 230–232
finite-set statistics, 212–221
mathematical derivations, 234–241
MTF, 200
multi-Bernoulli filters, 201
multitarget distributed fusion, 222–224
Pedigree techniques, 201
single-target distributed fusion, 202–212
XM implementation, 240–241
XM T²F formulas, 232–234
Threat analysis, distributed environments
action, event and reference point, 298–299
advantages, 320
analytical challenges, 321–322
capability indicators, 302
centralized and decentralized control, 319
collaboration challenges, 322
CPA, 301
data fusion model, 303–304
definitions, 297–298
dual perspective, 303
goal and plan recognition, 304–306
impact assessment, 300
intent indicators, 302
intentionality, 299–300
military operations, 315–318
and network-centric operation, 322–323
operational challenges, 320–321
opportunity indicators, 302
plan recognition, *see* Plan recognition
reasoning processes, 296
TML, *see* Transducer Markup Language (TML)
Track association
Bar-Shalom metric, 153

Chong-Mori-Chang metric, 153–154
CI metric, 153
comparison, metrics, 155–156
description, 150–151
expanded state metric, 154
problem definition, 151–152
sensor biases and the track
    association, 155
Singer-Kanyuck metric, 152
Track fusion
one-time, *see* One-time track fusion
repeated, *see* Repeated track fusion
Track-to-track fusion (T²F)
CPHD filter distributed fusion, 224–226
general multitarget distributed fusion,
    222–223
GM-PHD tracks, 231–232
independent sources, 204–205
known double-counting, 206–207
multitarget distributed fusion, 222–223
particle-PHD tracks, 232
PHD filter distributed fusion, 226
Trajectory information form equivalence,
    194–196
Transducer Markup Language (TML), 353
Tree network topology, channel cache
algorithm, 166
communicated terms, 166–167
DDF, 166
description, 165
disjoint subsets, 166
observation and communication cache, 167
transmission, communication term, 167
Trust and reputation
effective models, 35
evaluation
    covariance matrix and normalized
        error, 40
    estimated expected utility, 41
    expected information content *vs.* NSE,
        40, 41
    Kalman filter encode, 40
    reputation system, 39
    simulation run, 40
    single-dimensional trust models, 42
expected utility, contract, 35–37
heterogeneous contracts, *see* Heterogeneous
    contracts
interaction partners, 34–35

U

UAV, *see* Unmanned aerial vehicle (UAV)
UML, *see* Universal Modeling
    Language (UML)
Universal Modeling Language (UML), 330
Unmanned aerial vehicle (UAV), 282–283

V

Value chain
    concepts, 48–49
    individual nodes, 51
    ISS, 50
    multilayered process, 49
    NCW conceptual framework, 50, 51
    self-organization and self-synchronization,
        60–61
VEML, *see* Video event markup language
        (VEML)
VERL, *see* Video event representation language
        (VERL)
Video event markup language (VEML), 443
Video event representation language (VERL), 443
Visual sensor networks (VSNs)
    camera calibration, 437–438
    classification, 439
    communication, 437
    context-based approaches, HLIF, *see* High-
        level information fusion (HLIF)
    description, 436
    DETER system, 440
    DIF software architectures and techniques, 441

disadvantage, classical systems, 441
indoor surveillance, *see* Indoor surveillance,
        VSNs
monitoring and surveillance tasks, 436
multi-agent architectures, 441
multi-agent systems, *see* Multi-agent systems,
        VSNs
object detection, 438
object tracking, 438–439
process enhancement, 439
reliability and accuracy, 436–437
requirements and issues, 437
scalability, 440
third-generation surveillance systems, 440
third-generation video systems, 436
VSAM, 439
VSNs, *see* Visual sensor networks (VSNs)

X

XM fusion, single-target distributed fusion
    model, 210
    multidimensional Gaussian distribution, 209
    optimization approaches, 212
    Wasserstein distance, 211